THE ENCYCLOPAEDIA OF
UNDERWATER
LIFE

THE ENCYCLOPAEDIA OF UNDERWATER LIFE

Edited by Dr Keith Banister
and Dr Andrew Campbell

George Allen & Unwin
London, Sydney

Project Editor: Graham Bateman
Editors: Bill MacKeith, Robert Peberdy
Art Editor: Jerry Burman
Art Assistant: Carol Wells
Picture Research: Alison Renney
Production: Clive Sparling
Design: Chris Munday, Andrew Lawson
Index: Barbara James

 AN EQUINOX BOOK

Published by:
George Allen & Unwin
40 Museum Street
London WC1A 1LU

Planned and produced by:
Equinox (Oxford) Ltd
Littlegate House
St Ebbe's Street
Oxford OX1 1SQ

British Library Cataloguing in Publication Data

The Encyclopaedia of underwater life. — (The Unwin
animal library)
1. Marine Biology
I. Banister, Keith   II. Campbell, Andrew
574.92   QH91
ISBN 0-04-500035-2

Origination by Fotographics, Hong Kong; Alpha
Reprographics Ltd, Harefield, Middx.

Filmset by BAS Printers Limited,
Over Wallop, Stockbridge, Hants, England.

Printed in Spain by Heraclio Fournier S.A. Vitoria.

Advisory Editors

Professor Fu-Shiang Chia
The University of Alberta
Edmonton
Canada

Dr John E. McCosker
Steinhart Aquarium
California Academy of Sciences
San Francisco
USA

Dr R. M. McDowall
Ministry of Agriculture and Fisheries
Christchurch
New Zealand

Artwork panels and diagrams

S. S. Driver

Richard Lewington

Roger Gorringe

Mick Loates

Denys Ovenden

Norman Weaver

Left: Lion fish (Planet Earth Pictures); half-title: fan worm (Oxford Scientific Films); title page: Goosefish and crab (Oxford Scientific Films).

CONTRIBUTORS

RGB Roland G. Bailey BSc PhD
Chelsea College
University of London
England

GJB Gerald J. Bakus PhD
University of Southern
California
Los Angeles, California
USA

KEB Keith E. Banister PhD
British Museum (Natural
History)
London
England

RCB Robin C. Brace BSc PhD
University of Nottingham
England

BB Bernice Brewster BSc
British Museum (Natural
History)
London
England

AC Andrew Campbell BSc DPhil
Queen Mary College
University of London
England

JEC June E. Chatfield BSc PhD
ARCS
Gilbert White Museum
Selborne, Hants
England

GD Gordon Dickerson BSc
MIBiol
Formerly of Wellcome
Research Laboratory
Beckenham, Kent
England

GDi Guido Dingerkus BS MS
MPhil PhD
American Museum of
Natural History
New York
USA

PRG Peter R. Garwood BSc PhD
Dove Marine Laboratory
University of Newcastle
upon Tyne
England

GJH Gordon J. Howes MIBiol
British Museum (Natural
History)
London
England

JL-P Johanna Laybourn-Parry
BSc MSc PhD
University of Lancaster
England

JMcC John E. McCosker BA PhD
Steinhart Aquarium
California Academy of
Sciences
San Francisco
USA

RMcD Bob McDowall MSc PhD
Ministry of Agriculture and
Fisheries
Christchurch
New Zealand

LP Lynne Parenti
California Academy of
Sciences
San Francisco
USA

PSR Philip S. Rainbow MA PhD
Queen Mary College
University of London
England

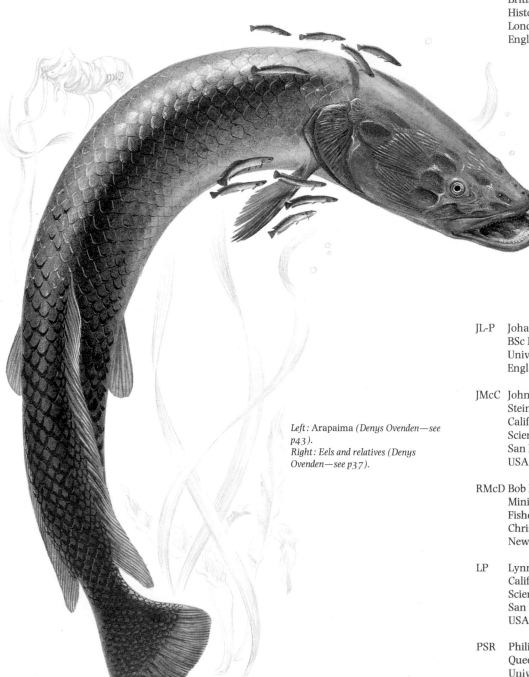

Left: Arapaima (Denys Ovenden—see p43).
Right: Eels and relatives (Denys Ovenden—see p37).

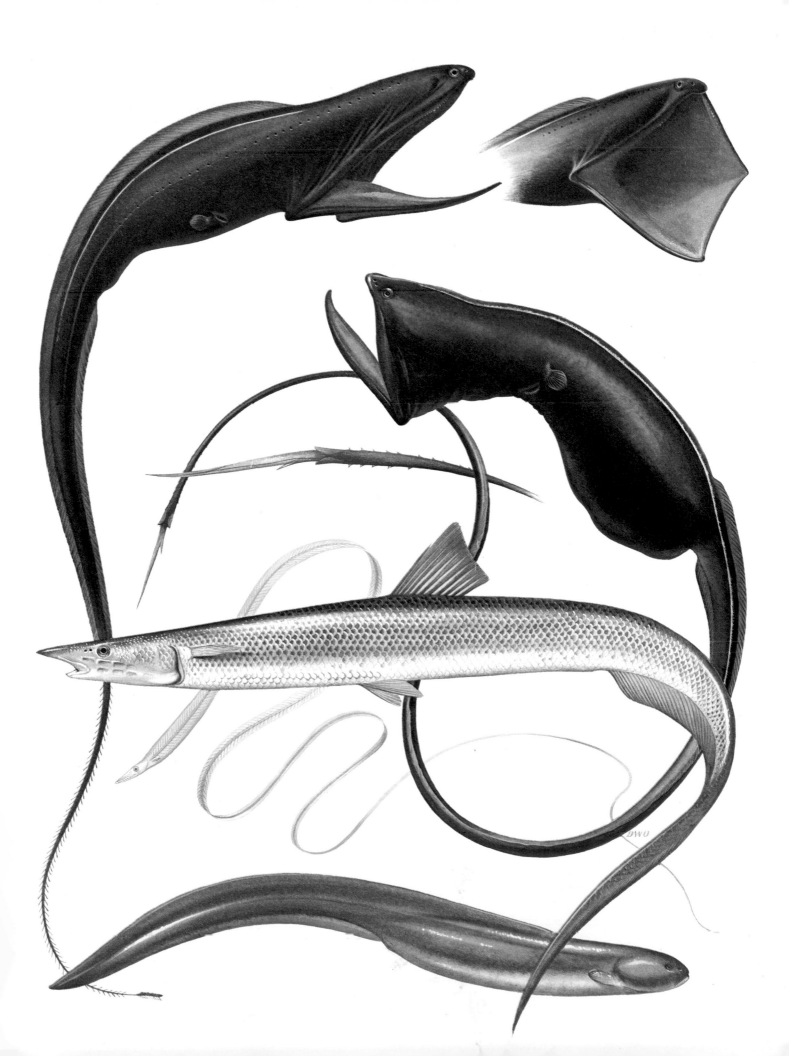

CONTENTS

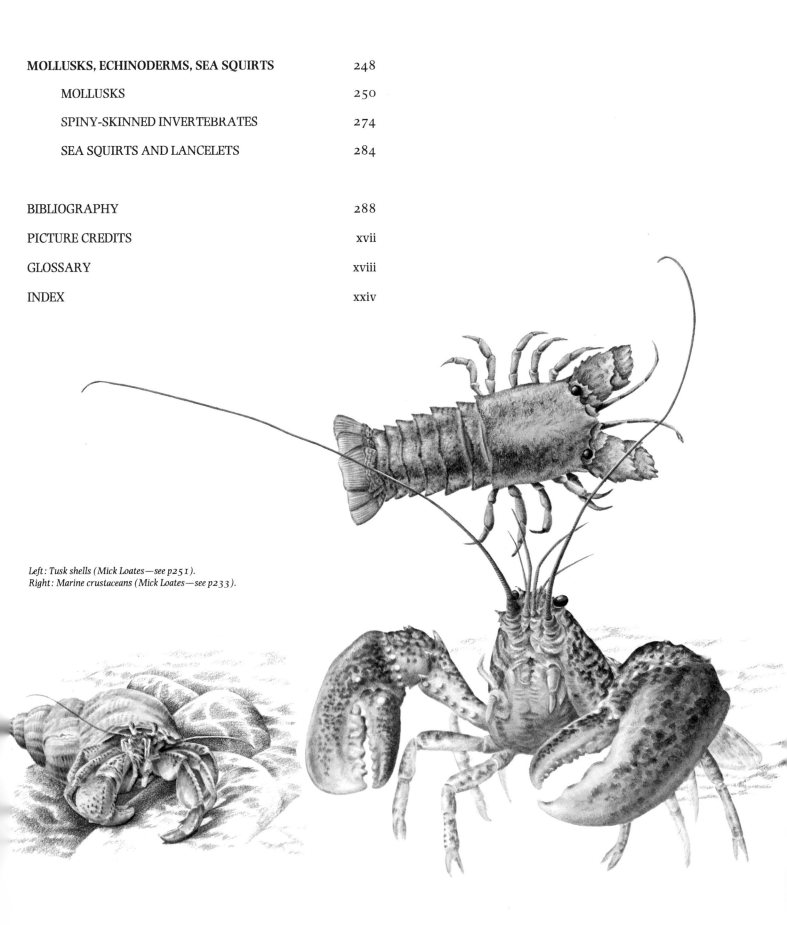

Left: Tusk shells (Mick Loates—see p251).
Right: Marine crustaceans (Mick Loates—see p233).

PREFACE

Life evolved in the primeval seas some 4,000 million years ago and, despite a much later major transition to land, has continued to evolve in our watery habitats since that time. The result of 4,000 million years of gradual change is a bewildering diversity of aquatic life, from the microscopic single-celled animals to the giant squid and monstrous sharks, and from beautiful and delicate sea anemones, corals and sponges to grotesque angler fish and other denizens of the deep sea. The underwater world is mostly invisible to humans, except for the lucky few, but in many respects it is more complicated and diverse than its more obvious terrestrial counterparts. The aim of this volume is to lift the watery covering and reveal the secrets of aquatic life.

The Encyclopaedia of Underwater Life is divided into two major sections: Fishes and Aquatic Invertebrates. Fishes form the most obvious fauna of our oceans, rivers and lakes and although many hundreds of millions of years ago they spawned the higher terrestrial animals—in the first instance amphibians—they remain today totally aquatic; with a few notable exceptions they do not come out of water onto land. The aquatic invertebrates contain an immense variety of often small and inconspicuous animals whose diversity relates to their often long evolutionary history. The purpose of this volume is to review the major taxonomic groupings that are essentially totally aquatic. It thus excludes such amphibious animals as frogs and newts, which return to water to breed, and seals and sea lions which leave the sea to breed on land. Also not covered here are the whales and their relatives and the sea cows which live out all their lives in rivers and oceans. All these groups are covered elsewhere in the series.

Fishes are found almost everywhere: from the cold lightless waters of the deepest oceans to lakes high in the Andes mountains, on land, in mud, underground, in the air, even in trees. There are over 20,000 species of fishes alive today. They do almost everything that other vertebrates (amphibians, reptiles, birds, mammals) do, and many things not done by any other vertebrates. Many show parental care; some have even paralleled the mammals in devising an equivalent to a placenta. Others feed their offspring with secretions from their bodies. The most avant-garde feminists would admire the fishes in which the male effectively "gives birth" and looks after the young. Some fishes change sex, in contrast to an Amazon guppy that comprises only females.

There are luminous fishes, transparent fishes and electric fishes. The size range is colossal, from species that are fully grown at 10mm (0.4in) to species reaching over 12m (40ft) in length. Some species are represented by countless millions of individuals, others just manage to survive precariously with a handful of individuals.

Because there are so many species of fishes there is an extensive literature reporting the research that has been conducted on many aspects of their biology, anatomy, behavior and relationships. What the contributors have aimed to do in this volume is to distill their knowledge to give the reader a flavor of the essence of fishes. It is inevitable in the space available that there are gaps, but these are largely of detail, not of import. As far as possible the major biological principles are each covered by an example. This may not necessarily be a familiar one, but will nonetheless illustrate the general point.

The bulk of the fishes section deals with the four major living classes of fishes (lampreys, hagfishes, jawed fishes, jawless fishes). There is no universally accepted classification for fishes. In some circumstances detailed studies have produced a classification for one level of hierarchy based on one philosophy whereas in other areas different philosophies have been used as the basis for the classification of a different category. Other groups of fishes have hardly been studied at all.

Within the world of fish classification these are exciting times, but producing more problems than solutions. Yet these have caused all professional ichthyologists to think deeply about the nature of classification and to try and produce a scheme that reflects the true genealogical relationships of fishes.

Rather than produce a misleading classification that might be assumed to be definitive, our text stresses many of the problems. We felt it better to admit the current state of uncertainty and confusion, to admit that there are profound disagreements between different schools of ichthyological thought and to admit that there is much about which we are ignorant.

Even the four living and two extinct classes of "fishes" mentioned in this volume cannot be arranged in a way that would not produce criticism from one group of researchers or another. There is no really satisfactory solution at the moment. Consequently the classification (arrangement might be a better word) adopted here is a neutral one, and is meant to be a convenient map to guide people around the many unfamiliar groups of fishes.

The present survey proceeds with articles covering a group of closely related species, the length of each reflecting the importance and interest of the group and to some extent the number of species in the group. Each article is prefaced by a fact panel which provides a digest of information, covering the number of species, genera and families as well as their distribution and size range. Outline drawings of examples from key families are also provided to help the reader visualize what the fishes look like. For large groups this information is consolidated in a separate table. Where possible, common names are used in the fact boxes or tables. However, some groups (and the great majority of species) have no common names, or they vary from country to country, hence the universally accepted scientific name for the group has to be used. This may seem unnecessarily cumbersome but it was the only way to provide a peg on which to hang the group. One value of scientific names is that the ending of many informs the reader of the hierarchy of the group being discussed; this system is explained in "What Is a Fish?"

ABOVE Angler fish and lure (Planet Earth Pictures). BELOW Salmon spawning (Denys Ovenden—see p49).

As well as the outline drawings there are spreads of color illustrations. Most of these have been drawn from actual specimens in an attempt to avoid perpetuating errors in the literature. Because fishes vary considerably in size it is not possible to show them to scale in the plates. An approximate size for each fish is given in the captions. Relevant line drawings of important anatomical features are also included. Color photographs drawn from sources worldwide complement and enhance the text and the other artwork. Here again, rarely illustrated species have been chosen.

In addition to the sequence of systematic articles there are the introduction ("What Is a Fish?") and numerous special features. These are multidisciplinary and deal with unrelated fishes. After all, a trawler does not selectively catch fishes of just one species! So concepts dealing with the associations of fishes with ourselves, or with habits common to different groups of fishes transcend the natural categories. Here are articles on the problems posed for fishes that live on land, and their solutions; on the difficulty of living in caves; and on bioluminescence. Aspects and consequences of our interaction with fishes as food and in captivity are also discussed. Whereas the subject of fish in captivity is largely historical, the treatment of endangered fishes and the future of fishes is concerned with what humans have done to fish species and whether we will learn from the mistakes of yesterday to make a better tomorrow.

Although fish have great intrinsic interest it is not so generally realized that research is being conducted on the use of the poison from fish spines in surgical techniques. We have problems of tissue rejection in heart and lung transplants—how much might we learn from the fusion of the male angler fish onto the female's body? They seem to have no rejection problems. Studies on a few small species of fish that exist as eyed surface forms and eyeless cave forms may help us understand the relationship between the genetic code and the environment.

It is hoped that the information in both the main text and in the special spreads will tempt the reader to plunge into the world of fish in more detail.

Aquatic invertebrates are invertebrate animals which live in the sea, fresh water or moist terrestrial habitats; they also include many parasites whose "aquatic environment" is that of the bodies of their hosts. The term invertebrates refers to the fact that none of these animals has a bony or cartilaginous backbone, a characteristic of most fishes and all reptiles, amphibia, birds and mammals. Over 30 major invertebrate subdivisions (phyla) of the animal kingdom are recognized by scientists and each one of these is "invertebrate." Even the phylum Chordata, which includes vertebrates such as fish, birds, mammals etc, has some members, eg sea squirts and lancelets, which lack a backbone, and these are dealt with here. The only invertebrates not covered are the terrestrial arthropods—the insects, millipedes and allies (phylum Uniramia), the spiders and other land-dwelling arachnids (most of the phylum Chelicerata) and the velvet worms (phylum Onychophora).

Some invertebrate phyla, while overwhelmingly aquatic in habits, contain groups that have terrestrial forms. For example, although segmented worms are mainly marine, the earthworms live in damp soil; also slugs and snails are terrestrial variations of the mainly aquatic mollusks. For completeness such terrestrial forms are also considered here. The diversities of form and biology are immense—a salmon and an elephant have more in common with each other than do many apparently related members of invertebrate phyla. Thus the subject is enormous.

In all some 20 percent of all species in the animal kingdom are in the aquatic phyla (if we take a somewhat arbitrary estimate of 1 million species of insects and their allies). It is hardly surprising that such a large portion of animal species includes many of scientific, medical, economic and ecological importance. For all this, most of these animals are unknown to the layman. Many households in the west have at least one book on birds; mammals too are popular, largely because of the prominence they play in our lives. Aquatic invertebrates may be held by some to be the poor relations. This is far from the case. For sheer beauty the microscopic architecture of the heliozoans or sun animalcules and the sea anemones is not easily surpassed. For devastating disease in man the malaria and bilharzia parasites are without equal. For complexity of structure and intelligence, the squids and octopuses come near to rivaling the fish in expertise in their mastery of the water and in their behavior. For enigmatic architecture and diverse solutions to evolutionary problems the starfishes, sea urchins and their allies are without equal in the animal kingdom.

Despite this wealth of evolutionary enterprise and extensive distribution, most people, other than the professional zoologist, have little knowledge of the aquatic invertebrates. For this reason it is difficult to refer to many of these animals without scientific terminology. In order to bring some simplicity to the extensive technical classification of so many phyla, the organisms treated here have been classified into a few subjective groupings which are: Simple Invertebrates, Sedentary and Free-swimming Invertebrates, Worm-like Invertebrates, Jointed-limbed Invertebrates, and Mollusks, Echinoderms and Sea Squirts. Each of these gross headings is introduced by an essay dealing with the ecological attributes which members of the grouping have in common, and indicating relationships. Then each phylum is dealt with, treating the classification, structure, development and ecology as appropriate. Each major entry starts with a fact panel which summarizes the biology of the group concerned and gives the scientific names of those "few" species that have valid common names. This work is not structured at a species level because the species it covers are far too numerous to be treated in this way. In the cases of large and conspicuous phyla like the Protozoa, Annelida, Crustacea and Mollusca, the main subdivisions of the phylum (often classes) are dealt with in classificatory order. In the smaller groups a general résumé of the whole is given. A number of topics worthy of special attention are drawn out as boxed features or in the case of some ecologically, economically or medically important subjects as full double-page spreads.

Illustrations play a major role in this encyclopaedia; they do not just show important variations in form and biology, but the photographs in particular, provide a backcloth of stunning colors, bewildering shapes and impressive waterscapes. The latest techniques of underwater and microscopic photography are shown at their best here and the world has been scoured for suitable picture material. Pictures have been selected and captions written, not only to illustrate the text, but to expand upon it.

Artwork comes in several forms. Firstly, there are a number of color panels illustrating the diversity of major groups. These have often been painted as ecological scenes and for this reason include some animals not relevant to the section in question, although covered elsewhere in the volume. Actual sizes of the animals depicted are indicated in the captions. Secondly, there are many simple line diagrams which, at a glance, show the form of major subdivisions of the larger groups. Thirdly, many life cycles, particularly those of parasites, are shown in simple line form. Finally, there are the diagrams showing, in particular, internal anatomy. Each phylum has its own special form, parts of which may be peculiar to it and have special names but not popular ones such as arms or legs. For this reason it has been very important to illustrate such form and anatomy in each phylum to enable the lay reader to find his way through inevitable scientific terminology—a glossary is also provided. A bonus to this approach is that the apparent external simplicity of many of these invertebrates can be shown to hide a complex internal structure.

Because of the diversity of aquatic invertebrates an international team of specialists have contributed to the volume, each one dealing with his or her special area from an individual standpoint. In this way particular impressions of the phyla are built up and valid zoological opinions developed. We have not attempted to subdue these but have allowed them to speak for themselves since the development of ideas and the criticism of theories is the very meat of zoological progress.

The degree to which a book of this type succeeds in attracting readership, and providing a usable source of reference, depends entirely on the enthusiasm and commitment of the contributors as well as on the skill of the editorial and publishing staff. It has been a great pleasure to work on this volume with friends old and new and we thank them all for their indulgence of our editorial views. A special mention must be made of Dr Graham Bateman and his team at Equinox (Oxford) Limited who have stoically borne the pangs of birth and adolescence of many of the articles. Their judgment has lead to a satisfactory maturation of the manuscript.

Keith Banister
BRITISH MUSEUM (NATURAL HISTORY)

Andrew Campbell
QUEEN MARY COLLEGE, UNIVERSITY OF LONDON

LEFT Community of mussels, hydroids, sea anemones, barnacles and algae (Planet Earth Pictures). OVERLEAF School of fish, Galapagos (Oxford Scientific Films).

FISHES

WHAT IS A FISH? 2

WHAT IS A FISH?

Jawless fishes
Classes: Cephalaspidomorphi, Myxini
Two or four families: 12 or 17 genera: 72 species.

Lampreys
Class: Cephalaspidomorphi
Order: Petromyzontiformes.
About forty species in 6 or 11 genera and
1 or 3 families.

Hagfishes
Class: Myxini
Order: Myxiniformes.
About thirty-two species in 6 genera of the family
Myxinidae.

Bony fishes
Class: Osteichthyes
Four hundred and twenty families: 3,700 genera:
about 20,750 species.

Sturgeons and paddlefishes
Order: Acipenseriformes
Twenty-seven species in 6 genera and 2 families.

Bowfin and garfishes
Families: Amiidae, Lepisosteidae
Eight species in 3 genera.

Tarpons, eels and notacanths
Superorder: Elopomorpha
About six hundred and thirty-two species in about 160
genera and about 30 families.

Herrings, anchovies and allies
Superorder: Clupeomorpha
About three hundred and forty-two species in
68 genera and 4 families.
Includes: **herring** (*Clupea harengus*).

Bony tongues and allies
Superorder: Osteoglossomorpha
About one hundred and sixteen species in
26 genera and 6 families.

Pike, salmon, argentines and allies
Order: Salmoniformes
Over three hundred species in 80 genera and
17 families.
Includes: **argentines** (family Argentinidae), **Brown trout**
(*Salmo trutta*).

Bristle mouths and allies
Order: Stomiiformes
About two hundred and fifty species in over
50 genera and about 9 families.

I N all probability life started in our planet's waters about 3,000 million years ago. For a very long time not much seems to have happened. The first known multicellular invertebrates occur about 600 million years ago. After a much shorter interval (in geological terms), about 120 million years, the first aquatic vertebrates, the fishes, appeared. From these innovative fish groups arose the animals most familiar to us today—birds, mammals and especially ourselves.

Over half of the vertebrates alive today are these masters of the water—the fishes. They do just about everything that the other vertebrates do and also have many unique attributes. Only fishes, for example, make their own light (see p64), produce electricity, have complete parasitism (see p98) as well as having the largest increase in volume from hatching to adulthood.

Different people have different impressions of fish. To some the epitome of the fish is the sharp-toothed shark, elegantly and effortlessly hunting its prey in the sea. To others fish are small, pretty animals giving some point to the aquarium in the home. For anglers, fish are a cunning quarry to be outwitted and caught. For the commercial fishermen fish are a mass of slimy silvery bodies being hauled on board the vessel: the more fish, the more money. Biologists regard fish as representing a mass of problems concerning evolution, behavior and form, the study of which provides more questions than answers.

It is the very diversity of fishes, the large number of species and the huge numbers of individuals of some species that make them so interesting, instructive and useful to us.

It is easy to see that the huge shoals (provided they are maintained by careful management) form a valuable food resource for us and other animals, but it is not so generally realized that research is being conducted on the use of the poison from fish spines in surgical techniques. In making transplants of hearts and lungs in humans there are problems of tissue rejection: how much might we learn from the fusion of the male angler fish onto the female's body? (See p98.) They seemingly have no rejection problems. Studies on a few small species of fish that exist as eyed surface forms and eyeless cave forms may help us understand the relationship between the genetic code and the environment (see p84). There are many similar examples.

Fish, then, are many things to many people and for millennia they have fascinated mankind in many ways, not least because they have conquered an environment that is to us alien. Yet all these associations between humans and fish beg the most vital question of all.

▲ **Elegant, effortless hunters,** sharks are a group of fishes that lack a bony skeleton. Their graceful, powerful bodies have a skeleton of flexible gristle-like substance called cartilage. Their gill slits open separately to the outside and are not covered by an operculum (cover). Their teeth are constantly replaced and there is a spiral valve in the intestine to increase the absorptive area. This is the Gray reef shark (*Carcharhinus amblyrhynchus*).

▶ **Calculating predators of the deep,** deep-sea angler fishes are commonest in lightless waters. Above the mouth is a modified dorsal fin ray (the illicium) which terminates in a luminous lure (the esca). With this they entice their prey. Some species also have an elaborate chin barbel (its function is uncertain). Most are small fishes (there is little food in the deep sea) but in one species, *Ceratias hoelboelli*, the females can grow to 1m (3.3ft) long.

Lizard fishes and lantern fishes
Orders: Aulopiformes, Myctophiformes
About four hundred and forty species in about 75 genera and about 13 families.

Characins, catfishes, carps and milk fishes
Superorder: Ostariophysi
About six thousand species in at least 907 genera and 50 families.
Includes: **carp** (*Cyprinus carpio*), **goldfish** (*Carrasius auratus*).

Beard fishes, cling fishes, cods, angler fishes and allies
Superorder: Paracanthopterygii
About one thousand one hundred species in about 200 genera, 30 families and 6 orders.
Includes: **rat-tails** or **grenadiers** (family Macrouridae).

Silversides, killifishes, ricefishes and allies
Series: Atherinomorpha
Superorder: Acanthopterygii.
About one thousand species in about 165 genera, 21 families, 2 orders and 1 division.
Includes: **Devil's Hole pupfish** (*Cyprinodon diabolis*).

Spiny-finned fishes
Series: Percomorpha
Superorder: Acanthopterygii.
About ten thousand five hundred species in 2,000 genera, 230 families and 11 orders.
Includes: **Nile perch** (*Lates niloticus*), **Red mullet** (*Mullus surmuletus*).

Bichirs, coelacanth and lungfishes
Orders: Polypteriformes, Coelacanthiformes; superorder: Ceratodontimorpha
Seventeen species in 6 genera, 4 or 5 families and 4 orders.

Cartilaginous fishes
Class: Chondrichthyes
Thirty-one families: 130 genera: about 711 species.

Sharks
Subclass Selachii
About three hundred and seventy species in 74 genera, 21 families and 12 orders.

Skates and rays
Order: Batiformes
About three hundred and eighteen species in 50 genera and 7 families.

Chimaeras
Subclass: Holocephali
About twenty-three species in 6 genera, 3 families and the order Chimaeriformes.

What is a fish?

There is no such thing as a fish. "Fish" is simply a shorthand notation for an aquatic vertebrate that is not a mammal, a turtle or anything else. There are four quite separate groups of fish, not closely related to one another. Lumping these four groups together under the heading fish is like lumping all flying vertebrates, ie bats (mammals), birds and even the flying lizard under the heading "birds" just because they all fly. The relationship between a lamprey and a shark is no closer than that between a salamander and a camel.

However, the centuries of acceptance of "fish" as a descriptive term dictate that, for convenience, it will be used here. But remember that using the word fish for the four different living groups is equivalent to referring to all other vertebrates as tetrapods, even if some have subsequently lost or modified their legs.

The four living groups consist of two groups of jawless fishes, the hagfishes and the lampreys, and two groups with jaws, the cartilaginous fishes (sharks and rays) and the bony fishes (all the rest). There are also two other groups that are now extinct.

The four living groups differ widely in their numbers of species. There are about 32 species of hagfish and about 40 of lamprey. Today the jawed fishes dominate: the sharks, rays and chimaeras comprise about 700 species while the greatest flowering is in the bony fishes with over 20,000 species.

A brief history of the major groups

The first identifiable remains of fishes are small, broken and crushed plates in rocks of the middle Ordovician era 460–480 million years ago. (Possible traces from the upper Cambrian era, more than 500 million years ago, have not yet been confirmed.) These plates represent parts of the bony external armor of jawless fishes. Although none of the *living* jawless fishes has any external protection, large defensive head-shields were not uncommon in the early forms. But it is not known what the overall body shape of the first known fishes was like.

About 150 million years after they first appeared, the jawless fishes radiated into many widely varying forms, quite unlike the eel-like forms alive today. In some species of the Devonian era (400–350 million years ago) the armor was reduced to a series of thin rods allowing greater flexibility of the body. In one poorly known group, represented now by mere shadowy outlines on rocks, the armor consisted solely of tiny isolated tubercles (nodules).

Most of these Devonian jawless fishes were small. An exception are the pteraspids, fishes with the front half of the body covered with a massive plate, often with a backward pointing spine. They could reach 1.5m (5ft) in length. The cephalaspids, with their shield-shaped head plates, are a group well known from fossils in the Old Red Sandstone. A fortunate discovery of some well-preserved cephalaspids buried in fine mud enabled, with careful preparation, the course of their nerves and blood vessels to be discovered.

The first recognizable fossil lampreys have been found in Carboniferous (Penn-

The Ranking of Fishes

The ranking of a fish or a group of fishes (a taxon) within the hierarchy is indicated by the ending of the group name. These endings are uniform up to and including order, but there is little standardization for higher categories, despite attempts having been made to introduce it. Taking as an example the now ubiquitous carp, *Cyprinus carpio*, we have:

carpio	the trivial name of that species
Cyprinus	genus, which includes other species
Cyprinidae	family, which includes many genera, eg *Cyprinus, Pimephales, Tribolodon*
Cyprinoidei	suborder, which includes other families, eg Cobitidae (loaches)
Cypriniformes	order, which contains the other suborders, eg Charcoidei (characins)
Ostariophysi	superorder, which contains other orders, eg Siluriformes (catfishes)

Although there are yet higher categories, as well as others inserted between the stages listed above, they tend to be used only in a rather specialized fashion.

Great care must be taken when expressing some of these categories by using informal names based on scientific names. Cyprinoids, for example, comprise all the species in six families of the suborder Cyprinoidei (Cyprinidae, Psilorhynchidae, Gyrinocheilidae, Homalopteridae, Cobitidae and Catostomidae) whereas cyprinids refers solely to the species in the one family Cyprinidae.

BODY PLAN OF FISH

The four extant classes of fishes are, in their way, as different from each other as are the classes of land vertebrates. Some of these differences, although fundamental, concern details of anatomy; other differences are more obvious. It must be remembered that in those classes containing large numbers of species (especially the bony fishes) there are fishes whose features make them exceptional to the generalizations given opposite. Such is the nature of the diverse creatures grouped together under the title "Fishes."

▼ **Gill structures in fishes.** (a) In the hagfishes water passes through a series of muscular pouches before it leaves through a single common opening. (b) In the lampreys each gill has a separate opening to the outside and the gills are supported by an elaborate cartilaginous structure called the branchial basket. (c) In the living chondrichthyes (sharks, skates and chimaeras) the gills (except in the chimaeras) primitively open directly to the outside via five slits. (In the chimaeras an operculum or cover is developed.) (d) In bony fishes the gills are protected externally by a bony operculum (cover).

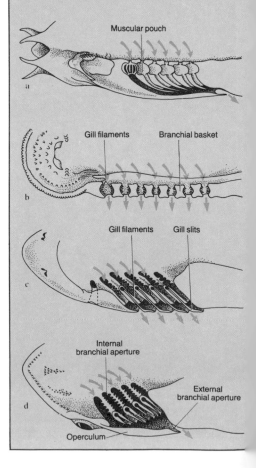

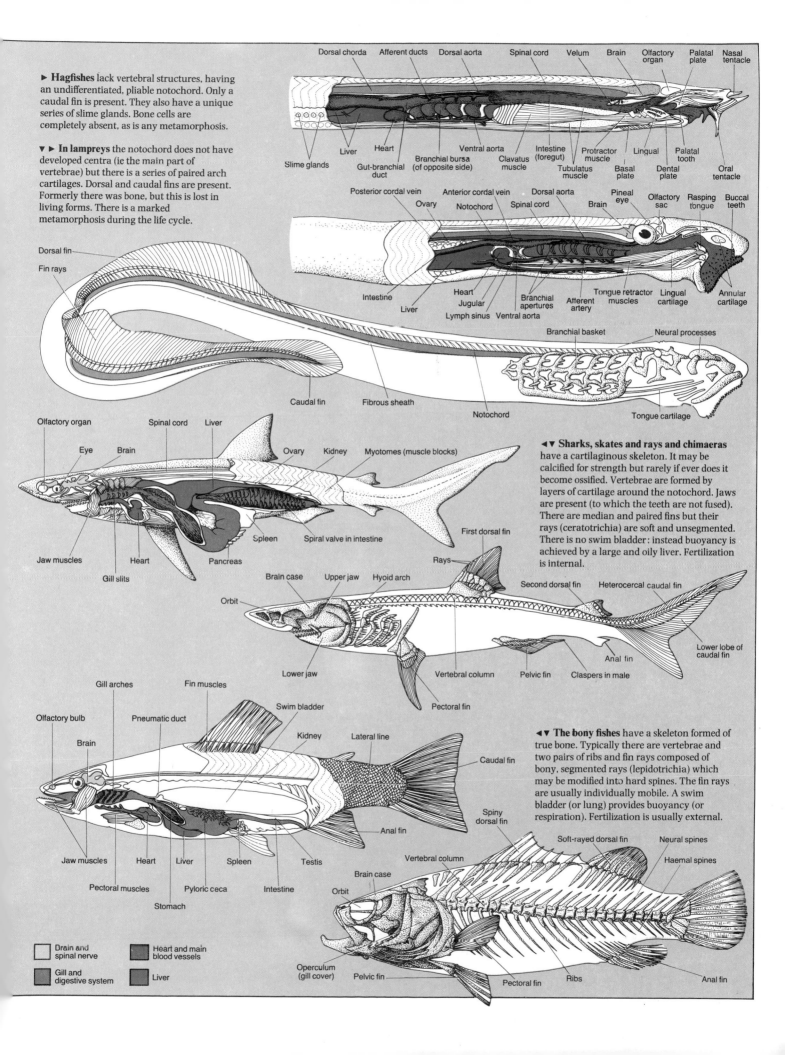

▶ **Hagfishes** lack vertebral structures, having an undifferentiated, pliable notochord. Only a caudal fin is present. They also have a unique series of slime glands. Bone cells are completely absent, as is any metamorphosis.

▼▶ **In lampreys** the notochord does not have developed centra (ie the main part of vertebrae) but there is a series of paired arch cartilages. Dorsal and caudal fins are present. Formerly there was bone, but this is lost in living forms. There is a marked metamorphosis during the life cycle.

◀▼ **Sharks, skates and rays and chimaeras** have a cartilaginous skeleton. It may be calcified for strength but rarely if ever does it become ossified. Vertebrae are formed by layers of cartilage around the notochord. Jaws are present (to which the teeth are not fused). There are median and paired fins but their rays (ceratotrichia) are soft and unsegmented. There is no swim bladder: instead buoyancy is achieved by a large and oily liver. Fertilization is internal.

◀▼ **The bony fishes** have a skeleton formed of true bone. Typically there are vertebrae and two pairs of ribs and fin rays composed of bony, segmented rays (lepidotrichia) which may be modified into hard spines. The fin rays are usually individually mobile. A swim bladder (or lung) provides buoyancy (or respiration). Fertilization is usually external.

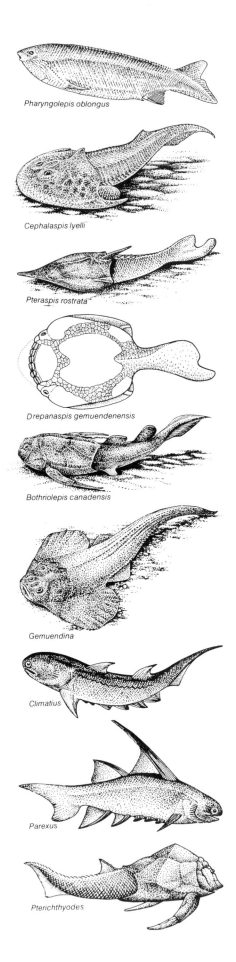

Pharyngolepis oblongus

Cephalaspis lyelli

Pteraspis rostrata

Drepanaspis gemuendenensis

Bothriolepis canadensis

Gemuendina

Climatius

Parexus

Pterichthyodes

sylvanian) rocks of Illinois, USA (dating from 325–280 million years ago). So far no indisputable fossil hagfishes have been found.

The earliest of the true jawed fishes were the acanthodians. Spines belonging to this group of large-eyed, scaled fishes have been reported from rocks 440 million years old. They probably hunted by sight in the upper layers of the waters. Some of the largest species which grew to over 2m (6.5ft) long have jaws that suggest they were active predators, much like sharks today. The majority of the acanthodians, though, were small fishes. The earliest acanthodians were marine; the later species lived in fresh water.

Acanthodians had bone, ganoid scales and a stout spine in front of each fin except the caudal. Most species had a row of spines between the pectoral and pelvic fins. The tail was shark-like (heterocercal), ie the upper lobe was longer than the lower. The tail shape and the presence of spines have led to them being called spiny sharks, despite the presence of bone and scales. Recent research has suggested that they may, however, be more closely related to the bony fishes. They never evolved flattened or bottom-living forms. One hundred and fifty million years after they appeared they became extinct.

The other group of extinct fishes is the placoderms, a bizarre class which may be related to sharks, or to bony fishes, or to both or to all other jawed fishes; no one knows for certain. The front half of their body was enclosed in bony plates, formed into a head shield which articulated with the trunk shield. Most had depressed bodies and lived on the bottom. One order, however, the arthrodires, were probably fast-swimming active predators growing to 6m (20ft) long. They probably paralleled the living sharks in the same way that the very depressed rhenanids paralleled rays and skates. Another group of placoderms, the antiarchs, are among the strangest fishes that have ever evolved. About 30cm (1ft) long, they had an armor-plated trunk which was triangular in cross section. The eyes were very small and placed close together on the top of the head. The "pectoral fins" are most extraordinary and unique among vertebrates. Whereas all other vertebrates had developed an internal skeleton, the "pectoral fin" of these antiarchs had changed to a crustacean-like condition. The "fins" resemble the legs of a lobster, ie a tube of jointed bony plates worked internally by muscles. The function of these appendages is unknown, but they may have been used

to drag the fish slowly over mud or rocks. The antiarchs also have a pair of internal sacs, which have been interpreted as lungs. In another group of placoderms, the ptyctodontids, a little can be surmised about their reproduction. The pelvic fins show sexual dimorphism, those of the males being enlarged into shark-like claspers. From this it is assumed that fertilization was internal.

Most of the early, and a few of the later, placoderms lived in fresh water; the rest, including the delightful antiarchs, were marine. This enigmatic group appeared about 400 million years ago and died out about 70 million years later.

History of fish in captivity

Seventy percent of the earth's surface is covered with water which, unlike the land surface, offers a three-dimensional living space, very little of which is devoid of fish life. The overall contribution that fish make to the total vertebrate biomass is therefore considerable. Yet because they are not as obvious a part of the environment as are birds and mammals they have not been as generally appreciated as they deserve, though man has been interested in them for at least 4,000 years.

It is not known when fishes were first kept in captivity; nor is the motive for doing this known, although it may be surmised that it was to provide an easily obtainable source of fresh food rather than for aesthetic purposes. Around 4,000 years ago the Middle East in general, and the fertile crescent of the Tigris and Euphrates in particular, were much wetter and more fertile than now. It was in this period that the first identifiable fish ponds were built by the Sumerians in their temples. The Assyrians and other races followed a little later. It is possible that fishes left behind and surviving in natural depressions after a flood may well have suggested the idea. It is not known which fishes were kept in the fertile crescent ponds. The Assyrians depicted fishes on their coins, but diagrammatically, so we cannot identify them.

The story is quite different with the Egyptians. Their high standard of representational art has enabled the fishes in their ponds to be identified. Even more conveniently they mummified some of their important species so that the accuracy of the drawings can be checked. Various species of "Tilapia" (still a much valued food fish in the region), Nile perch and *Mormyrus* are among those present. The Egyptians added a new dimension to the functional aspect of fish ponds, that of recreation. Murals depict

fishing with a rod and line, which must have been for fun because it is not as efficient as netting for catching fish in commercial quantities. The Egyptians also worshiped their fish.

The Roman Marcus Terentius Varro (116–27 BC) wrote in his book *De re rustica* of two kinds of fish ponds: there were fresh-water ponds (*dulces*) kept by the peasantry for food and profit, and salt-water ponds (*maritimae* or *sales*) which were only owned by wealthy aristocrats who used them to entertain guests. Red mullet were especially favored as the color changes of the dying fish were admired by guests before the fish was

cooked and eaten. Large moray eels were also kept and in the most extreme examples they were decorated with jewelry and fed on unwanted or errant slaves. There is also a record of 6,000 morays being on show at an imperial banquet. Although the Romans had a glass technology there is no record of any form of aquarium having been constructed. The Romans' involvement with fish was not totally for show. They explored fish culture methods and were known to have transported fertilized fish eggs. They may well have externally fertilized the eggs by stripping the fish.

Records of fishes are few in the western

world after the fall of the Roman Empire. Indeed, during the Dark Ages there are few records of anything. Cassiodorus (*c*.490–*c*.585 AD) mentioned that live carp were taken from the Danube to the Goth king Theodoric at Ravenna in Italy. Charlemagne marketed the live fishes he kept in ponds.

The tradition had doubtless been kept alive, however, by clergy and nobles. It is stated in Domesday Book (1086) that the Abbot of St Edmund's had fish ponds providing fresh food for the monastery table and that Robert Malet of Yorkshire had 20 ponds taxed to the value of 20 eels. Stew ponds were common in monasteries during the Middle Ages and were regarded as essential as the church forbade the eating of meat on Friday. The word stew comes from the Old French *estui*, meaning to confine, and does not allude to the means of preparation of the fish for consumption.

Modern-style aquaria were developed in the first half of the 19th century. At a meeting of the British Association for the Advancement of Science in 1833 it was shown that aquatic plants absorbed carbon dioxide and emitted oxygen thereby benefiting the fish. The first attempt at keeping marine fish alive and the water healthy by the use of plants was made by Mrs Thynne in 1846. Only six years later came the first public aquarium of any size, built in the Zoological Gardens in London. In late Victorian times, as now, many homes had aquaria and the invention of the heater and thermostat latterly allowed more exotic fishes to be kept.

But it is to the humblest of all aquarium fishes, the goldfish, that we now return. This species is native to China and has been bred for its beauty for over 4,500 years. In 475 BC Fan Lai wrote that carp culture had been associated with silkworm culture—the fish feeding on the feces—since 2689 BC. About 2000 BC the Chinese were, according to fishery experts, artificially hatching fish eggs. Red goldfish were noted in 350 AD and during the T'ang dynasty (about 650 AD) gold-colored fish-shaped badges were a symbol of high office. By the 10th century elementary medicines for fish were available. Poplar bark was advocated for removing fish lice from goldfish.

In the wild state goldfish are brown, but when first brought into Britain (probably in 1691) gold, red, white and mottled varieties were available. They certainly were in 1728 when the merchant and economist Sir Matthew Dekker imported a great number and distributed them to many country houses.

They reached America in the 18th century and are now one of the most familiar fishes. Next time you see them in suspended plastic bags as prizes at a fair, think on their long history, for they are not actually such humble fish but fish of a long pedigree.

Endangered fishes

In 1982 the Fish and Wildlife Service of the USA proposed the removal of the Blue pike (*Stizostedion vitreum glaucum*) and the Longjaw cisco (*Coregonus alpenae*) from the American "List of Endangered and Threatened Wildlife." Not because their numbers had recovered to their former abundance but because they were deemed extinct. The Blue pike, which formerly lived in the Niagara River and Lakes Erie and Ontario, had not been seen since the early 1960s. The Longjaw cisco, from Lakes Michigan, Huron and Erie, was last reported in 1967. What caused their final disappearance after thousands of years of successful survival? Both species, directly and via their food chain, were severely hit by pollution and, on top of that, the Longjaw cisco in particular suffered from predation by parasitic lampreys that came into its habitat when the Welland Canal was built (see p18).

In South Africa a small minnow, now called *Oreodaimon quathlambae*, was des-

◄▲ Sea fish for public view. During the latter half of the 19th century public aquaria were popular spectacles as well as meeting places. The public aquarium LEFT, was constructed for the 1867 Paris exhibition.

In contrast a modern aquarium, such as the Fish Roundabout at the Steinhart Aquarium in San Francisco ABOVE, has a totally different appearance and function: there is a strong emphasis on facilitating the conservation and reproduction of endangered species.

◄ Ornamental fishes, some of the cultivated varieties of goldfish. Centuries of intensive breeding have led to the development of varieties of fish that could not live in the wild.

▼ Danger in Death Valley. Some fish now live throughout the world. Others survive only in particular places and habitats. The Devil's Hole pupfish seen here is restricted to a tiny water-filled hole in the floor of Death Valley, USA. Its future depends on the maintenance of the water table. Further pumping of water in the region will destroy its food supply.

cribed in 1938. A few years later it was extinct in its original locality in Natal. No canals had been dug, pollution was minimal, but an exotic species, the Brown trout (*Salmo trutta*), had been introduced to provide familiar sport for the expatriates. Small trout ate the same food as *Oreodaimon* and large trout ate the *Oreodaimon*; result, no *Oreodaimon* in the Umkomazana river. Luckily late in the 1970s a small population was discovered living above a waterfall in the Drakensberg mountains. The fall had prevented the spread of trout, but they had recently been transplanted above the falls. Although this population is protected as far as possible from predation, a more serious threat to its survival is the overgrazing of the land adjoining the river, causing the silting up of the river and changes in the river's flow.

In parts of Malaysia, Sri Lanka and in Lake Malawi some of the more brightly colored freshwater fishes are becoming harder and harder to find. In these cases, on top of any other pressures that may exist, the numbers have fallen because of extensive collecting for the aquarium trade and concomitant environmental damage.

Formerly, large parts of the American southwest were covered with extensive lakes. With post-Pleistocene desiccation (ie about 100,000 years ago) the lakes, with their associated fishes, dwindled, and now some of the pupfish only survive in minute environments. The Devil's Hole pupfish (*Cyprinodon diabolis*) lives only in a pool 3 × 15m (10 × 50ft) located 18m (60ft) below the desert floor in southern Nevada. It depends for its food on the invertebrates living in the algae on a rock shelf 3 × 6m (10 × 20ft) just below the surface of the water. Although in a protected area in the Death Valley National Monument, distant pumping of subterranean water lowered the water table, threatening to expose the shelf and deprive the fish of its only source of food. Attempts to transplant some of the fish to other localities failed. Hastily a lower, artificial shelf was installed and the case was brought up before the US Supreme Court. They ruled in favor of the fish, the pumping was stopped, the water level stabilized and the Devil's Hole pupfish was saved.

The examples given above illustrate what is happening and what can be done. All the fishes mentioned live in fresh waters, where the area occupied is small enough for low-level population changes to be monitored. The same detailed knowledge about marine fishes is lacking because they can occupy much greater territories, precluding the col-

lection of such information.

For whatever reason, far too many species of fish are endangered, ie in danger of extinction. What is being done to secure their future? Many countries are signatories to an organization whose acronym is CITES (Convention on International Trade in Endangered Species). For this the member countries produce lists of their endangered animals and plants and all agree that there shall be no unlicensed trade in certain of the species. Unfortunately as there is little general awareness about fish, and hence little political capital to be gained, fish do not feature strongly on the CITES lists. Mammals and birds occupy the great majority of the pages. To complicate the issue, would a customs officer (who is largely responsible for the implementation of the rules by ascertaining that the species in the consignment are what they purport to be) be able to identify accurately the 5,000 species he might be confronted with? Even if he did (and many a professional fish specialist finds living fish difficult as the diagnostic characters may be internal) what would then happen to the fish? Even if it were possible to ship them back to their native land before they died, there is no guarantee they would be released into their original habitat.

Some countries follow the spirit of the laws of conservation better than others. In America, dam construction on the Colorado River prevented many of the endemic species from breeding. Many species quickly became very rare and were in danger of extinction. After prompting from scientists, the Dexter National Fish Hatchery was constructed in New Mexico and the endangered species were taken there and allowed to breed. The breeding program has been successful and many young are put back into suitable sites each year. But even this venture is only a partial answer. Pity poor *Gambusia amistadensis* that now lives and breeds in a reed-fringed pool at Dexter. It cannot be reintroduced. The Goodenough Spring, in which it lived, is now at the bottom of a reservoir.

A cynic might ask "Why bother to save a small fish from extinction?" There are two types of answer. The altruistic one is "because they are there and have as much right to be there as we have." The selfish answer is "because they could be of use to us." Living in a cave in Oman is a small, eyeless fish whose total population is probably 1,000. Recently it has been discovered that it can regrow about one-third of its brain (the optic lobes)—the only vertebrate known to do so. Such a discovery could be

of great importance in human neuro-
surgery. Luckily this fish was discovered;
how many more useful attributes were there
in fishes that have been allowed to die out?

The status of some species is more critical
than others, but what happens to them all
is almost entirely our decision. If the
endangered species are to survive, the right
decision must be taken—soon.

The Future of Fish

For a few years prior to 1882 there was an
extensive fishery for a tile fish (*Lopholatilus
chamaelionticeps*) off the east coast of
America. This nutritious, tasty species
reached about 90cm (3ft) in length and
18kg (40lb) in weight. Tile fishes lived in
warm water near the edge of the continental
shelf from 90 to 275m (300–900ft) down.
During March and April 1882 millions and
millions were found floating, dead, on the
surface. One steamer reported voyaging for
two days during which time it was never out
of sight of tile fish corpses. For over 20 years
no more were caught, but by 1915 the
numbers had recovered enough to sustain
a small fishery.

Despite the long gap in records, enough
individuals must have survived to be able to
rebuild the population, albeit to a lower
level. The cause of this carnage is uncertain,
but it is believed that ocean currents
changed and the warm water in which the
tile fish lived was rapidly replaced by an
upwelling of deep, cold water in which they
died.

Although many fish species undergo
natural variations in abundance, for no
matter what reason, a major natural disaster can affect them all. Should the population be at a low level at such a time, the
chance of extinction, and the loss of a
resource, could occur. Luckily the tile fish
population was high when the currents
changed.

It is largely axiomatic that where industries are few, rivers are clean. This is certainly true of parts of Scotland at least,
where communities depend for financial
viability on the spawning runs of the Atlantic salmon. For over a century the mainstay
of the tourist/hotel business has been the
anglers who stay in the hope of hooking a
salmon. Nearer the coast are the netsmen
who catch salmon as they begin their
upstream migration and sell their luxury
food in the markets further south. These
communities remained stable and thrived,
until an accidental observation was made in
the late 1950s.

Before that time nobody knew where the

Herrings and History

Factors, both natural and unnatural, can
combine on the most abundant species to
produce the most alarming and unforeseen
consequences. The Hanseatic states on the
south coast of the Baltic prospered until the
15th century, because not only did they have
a remarkably rich supply of herring on their
doorsteps but they also knew how to preserve
them and export them all over Europe. Then
the herring suddenly left the Baltic for the
North Sea. The Hanseatic states collapsed
economically and politically and the Dutch
rejoiced because they now had access to the
herring. The Dutch had a good fleet and set
about catching herring, frequently in British
waters. Charles I of England decided that a
good way of boosting his exchequer would be
to tax the Dutch for fishing in his waters. To
enforce the payment of this somewhat
unwelcome tax and to keep British herring for
the British, Charles had to build up the
declining British Navy. The Dutch decided to
protect their fishing boats and before too long
the two countries were at war. Although that
particular series of skirmishes was finally
resolved, the dispute over the ownership of
herrings (among others) continues. The most
recent wrangle over herring quotas within the
member states of the EEC is merely the
modern, politely political equivalent of firing
cannon balls at one another.

Parenthetically, one should note that "the
battle of herrings" occurred in 1429 when Sir
John Fastolf routed the French who tried to
prevent him taking herrings to the Duke of
Suffolk who was besieging Orleans.   KEB

The examples given above (and in the box) demonstrate how valuable are the fish stocks. Fish should not be just a short-term source of profit for a few but a continuing self-renewing resource; and they would be if used properly. When herring were caught by drift nets, before the Second World War, there were over 1,600km (1,000mi) of net out in the North Sea each night. Yet because the shoals moved up and down in the water, and the nets only occupied the top 6m (20ft), enough herring escaped to spawn and thereby maintain the stocks. Changing fishing techniques, using echo sounders and adjusting the depth of the net, resulted in a much greater yield for effort ratio, but also a rapid destruction of the stocks. It was only a few years ago that herring fishing was banned in an effort to conserve stocks. Although this was laudable, were one of the few remaining shoals caught by accident it would be too late by the time the error was discovered.

It is not only prime food fish that are under threat. Species living in deeper waters, such as rat-tails and argentines, are now caught; their flesh is shredded, breaded, molded and served as fish fingers. When it becomes uneconomical to catch these species, what happens? Already there is random catching of nonfood fish which are turned into fertilizer. Add to all these pressures the unnatural, ie man-made, catastrophes, which are becoming commoner than heretofore. The dumping of industrial waste may have poisoned fish locally, but more importantly the toxic substances become concentrated up the food chain and can reach deleterious levels in the fish that we eat from the top of the chain. Oil spillages have prevented eggs near the surface from hatching, yet for the comfort of our way of life we want the waste disposed of and the oil transported. The nets used to catch fish for fertilizer have a small mesh so they also catch the young of larger species before they can spawn. The trawls destroy the spawning grounds.

◄◄ ▼ Nature's bounty in human hands. In areas of the world where marine food supplies are rich fish are so numerous that many tonnes can be caught LEFT. However, no resource is so permanent that it can be taken as a perennial cornucopia. All resources have to be managed. Irresponsible overfishing, usually done to make a quick profit, is a short-sighted, destructive process.

One way of utilizing fish sensibly is to farm them. Here, ABOVE, the luxury fish salmon is farmed in the USA. If there is persistent overfishing it might also be viable to farm essential sources of fish flesh.

In the tropics BELOW, many fishes are sun-dried. This low-level technology enriches the diet of people far from water.

salmon passed their growing period in the sea, but a US nuclear submarine spotted vast shoals of salmon under the ice in the Davis Strait (between Greenland and Baffin Island). When this information became generally available commercial fishing boats went after this valuable catch. The stocks were rapidly decimated, leaving progressively fewer fish to return to their natal rivers to spawn. There were fewer fish for the netsmen to catch and their livelihood suffered; fewer still for the angler to catch and even fewer to spawn and replenish the stocks.

On two famous pools on the River Tay anglers normally landed about 500 fish before the end of May. In 1983 only 36 were caught. Anglers hardly bothered to come. Hotels that used to open shortly after Hogmanay did not bother to open until Easter. Result: fewer jobs and increasing economic decline in an already hard-pressed area devoid of other gainful employments.

Two further factors emphasize the ridiculous artificiality of the situation. Firstly, there is a slow climatic change that tends to reduce the number of fishes involved in the spring run. Secondly, salmon farming has reached the point at which it could provide such a large proportion of the market's demands that there is little necessity to continue cropping wild stocks at sea. Biologically, it is better to conserve vigorous and healthy wild populations as a genetic reserve in case inbreeding in farmed fish weakens their constitution. As salmon farming is fairly new this may seem a remote contingency, yet already the number of young born with foreshortened jaws, and hence difficulty in feeding and therefore slow growth rate, is already higher in some farmed stocks than in the wild.

Do all these factors, of which many people are unaware, necessarily spell out a recipe for disaster? The answer is probably "yes," unless an international compromise can be effected and adhered to. At the moment no such compromise is even planned. The seas can provide a sustainable crop if treated with care and thoughtfulness. The future of fishes is our responsibility and is intimately linked with the quality of human life. It is up to us to realize this and act accordingly to ensure that the blackest picture of all is never painted. KEB

Fishy Tales

Four thousand years of myth and mystery

When one thinks about the matter, humans have an enigmatic relationship with fish. To start with, fish live naturally in a medium that we cannot enter without the assistance of technology to take a microcosm of our medium with us. For most people an awareness of fish is bounded by fried cod on a plate (or smoked salmon at a party), goldfish (or Neon tetras) in an aquarium, anglers' tales of the one that got away (or failed so to do), thriller films about killer sharks and governments arguing over fishing rights.

It was not always so. Fish were sacred to the ancient Egyptians—one of the forms of the senior deity Isis has a fish head. A giant fish, the Nile perch, was deified and the town of Esueh was renamed Latopolis, from the Greek name for the fish, by Greeks who found thousands of these fish embalmed there. Latterly Latopolis was used as the root for the scientific name for the Nile perch, *Lates niloticus*.

The myth and mystery of fishes persist. In the 1970s a story appeared in an English newspaper of an angler in Ireland who hooked a pike of such size and power that it towed his boat along the lough before breaking the line. There is nothing new in our wonderment about the size of fishes, especially of the pike. Conrad Gesner, in 1558, wrote that a man took his mule to a lake to drink and there a pike bit the mule's lips and hung on and it was only after a struggle that the mule was able to pull the pike from the water. The size of the pike and the size of the mule were not given, but there is an implication that they were of the same order of magnitude and therefore the pike was giant. Somewhat better documented is Gesner's account of the Emperor's pike of which several paintings exist. Apparently in 1497 a pike was caught in Germany.

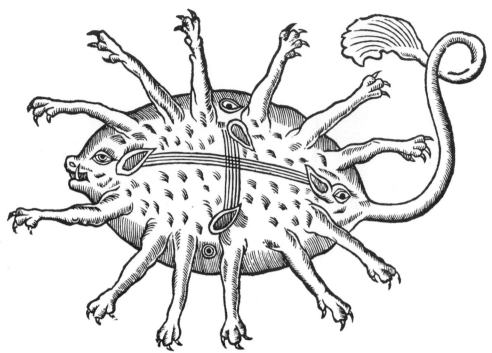

Around its "neck" was a copper ring bearing the message that the ring had been placed there by Frederick II in 1230. The pike was 5.8m (19ft) long and weighed about a quarter of a tonne. Furthermore, to prove it was not just an angler's tale, the skeleton was respectably housed in Mannheim cathedral. When the aforementioned skeleton was examined in the 19th century it was found that the skull came from one pike, but the body was compounded from several pike bodies joined together. The truth about the pike is less spectacular: it can weigh up to about 18kg (40lb) and live for some 30 years. Even so an 18kg pike would have left enough of an impression on a medieval peasant to start a myth about it.

▲ **Fact and fiction.** One of the first major animal encyclopedias based on careful study was that of the meticulous Conrad Gesner, published in 1551–58. Even he however could not entirely sift fact from fiction, especially where aquatic vertebrates were concerned. This picture from his work is derived from imagination.

▶ **Object of vain affection.** ABOVE This is the classic sailor's idea of a mermaid, with long hair (doubtless blond), looking glass, and comb. Regrettably there are no mermaids.

▶ **An attested improbability.** There are many well-documented accounts of fish falling from the sky, but no convincing explanations.

▼ **The ambition of every fisherman** is to catch a fish of this magnitude, which some claim to have done.

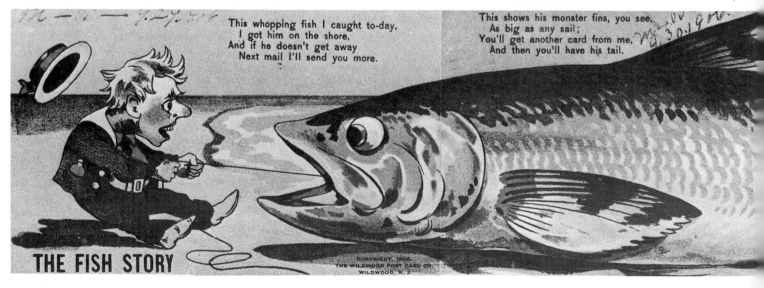

This whopping fish I caught to-day,
I got him on the shore,
And if he doesn't get away
Next mail I'll send you more.

This shows his monster fins, you see,
As big as any sail;
You'll get another card from me,
And then you'll have his tail.

THE FISH STORY

COPYRIGHT, 1906.
THE WILDWOOD POST CARD CO.
WILDWOOD, N.J.

Some myths, such as those about mermaids, can never be solved; their origins are too remote in both time and fact. The mermaids' mythological compatriots, the sea serpents, could have had their origins in misinterpretations of rarely seen fishes. The inquiring reader should consider the Oar fish (*Regalecus glesne*), the sun fishes (*Mola*), the Whale shark (*Rhincodon typus*) and "megamouth."

Shark attacks have been less dangerous to humans than fish poisons. A delicacy in Japanese restaurants is *fugu* made from the extremely poisonous liver of a porcupine fish. If cooked correctly it is edible; if not, more people can die in one restaurant in one night than are taken by sharks in a year. Edible fish can also prove fatal. Regicide can be claimed by lampreys: it is alleged that a surfeit of lampreys killed King Henry I of England.

Rains of fishes (and of frogs) have been witnessed many times in the last thousand years. Despite this we can only surmise about their causes. It is thought that a tornado passed over water, turned into a waterspout, sucked up the fish with the water and when the wind slackened the fish dropped. Most fish subject to such indignity are small, but in India there was a rain with fishes weighing up to 3kg (6.6lb). Sometimes small fish were carried up to the height at which hail formed and came down embedded in ice. In Europe some ten species of fish, including pike, herring, trout and sticklebacks, have fallen, but so far neither mermaids nor sea serpents have been recorded. KEB

His tail it lashed the briny sea,
I dropped my line to run,
And then he got away from me,
And thus my story's done.

LAMPREYS AND HAGFISHES – JAWLESS FISHES

Classes: Cephalaspidomorphi, Myxini
About seventy-two species in 12 or 17 genera
and 2 or 4 families.
Distribution: worldwide in cool marine and
fresh waters.

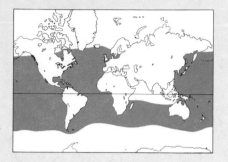

Size: length 8–90cm (8–35in).

Lampreys
Class: Cephalaspidomorphi
Order: Petromyzontiformes.

Family: Petromyzontidae or Petromyzonidae
(or subfamily: Petromyzontinae)
About thirty-six species in 4 or 9 genera.
Northern hemisphere.
Species include: **Brook lamprey** (*Lampetra
planeri*), **Freshwater lamprey** (*Lampetra
fluviatilis*), **Sea lamprey** (*Petromyzon marinus*).

Family: Geotriidae (or subfamily: Geotriinae)
Geotria australis.
Southern Australia, Tasmania, New Zealand,
Chile, Argentina, Falkland Islands and South
Georgia, possibly Uruguay and southern
Brazil.

Family: Mordaciidae (or subfamily:
Mordacinae)
Three species of the genus *Mordacia*.
Chile, SE Australia, Tasmania.

Hagfishes
Class: Myxini
Order: Myxiniformes.
Family: Myxinidae.
About thirty-two species in 6 genera.
Atlantic, Pacific and Indian oceans.

Dʊʀɪɴɢ the Devonian era (about 400 million years ago) the world's rivers and seas were occupied by heavily armored jawless vertebrates. Today their bony plates and head shields are common as fossils in rocks of the period, testifying to the success of their radiation. With the arrival of the jawed fishes these jawless forms became less common and are now represented by just two groups, lampreys and hagfishes.

Neither group bears much resemblance to the Devonian forms; the living forms are eel-like and lack bony plates in the skin. Indeed, the relationship of the two living groups to each other and to the fossil forms is the subject of much debate and different viewpoints are staunchly defended. It is true that they are both at a similar level of organization, ie at a pre-jaws stage, but both have had a long, separate history and there are many important internal differences. The association of the two groups here, and in many other works (often under the name of Cyclostomata) is one of convenience and should not be taken as necessarily implying a close relationship.

Lampreys are found in the cooler waters of both hemispheres. All have a distinct larval stage during which they little resemble the adults in structure or life-style. The adult is eel-like, with one or two dorsal fins and a simple caudal fin. There are no paired fins. The mouth is a disk adapted for sucking, bearing a complex arrangement of horny teeth, the exact arrangement of which is specific to a particular species and heavily relied upon in classification. There are seven gills, each of which has a separate opening.

Water enters each side through an opening near the mouth and then flows into any of seven ducts, each of which leads to one of seven external gill openings. Each duct has a muscular pouch which encourages water flow. Between the eyes on the top of the head, and just behind the single median nostril is a small patch of translucent skin covering the pineal organ. In lampreys this organ is light sensitive and the light levels it receives control hormone levels. (In higher vertebrates the sensitivity to light is lost but hormonal control is still exercised by the remaining part of the organ.) The lamprey's skin has glands that secrete a mucus with toxic properties; it is thought to discourage larger fish from eating the lamprey. (The mucus should be removed before using lampreys for human consumption as it can cause severe stomach upsets.)

The inner ear of bony fish and the higher vertebrates consists of three semicircular canals at right angles which maintain balance as well as being involved in the hearing process. In lampreys only the two vertical canals are present; the horizontal canal is absent.

From a biological point of view there are two interconnected groups of lampreys: those that spend most of their adult life at sea and enter fresh water to breed (compared with those that spend their entire life in fresh water) and those that feed parasitically on fish as adults (compared with those that feed on small invertebrates).

To some extent all lampreys move upstream to spawn. The marine species, which are larger than the freshwater species, are anadromous, ie they leave the sea to migrate up rivers. A problem immediately encountered by these species is the need to adjust the concentration of salts in their blood and body fluids as they move from salt water to fresh water.

The timing of the spawning run varies with species and locality. In northwest

▲ **The sucking mouth** of lampreys is ringed by horny teeth which rasp away the flesh of their prey.

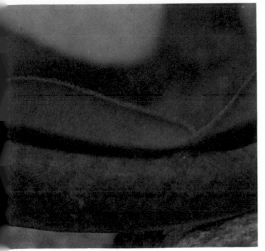

◄ **At rest, a Brook lamprey** lying near the bed of a stream. The row of seven gill openings, an eye and the nostril led to lampreys once being known as "nine-eyes." Young Brook lampreys, unlike those of other species, do not migrate to sea but move straight to the spawning area where spawning begins immediately.

Europe the Brook lamprey begins its run in September or October but in the Adriatic region the peak comes in February and March. In northwest Russia there are spring and fall runs. Lampreys in the southern hemisphere spend much longer on the spawning run and may not spawn until a year or more after they enter fresh water. The spawning grounds may be several hundred kilometers from the estuary and lampreys may, in slow-flowing stretches, be able to travel at 3km (about 2mi) each day. Weirs and rapids may be overcome, and lampreys cope with weariness by temporarily holding on to rocks with their suckers. Lampreys do not feed during the spawning run.

During the upstream migration and the subsequent spawning some changes in body shape occur. These may be as minor as the changes in the relative positions of the two dorsal fins and the anal fin or as substantial as the enlargement of the disk (sucker) in males of the southern genera *Geotria* and *Mordacia*. The males of *Geotria australis* and *Mordacia lapicida* (but not of other species of *Mordacia*) develop a large, spectacular, sac-like extension of the throat. Its function is unknown.

The spawning grounds, often used year after year, are chosen for particular characteristics of which gravel of the right size for the larvae to live in seems to be the most important. Other features are less important, although water about 1m (3ft) deep and of moderate current is favored. The males of the Sea lamprey arrive first and start building a nest by moving stones around to make an oval, shallow depression with a gravelly bottom. Large stones are removed and the rest graded, with the biggest upstream. All species of lamprey build similar nests but in different species the sex of the nest-builder differs and may not be consistent.

Spawning is usually a group activity, in groups of 10–30 in the Brook lamprey, in small numbers in the Sea lamprey, often a pair to a nest. In the Freshwater lamprey there is courtship. While the male is building the nest the female swims overhead and passes the posterior part of her body close to the male's head. It is suggested that stimulation by smell may be involved. Fertilization occurs externally, the male and female twining together and shedding sperm and eggs, respectively, into the water. Using her sucker the female attaches herself to a stone and the male then sucks onto her to remain in position during spawning. In the Brook lamprey two or three males may attach to the same female. Only a few eggs are extruded at a time, so mating takes place at frequent intervals over several days. The eggs are sticky and adhere to the sand grains, and the continued spawning activity of the adults covers the eggs with more sand. After spawning the adults die.

The eggs hatch into burrowing larvae (called ammocoetes) which are structurally unlike the adults. Their eyes are small and hidden beneath the skin, and light is detected by a photosensitive region near the tail. The sucking mouth with its horny teeth is not yet developed. Instead there is an oral hood, rather like a cowl or a greatly expanded upper lip, at the base of which is the mouth surrounded by a ring of filaments

(cirri) which act as strainers. Water is drawn through the oral hood by the action of the gill pouches and of a valve-like structure behind the mouth called the velum. On the inner surface of the hood are rows of minute hairs (cilia) and much sticky mucus. Particles in the water are caught on the mucus and the action of the cilia channels the food-rich mucus through the mouth into a complex glandular duct (the endostyle) at the base of the pharynx. The ability of the ammocoetes to secrete mucus is important, for not merely does it enable them to feed but it also helps them to cement the walls of their burrows and stop them from collapsing.

Ammocoetes rarely leave their burrows, and then only at night, but they frequently change position. In the burrows they lie partially on their backs, tail down, with oral hood facing into the current. In this position the importance of the photosensitive tail region can now be appreciated: it helps the ammocoete to orient itself correctly. The longest phase in the life history of lampreys is spent as ammocoete larvae (often called "prides"). The Sea lamprey can pass seven years of its life in this stage, *Ichthyomyzon fossor* six, the Brook lamprey three to six years and *Mordacia mordax* three years.

Although lampreys live successfully as ammocoetes and as adults, the change from one state to the other is a dangerous period, from which high mortality results. The changes involved are profound: the entire mouth and the feeding and digestive systems have to be restructured, eyes have to be developed and the burrowing habit abandoned for a free-swimming one. During this time, which may last for eight months, lampreys do not and cannot feed. It was discovered in Russia, for example, that large numbers of metamorphosing lampreys found dead during early spring had been

◄ **Anchored against the flow,** a Brook lamprey uses its sucker to hold itself fast to a rock. The lamprey mouth is efficient and effective, being used for carrying stones for the nest and in mating.

▼ **Hagfish are normally sedentary,** living in burrows in mud, the walls of which are partly reinforced by secretions of slime. Hagfish eat rarely; they have a low metabolic rate and store a lot of fat. When they are hungry their sedentary habits cease. They swim around leisurely, tasting odors in the water by lifting their heads and spreading their barbels.

effectively suffocated by the breakdown of the velum, after the skin of the gills and the mouth had been blocked by mucus.

Metamorphosis in northern hemisphere lampreys usually starts in late summer. In any given area the ammocoetes start their transformation almost simultaneously, mostly within a couple of weeks. Environmental conditions seem to be the trigger: it has been noted that the metamorphosis starts earlier in colder waters. Initially the newly eyed adults are inactive but then a downstream migration begins. It is now that the difference between the parasitic and nonparasitic lampreys becomes important. After metamorphosis nonparasitic lampreys do not feed: they breed and die. Hence all their growth is spent as a larva. Parasitic lampreys, however, feed on the blood and fluids of fishes for up to two years, sometimes with spectacular effect.

During the parasitic phase lampreys travel extensively. Species that go to sea have been caught many kilometers off the coast and at depths as great as 1,000m (about 3,300ft). Lampreys detect their prey by sight and usually attach to the lower side surface in the central third of the body. They move in with the sucker closed to reduce water resistance, but open it just before attack. After attachment they may move to a more favorable position. The opening of the disk brings the teeth into play. Only rarely do fish free themselves of lampreys, by coming to the surface and turning over so that the lamprey's head is in the air. During the course of the parasitic phase it has been calculated that 1.4kg (3lb) of fish blood will suffice to feed the lamprey from metamorphosis to spawning. Scarcely any fish is immune from attack, not even the garfishes which are covered with diamond-shaped scutes. Several lampreys may attack the same fish and newly metamorphosed individuals are the most voracious.

Any change in the environment may have unforseen consequences. There used to be an extensive and profitable fishery for trout in the Great Lakes of North America. When the Welland Canal was built to allow ships to sail from Lake Ontario into Lake Erie (opened in 1828) it also let through sea lampreys. From Lake Erie they traveled into lakes Huron, Michigan and Superior. By the mid 1950s sea lampreys were well established and the fishery was decimated. Considerable attempts were made to control the lampreys and restock the trout, but not only were these attempts expensive, they were also unsuccessful. The only real hope now is that a biological balance will be achieved before pollution wipes out all the fish life.

Hagfishes are superficially undistinguished and unprepossessing animals. They are eel-like, white to pale brown, with a fleshy median fin on the flattened caudal region and four or six tentacles around the mouth. But their superficial simplicity hides a number of extraordinary, even unique, features. Hagfishes have neither jaws nor stomach, yet they are parasites of larger fishes. They have few predators but defend themselves by exuding large amounts of slime. They can tie their bodies into knots. They also have several sets of hearts.

There are 32 species, which rarely exceed

70cm (about 28in) in length and live in cold oceanic waters at depths from 20 to 300m (65 to about 1,000ft). The exact number of species is not known for certain because of the paucity of agreed, tangible characters and the unknown degree of individual variability. (The two best known genera are *Eptatretus* and *Myxine*.) The relationship of the hagfishes to the other extant jawless fishes, the lampreys, is uncertain and the subject of much debate.

The mouth is surrounded by four or six tentacles according to species. There are no eyes; instead there are two pigmented depressions on the head and other, unpigmented, regions on the head and around the cloaca that can detect the presence of light. The senses of touch and smell are well developed. There is a single median nostril above the mouth which leads into the pharynx; a blind olfactory sac is also present which detects odors. During breathing, water enters the nostril from where it is pumped by the velum to the 14 or so pairs of gill pouches leading from the pharynx to the outside. In the gill pouches oxygen is taken from the water by the blood which simultaneously loses its carbon dioxide. It is thought that the skin also absorbs oxygen and excretes carbon dioxide. Externally the gill pouches are visible as a series of small pores, increasing in size towards the rear. This row is continued by small openings of the slime glands. The only teeth of the hagfish are on the tongue.

The main prey of hagfishes are dying or dead fish and when these are detected the hagfishes swim rapidly upcurrent to reach them. Now the tongue and flexible body come into play. The toothed tongue quickly rasps a hole in the side of the fish. With a large fish extra leverage is needed, so hagfishes loop their bodies into a half hitch and use the fore part of this loop as a fulcrum to help thrust the head into the body of the fish. Hagfishes feed quickly and may soon be completely inside the fish, voraciously eating the flesh. It is not rare for a trawler to catch the remains of a fish almost hollow save for a sated hagfish inside. Worms and other invertebrates are also eaten.

The ability to knot the body has other uses. It can be used to evade capture, especially when this is coupled with slime secretion. Knotting is also used to clear a hagfish of secreted slime which might otherwise clog gill openings and cause suffocation. In this case it simply wipes itself free of slime by sliding through the knot. The nostril cannot be cleared in this maneuver, so a powerful "sneeze" has been developed to reestablish normal nasal functions. Slime is secreted into water as cotton-like cells which, on contact with sea water, rapidly expand, coagulate and form a tenacious mucous covering. A single North Atlantic hagfish (*Myxine glutinosa*) 45cm (18in) long can, when placed in a bucket of sea water, turn the entire contents of the bucket into slime.

Like most fish hagfishes have a heart to pump the blood through the fine vessels of the gills. However, they also have a pair of hearts to speed the blood after it has passed through the gills. Another heart pumps the blood into the liver, and there is a pair of small hearts, like reciprocating pumps, near the tail. These accessory hearts are necessary because in hagfishes alone the blood is not always contained in restraining blood vessels. Hagfish have a series of open spaces or "blood-lakes," called sinuses, in which there is a great drop in blood pressure. An accessory heart is found after each sinus to increase the pressure and push the blood on to the next part of the circulatory system. The tail hearts are a puzzle. They are

▼ ► **Lampreys and hagfishes.** (1) Juvenile lampreys (ammocoete larvae) on the bottom, filtering detritus from water (8–10cm, 3–4in). (2) Head of a juvenile Freshwater lamprey (*Lampetra fluviatilis*) showing the oral hood. (3) Mouth of an adult Freshwater lamprey showing horny teeth. (4) River lampreys feeding on a sea trout (50–60cm, 20–24in). (5) A Sea lamprey building a nest (1m, 3.3ft). (6) Head of a hagfish (a species of the genus *Myxine*) showing the mouth and nostril surrounded by tentacles. (7) A hagfish of the genus *Myxine*, showing its eel-like form (60cm, 24in). (8) A hagfish about to enter the body of its prey. By twisting into a knot it can gain extra leverage for thrusting itself into the fish.

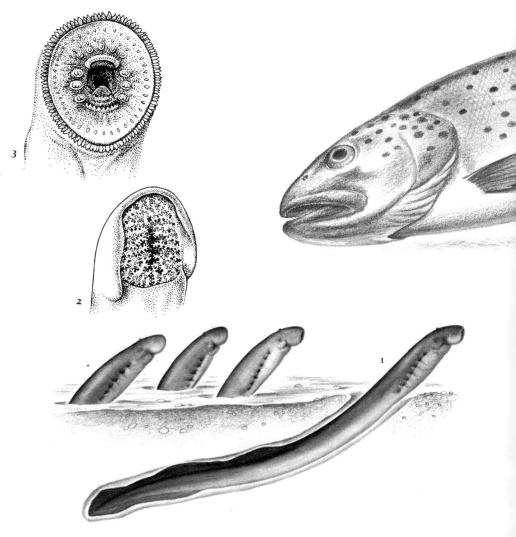

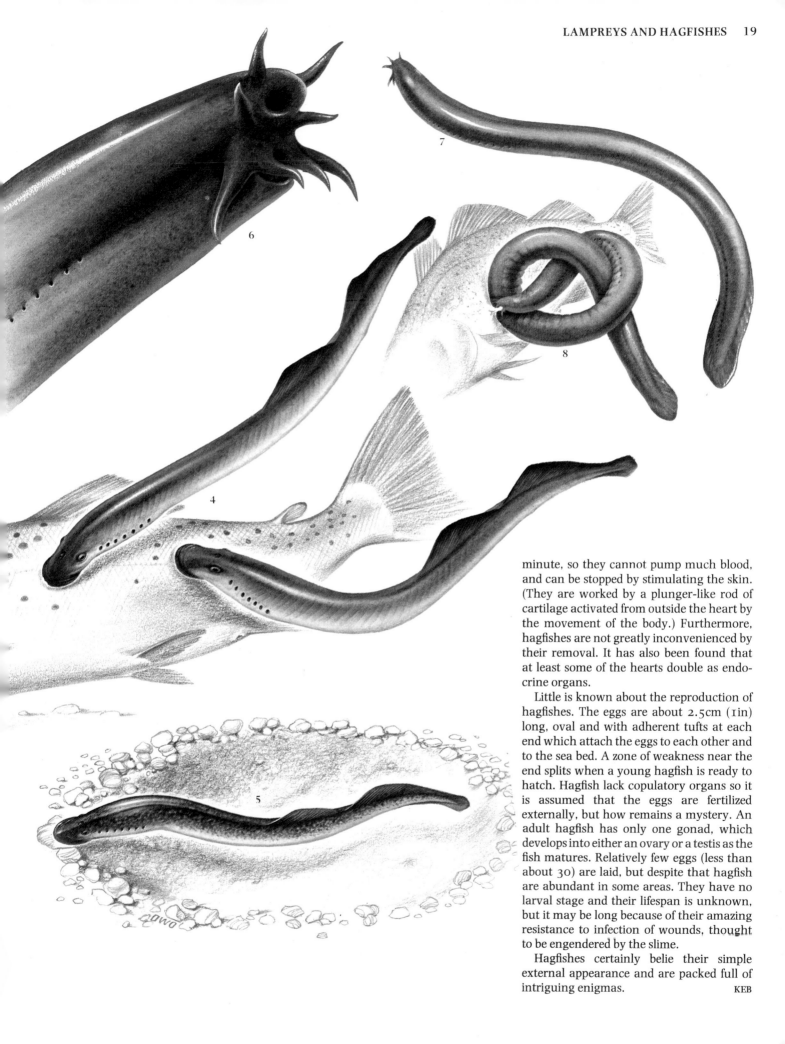

minute, so they cannot pump much blood, and can be stopped by stimulating the skin. (They are worked by a plunger-like rod of cartilage activated from outside the heart by the movement of the body.) Furthermore, hagfishes are not greatly inconvenienced by their removal. It has also been found that at least some of the hearts double as endocrine organs.

Little is known about the reproduction of hagfishes. The eggs are about 2.5cm (1in) long, oval and with adherent tufts at each end which attach the eggs to each other and to the sea bed. A zone of weakness near the end splits when a young hagfish is ready to hatch. Hagfish lack copulatory organs so it is assumed that the eggs are fertilized externally, but how remains a mystery. An adult hagfish has only one gonad, which develops into either an ovary or a testis as the fish matures. Relatively few eggs (less than about 30) are laid, but despite that hagfish are abundant in some areas. They have no larval stage and their lifespan is unknown, but it may be long because of their amazing resistance to infection of wounds, thought to be engendered by the slime.

Hagfishes certainly belie their simple external appearance and are packed full of intriguing enigmas.   KEB

STURGEONS, PADDLEFISHES

Order: Acipenseriformes
Twenty-seven species in 6 genera and
2 families.

Sturgeons
Family: Acipenseridae
Twenty-five species in 4 genera.
Distribution: North Atlantic, North Pacific and
Arctic oceans and associated feeder rivers.
Size: adult length 0.9–9m (3–20ft).
Species include: **Baltic sturgeon** (*Acipenser
sturio*), **beluga** (*Huso huso*), **Kaluga sturgeon**
(*H. dauricus*), **Lake sturgeon** (*Acipenser
fulvescens*), **sterlet** (*A. ruthenus*), **White sturgeon**
(*A. transmontanus*).

Paddlefishes
Family: Polyodontidae
Two species in 2 genera.
Distribution: Mississippi and Yangtze rivers.
Size: maximum adult length about 3m (10ft).
Species: **Chinese paddlefish** (*Psephurus gladius*),
paddlefish (*Polyodon spathula*).

STURGEONS are amazing. They are the largest freshwater fishes, and the longest lived. They also provide the expensive delicacy caviar.

Sturgeons and paddlefishes are the only survivors of an ancient group of fishes known from the Upper Cretaceous period (135–100 million years ago). (Together with five extinct orders they form the infraclass Chondrostei.) Today they are confined to the northern hemisphere, within which there are two distributions, one centered on the Pacific, the other on the Atlantic.

Sturgeons have heavy, almost cylindrical bodies which bear rows of large ivory-like nodules in the skin (scutes or bucklers), a ventral mouth surrounded by barbels, a heterocercal tail and a cartilaginous skeleton.

Some sturgeons live in the sea but breed in fresh water, others live entirely in fresh water. Little is known of the life at sea of anadromous sturgeons. They appear to take a wide variety of food including mollusks, polychaete worms, shrimps and fishes. Adults have few enemies, though they are known to be attacked and even killed by the sea lamprey. The Baltic sturgeon has been found at depths of over 100m (330ft) in submarine canyons off the continental shelf. Although it was known that the Kaluga sturgeon inhabited fishing grounds off

Sakhalin Island (USSR), it was not until 1975 that the first specimens were caught in the well-fished grounds off Hokkaido, northern Japan. The White sturgeon is known to travel over 1,000km (625mi) during its time at sea. Freshwater sturgeons usually remain in shoal areas of large lakes and rivers feeding on crayfish, mollusks, insect larvae and various other invertebrates, but rarely on fishes. Seasonal movements are from shallow to deeper waters in the summer and a return to the shallows in winter. In the Volga sturgeons overwinter along a 430km (270mi) stretch, aggregating in bottom depressions.

Anadromous sturgeons spawn during spring and summer months, though in some species there are "spring" and "winter" forms which ascend the rivers in their respective seasons. The spring form spawns soon after going up the river whereas the winter form spawns the following spring. Some adults spawn every year, others intermittently. Freshwater sturgeons make their way from their home streams and lakes into the upper or middle reaches of large rivers. The North American Lake sturgeon will spawn over rocks in wave conditions when more suitable quiet areas are unavailable. Courtship involves the fish leaping and rolling near the bottom.

Both anadromous and freshwater species cease feeding during the spawning period. Eggs (caviar) are produced in millions— over 3 million in a female Atlantic sturgeon 2.65m (8.7ft) long. They are adhesive, attaching to vegetation and stones. Hatching takes about a week. Few data are available on the development of the young but growth is generally rapid in the first five years, approximately 50cm (20in).

The size and age of sturgeons are impressive; the White sturgeon of North America and the Russian beluga are the world's largest freshwater fishes. An 800kg (1,800lb) White sturgeon caught in Oregon in 1892 was exhibited at the Chicago World Fair, but the only great White sturgeon actually weighed and measured came from the Columbia River (Canada and USA). Caught in 1912, it was 3.8m (12.5ft) long and weighed 580kg (1,285lb). When 1.8m (6ft) long the White sturgeon is between 15 and 20 years old. The largest Lake sturgeon, caught in 1922, weighed in at 140kg (310lb), and the greatest recorded age was 154 years for a specimen caught in 1953. A beluga caught in 1926 weighed over 1,000kg (2,200lb) and yielded 180kg (396lb) of caviar and 688kg (1,500lb) of flesh; it was at least 75 years old.

▲ **An indoor sturgeon.** The sterlet has recently proved to be a popular though expensive aquarium fish. It is an attractive animal, but not easy to feed. It favors finely ground mussels, but unless great care is taken it will not live long.

◀ **Eating apparatus.** The paddlefish is a filter-feeder, that is it sieves food out of the water using gill rakers, which are prominent in this picture. The long rostrum may enable the fish to detect the zooplankton on which it feeds.

▼ **Treasure from a frozen wilderness**—fishing for sturgeon in Siberia *c.* 1870. Today caviar (sturgeon eggs or roe) is one of the USSR's most famous exports.

There are two **paddlefishes**, the American paddlefish of the Mississippi and the Chinese paddlefish of the Yangtze, but fossils are known from North America from the Cretaceous and Eocene periods (135–38 million years ago).

Paddlefishes are instantly recognized from their long, flat broad snout. The mouth is sack-like and the gapes open as the fish swims, scooping up crustaceans and other plankton. The American paddlefish occurs in silty reservoirs and rivers; in length it exceeds 1.5m (5ft) and weighs up to 80kg (175lb). The fish is nocturnal, resting at the bottom of deep pools during the day. Spawning was first observed only in 1961. When

water temperature reaches 10°C (50°F) adults are stimulated to move upstream to shallows.

The function of the paddle-like upper jaw is uncertain. It is suggested that it is an electrical sensory device for detecting plankton swarms; a stabilizer to balance the head against downward pressure in the huge mouth; a scoop; a mud-digger; or even a beater to release small organisms from aquatic plants (though individuals that lost their paddles in accidents are known to feed perfectly, from the evidence of their full stomachs).

Virtually nothing is known of the biology of the Chinese paddlefish.   GJH

Sturgeons and Man

To Longfellow's Hiawatha (1855) the Lake sturgeon was "Mishe-Nahma, King of Fishes"; on the early explorers of North America the mighty fish also made a deep impression: witness the numbers of Sturgeon rivers, Sturgeon bays, lakes, falls etc on the map of North America. To ancient man in both America and Russia sturgeons were a valuable resource. Scutes were used as scrapers, oils as medicine, flesh as food, eggs, latterly, as a caviar. Horrendously, steamboats in North America could trawl large numbers of sturgeons and used their oil as fuel, thereby speeding their demise. Thousands more were butchered by 1885— when over 2 million kg (5 million lb) of smoked sturgeon were sold at Sandusky, Ohio, and distributed to cities.

The eggs or roe provide the sturgeon's most prized product: caviar. The center of the caviar trade is the Russian Caspian Sea basin. Here sturgeons were fished intensively for 200 years, and by 1900 stocks had declined dramatically. The First World War and

internal strife allowed some recovery of stock but by 1930 feeding grounds were intensively overfished and dam construction depleted the fishery even more by precluding breeding migrations. The development of sturgeon hatcheries in the 1950s conserved and increased populations; the hatcheries released millions of young (45 million in 1965) into the rivers. Fishing for sturgeon is now banned in the USSR. A product which also led to a reduction in sturgeon stocks was isinglass.

Procured from the sturgeon's swim bladder and vertebrae it was used as a clarifier of wines and a gelatinizing agent for jams and jellies. In 1885 about 1,350kg (3,000lb) of isinglass were exported (derived from approximately 30,000 sturgeons).

Today in both the USSR and the USA the plight of the sturgeon is recognized. However with few biological data and constant modification of the environment the situation of many species is precarious.

BOWFIN, GARFISHES

Families: Amiidae, Lepisosteidae
Eight species in 3 genera.

Bowfin
Family: Amiidae
Sole species *Amia calva*.
Distribution: N America from the Great Lakes
S to Florida and the Mississippi Valley.
Size: length 45–100cm (18–39in), maximum
weight 4kg (9lb).

Garfishes
Family: Lepisosteidae
Seven species in 2 genera.
Distribution: C America, Cuba, N America as
far N as the Great Lakes.
Size: length 75cm–4m (2.5–13ft), weight
7–137kg (15–62lb).
Species: **Alligator gar** (*Atractosteus spatula*),
Cuban gar (*A. tristoechus*), **Florida spotted gar**
(*Lepisosteus oculatus*), **Longnose gar** (*L. osseus*),
Shortnose gar (*L. platostomus*), **Spotted gar**
(*L. platyrhincus*).

THE **bowfin** is a renowned fish, its fame being due to its size, predatory habits, ability to survive out of water and its status as the only living representative of a group of primitive bony fishes (the Halecomorphi) whose ancestral lineage stretches back to the Jurassic period (195–135 million years ago). Its systematic position, however, has been contentious for many years, and among bony fishes its anatomy is probably the most thoroughly described. The cranial anatomy alone forms the subject of a 300-page monograph (by Edward Allis) published in 1897. (Present opinion is that bowfins are most closely related to the Teleostei.)

During the Tertiary period (65–7 million years ago) bowfins were widely distributed in Eurasia and North America. Nowadays only one species survives, in central and eastern North America.

The common name alludes to the bowfin's long, undulating dorsal fin. In the Great Lakes it is known as the dog-fish, and in the southern states as the grindle. The other distinctive features of the bowfin are its massive blunt head and cylindrical body. At the upper base of the tail is a dark spot, edged in orange or yellow in males (probably a sign for recognition). Females lack the edging and sometimes the spot itself. Most bowfins grow to 45–60cm long (18–24 in) but a few will reach 1m (3.3ft) and weigh 4kg (9lb).

Bowfins can use their air bladder as a "lung" and can survive a day out of water. This feature enables them to exist in swampy, stagnant waters unusable by other predatory fish. There is also evidence that during dry conditions individuals enter a state of torpor (aestivate). Bowfins spawn in spring; males move to shallow water and each prepares a saucer-shaped nest, 30–60cm (12–24in) across, by biting away plants growing in the bed of the river or lake. The males vigorously defend their nesting sites against other males but often spawn

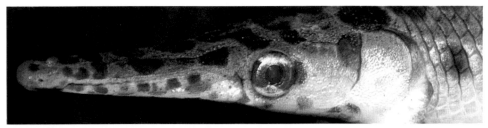

▲ **The long dorsal fin,** to which the bowfin's name alludes, allows it to remain stationary while it watches for prey. It catches its food by surging forward, fast acceleration being provided by the powerful sweeps of its muscular body.

◄ **This alligator-like snout** and the heavily enameled diamond-shaped ganoid scales are those of the Cuban gar. The speckled color pattern is part of the camouflage of this lurking predator.

▼ **Sleek and sleepy,** a garfish of the genus *Atractosteus.* Its three extant species are restricted to northern America; fossil species are also known from Europe and India.

with several females. Females lay about 30,000 adhesive eggs, which the male guards and fans. In 8–10 days the eggs hatch and the young fish cling to vegetation by an adhesive snout pad. The male continues to guard the brood until the young are about 10cm (4in) long.

Bowfins are predators; they feed mostly on other fish (sunfish, bass, perch, pike, catfish and minnows), frogs, crayfish, shrimps and various water insects. Bowfin flesh itself is unsavory, and its plundering of sport fishes and comparative abundance make for few friends amongst fishermen and conservationists.

Garfishes are long-bodied predators with the habit of lurking alongside submerged branches. They are characterized by their long jaws with many teeth and by their heavy armor-like scales. Their swim bladder is connected to the esophagus and acts as a lung enabling them to breathe atmospheric air. The two living genera of garfishes are also known as fossils from the Cretaceous and Tertiary deposits of Europe, India and North America but nowadays they exist only in North and Central America. Members of the genus *Atractosteus* live in Central America and the southern USA.

The Alligator gar is one of the largest North American freshwater fishes. One weighing over 135kg (300lb) and 3m (10ft) long was caught in Louisiana. It is now scarce throughout its range (an arc along the Gulf coast plain from Veracruz to the Ohio and Missouri rivers). It has been "accused" of eating game fishes and waterfowl. However, careful studies have shown that the Alligator gar rarely feeds on these animals and preys mostly on forage fishes and crabs. The other two *Atractosteus* species are *A. tropicus* of Nicaragua and the Cuban gar. Little is known of their biology; *A. tropicus* inhabits the shallow, protected areas of Lake Nicaragua, where specimens can grow to over 1.1m (3.6ft) and weigh over 9kg (20lb).

The genus *Lepisosteus* has a wider distribution, from the northern Great Lakes to Florida and the Mississippi basin. The Spotted gar occurs throughout the Mississippi drainage and grows to over 1.1m (3.6ft) and a weight of 3kg (6.6lb). The distribution of the Longnose gar is wider than that of the Spotted gar, and it too inhabits brackish water in its coastal range. The Longnose gar can grow to over 1.8m (5.9ft) and weigh 30kg (66lb). The females are longer than the males, the difference being as much as 18cm (7in) in their tenth or eleventh year. Males rarely survive beyond this period, but females may live for up to 22 years. Group spawning occurs from March to August, according to locality, in shallow warm water over vegetation. The adhesive eggs, about 27,000 per female, are scattered randomly and hatch within 6–8 days. As in the bowfin the young cling to vegetation by means of an adhesive pad on their snout. Growth is rapid, 2.5–3.9mm (0.1–0.15in) per day. The Shortnose gar covers an area embracing northeastern Texas, Montana, southern Ohio and the Mississippi, whilst the Florida spotted gar occurs there and in the lowlands of Georgia.   GJH

TARPONS, EELS, NOTACANTHS

Superorder: Elopomorpha
About six hundred and thirty-two species in about 160 genera and about 30 families.

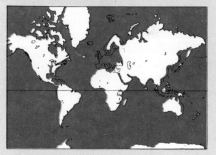

Tarpons and allies
Order: Elopiformes
About twelve species in 4 genera and 3 families.
Distribution: most tropical seas. Rarely brackish or fresh water.
Size: maximum length about 2m (6.5ft), maximum weight 160kg (350lb).
Species include: **Bone fish** (*Albula vulpes*), **Pacific tarpon** (*Megalops cyprinoides*), **tarpon** (*Tarpon atlanticus*), **tenpounder** (*Elops saurus*).

Eels
Order: Anguilliformes
About six hundred species in about 150 genera and 24 families.
Distribution: all oceans, N America, Europe, E Africa, Madagascar, S India, Sri Lanka, SE Asia, Malay Archipelago, N and E Australia, New Zealand.
Size: maximum length about 3m (10ft).
Families, genera and species include: **anguillid eels** (family Anguillidae), including **American eel** (*Anguilla rostrata*), **European eel** (*A. anguilla*), **Japanese eel** (*A. japonica*); **curtailed snipe eels** (family Cyemidae); **garden eels** (family Heterocongridae); **Gulper eel** (family Eurypharyngidae, sole species *Eurypharynx pelecanoides*); **monognathids** (family Monognathidae); **moray eels** (family Muraenidae), including **moray eels** (genus *Muraena*); **saccopharyngids** (family Saccopharyngidae); **snipe eels** (family Nemichthyidae); **Snub-nosed parasitic eel** (family Simenchelyidae, sole species *Simenchelys parasiticus*); **synaphobranchid eels** (family Synaphobranchidae); **worm** or **spaghetti eels** (family Moringuidae); **xenocongrid eels** (family Xenocongridae).

Notacanths
Order Notacanthiformes
About twenty species in 6 genera and 3 families.
Distribution: all oceans.
Size: maximum length about 2m (6.5ft).
Families include: **halosaurs** (Halosauridae), **notacanthids** (Notacanthidae).

▶ **Scaling a tarpon** in Northeast Brazil. The Atlantic tarpon is a large and prized food fish. Its scales have been incorporated into jewelry.

▶ **A shoal of tarpon** OPPOSITE, feeding around a rocky outcrop.

EELS have long been prized and in the Middle Ages were a staple food. But it was then believed that eels were different: unlike other freshwater fish they did not produce eggs and sperm at the start of the breeding season. Where did eels come from? The problem produced ingenious answers but no solution until the end of the 19th century. Even today eel behavior retains many mysteries.

The superorder of tarpons, eels and notacanths comprises three orders, the members of each of which seem to be unlike the others, apart from certain anatomical similarities. They are united in having a larva that is quite unlike the adult: it is transparent, the shape of a willow leaf or ribbon and is graced with the name of leptocephalus. The three orders are the Elopiformes, containing the tarpons, tenpounder and the Bone fish; the Anguilliformes with about 24 families of eels; and the Notacanthiformes which lack any accepted common name—sometimes they are called deep sea spiny eels, but as this is abbreviated to "spiny eels" they could be confused with the unrelated family that aquarists call "spiny eels"; confusion will be avoided by using the name "notacanths."

The un-eel-like **tarpons** and allies represent an ancient lineage. Fossils belonging to this group occur in the Upper Cretaceous deposits of Europe, Asia and Africa (135–100 million years ago). In addition to developing from a leptocephalus larva they all possess gular plates—a pair of superficial bones in the skin of the throat between each side of the lower jaw. Gular plates were much more common in ancient fishes than they are in living fishes. All are marine fish.

The tarpon is the largest fish in the group, weighing up to 160kg (350lb). It is a popular angling fish because it gives the angler a fight, leaping into the air, twisting and turning to dislodge the hook. This species lives in the tropical waters on both sides of the Atlantic: the adults breed at sea and the larvae make their way inshore where they metamorphose. The young live and grow in lagoons and mangrove swamps. These swampy regions can often be low in oxygen, but the tarpon can breathe atmospheric oxygen and does so at such times.

The Pacific tarpon is very similar in appearance and even has the last ray of the dorsal fin similarly prolonged. Its life cycle is also like that of the tarpon. Apart from some minor anatomical differences (in the way that processes from the swim bladder come to lie closely against the region of the inner ear in the skull) the most obvious difference is that the Pacific tarpon is smaller, rarely reaching 50kg (110lb).

The tenpounder is another warm-water species, known from the tropical Atlantic. What may be a different species is widespread in the Indo-Pacific. Despite its common name it can reach 6.8kg (15lb) in weight.

The Bone fish is widespread in shallow tropical marine waters. It is a shoaling fish and feeds on bottom-living invertebrates, often in water so shallow that the dorsal fin and upper lobe of the caudal fin stick out of the water. Although rarely exceeding 9kg (20lb) it is, despite its poor flavor, a much sought after angling fish because of its fighting qualities.

The exact number of families of **eels** is uncertain. There are differences of opinion because some groups are poorly known—sometimes on the basis of just one or a few specimens. In addition there is the problem of matching up the known leptocephali with the known adults.

The larvae of European eels hatch in the Sargasso Sea (see box) and then drift back in the Gulf Stream and take about three years to reach the colder, shallower and fresher European coastal waters. Here they shrink slightly, metamorphose into elvers and move upstream, where they grow and feed until some years later the urge to migrate comes upon them. The American

eel breeds in the western part of the Sargasso and its leptocephali take only one or two years to reach fresh waters.

The breeding grounds have not been found for all anguillid eels. The commercially significant Japanese eel *Anguilla japonica*, for example, has an unknown spawning ground. Eels from the eastern part of southern Africa (*A. nebulosa*, *A. bicolor* and *A. mossambica*) breed at depth to the east of Madagascar. As adults, the European, American and Japanese eels are widespread. A fourth species of eel in eastern Africa (*A. marmorata*) also occurs on many Pacific islands and as far east and north as Hong Kong and southern Japan. In complete contrast, *Anguilla bornensis* lives in the river Bo in Borneo, *A. anterioris* in mountain streams in New Guinea and *A. ancestralis* is found only in the northeastern part of the Celebes.

Other species have intermediate ranges. But all have to find their way to their breeding grounds; how and why? We do not know the answers, but a few interesting possibilities can be suggested.

In the European eel, as the larvae are carried by the Gulf Stream, might not the adults also follow it back to the breeding grounds? However, if the breeding grounds are characterized by three parameters—temperature, salinity and pressure—could these be guides? But these latter possibilities present difficulties. An adult eel leaving northern Europe could follow a temperature gradient and reach the Sargasso, but an eel leaving Italy would not (the Mediterranean is warmer than the adjacent parts of the Atlantic). The appropriate pressure (as a measure of depth) could be reached very much closer to European shores than the other side of the Atlantic. Also it should not surprise anyone that adult eels are not particularly sensitive to salinity changes as they have already had to cope in their earlier migration with change from fresh water to salt water.

Our ignorance is reiterated, but we cannot leave the problems without some potentially exciting speculations. The first is that many animals are known to be able to navigate using the earth's magnetic field and although no magnetic-field detectors have yet been found in the eel the possibility remains. A second possibility involves the movements of continents—continental drift. Most of Europe and North America were once joined together as one continent, as were Africa and South America (look at their opposing coastlines). At one stage in the shaping of the present continents there

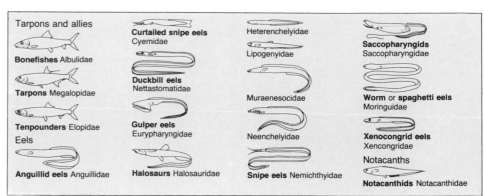

Tarpons and allies

Bonefishes Albulidae

Tarpons Megalopidae

Tenpounders Elopidae

Eels

Anguillid eels Anguillidae

Curtailed snipe eels Cyemidae

Duckbill eels Nettastomatidae

Gulper eels Eurypharyngidae

Halosaurs Halosauridae

Heterenchelyidae

Lipogenyidae

Muraenesocidae

Neenchelyidae

Snipe eels Nemichthyidae

Saccopharyngids Saccopharyngidae

Worm or **spaghetti eels** Moringuidae

Xenocongrid eels Xencongridae

Notacanths

Notacanthids Notacanthidae

was a narrow sea between the two. Is it possible that when eels first appeared they bred in this sea and still need its physical conditions to ensure successful reproduction even though it involves a journey of thousands of kilometers by an inefficient swimmer?

From tagged migrating eels it has been extrapolated that they take 4–7 months to make the journey and during that time they do not eat.

Further enigmas occur with the allegedly passive journey of the leptocephali in the water currents that bring them back to the feeding grounds.

It is likely that they do not eat. No food has been found in the guts and they develop peculiar forward-pointing teeth, mostly on the outside of their jaws, which are ill adapted to catching food. However, they grow, until at metamorphosis they shrink and lose their teeth.

A few clues are available to hint at partial solutions to these conundra. A conger eel larva that was studied was shown to be able to absorb nutrients and vital minerals from the sea water. The mouth of this species is lined with minute projections (villi) which are thought to absorb the nutrients. However, this has not yet been shown to be the case in anguillid eels. The metamorphosis is rapid, and a lot of calcium must be made available very quickly to mineralize the bones and convert them from the juvenile to the adult state. The forward-pointing teeth of the leptocephalus, which disappear at metamorphosis, contain a lot of calcium and these may act as the necessary calcium reservoirs.

As the leptocephalus metamorphoses it modifies its compressed body into the adult's shape and develops pigment, pectoral fins and scales. Before pigmentation develops it is called a glass eel and latterly an elver. Depending upon local conditions it is either the glass eels or the elvers that ascend the rivers in vast numbers. In the 19th century,

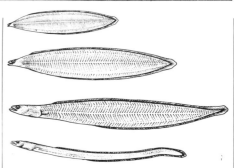

▲ **Outlines of representative species** TOP of tarpons, eels and notacanths.

▲ **Preparation for change.** The transparent, willow-leaf-shaped leptocephalus larva of an eel shrinks before it metamorphoses into its adult form.

▶ **During metamorphosis** ABOVE leptocephali are called glass eels. In some areas of the world they are a valued but seasonal delicacy.

▼ **Exposed on the ground,** a European eel. If left stranded by receding flood waters, eels can move back to water overland.

before estuarine pollution had its effect, millions of elvers were caught and eel fairs were held. Elvers caught on the River Severn in England were particularly famous. There elvers were used principally for food but in Europe they were also used to make preservatives and glue. Elvers are still caught today and their fate is to be taken to eel farms and grown to maturity in less time than it would take in the wild. Adult eels are very nutritious and are popular in many parts of the world, smoked or stewed. Jellied eels are a famous delicacy of the eastern part of London.

Eel blood contains an ichthyotoxin which can be dangerous. It is important not to let eel blood come into contact with eyes or any other mucous membrane. The poison is quickly destroyed by cooking.

Muraena, the generic name of the Mediterranean moray, commemorates a wealthy and powerful Roman, Licinius Muraena, who lived near the end of the 2nd century BC. Licinius, so we read in Pliny the Elder, kept moray eels in captivity to demonstrate to the world that he was wealthy. Gaius Herrius built a special pond for his morays and they were such splendid fish that Julius Caesar offered to buy them from him. Gaius refused but, presumably to maintain his position in society and not incur Caesar's displeasure, lent them to Caesar to display

Where Do Eels Come From?

The best-known eel is the European eel (family Anguillidae). Although anguillid eels are the only family that spend most of their life in fresh water, the European eel's life history will serve as an illustration for all the other families, not least because the eel's familiarity has over the centuries occasioned so much curiosity about eel reproduction.

From before the days of the Greeks and Romans this eel has been an important source of food. Aristotle and Pliny wrote about large eels going down to the sea and small eels coming back from the sea. Other freshwater fish had been noticed to have eggs and sperm at the start of the breeding season, but not the eel; it was therefore concluded that the eel was "different" and two millennia of speculation were born. Aristotle was certain that baby eels sprang out from "the entrails of the earth," Pliny concluded that young eels grew from pieces of the adult's skin scraped off on rocks. Subsequent suggestions continued to be as unrealistic. In the 18th century the notion of the hairs from horses' tails giving rise to eels was in vogue and in the 19th a small beetle was advocated as the natural mother of eels. The truth, when finally resolved, like the denouement of a detective story, was hardly more credible.

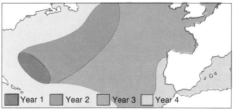
Year 1   Year 2   Year 3   Year 4

Countless millions of eels had been caught and gutted for eating over the centuries, but not until 1777 were developing ovaries identified by Professor Mondini of Bologna. In 1788 his finding was challenged by Spallanzani who pointed out that none of the 152,000,000 eels from Lake Commachio ever showed such structures. Nonetheless Spallanzani was struck with the determination of the eel to reach the sea, traveling overland on damp nights. In 1874 indisputable testes were found in an eel in Poland. Not until 1897 was an indisputably sexually mature female caught in the Straits of Messina. The beetle myth was laid to rest; eels must lay eggs in the sea: but where? When the eels reappear on the coasts they are about 15cm (6in) long. Why were smaller specimens never caught?

The answer had been available, but unrealized, since 1763. In that year the zoologist Theodore Gronovius illustrated and described

a transparent fish like a willow leaf, which he called leptocephalus. One hundred and thirty-three years later (1896) Grassi and Calandruccio (the two biologists who found the sexually mature female eel) caught two leptocephali and kept them alive in aquaria. The leptocephali, being caught near the coast, were on the point of metamorphosing; their transformation in the aquaria tied up a part of the story.

The hunt was now on for leptocephali and the breeding ground. Johannes Schmidt took leptocephali samples and followed their decreasing size. Finally he found the smallest of all, 1cm (0.4in) long, in an area of the west Atlantic between 20 and 30 degrees N, 48 and 65 degrees W: the Sargasso Sea.

On the basis of much subsequent research we now know that the European eel probably breeds at moderate depths (the eyes of the adults enlarge on their 4,000 mile journey) and at a temperature of about 20°C (68°F). The Sargasso Sea is one of the few areas where such a high temperature extends to such a depth. There are throughout the world 16 species of the genus *Anguilla* and all are thought to breed in deep warm waters, although only two, the European and the American eel, in the Sargasso.

at a banquet; 6,000 morays were present. Moray owners, those that had money left over after paying for their pets' expenses, decorated the eels with jewels. Vedius Pollo, on the other hand, entertained his dinner guests by letting them watch his pet morays eat recalcitrant or superfluous slaves. It is possibly from such actions that legends about morays' savagery stem, culminating in the current rumors that morays find skin divers irresistible. At the après-dive parties of this fish-imitating subculture tales are told and retold of moray eels attacking divers. How true are such stories? The answer seems to be, not very. There are two aspects to the stories, one is that the bite is poisonous, the second that the moray does not let go once it has bitten. The moray's bite is not poisonous; any infection of the wound is secondary. Moray eels' teeth are adapted for holding small prey, ie they are long and thin and the larger teeth are hinged to permit the smooth passage of prey into the stomach.

Most morays have teeth like those of the Mediterranean moray, but species of the tropical genus *Echidna* have blunt, rounded teeth adapted for crushing crabs which are its major food item. The genus *Rhinomuraena* is characterized by having the anterior nostrils expanded into tall, leaf-like structures. *Rhinomuraena quaesita* is brilliantly colored and much prized by aquarists. The body is turquoise, the dorsal fin bright yellow with a white margin, the front and underside of the head yellow and the rest blue.

Moray eels are found in all tropical seas, usually in shallow waters. They occasionally enter fresh water.

Xenocongrid eels have no common names. They are probably close relatives of the morays. Like them they lack scales and have thick skin covering the continuous median fin. The pectoral fins are reduced or, in at least one case, lost. The coloration is mottled like that of morays. That the pores on the head are much smaller than in morays and the body is much thinner has been deemed sufficient justification to consider them a separate family. Representatives are found worldwide but little is known of them because they are secretive fishes.

The leptocephalus of *Chlopsis bicolor* is very common in the Mediterranean but until recently the adults were poorly known. This species was described over a hundred years ago, but only a very few specimens were in museum collections until divers started using underwater narcotics to drug fish and succeeded in catching large num-

bers. *Chlopsis bicolor*, which also lives off Florida, is a small fish (shorter than 25cm, 10in), dark brown above and paler below. *Xenoconger olokun* was known from just one specimen captured by the research vessel *Pillsbury* off the estuary of the St Andrews River, Ivory Coast, at a depth of 50m (165ft). (The species name is derived from Olokun, a sea deity in the Yoruba religion of the region.) *Xenoconger fryeri* is widespread in the Pacific but *X. olokun* was the first member of this genus to be found in the Atlantic.

The common names of the family of worm eels or spaghetti eels are particularly apt. "Moringuids" are long, thin and scaleless, usually with poorly developed dorsal and anal fins. They are commonest in the Indo-Pacific but a few species are found in the Atlantic, mostly in the west. Most unusually for eels they have a distinct caudal fin which can vary from having distinct forks to being trilobed.

In some species there are marked differences between the sexes. In *Moringua edwardsii* from the West Indies, for example, the females are nearly twice as long as the males, have more vertebrae and the heart lies much further back in the body. These differences led to members of this species having previously been described under several different names.

Worm eels are burrowing eels but unlike most others they burrow head first, not tail first, preferring a bed of sand or fine gravel. Their life history is exemplified by the Pacific species *Moringua macrochir*. As the leptocephalus metamorphoses it starts burrowing. With its minute eyes and reduced fins it looks very worm-like. It rarely leaves the burrow and then only at night, until the onset of sexual maturity. Then the eyes and fins enlarge and the metamorphosing forms swim around at night to find a mate.

Garden eels have acquired their common name because they live in underwater colonies (gardens), rooted in the sea bottom where they wave around like plants. They are closely related to conger eels—some authorities place them in the conger eel family. Garden eels are characterized by having a long thin body, small gill openings and a hard fleshy point at the rear of the body with the caudal fin hidden below the skin. They burrow tail-first, their modified

▶ **The silent spy.** The moray eel is a territorial carnivore that spends some of its time looking out of a crevice. Its fins are covered in a thick skin, so the fin rays are inconspicuous.

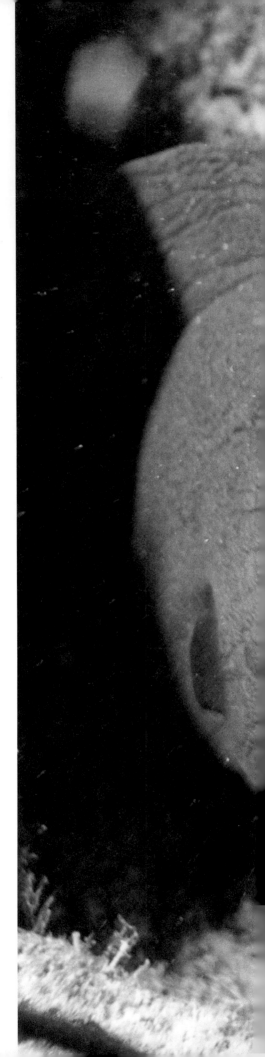

The Colonies of Garden Eels

There are about twenty species of garden eels grouped into three genera, and it seems that the daily cycle and life histories of all are very similar. Colonies usually live in shallow water where there is sufficient light for the eels to see their food and a reasonable current. There the eels give a passable imitation of prehensile walking sticks facing into the current. The densest colonies are deepest, with perhaps one eel every 50cm (20in). The distance apart depends upon the length of the eel. Each eel occupies a hemisphere of water with a radius similar to the eel's body length, so that it is just separated from its neighbor.

A colony of *Gorgasia sillneri* in the Red Sea has been studied intensively and has produced some interesting information. The eels' burrows themselves are twisted and lined with mucus secreted by the skin (during the period of study, no eel ever left its burrow). The eels' day starts about half an hour before sunrise when they emerge. By sunrise all are out eating. Any disturbance causes the whole colony to disappear. For a period either side of noon the eels rest in their burrows, but from mid afternoon to sunset they are out feeding again. An interesting phenomenon is that while the eels are in their burrows their feeding space is taken over by another fish (*Trichonotus nikii*), which feeds at the level of where the eels' heads would be. When the eels are out the fish hides in the sand.

Even mating is conducted from the security of the burrow. The male stretches out and ripples his iridescent fins at the adjacent female. She stretches out of her burrow and they intertwine. A few eggs are shed and fertilized at a time and a couple may mate 20 times a day. Detailed study of a plot within a colony showed that females never moved their burrows but some males did, to be near females. As this movement was never observed it was presumed to take place at night. The males also moved away from the females after breeding.

Small eels settle down together until they are about 25cm (10in) long when the colony breaks up and the eels try to find a place in an adult colony. The adults resent this intrusion and unless the juvenile can stand up to aggression it moves away to a new area.

A major predator of garden eels is the ray. The effect of a ray swimming over a colony has been likened to a mower cutting grass. Whether the eels retreat in time or are caught has not been satisfactorily ascertained.

◄ **A congregation of eels.** The Conger eel (*Conger conger*) is an Atlantic species. Largely nocturnal, it hides away by day. Although it resembles the European and American anguillid eels it never enters fresh waters.

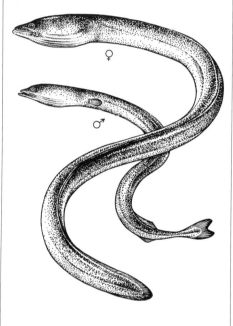

▲ **Dimorphic duo.** Some eels, such as these worm eels of the species *Moringua edwardsii*, show marked differences between the sexes. Here the size and nature of the fins are very different.

▼ **The mark of a family.** Synaphobranchid eels are sometimes called cut-throat eels: they have a single ventral gill-slit. They are found in most oceans and some are parasitic.

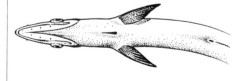

tail regions being used to penetrate the sand. Their distribution is similar to that of the worm eels.

Three further families of eels (Dysommatidae, Synaphobranchidae and Simenchelyidae) are here grouped together because the present division into three families may not be justified. All have leptocephali with an unusual characteristic: telescopic eyes. This may well indicate a very close relationship.

The genus *Dysomma* includes about six species characterized by enlarged teeth in the roof of the mouth (vomerine teeth). There is also a tendency towards the loss of the pectoral fin. The species tend to be laterally compressed and to taper evenly towards the tail. The swim bladder is very long and the nostrils are tubular. The main reason for including this otherwise unremarkable Indo-Pacific family is because of the history of the genus *Media* which exemplifies some of the problems faced by ichthyologists. A fish was bought in a fish market near Kochi in Japan. No one knew what it was and it became the type of the new species (and genus) *Media abyssalis*. During the Second World War this specimen was destroyed by fire. Not until 1950 were more caught off Japan and off Hawaii. Although the search for this brownish, scaleless eel was not particularly assiduous, it does show how fate can destroy, so easily, the only representatives available to the scientific world of a species that is probably common in its own particular habitat.

The thought of fire and the dysommatid eels brings us to fire and the synaphobranchid eels. Uncommon in collections, synaphobranchids are deep-sea eels with the gill slits almost confluent ventrally, giving them the vernacular name of cut-throat eels. In the course of one expedition a deep-sea research vessel, *Galathea*, caught 12 specimens. These were sent, in three separate parcels for safety, to a researcher in New Zealand. The second parcel, with five specimens, was destroyed in a fire in a mail-storage warehouse in Wellington in July 1961.

Cut-throat eels live in cold, deep waters, where temperatures average 5°C (41°F). They feed on crustaceans and fishes. (One specimen of *Synaphobranchus kaupi*, trawled at a depth of 1,000m (3,300ft), had eaten octopus eggs.) They are thick set, scaled eels with large jaws and small teeth. The scale pattern is like that of the following family. Several species have a characteristic nick in the ventral outline of the body just below the pectoral fin. The genus *Haptenchelys* is odd in lacking scales. Their distribution is

world wide from 2,000m to 400m deep (6,600–1,300ft). The adults are bottom dwellers, the larvae have telescopic eyes, but beyond that little is known about their life history.

The third family in this assemblage (Simenchelyidae) probably has just one species, the Snub-nosed parasitic eel, but as in many other instances there is no agreement on the matter. This deep-sea eel has been caught off the eastern coast of Canada, in the western and northern Pacific, off South Africa and off New Zealand. Whether these discrete collecting localities reflect different local species, or just the collecting effort, is impossible to say. It seems to be plentiful because it is caught in large numbers. It may be a gregarious species or it may need a particular habitat. It has a blunt head with a small transverse mouth. Jaws and jaw muscles are strong. The gill slits are short and lie horizontally below the pectoral fins. The dorsal fin starts well in advance of the anal fin. The body is cylindrical until the anus and compressed thereafter. The small scales are grouped and angled acutely to the lateral line.

The young eat small crustaceans but the adult is probably at least partly parasitic. The stomachs of the first specimens found were full of fish flesh, as are those of most of the subsequent specimens. Evidence from the North Atlantic suggests that they feed on the flesh of halibut and other large fish, but probably only injured or moribund specimens. As *Simenchelys* only reaches a length of 70cm (28in) it is unlikely to be an active predator.

One further eel family (Aoteidae) may not exist at all. It was erected on the basis of one fish found inside the stomach of a snapper caught in Cook Strait (New Zealand) in 1926 and called *Aotea acus*. It was damaged and slightly digested, but could not be related to any previously described species. The corpse is elongate, scaleless, cylindrical, finless and has a wide mouth and a pointed snout. It is not known where the snapper had eaten it but it was presumed to be in deep water. No more have ever been found. *Aotea acus* has not been reexamined, but it may be the partially decomposed corpse of something else.

The remaining families in this order are all deep-sea eels and are modified for their environment in spectacular if different ways.

Members of the family of snipe eels are large-eyed, extremely elongated animals. The rear of the body is little more than a skin-covered continuation of the spine. Here

the vertebrae are weak and poorly ossified, hence few complete specimens are known. One undamaged specimen, a lucky catch, has 670 vertebrae, the highest number known.

This family's common name alludes to its long and widely flaring jaws, fancifully seen as resembling the bill of the similarly named bird. It had been thought that there were two groups of snipe eels, those with extremely long divergent jaws and those with short divergent jaws. Recently enough specimens have been caught to show that mature males are short jawed and mature females and juveniles of both sexes are long jawed. Before maturity both the inside and outside of snipe eels' jaws are covered with small, backward-pointing teeth presenting a passable imitation of sandpaper. It seems likely that because adults have lost many teeth they eat little; possibly they die after they have reproduced.

Even when they have their granular teeth, how do snipe eels feed? The jaws can only close posteriorly. A few observations of living eels from deep-sea research submarine vessels have shown that they spend their time hanging vertically, mouth down, in the water, either still or with bodies gently undulating. A few specimens caught with food still in their stomachs revealed that the major food source is deep-sea shrimps. Characteristic of these are very long antennae and legs. It is suggested that snipe eels feed by entanglement, ie that once the long antennae or legs of the crustacean become embrangled with the teeth inside or outside the jaws they are followed down to the body which is then consumed.

There are three recognized genera of snipe eels (*Avocettina*, *Nemichthys* and *Labichthys*). They are widely distributed in warmer parts of the oceans at depths down to 2,000m (6,600ft). The leptocephali are easily recognized by their thinness and the long caudal filament—a precursor of the adult's prolongation of the caudal region. They metamorphose when some 30cm (12in) long.

One of the most distinctive eels is the Curtailed snipe eel, probably a single-species family (Cyemidae). It looks somewhat like a dart with a long thin point divided into diverging dorsal and ventral parts. This eel occurs in all tropical and subtropical oceans at depths varying from 500 to 5,000m (1,600–16,500ft), but rarely as deep as the deepest figure. It is a laterally compressed, small fish, rarely exceeding 15cm (6in) long. It has a dark velvety skin and minute, but functional, eyes. Its biology is poorly known, though what is known of its breeding has some interesting aspects. Many eels breed in clearly circumscribed areas, where the physical conditions meet the stringent requirements of their physiology. By contrast in the Atlantic the Curtailed snipe eel spawns over large stretches of the warmer parts of the north Atlantic.

This eel's leptocephalus is quite different

▲ Like an underwater snake. There are about two hundred species of ophichthid or snake eels. They live in shallow, warm seas. Some have conspicuous colorings. Many burrow tail-first into the sea floor. This is *Myrichthys oculatus*.

◄ Danger in the wings. Moray eels are territorial and spend most of the day hidden in crevices with the head poking out to see what is happening. Normally if approached an eel retreats. If, however, the eel is provoked it may well bite. Some divers have been known to hand-feed morays for a while and then suddenly get bitten. An explanation for this is that the eel has mistaken the diver's hand for a favorite food item—the octopus. The Mediterranean moray can grow to over 2m (6.5 ft) long and can inflict a nasty bite which, if by chance it severed an artery, could be fatal. Speared morays are very likely to attack because to a moray being speared counts as provocation. This is the Spotted moray (*Gymnothorax moringa*).

from that of others. Whereas the typical leptocephalus is a willow-leaf shape (save for minor variations), the leptocephalus of the Curtailed snipe eel can be nearly as deep as it is long. The little evidence available suggests that it spends at least two years in the larval phase.

On account of their mouths three families of deep-sea eels (Monognathidae, Saccopharyngidae and Eurypharyngidae) are sometimes grouped together under the name gulper eels. (Although three families are often recognized in the scientific literature there is growing evidence that fishes purportedly belonging to the Monognathidae may be young stages of fishes placed in the family Saccopharyngidae.) Monognathids are known from six species found in the Atlantic and Pacific oceans. Their scientific name, which means "one jawed," alludes to the fact that they lack upper jaw bones and have a conspicuous lower jaw which can be longer than the head. All known specimens are small. The largest known individual of

Monognathus jespersensi is only 11cm (4in) long but is only recently metamorphosed. Only one specimen, referred to *Monognathus isaacsi*, has developed any pigment. It seems likely that monognathids are juvenile saccopharyngids but adult saccopharyngids have many more vertebrae than the monognathids. However, the largest monognathids have the most vertebrae so perhaps vertebral formation continues as the fish grows. The conundrum will only be answered when either a sexually mature monognathid is discovered or when a clear monognathid–saccopharyngid transition series is built up.

Saccopharyngids are archetypal deep-sea fishes. They have huge mouths, elastic stomachs, toothed jaws and a luminous organ on the tail. An immediate enigma is that only four species are generally recognized whereas six species of monognathids are usually admitted. Therefore *if* monognathids are juvenile saccopharyngids then there are two more saccopharyngids awaiting discovery or

two monognathid species have been erroneously described.

Only one species from the western Atlantic has been seen alive (*Saccopharynx harrisoni*). This fortunate event happened because the fish, although trawled in 1,700m (5,600ft) of water, was not in the net but entangled by its teeth in the mouth of the trawl and so escaped being crushed by the weight of fish during the haul to the surface. Also, lacking a swim bladder, it escaped the fate of many deep-water fishes—that of the considerable expansion of the gas in the swim bladder as the water pressure reduces towards the surface.

For deep-sea fishes saccopharyngids are large, growing to over 2m (6.5ft) long, but most of the body is tail. The huge mouth has necessitated some morphological changes. The gill arches are a long way behind the skull and dissociated into separate lateral halves. The opercular bones are not developed and the gill chambers are incompletely covered by skin. These modifications

▲ **The bright colors** and flaring, saucer-like nasal palps (the sensory appendages on the mouth) make the members of the genus *Rhinomuraena* popular with aquarists.

▶ **The remarkable jaws** of a deep sea snipe eel. They bear file-like teeth which entangle with the long legs and antennae of deep-sea crustaceans.

mean that its respiratory mechanism is unlike that of other fishes. Another peculiarity of these fishes, apart from the absence of a pelvic girdle, is that the lateral line organs, instead of lying in a subcutaneous canal, stand out from the body on separate papillae. It is surmised that this adaptation makes the fish more sensitive to vibrations in the water and enhances its chance of finding a suitable fish to cram into the distensible stomach. The escape of the prey, once in the mouth, is prevented by two rows of conical and curved teeth on the upper jaw and a single row of alternating large and small teeth on the lower jaw.

At the end of the long, tapering tail of all saccopharyngids there is a complex luminous organ whose function is unknown. Indeed, the whole arrangement of the luminous organs is odd. On the top of the head are two grooves that run backwards towards the tail. These contain a white, luminous tissue that glows with a pale light. The grooves separate to pass either side of the dorsal fin, each ray of which has two small, angled grooves containing a similar white tissue. The tail organ is confined to about the last 15cm (6in) of the body. Where the body is shallowest there is a single, pink, club-shaped tentacle on the ventral surface. Further back, where the body is more rounded, there are six dorsal and seven ventral scarlet projections (papillae) surmounting depigmented mounds. The main part of the organ lies behind these and is a transparent leaf-like structure with an ample supply of blood vessels. Its dorsal and ventral edges are prolonged and scarlet whereas the organ is pink because of the blood vessels. The main organ is split into two zones by a band of black pigment with red spots. Further finger-like papillae are even nearer the end of the body where the tail narrows and the black of the rest of the body is replaced by red and purple pigments. The small papillae produce a steady pink light. The leaf-like tail organ can produce flashes on top of a continuous reddish glow.

It is unlikely that the organ is a lure because the contortions required to place the organ where it will act as a lure, even for such a long, thin and presumably prehensile fish, would leave the body in a position where it would be unable to surge forwards to grasp the prey.

Equally bizarre is the last member of this group of deep-sea eels, the Gulper eel, the only member of its family. The mouth is bigger than that of the saccopharyngids, and the jaws can be up to 25 percent of the body length. The jaws are joined by a black elastic membrane, the overall appearance of which prompted the trivial name *pelecanoides*, ie pelican-like. The eyes and brain are minute and confined to a very small area above the front of the mouth.

Almost nothing is known of its biology. The teeth are minute so it probably feeds on very small organisms. A small complex organ is present near the tail but it is not known if it is luminous. As in the saccopharyngids there is no swim bladder but there is an extensive liquid-filled lymphatic system which may aid buoyancy. The lateral line organs are external and show up as two or three papillae emerging from a small bump.

The larvae metamorphose at a small size, less than 4cm (1.5in), but even at this diminutive size have already developed the huge mouth. Interestingly the larvae live much nearer the surface of the sea (100–200m, 330–660ft) than the adults which live at great depths.

The order of **notacanths** contains three families: Halosauridae, Notacanthidae and Lipogenyidae. All are scaled, deep-sea fishes rarely exceeding 2m (6.5ft) in length. They have a snout that extends in front of the mouth and which may be sharply pointed. The head is usually the deepest part of the body; the latter tapers away to a rat tail. In some species a minute caudal fin is present, in others the anal and caudal fins are confluent. The dorsal fin, if present, is short and placed well forward. The notacanths have a series of isolated spines on the back and in front of the anal fin.

The halosaurs live world wide, mostly at depths down to 1,800m (5,900ft), but one species, *Aldrovandia rostrata*, was caught in 5,200m (17,000ft) of water in the North Atlantic. They feed on a wide range of invertebrates, mostly deep-dwelling forms which some notacanths are thought to dislodge from the sea bottom with their snouts. Other species include small squid in their diet and one species, *Halosauropsis macrochir* from the Atlantic, has a row of what are thought to be taste buds across the top of its head. They are mostly dark-hued fishes but *Halosaurus ovenii* is pinkish with silvery sheens. This species is also one of several in which the roof of the mouth has alternate light and dark stripes.

The notacanthids are stockier than the halosaurs and have a series of dorsal spines instead of a dorsal fin. They are distributed worldwide and feed on echinoderms, sponges and sea anemones from the sea floor.

The family Lipogenyidae contains only one species, *Lipogenys gilli* from the western North Atlantic. The main interest of this species is that it has a toothless mouth that functions like a vacuum cleaner and sucks up vast quantities of ooze. The amount of organic material in ooze is probably small but as this species has a long intestine, permitting maximum absorption of nutrients, it apparently survives on such an unpromising diet.

During a research voyage of 1928–30 the research vessel *Dana* caught, off South Africa, a leptocephalus larva 1.84m (6ft) long. Much speculation appeared in the popular and pseudoscientific press along the following lines: "if a 10cm (4in) conger eel leptocephalus produces an adult 2m (6.5ft) long, then this larva will produce an adult over 30m (100ft) long; hence it is a baby sea serpent and the sea serpent is an eel." Other giant leptocephali have been caught, and were named *Leptocephalus giganteus*. In the mid 1960s luck came to the aid of science. Another giant leptocephalus was caught but this time it was in mid metamorphosis and it could be established that the adult form was a notacanth. Reexamination of the other giant larvae showed they too could be referred to notacanths, thereby establishing the relationships of the notacanths with the eels and dispelling one set of rumors about sea serpents. Unlike other eels notacanths hardly grow after metamorphosis and the 30m (100ft) adult—the sea serpent—is merely a product of imaginative extrapolation.   KEB

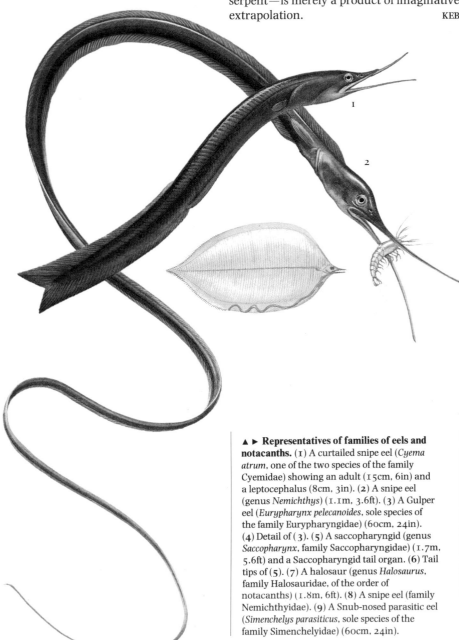

▲ ► **Representatives of families of eels and notacanths.** (1) A curtailed snipe eel (*Cyema atrum*, one of the two species of the family Cyemidae) showing an adult (15cm, 6in) and a leptocephalus (8cm, 3in). (2) A snipe eel (genus *Nemichthys*) (1.1m, 3.6ft). (3) A Gulper eel (*Eurypharynx pelecanoides*, sole species of the family Eurypharyngidae) (60cm, 24in). (4) Detail of (3). (5) A saccopharyngid (genus *Saccopharynx*, family Saccopharyngidae) (1.7m, 5.6ft) and a Saccopharyngid tail organ. (6) Tail tips of (5). (7) A halosaur (genus *Halosaurus*, family Halosauridae, of the order of notacanths) (1.8m, 6ft). (8) A snipe eel (family Nemichthyidae). (9) A Snub-nosed parasitic eel (*Simenchelys parasiticus*, sole species of the family Simenchelyidae) (60cm, 24in).

3

4

5

6

7

8

9

DWO

Fish Out of Water

How fishes are adapted for life on land

That humans are here reflects the fact that over 350 million years ago fish came out of water, adapted progressively to a terrestrial environment, evolved, and one of the results is us. Today various bony fishes leave the water for various lengths of time, although it is not suggested that the end point of their excursions will necessarily be as dramatic as that of their ancestors.

It must be pointed out that the ability of a fish to breathe air does not equate with its ability to leave water. The spectacular attributes of the lungfishes, among others, are dealt with elsewhere (see p126). Here we are concerned with fishes that actively travel out of water. Naturally there are degrees of extra-aquatic activity from small fishes that skitter along the water's surface momentarily to escape predation to the Pacific moray eel which may spend up to ten minutes out of water and others that spend far longer on land. What problems do they face?

Their problems are several. They include (*not* in order of severity): (1) respiration, (2) temperature control, (3) vision, (4) desiccation and (5) locomotion.

An initial problem facing fishes leaving water is that the surface area of the delicate gill filaments is reduced since they are no longer kept separate by the buoyancy of water. Drying out of the gills also occurs. Consequently other means of respiration have to be present and ways of protecting the gills. This usually involves the development of a sac into which the air is sucked, with a moist lining richly endowed with blood vessels (vascularized). As is always the case with fish, there are exceptions. The Chilean cling fish *Sicyaces sanguineus*, for example, which spends a substantial part of its life out of water, has a vascularized layer of skin in front of the sucker formed from its ventral fins. When it needs oxygen it raises the front of its body off the rocks and exposes this skin patch to the air. More "orthodox" are the anabantids (which include the Climbing perch) and the walking catfish (family Clariidae), which have pouches above the gill chamber, with linings expanded into convoluted shapes to increase the surface area. The Electric eel (*Electrophorus electricus*) which spends only a small part of its life out of water, can use its gills for aerial respiration, and the mud skippers (*Periophthalmus* species), which spend a lot of their life out of water, can use their skin.

Fishes on land have a problem in keeping cool. The difficulties are compounded because most "semiterrestrial" fish live in

▲ **An athletic fish,** the Climbing perch (*Anabas testudineus*). It can struggle across land and climb over rocks and stumps.

◄▲ **Climbing gear.** Mudskippers (genus *Periophthalmus* of the goby family of spiny-finned fishes) spend much of their time out of water. Their eyes, placed high on the head on short stalks, are adapted for aerial vision but have to be kept moist. Their pelvic fins, typical of gobioids, are fused into an adhesive disk TOP. The pectoral fins, used in their agile movements out of water, have short, muscular peduncles.

▲ **Fish on the highway** TOP RIGHT. Clariid catfishes are widespread in the old world. They have tree-like accessory breathing organs in the top of the gill chamber although in some deep-living species in Lake Tanganyika these are reduced. The Indian species *Clarias batrachus* was introduced into Florida (USA); it flourished and alarmed inhabitants when shoals left the waters and wriggled across roads.

tropical areas. Although living in a coolish climate, one technique for keeping cool is displayed by the Chilean cling fish, that of not producing heat by muscular activity. At its best, it has been described, when out of water, as "inactive and difficult to tell when dead." It also lives in a shore zone where waves will, from time to time, wash over it. The length of time it can stay on land depends very much on the weather. When cloudy it can stay out for about two days and lose only 10 percent of its body water. In captivity it has been shown that it takes about a 25 percent loss of body water to cause death. Chilean cling fish that die in the wild have lost little water, but the sun has driven their body temperatures up to some 24°C (75°F) which is fatal. Adult Chilean cling fish do not breathe continuously through the vascularized skin in

front of the sucker but alternate this with holding air in the mouth and pharynx and exchanging gases through the moist lining.

When on land fish have to see, and unless modifications occur they would be short-sighted. Two physical responses are possible, first to change the shape of the lens, and/or second to keep the lens spherical and change the shape of the cornea. The mud-skippers have done both and have such excellent vision that they can catch insects on the wing.

Eyes are delicate and must be protected on land. Mudskippers have a thick layer of clear skin to protect the eye against physical damage; to keep the eyes moist (as fish lack tear ducts), they roll them back into the moist sockets. The walking catfishes (*Clarias* species), which probably cannot see more than 2m (6.5ft) on land, protect their eyes by a thick layer of clear skin and by confining them to within the body contour. It is brighter on land than in water, so to cope with excess light hitting the retina some fishes, eg those of the blenny genus *Mnierpes*, have developed a layer of pigment—the fishes' equivalent of sunglasses.

For an eel, locomotion on land is just like swimming. *Clarias* species supplement the waving movements of the body with the use of the pectoral fin spines as bracers. An African species, *Clarias lazera*, burrows in drying pools and comes out at night to feed. Migrations of up to a thousand individuals have been recorded and it is suggested that the whiskers are used to keep in touch. *Clarias batrachus*, originally from Southeast Asia, was introduced into Florida where it flourished and its ambulatory habits have, not surprisingly, caused much disquiet among many residents of that state.

KEB

HERRINGS, ANCHOVIES...

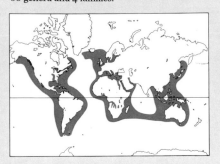

Superorder: Clupeomorpha
Order: Clupeiformes.
About three hundred and forty-two species in
68 genera and 4 families.

Anchovies
Family: Engraulidae
About one hundred and forty species in
16 genera.
Distribution: oceans worldwide.
Size: maximum length about 50cm (20in).
Species include: **anchovetta** (*Engraulis ringens*),
anchovy (*E. encrasicolus*).

Family: Denticipitidae
Denticeps clupeoides.
Distribution: rivers in W Africa (near border of
Nigeria with Benin).
Size: maximum length 8cm (3in).

Herrings
Family: Clupeidae
About two hundred species in about 50 genera.
Distribution: oceans worldwide, fresh water in
Africa.
Size: maximum length 90cm (35in).
Species include: **alewife** (*A. pseudoharengus*),
Allis shad (*Alosa alosa*), **herring** (*Clupea
harengus*), **Pacific sardine** (*Sardinops sagax*),
pilchard (*Clupea pilchardus*), **sprat** (*C. sprattus*),
Twaite shad (*Alosa fallax*).

Wolf herring
Family: Chirocentridae
Sole species *Chirocentrus dorab.*
Distribution: Indian and W Pacific oceans.
Size: maximum length 3.5m (11.5ft).

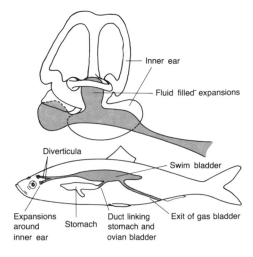

- Inner ear
- Fluid filled expansions
- Diverticula
- Swim bladder
- Expansions around inner ear
- Stomach
- Duct linking stomach and ovian bladder
- Exit of gas bladder

Sᴏᴍᴇ of the most familiar fish in the world belong to an order comprising just four families: herrings, anchovies, the Wolf herring and the Denticipitidae. Their influence has been enormous. The North Atlantic herring alone has been the subject of wars and its migrations have brought down governments and caused the dissolution of states.

Herrings and their relatives are a cosmopolitan, largely marine group of fishes. Compared with many groups they are beautifully coherent and easy to diagnose.

Freshwater representatives of the herrings and shads occur in the eastern USA, the Amazon basin, West and Central Africa, eastern Australia and occasionally and sporadically elsewhere. The anchovies are coastal forms in all temperate and tropical regions, with freshwater species in the Amazon and Southeast Asia. The Wolf herring is solely marine and the Denticipitidae live only in fresh water, the single living species occurring in a few rivers in West Africa.

In this group the **herrings** and shads family contains the largest number of species (about 200). They are of great commercial importance. Thanks to their former abundance, highly nutritious flesh and shoaling habits they became a prime target for fishing fleets. In 1936–37 members of this family made up 37.3 percent by weight of all the fish caught in the world. About half of this weight was from one species, the Pacific sardine.

In the North Atlantic the herring has long been exploited. In AD 709 salted herring were exported from East Anglia in England to the Frisians; the fisheries even appear in Domesday Book (1086). The great advantage of this herring (and others) was that it could be preserved in so many ways: pickled in brine, salted, hot and cold smoked (either salted first and split or not—resulting variously in kippers and red herrings). All were good techniques before the days of freezing and canning.

Herrings spawn in shoals in the warmer months, laying a mat of sticky eggs on the sea floor. After hatching the young become pelagic. Their food, at all stages in their life history, consists of plankton, especially small crustaceans and the larval stages of larger crustaceans. Herrings can swim at a maximum speed of 5.8km/h (3.6mph).

Sprats are small relatives of herrings. The fish often marketed as sardines are the young of the pilchard, another herring species. Whitebait are the young of both the herring and the sprat.

Shads are larger members of the herring clan. An American Atlantic species (*Alosa sapidissima*) reaches nearly 80cm (31in) in length. In 1871 they were introduced into rivers draining into the Pacific and are now widespread along the Pacific coast of the USA where they grow to 90cm (35in). In the European North Atlantic there are two species, both now scarce, the Allis shad and the Twaite shad. The latter species has some non-migratory dwarf populations in the Lakes of Killarney (Eire) and some Italian lakes. Freshwater populations of the American alewife are also known.

The external appearance of members of the herring family is very similar. Most are silvery (darker on the back), lack spines in the fins and have scales that are very easily shed (deciduous scales).

Species in the **anchovies** family are longer and thinner than many clupeids. They have a large mouth, a pointed, overhanging snout and a round belly without the scute-covered keel present in clupeids. *Engraulis mordax* from the eastern Pacific grows to about 18cm (7in). It is in great demand, more as a source of oil and as meal than as a delicacy. Only a very small proportion of the catch is canned or made into paste. It was once far commoner than now. There is a record of a single set of purse seine nets catching over 200 tonnes in November 1933.

▲ **An easy catch.** Because of their shoaling
habits and former abundance herring shoals
have long been an important source of
nutritious food in the North Atlantic region. A
major factor in reducing their numbers has
been irresponsible overfishing.

▶ **A herring of the southern hemisphere,**
Nematalosa erebi. It sometimes enters fresh
water.

◀ **Herring hearing apparatus.** So members of
the superorder Clupeomorpha can pick up
sound waves in the water they have a pair of
thin sacs (diverticula) running from their swim
bladder which penetrate the rear of the skull
and expand around part of the inner ear.

Further south in the Pacific the
anchovetta, which is abundant in the food-
rich Humboldt Current, is one of the major
resources of South American countries,
especially Peru. The anchovy is found in the
warmer parts of the North Atlantic and the
Mediterranean. Before being sold, in whatever form, anchovies are packed into barrels
with salt and kept at 30°C (86°F) for three
months until the flesh has turned red.
Anchovies rarely grow to more than 20cm

(8in) long, may live for seven years, but are
usually sexually mature at the end of the
first or second year. They are plankton
feeders and eat the edible contents of the sea
water filtered out by their long, thin gill
rakers.

The family Chirocentridae houses only
one genus, and probably only one species,
the **Wolf herring**. This giant among the herrings grows to over 3.5m (11ft) long and
lives in the tropical parts of the Indo-Pacific
Ocean. It is an elongate fish, strongly compressed and with a sharp belly keel. It has
large, fang-like teeth on its jaws and smaller
teeth on the tongue and roof of the mouth.
An avid hunter, it makes prodigious leaps.
It is not used for food because it has a lot
of small intermuscular bones and also, if
caught, it struggles violently and snaps at
anything—including fishermen. Within the
intestine is a spiral valve whose function is
to increase the absorbtive surface. Spiral
valves are very rare in bony fishes.

The family Denticipitidae has only one living member, a fish 8cm (3in) long found
only in the fastest flowing parts of a few
rivers near the Nigerian border with
Dahomey. It is silvery with a dark stripe
along the side. A seemingly inconspicuous
and insignificant fish, it is noteworthy for
the presence of a large number of tooth-like
denticles over the head and front part of the
body. Their significance and function
are unknown. Fossils, hardly distinguishable from the living species, have been found
in former lake deposits in Tanzania (formerly Tanganyika) and named *Palaeodenti-
ceps tanganikae.* The living *Denticeps* is
believed to the most primitive member of the
clupeomorphs.   KEB

BONY TONGUES AND ALLIES

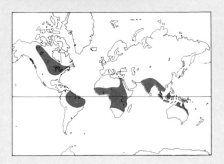
Bony tongues, butterfly-fish, mooneyes, featherbacks and elephant-snouts make up the diverse group collectively known as osteoglossomorphs. They are essentially tropical freshwater fishes. Although they typically possess toothed jaws, in all of them the main bite is exerted by teeth on bones in the tongue pressing against teeth on the roof of the mouth. They share a number of other structural characteristics, including a complex ornamentation of the scales.

The **bony tongues** are moderate to very large fishes with prominent eyes and scales and dorsal and anal fins placed well back on their long bodies. They cruise gracefully but can produce sudden powerful bursts when occasion demands. The almost fabulous Amazonian arapaima is perhaps best known. If reputed lengths of 5m (16ft) and weights of 170kg (375lb) were authenticated then this species would rank as *the* giant of strictly freshwater fishes. There is no doubt that it grows to 3m (10ft) and 100kg (220lb) in the wild but these modest achievements can be matched by other record-breaking contenders.

The arapaima is an air-breather. Its swim bladder is joined to the throat and has a lung-like lining. A similar bladder condition prevails in the African species *Heterotis niloticus* which also has an accessory respiratory organ above the gills. How much they use air is uncertain but they penetrate inhospitable swamps for spawning. Both fishes build nests and guard their young. A modest hollow about 0.5m (1.6ft) across in sand suffices for the arapaima, whereas *Heterotis* constructs a walled nest of broken vegetation about 1m (3.3ft) in diameter. By contrast *Osteoglossum* (South America) and *Scleropages* (Asia and Australia) species brood eggs and young in their mouths.

Bony tongues are carnivores that eat insects and other fish, except for *Heterotis* which consumes mud and plankton.

At the other end of the size spectrum of osteoglossomorphs lies the curious West African **butterfly-fish** which is only about 6–10cm (2.4–4in) long. This species, also an air-breather, inhabits grassy swamps where it swims close to the surface, trailing below the amazingly long and separated rays of its pelvic fins as it feeds on floating insects. It is capable of leaping high out of the water with wing-like pectoral fins extended or of simply skittering over the surface. It was originally discovered by a butterfly collector who caught one in his net.

Mooneyes are herring-like fishes of modest size which extend the distribution of osteoglossomorphs into North America. **Featherbacks** are laterally compressed fishes which swim by undulating a very long anal fin, extending from the rudimentary pelvic fins to the tip of the tail. *Notopterus chitala,*

3

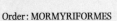

1

2

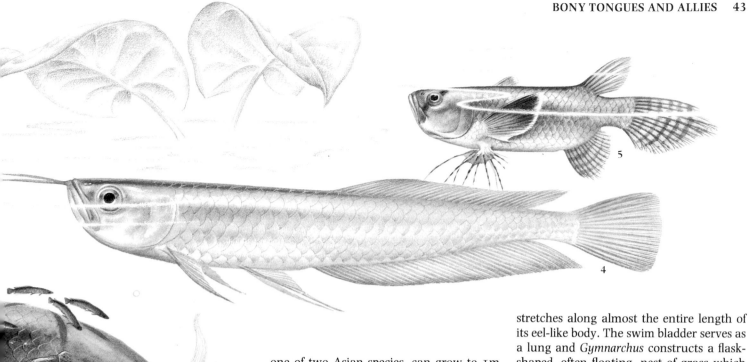

◄ ▲ **Representative species of bony tongues
and allies.** (**1**) *Petrocephalus catostoma*, a species
of elephant-snout (10cm, 4in). (**2**) An Elephant
mormyrid (*Campylomormyrus rhynchophorus*)
(80cm, 31in). (**3**) An adult arapaima (*Arapaima
gigas*) (2.2m, 7.2ft) surrounded by juveniles.
(**4**) A juvenile bony tongue (genus
Osteoglossum) at the surface (10cm, 4in; adult
size 1m, 3.3ft). (**5**) A butterfly-fish (*Pantodon
buchholtzi*) (10cm, 4in).

one of two Asian species, can grow to 1m
(3.3ft) in length and shows some parental
care of spawned eggs. Other featherbacks
are smaller. Those of Africa, *Papyrocranus*
and *Xenomystus*, have accessory respiratory
structures above the gills and can inhabit
swampy pools. All of these fishes have large
mouths and predatory habits.

The **elephant-snouts** or mormyrids make
significant contributions to the fish stocks of
African lakes, rivers and floodpools. Most
are bottom dwellers feeding on worms,
insects and mollusks. Some are elephant-
snouted, but although there is a tendency
in others for a forward and downward prolongation of the snout region, within the
family head-shapes are highly variable.
Small to medium in size, all have small
mouths, eyes, gill-openings and scales. Dorsal and anal fins are set well back and the
deeply forked caudal fin has an exceptionally narrow peduncle. The muscles are
modified to form electric organs which produce a continuous field of weak discharges
at varying frequencies around the fish and
there are electro-reception centers in part of
the enlarged cerebellum of the brain (giving
mormyrids the largest brains to be found
among lower vertebrate animals). The
system as a whole acts as a sort of radar,
sensing distortions in the electrical field from
objects coming within it. This would seem
to be an ideal adaptation for nocturnal
activities, including social interactions, in
murky waters.

Much early research into electrogenic
activity was conducted on *Gymnarchus
niloticus*. Recently separated from the mormyrids by taxonomists, it is another large,
predaceous osteoglossomorph remarkable
for its shape. It lacks anal and caudal fins
and moves by undulating a dorsal fin which

stretches along almost the entire length of
its eel-like body. The swim bladder serves as
a lung and *Gymnarchus* constructs a flask-
shaped, often floating, nest of grass which
it is reputed to defend with vigor.

Apart from their scientific interest man is
involved with osteoglossomorphs at various
levels. There are capture fisheries for
Arapaima and *Osteoglossum* in South
America and for *Heterotis* and *Gymnarchus*
in the seasonal swamps and floodplains of
West Africa and the upper White Nile.
Larger mormyrids are more widely fished in
Africa, although they are not universally
acceptable as food. For example East African
women may forego them lest their children
are born with elephantine snouts. Bony
tongues, mooneyes and *Gymnarchus* are
rated as good sport fish by anglers and so
far as pond culture is concerned some progress has been made with representatives of
the group in each continent, namely the
arapaima, *Heterotis niloticus* and *Notopterus
chitala*. Good growth rates have been
achieved but in the case of *Heterotis* harvesting may be thwarted by their superb ability
to leap an encircling seine net.

Pantodon, featherbacks and a number of
small mormyrids are particularly interesting for aquarists on account of their specialized features, but they are not
universally popular because breeding them
in captivity is either difficult or impossible.
Many public aquaria in the USA have
imported arapaima which at first caused
problems by attempting to swim through
the walls of tanks. In the Far East large specimens of *Scleropages* are also in captivity and
command astounding prices because of the
symbolism attached to them. In this connection it can be noted that *Gymnarchus* is an
important item in ceremonial feasts of the
Hausa people, and carved and painted mormyrids feature in the tombs of ancient
Egyptians.   RGB

PIKE, SALMON, ARGENTINES...

Order: Salmoniformes
Over three hundred species in 80 genera and 17 families.

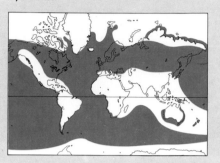

► **A small relative of the pike,** the European mudminnow. Mudminnows feed on invertebrates and small fish. Like the pikes they have dorsal and anal fins set well back on the body.

▼ **A lurking predator,** a Northern pike. This species is widely distributed throughout the cool, slow-flowing waters of the northern hemisphere. Its staple prey is small fishes but it also feasts on birds, frogs and other small vertebrates. It is usually an inconspicuous green but golden specimens are occasionally found.

MEMBERS of the order of salmon, pike and allies are of great interest to many people. It contains prize angling fishes, honored food fishes and fishes of great interest to biologists for their migratory habits.

The superorder Protacanthopterygii was originally erected to contain primitive teleostean fishes such as salmon, pike, lantern fishes, whale fishes and galaxiids. Research in the last few years has shown that these groups were "false friends," largely united on primitive characters which do not indicate relationships. Consequently the superorder has been redefined (though is probably still liable to change) and is now limited to the old order Salmoniformes (see table).

The **pike, muskellunge and pickerels** are famous game fishes, some of which grow to a very large size and are known for their ferocity once hooked. They are powerful and aggressive predators, mostly on other fish, and generally live solitary lives. They perform an important role in coarse-fish populations by controlling the numbers of abundant, fast-breeding fish species of little importance to anglers.

The distribution of pike is basically circumpolar. Of the five species the Northern pike is widespread in North America, Europe and Asia, the others more local in range, with one in Siberia and three in North America. The biggest of them is the muskellunge or muskie which may reach more than 30kg (66lb) and 1.5m (5ft) in length, while the Northern pike may exceed 20kg (44lb) and 1m (3.3ft). Legends of giant pike abound (see p12).

The North American pickerels are regarded by some experts as three species, the Redfin pickerel, the Grass pickerel and the Chain pickerel. Other authorities, however, regard the last two as subspecies of the Redfin pickerel. All are small fishes, the Redfin and Grass pickerels rarely growing to more than 30cm (1ft) long. The Chain pickerel is somewhat larger and in areas where it cohabits with the others may hybridize and produce fishes that are often claimed to be Redfin or Chain pickerels of record size. It is not always easy to distinguish the two or three species, and even more difficult to establish the true nature of the hybrids.

All pike species are similar in appearance, being slender, elongate fish somewhat laterally compressed with a long, flattened, almost alligator-like snout. The mouth itself is also long and has large pointed teeth. Perhaps the most distinctive feature of pike is the clustering of their dorsal and anal fins. This concentration of finnage at the rear enables them to accelerate rapidly and has endowed them (and also other fish with similar fin arrangements) with the name

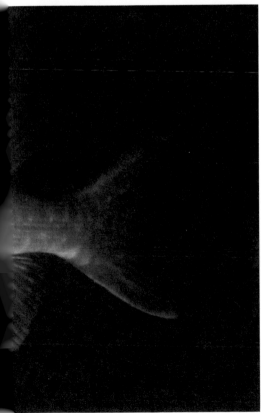

junct distribution is a relict one: fossils found in Europe and North America show that their former distribution was much like that of the extant pike. Today they occur in Europe, to the west of the Caspian Sea (European mudminnow); on the eastern side of North America (*Umbra pygmaea*, which has been introduced to Europe); in the Chelialis River, Washington, USA (*Novumbra hubbsi*); throughout the Great Lakes and Mississippi drainage (*Umbra limi*); and from Alaska and Eastern Siberia (the Alaskan black fish).

Mudminnows are small fishes, rarely exceeding 15cm (6in) in length. The caudal fin is rounded and the dorsal and anal fins are set far back, as in the pike. Their coloring is cryptic—mottled dark brown or olivaceous. All are carnivorous and feed on small invertebrates and larval fishes which they seize by making rapid lunges. They are sluggish, retiring fishes, hiding away among weeds waiting for the prey to come within striking distance.

Mudminnows are capable of living in high densities in poorly oxygenated swampy areas as they can utilize atmospheric oxygen. They are tolerant of drought, which they escape by burrowing into soft mud and ooze, and also tolerate cold, especially the Alaskan black fish, much as one might expect from where it lives. In many books there are accounts of the Alaskan black fish being able to withstand freezing. This frequently repeated untruth seems to stem from a book written by L.M. Turner in 1886. In *Contributions to the natural history of Alaska*... he wrote: "The vitality of this fish is astonishing. They will remain in ... grass buckets for weeks, and when brought in the house and thawed out they will be as lively as ever. The pieces which are thrown to the ravenous dogs are eagerly swallowed; the animal heat of the dog's stomach thaws the fish out, whereupon its movements cause the dog to vomit it up alive. This I have *seen*..." Sadly for sensation, properly controlled experiments have shown that although the Alaskan black fish is capable of living at very low temperatures, it cannot survive freezing or being ice-bound.

The **galaxiids** (they have no common name) are distinctive small fishes found in all the major southern land masses (Australia, New Zealand, South America and South Africa) and also on some of the more remote southern islands like New Caledonia, Auckland and Campbell and the Falklands. The first galaxiid was collected by naturalists with Captain Cook in New Zealand during 1777. The name *Galaxias* was given to the

"lurking predators." They skulk among vegetation around the margins of lakes and streams and surge out to catch passing prey.

Pike are mainly freshwater fish, but a few venture into mildly saline waters in Canadian lakes and the Baltic. Spawning takes place among vegetation in still or gently flowing marginal shallows. The female pairs with a male and the two spend several hours together releasing and fertilizing small batches of quite large eggs (2.5–3cm, about 1in, long). A large female may lay several hundred thousand eggs. From the time they hatch pike are predators, initially eating insects and small crustaceans but very soon becoming fish-eaters like the adults. Large pike may also occasionally take small mammals and birds.

Pike are prized by anglers, especially in Europe. In North America, where salmon species are more diverse, abundant and freely available the popularity of pike is not as great though many fishermen have the aspiration of catching a large muskellunge.

The **mudminnows** are closely related to pike. They were formerly included in the families Daliidae and Novumbridae, but are now united in one family. Their present dis-

fish, because it was dark, black-olive and covered with a profusion of small gold spots resembling a galaxy of stars. Galaxiids are fish of distinctive appearance. Scaleless, they have smooth leathery skins. Unlike some northern relatives they have no adipose fin but have the single dorsal fin well back towards the tail, over the anal fin. Most are small, 10–25cm (4–10in) in length, but one reaches 58cm (23in), while there are several tiny species, 3–5cm (1–2in) long. Most species are tubular, cigar-shaped fish with blunt heads, thick fleshy fins and a truncated tail. A few are stocky, thick-bodied fish and most of these are secretive fish that skulk among boulders, logs or debris in streams or lakes. Many are solitary but a few are shoaling fish. All but about six species are freshwater fish and a few are marine migratory fish in which the larval and juvenile phases are spent at sea.

Migratory species spawn mostly in fresh water, rarely in estuaries. When their eggs hatch tiny larvae, about 1cm (0.4in) long, are swept to sea. Their ability to cope with a sudden transition from fresh to sea water shows remarkable flexibility. They spend 5–6 months at sea before migrating back into fresh waters during spring, as elongate transparent juveniles. It will be months before these fish reach maturity, at about a year in some species, two to three years in others.

What is known about the spawning of these fish is equally remarkable. One species, *Galaxias maculatus*, is known to spawn in synchrony with the lunar or tidal cycle, spawning over vegetated estuary margins during high spring tides. When these high tides retreat the eggs are stranded in the vegetation and protected from dehydration only by humid air. Development takes place out of water, the eggs not being reimmersed until the next set of spring tides two weeks later. Then the eggs hatch, releasing the larvae which are swept quickly out to sea. Another species, which lives in small heavily forested streams, spawns during floods. This fish, *Galaxias fasciatus*, deposits its eggs in leaf litter along stream margins. When the floods dissipate the eggs are left stranded among rotting leaves, where they develop. Hatching cannot occur until there is a further flood and then the larvae are swept downstream and out to sea.

It is clear that both these modes of spawning involve substantial risks. The breeding habits described above are exceptional. As far as is known most galaxiid species lay their eggs in clusters between rocks and boulders. A few very small species pair up

for spawning and lay their eggs on the leaves of aquatic plants.

Although the southern lands in which these fish occur tend to have moist climates, some species have become adapted to surviving droughts and therefore aestivate. Some live in pools that lie on the floor of wet podocarp forests in water usually only a few centimeters deep which covers leaf litter shed from the towering forests above. In the late summer and fall these pools frequently dry up and the fish disappear into natural hollows around the buttresses of trees. They survive for several weeks in these damp pockets until rainfall restores water to the forest floor. The fish wait for the return of water before spawning. As the pools are increasingly replenished larvae are enabled to disperse around the forest floor and invade available habitats.

Historically the very broad distribution of the galaxiid fishes has attracted intense interest. Long ago zoologists asked how a group of freshwater fish could be so widely distributed around southern lands. Not only is the family widely dispersed but one species, *Galaxias maculatus*, is found in Australia, Tasmania, Lord Howe Island, New Zealand, Chatham Islands, Chile, Argentina and the Falkland Islands. With such a broad range this fish is one of the naturally most widely distributed freshwater fish known. Noting the remarkable range of this species, zoologists of the late 19th century suggested that there must have been former land connections between the areas where the fish are present. Some thought that Antarctica might have been involved.

In recent years it has been suggested that the distribution of galaxiid fishes results from their former presence on the southern continent Gondwanaland. This huge land mass was breaking apart some 70 million years ago. If galaxiid fishes were then present their distribution could date back to these ancient times, but this view is debatable. An alternative interpretation is that they dispersed through the sea to attain their present distribution. The latter viewpoint becomes more acceptable when it is remembered that some species spend up to six months at sea. It is probably no coincidence that the widely distributed species are also the sea-going ones.

That such small fish should have attained importance for commercial fisheries may seem remarkable. However, when European settlers arrived in New Zealand in the mid 19th century they found that vast populations of the sea-living juveniles of several *Galaxias* species migrated into rivers during

▲ **Scattered across the southern hemisphere** are the galaxiids, about 40 species of mostly freshwater fishes. They are the southern relatives of the salmons and trouts. Their enigmatic distribution has caused much speculation. One school favors the theory that they dispersed along ocean currents. Another argues that they developed from fishes that were separated when the great southern continent Gondwanaland broke up some 70 million years ago. This is one of the 20 Australian species, the Flathead galaxias (*Galaxias rostratus*).

▶ **The markings on some galaxiids,** like the Golden galaxias (*Galaxias auratus*) shown here, resemble the specklings on the northern trouts. This species is restricted to Lakes Crescent and Sorell and associated streams in Tasmania.

spring. The Maoris of New Zealand exploited huge quantities of them and not surprisingly the Europeans followed their example, calling the tiny fish whitebait on account of their similarity to fish they knew at home in England. Quantities caught today do not compare with those of the early years, but a good fishery persists in some rivers. Colossal numbers of fish are caught (each weighing only about 0.5g—there are about 1,800 fishes to a kilogram). The catches of individual fishermen vary from just a few to many kilograms, exceptional catches reaching several hundred kilograms in a day. Not only are these a delicious gourmet seafood, but they may sell for up to 50 Australian dollars a kilogram in the shops.

The family of **salmon, trout, charrs and their relatives** contains some of the most prestigious and influential fishes of the northern hemisphere. Included here are the Atlantic and Pacific salmon, trout, charrs, grayling and whitefish.

Of this august company the Atlantic salmon is the best known. Many people pay high prices for its flesh and even higher prices for the pleasure of trying to catch it by inefficient means. It is caught more efficiently by commercial fishermen (ensuing problems are discussed on pp10–11). Although salmon is now a luxury item, in the 19th century apprentices in London protested about being fed salmon six days a week.

To emphasize the significance of the only salmon species native to the Atlantic, just think of the different names applied to each stage of its life history. On hatching they are called fry and alevins. Later, when they are a few centimeters long and have developed

dark blotches on the body, they are parr (the blotches are parr marks). When they make for the sea the marks become covered by a silvery pigment and they are called smolt. When the adult male returns early from the sea it is a grilse. The adults, spent after spawning, and drifting back to sea, are called kelt.

After adult salmon have spawned in their home streams (see box) the eggs remain in the gravel, for much of the winter in the case of the early run salmon. Hatching time can be calculated on the 440 day-degree principle, ie when the average water temperature in degrees Celsius multiplied by the number of days since spawning equals 440, the eggs hatch. This is usually in April–May in northern Europe.

The young are about 2cm (0.8in) long on hatching and for the first six weeks or so live in the gravel, feeding on their yolk sacs. With the exhaustion of the yolk supply they emerge from the gravel and start feeding on insect larvae and other invertebrates. As they grow they develop into parr and their markings give them camouflage for hunting actively. The length of time spent in fresh water varies from five years in the north of the range to one year in the south.

Not all members of the year class migrate to the sea. A few males stay in fresh water and become precociously mature. These have been seen shedding sperm in the company of mating adults and it is argued that the precocious males act as a security in case no males return from the sea. The migrating young spend some time in estuaries, acclimatizing themselves for coping with salt water. In the sea they grow rapidly and can reach 14kg (30lb) in 3 years. They feed on fish and spend up to four years in the sea building up strength for the rigors of the spawning migration. When returning to the natal stream for spawning they can travel 115km (70mi) a day.

There are a few populations of land-locked salmon in lakes in the far north of America and Europe. They never grow as large as the others but still run up streams to spawn. It is thought that their access to the sea was stopped after the last ice age.

There are probably seven species of the Pacific salmon, whose generic name (*Oncorhynchus*) means hooked snout. Their life history is very similar to that of the Atlantic salmon. Two species have land-locked forms, first the Sockeye salmon, which is found from southern Oregon to Japan, and the somewhat doubtful species *Oncorhynchus rhodiurus* from Lake Biwa in Japan.

▲ **Driven by an irresistible compulsion,** adult salmon returning to the spawning ground can make prodigious leaps to overcome obstacles.

◄ **At the spawning ground** TOP the female salmon (here a Sockeye salmon) scrapes out a nest in the gravel of the river bed.

◄ **The large yolky eggs** CENTER are normally hidden in the gravel to hide them from predators.

◄ **Newly hatched salmon young** BELOW LEFT are called alevins or fry.

▼ **Young salmon** feed on their yolk sac at first. When this is almost exhausted, as shown here, they start to feed on small invertebrates.

► **A silvery school** OVERLEAF of Atlantic salmon. It was once one of the most widespread species in river systems draining into the Atlantic, until the rise of pollution, poaching and gross overfishing.

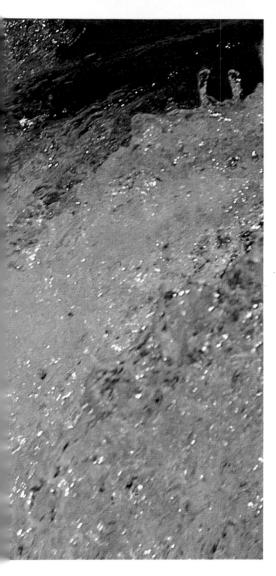

In the Atlantic Ocean the Atlantic salmon grows to about 32kg (70lb) in weight. The Chinook salmon has been recorded at 57kg (126lb). The Humpback or Pink salmon rarely grows to more than 9kg (20lb) but only takes two years to mature.

A Eurasian genus is *Hucho*. It includes the huchen, a slender species from the Danube; others occur in Central Asia. Attempts were made to translocate the huchen to the Thames in the 19th century. Despite rumors of it surviving into the early 20th century there is no reliable evidence that the introduction succeeded. The introduction of the Pink salmon into the northwest Atlantic, however, was an accident and the range of this Pacific species is continually expanding.

The charrs come from the cold deep lakes and rivers of Europe and North America. Only in the very north of the Atlantic are they migratory. The sole European species, *Salvelinus alpinus*, is very variable. Until relatively recently a different species was named from almost each lake. In the breeding season the males sport a spectacular deep red on the breast. If anything their flesh exceeds that of the salmon in quality and those lucky enough to have the chance to sample it should have charr steaks, lightly boiled with bay leaves, cold on toast.

Despite its common name of Brook trout, this eastern American species is a charr. In its native haunts the migratory form can grow to nearly 90cm (3ft) long but the European introductions rarely reach half that length. In Europe, the Brook trout hybridizes with both the native Brown trout and the introduced Rainbow trout. The offspring of both mismatches are a striped fish called the tiger or zebra trout. They are sterile.

The European brown trout has caused much confusion in the past among anglers and fishmongers. It is a very variable species, both in form and behavior, so consequently has been given many common names: eg trout, River trout, Lake trout, Salmon trout and Sea trout. Brown trout that live in lakes may become very large and cannibalistic—these are Lake trout. Brown trout that migrate to the sea to feed become very silvery and are called Sea trout or Salmon trout (*not* a hybrid between a salmon and a Brown trout). As a result of the richer feeding possibilities in the sea, the migratory Brown trout reach nearly twice the size of their nonmigratory brothers. What is not understood is why, when some populations of Brown trout could migrate to sea, they do not.

A similar situation occurs on the western coast of North America with the Rainbow trout. This commonly introduced denizen

Return to Base

The life history of the Atlantic salmon exemplifies that of the other anadromous species (ie those that migrate from the sea to fresh water to spawn). Larger salmon run up rivers in winter, smaller ones in summer, overcoming most obstacles to return to the stream in which they were hatched (this may be a journey of thousands of kilometers for the Pacific salmons). The key to this remarkable ability is memory and a sense of smell. It has been demonstrated that the adults remember the smell of their birth stream (a combination of chemicals in the water contributed by the rocks, soils, vegetation etc) and can find it again. Naturally, occasional errors are made, or the return route is blocked as a result of dam construction or drought, which enables the species to extend its range.

The males usually arrive back first, but the females normally select the spawning site, where they are aggressively courted by the males. During maturation the colors of the males are enhanced and the lower jaw bends upwards to form a hook called the kype. The spawning site (called a redd) is excavated in gravel by the vigorous movements of the female's tail until it is up to 3m (10ft) long and 30cm (12in) deep. The pair lay alongside each other for spawning and accompany the act

with much trembling and jaw opening. Each act of spawning continues for about five minutes and the whole process may continue for a fortnight. In between, the adults rest in deep holes in the river bed. After each spawning session the redd is filled in and another excavated.

When all the eggs are shed the adults drift slowly back to sea, exhausted and prone to many infections. One of the mysteries of Atlantic salmon migrating upstream to spawn is that they do not feed in fresh water (hence the understandable exhaustion and loss of up to 40 percent of body weight), yet they are caught on anglers' spoons and spinners. Why they are tempted to ingest these flashing objects is unknown. Very few males survive spawning, but those that do will recover rapidly in the rich feeding grounds in the sea. Only a tiny number of females have survived four spawnings.   KEB

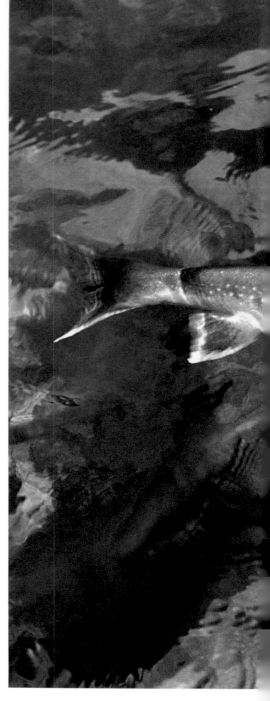

of fish farms in Europe has an extensive natural range from southern California to Alaska. In the northern part of its range it is migratory, and much larger and more intensively colored than in the south. The Canadians call it the Steelhead trout (migratory form) whereas the nonmigratory is the Rainbow. Some years ago a water authority in the United Kingdom bought from North America a large number of Rainbow trout to enhance the trout fishing in their waters. Unfortunately they were provided with the migratory form, and very little of their investment ever returned.

Because of their sporting and palatable qualities salmon and trout have been introduced into many countries. The Brown trout is now found in the North American west, doubtless in exchange for the Rainbow trout, and in almost all southern hemi-sphere countries, even in some tropical ones where there are cool streams at high altitudes in which they can thrive.

If the naming of trout species formerly caused chaos it is nothing compared with the current chaos concerning the whitefish. These are relatively plain, silvery salmonoids, the great majority of which live in cold deep lakes of Asia and northern America. (A few species are also found in Europe.) There are probably three genera. Many species are highly variable. For example populations in European lakes have been grouped into any number from three to eight species. They are of economic importance, especially in North America where they are smoked. A few species are migratory, but most are landlocked as a result of hydrological changes and changes in sea level since the last ice age.

▲ **The huchen** LEFT is a species of salmon from the River Danube and its tributaries. Attempts to introduce it to other rivers have been unsuccessful. It eats aquatic vertebrates and invertebrates but will also feed sporadically on other food, eg small mammals.

▲ **The charrs** ABOVE live in cold northern waters. In the breeding season the underside turns a bright red. This is the European charr (*Salvelinus alpinus*).

◄ **The Rainbow trout** derives its common name from the stripe on its flanks. Native to North America it has now been introduced to many parts of the world, often to the detriment of the local fauna.

► **The Brook trout,** a native of North America now introduced to Europe, is despite its common name a charr.

The graylings derive their generic name (*Thymallus*) from the thyme-like smell of their flesh. They are fishes of cool, swift-flowing rivers. Occasionally they enter brackish water. *Thymallus thymallus* is widespread in Europe and Asia. The slightly more colorful *Thymallus arcticus* occurs in northern America. In northern Asia some dubious species have been described.

To complete this important family two more poorly known species of uncertain affinities must be mentioned. In the lakes of the Ohrid region on the borders of Albania, Yugoslavia and Greece live salmonoids placed in the genus *Salmothymus*. They may be southern landlocked salmons. The genus *Brachymystax* comes from rivers of Mongolia, China and Korea. No satisfactory conclusions have been reached on its relationships—very few specimens have ever been available for study.

The family of scaled, cylindrical, southern hemisphere salmonoids are commonly known as **southern graylings**. Only two species are included in this family: *Prototroctes oxyrhynchus* from New Zealand and *P. maraena* from southeastern Australia and Tasmania. The New Zealand species was first described and named in 1870 when it was abundant. The last known specimens were caught by accident in a Maori fish trap in 1923. It is presumed to be extinct.

The Australian species appears to be following the same path; it is now described as severely endangered. Why these two remarkable fish species should have suffered so is unknown. In New Zealand they disappeared from regions completely untouched by human development, as well as from those areas where the cause was evident. When Europeans first settled in New Zealand in the 1860s *Prototroctes* was abundant and regarded as a good source of food, but by the late 1870s there was already concern about the decline of this widespread fish. In 1900 they were described as "thinned out." Twenty-three years later the last known specimens were caught. Once they existed in their millions; today there are less than 40 bodies preserved in the great national museums of the world—and a stuffed specimen in the Rotoiti Lodge of the New Zealand Deer Stalkers' Association, Nelson.

There are only a few records describing the New Zealand *Prototroctes* alive. Like its Australian congener it had only one ovary or testis. The early reports of its body color are inconsistent. One suggests that it was slaty on the back, merging to silvery on the sides and belly with patches of azure. The

fins were orange, slaty dark at the tips and the cheeks had a golden tinge. It seemed to have migrated regularly from the sea to fresh water where it spawned.

Aplochiton zebra is a fairly widespread fish, occurring in rivers of Patagonia and the Falkland Islands where it is called trout. Unlike the inconspicuously colored *Prototroctes*, *Aplochiton* has dark vertical stripes over the back and sides of its scaleless body. Its persistence in the Falkland Islands has been placed in jeopardy by the introduction of the Brown trout from Europe—a sad fate for a species first collected by Charles Darwin when he visited the Falklands. Very little is known of the biology of this fish.

The **osmerids** or smelts may well have acquired their alternative common name from the fact that a fresh *Osmerus eperlanus*, the northern European species, has a strong smell (like that of a cucumber). Even if apocryphal, it is a pleasing idea. Mostly small, silvery fish, smelts live in coastal and brackish cool waters in the northern hemisphere. Numerous, probably landlocked populations are known. They are carnivores, feeding on small invertebrates seized with their sharp conical teeth.

Their importance to subsistence fisheries in the far north is significant. They can be numerous and have a high fat content. When not eaten by the original inhabitants of the British Columbian coast, they were dried, and because of their fat content could be set alight and used as a natural candle. Hence *Thaleichthys pacificus* now has the common name of Candle fish.

The **ayu** is the only species in its family. This very peculiar salmonoid lives in Japan and adjacent parts of Asia, where it is of great economic importance. Its body is olive brown with a pale yellow blotch on the side. The dorsal fin is expanded and, like the other fins, has a reddish tint. When these colors, especially the reds, are enhanced in the breeding season, the Japanese name for the fish changes from *ayu* to *sabi*, which means rusty. Instead of the male developing an extended snout (kype) as many salmonoid males do, both sexes become covered in warty "nuptial tubercules" (warty growths) at the onset of breeding. The upper jaw of the male shortens and the female's anal fin expands. These changes start during the summer and the fish breed in the fall.

The fish mature in the upper reaches of rivers and, unlike most other salmonoids, move downstream towards the sea to breed. Little is known about their breeding behavior. Some 20,000 adhesive eggs are produced which hatch in about three weeks,

A Pike in the Southern Hemisphere?

If the galaxiids are the salmonoids of the southern hemisphere, and salmonoids are related to pike, is there a southern hemisphere pike? The answer is probably yes.

Described as recently as 1961, it is a small fish found in numerous localities in southern Australia and called *Lepidogalaxias salamandroides*. It was originally thought to be a galaxiid, but unlike galaxiids it has scales, dorsal and anal fins set much further forward and the anal fin of the mature male is highly modified with gnarled and hooked rays and peculiar dermal flaps.

This tiny fish, about 4cm (1.5in) long, has recently been much studied and the general, but not universal, conclusion is that it is more closely related to the northern hemisphere pikes than to the southern hemisphere galaxiids.

Little is known of its biology. It is apparently capable of surviving drought by burrowing in mud or under damp leaves. Its distribution appears to be confined to small pools and ditches mostly in the sand-plain area between the Scott River and Lake Powell. It is usually found in acid water (with a pH up to 4.5).   KEB

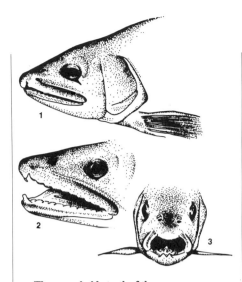

▲ **The remarkable teeth of the ayu.**
(1) Juxtaposing comb teeth on the outside of the jaw. (2) The fleshy snout with the few canine teeth behind it. (3) The canine teeth and the elaborate skin folds inside the mouth. The ayu's actual method of feeding is only known from supposition.

▼ **Named for its smell.** The scientific name of the European grayling (*Thymallus thymallus*) alludes to the thyme-like smell of its flesh, which is delicious though it soon deteriorates after the fish's death. For this reason graylings have virtually no commercial importance.

depending on temperature. The larvae stay in the river until they are about 2.5cm (1in) long when they move into the sea.

This seaward migration of the larvae is part of an interesting survival strategy. If the young stayed in the river, having spawned in the fall, they would have to endure the cold and compete with larger young of the species that spawned in spring (when most fishes spawn). However, during winter the temperature of the sea is more stable than that of the river and at sea food is more abundant. On the other hand, the young ayu have to have developed a physiological (osmotic) mechanism to enable their small bodies to cope with the shock of transferring from fresh to salt water. They must be able to cope with the transfer but it is not known how. During winter they feed on zooplankton and small crustaceans and, by spring, when they reenter fresh water, they have grown to about 8cm (3in) long. Then they migrate upstream in huge shoals, when thousands are caught and taken to capture ponds to facilitate a rapid growth rate and to provide an easily accessible source of food. The fish that escape continue up to the fast-flowing upper reaches, where each fish establishes a territory for itself among rocks and stones. Here they feed on diatoms and algae until summer or fall when they move downstream to spawn. The ayu is an annual fish as almost all adults die after spawning. The few adults that survive spawning (a very small percentage) spend the winter at sea and repeat the cycle.

Concomitant with the change of diet from young to adult and the move from salt to fresh water, the teeth change drastically. While the ayu is at sea its diet is typically salmonoid: using its conical teeth it catches small crustaceans and other invertebrates. Adults, by contrast, feed on algae and have a whole series of groups of teeth forming comb-like structures. Even more unusual is the fact that the teeth lie outside the mouth.

The comb-teeth develop under the skin of the jaw and erupt, shedding the conical teeth, when the fish enter fresh water. Each comb-tooth consists of 20–30 individual teeth, each one shaped like a crescent on a stick: narrow in the plane of the fish but very broad transversely. The gutter of the crescent faces inwards and, because of the different lengths of the arms of the crescent on each tooth, forms a sinuous gutter across the width of the comb-teeth. The combs of the upper and lower jaws juxtapose outside the mouth when it is closed. At the front of each lower jaw is a bony, pointed process that fits into a corresponding recess in the

upper jaw. On the mid-line of the floor of the mouth is a flange of tissue, which is low at the front but higher at the back where it branches into two. Each branch bends back on itself to run forwards, parallel to the sides of the jaw and decreasing in height towards the front. Muscles link this device with a median bone in the branchial series.

It is known that adult ayus eat algae, but how? "Grazing marks" have been seen on stones covered by algae in an ayu's territory. It is usually stated that they are formed by the comb-like teeth, but as these teeth are outside the jaw, any algae so scraped off would be washed away and not eaten. Even to make the grazing marks would necessitate the fish lying on its side, a maneuver that it is most unlikely to make in a swift current.

It is suggested here that the ayu is a filter-feeder in a manner analogous to that of the baleen whales. The ayu's snout is fleshy and slightly overhangs the front of the upper jaw. Behind the snout, at the front of the upper jaw, is a row of about eight small conical teeth. Hanging down from the palate is a complex series of curtain-like structures which have a relationship to the various flanges on the floor of the mouth. It is possible that if the ayu were to face into the current and scrape the algae off the rocks with the conical teeth, the algae so scraped would wash into the mouth. This would at least offer a reasonable explanation for the grazing marks and, with its large mouth and wide gape, the ayu is theoretically capable of performing this maneuver. With enough algae inside, the mouth could be tightly closed. (Certainly with the peg and socket fitments at the front and the clean, straight toothless abutting parts of the upper and lower jaws the mouth is designed to close firmly.) When sealed, if the muscles mentioned above were contracted, the floor of the mouth would be brought up to the palate, thus squeezing out the water. The algal particles would be trapped on the hanging fringes and could be swallowed. How they are swallowed is unknown. If the comb-teeth are a part of the filtration system they can only act as a long-stop, but if so, how the entrapped particles are removed is a mystery.

The shape of the teeth suggests that they ought to have some filtering function, but if it turns out that they do not, then we have to find another explanation for the presence of these peculiar yet highly evolved structures. Indeed, much work remains to be done on this enigmatic, yet commercially important, species.

THE 17 FAMILIES OF PIKE, SALMON, ARGENTINES AND ALLIES

Superfamily: ESOCOIDEA

Pike, muskellunge, pickerels and their relatives

Family: Esocidae
Five species of the genus *Esox*.
Distribution: fresh waters of Northern Hemisphere mostly N of 40 degrees N, but largely absent from USA W of Mississippi basin.
Size: maximum length 1.5m (5ft), maximum weight 30kg (66lb).
Species: **Chain pickerel** (*Esox niger*), **Grass pickerel** (*E. vermiculatus*), **muskellunge** or **muskie** (*E. masquinongy*), **Northern pike** (*E. lucius*), **Redfin pickerel** (*E. americanus*).

Mudminnows

Family: Umbridae
Five species in 3 genera.
Distribution: N America, E Europe, USSR W of Caspian Sea and E Siberia.
Size: maximum length 15cm (6in).
Species include: **Alaskan black fish** (*Dallia pectoralis*), **European mudminnow** (*Umbra kraemeri*).

Superfamily: LEPIDOGALAXOIDEA

Family: Lepidogalaxiidae
Lepidogalaxias salamandroides.
Distribution: S Australia, in sand plain area between R Scott and Lake Powell.
Size: length 4cm (1.5in).

Superfamily: SALMONOIDEA

Galaxiids

Family: Galaxiidae
About forty species in 6 genera.
Distribution: S America, S Africa, Australia, New Zealand, southern islands, eg New Caledonia, Falkland Islands.
Size: length 3–57cm (1–22in).

Salmon, trout, charrs and their relatives

Family: Salmonidae
About seventy species in 10 genera.
Distribution: northern temperate and subarctic fresh waters.
Size: maximum length 1.5m (5ft).
Species and genera include: **Atlantic salmon** (*Salmo salar*), **Brook trout** (*Salvelinus fontinalis*), **Brown trout** (*Salmo trutta*), **charrs** (genus *Salvelinus*), **Chinook salmon** (*Oncorhynchus tshawytscha*), **graylings** (genus *Thymallus*), **Humpback** or **Pink salmon** (*Oncorhynchus gorbuscha*), **Pacific salmon** (genus *Oncorhynchus*), **Rainbow trout** (*Salmo gairdneri*), **Sockeye salmon** (*Oncorhynchus nerka*),

whitefish (genera *Coregonus*, *Prospium*, *Stenodus*).

Southern graylings*

Family: Prototroctidae
Two species of the genus *Prototroctes*.
Distribution: SE Australia, Tasmania, New Zealand.
Size: maximum adult length 35cm (14in).

Family: Aplochitonidae*
Probably four species in 2 genera.
Distribution: Patagonia, Falkland Islands.
Size: length 30cm (12in).

Superfamily: OSMEROIDEA

Smelts or osmerids

Family: Osmeridae
Ten or eleven species in probably 6 genera.
Distribution: Northern Hemisphere in Atlantic and Pacific oceans in coastal and brackish waters.
Size: maximum adult length 40cm (16in) but most are less than half this size.
Species include: **Candle fish** (*Thaleichthys pacificus*).

Ayu

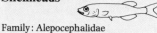

Family: Plecoglossidae
Plecoglossus altivelis.
Distribution: Japan and adjacent parts of Asia in marine and fresh waters.
Size: about 30cm (12in).

Southern smelts

Family: Retropinnidae
Four species in 2 genera.
Distribution: southern Australia, Tasmania, New Zealand in fresh, coastal and marine waters.
Size: maximum adult length 15cm (6in).

Noodle fishes or ice fishes

Family: Salangidae
About fourteen species in 6 genera.
Distribution: SE Asia and offshore areas in estuary and marine waters.
Size: 18cm (7in).

Family: Sundasalangidae
Two species of the genus *Sundasalanx*.
Distribution: southern Thailand and Borneo in fresh water.
Size: probably less than 3cm (1in).

Superfamily: ARGENTINOIDEA

Argentines or herring smelts

Family: Argentinidae
Twenty species in 2 or 3 genera.
Distribution: Atlantic, Indian and Pacific oceans.
Size: maximum length about 60cm (24in).

Family: Bathylagidae
Probably thirty-five species in 3 genera.
Distribution: Atlantic and Pacific oceans in deep water.
Size: 25cm (10in).

Superfamily: ALEPOCEPHALOIDEA

Slickheads

Family: Alepocephalidae
At least sixty species in about 22 genera.
Distribution: all oceans.
Size: maximum adult length 76cm (30in) but many are smaller.

Family: Bathyprionidae
Bathyprion danae.
Distribution: N and SE Atlantic Ocean, S Indian Ocean.
Size: 45cm (18in).

Family: Searsiidae
About thirty species in about 12 genera.
Distribution: all oceans.
Size: maximum length probably 38cm (15in).

*Some authors place these families within the southern smelts (Retropinnidae).

▶ **Top sport** ABOVE, a giant Pacific salmon caught by a North American fisherman.

▶ **An uncertain future** BELOW faces the Australian species of southern grayling. During the last 40 years its numbers have fallen dramatically. Now it is severely endangered. Its New Zealand relative has not been seen for well over 50 years and is deemed extinct.

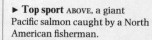

The **southern smelts** are slender, small-mouthed fishes from southeastern Australia, Tasmania and New Zealand. Some populations are migratory, others landlocked. Because they are very variable in form their taxonomy is far from clear. There may be three Australian species, or just one, which may or may not be the same species as *Retropinna retropinna* of New Zealand.

Noodle fishes are very small, slender transparent fishes from the western Pacific. Although marine they move into estuaries to spawn. Their head is tiny and pointed and the deepest part of the body is just in front of the dorsal fin. The only coloration on the 10cm (4in) long *Salangichthys microdon* is two rows of small black spots on the belly. Although small, they are sometimes so abundant that they can easily be caught for food. To the Japanese they are a delicacy called *shirauwo*.

The shape and transparency of the adult is very similar to that of the larvae of other, larger salmonoids. If it were not for the fact that sexually mature examples are known, they would have been thought to be larvae. It is not unknown in lower vertebrates for a species to evolve by a process called neoteny or paedomorphosis in which the body development is curtailed whilst sexual development is not. Both this family and the following one are neotenic. The best known example of neoteny occurs in the axolotl, which is the sexually mature, reproducing, former larva of a salamander. With time the ability to "grow up" is lost and a new species formed.

The two very small freshwater species that form the family Sundasalangidae, from Thailand and Borneo, were described as recently as 1981. The body is scaleless and transparent. Although they have the posteriorly placed dorsal and anal fins typical of salmonoids they lack the characteristic adipose fin. They also lack several bones in the skull. Superficially, they resemble the noodle fish but have particular features of the paired fin girdles and gill arches which make them unique. They are the smallest of all salmonoid species, one of them becoming mature at 1.5cm (0.5in) long. If such a phenomenon were possible, they are "even more" neotenic than the Salangidae.

The **argentines** (their name reflecting their silvery sheen, not their geographical distribution) are completely marine salmonoids. Also called herring smelts they reveal their osmeroid relationships by having an adipose fin and by superficially resembling the anadromous freshwater osmerid smelts. They are mostly small, usually less than 30cm (12in). They are elongate, slender, silvery fish, darker on the back, usually lacking distinctive markings or coloration. They have scales, a well-developed, rather flaglike dorsal fin high on the back, usually in front of the ventral fins, which are in the abdominal region. The head is longish with a pointed snout, the mouth small and terminal. The eyes are very large, a common feature of fish living at some depth in the sea. Although not well known, argentines are widespread in most oceans of the world; they are found down to about 1,000m (3,300ft), mostly a few hundred meters down, where they probably live in aggregations, if not in coordinated schools. From their teeth and stomach contents we know that they are carnivores, living on small crustaceans, worms and other prey. Their small size and the depths at which they occur mean that they are not of prime importance to commercial fisheries, but are taken for processing. They also have a role as a forage fish for larger and more economically useful food fish of deep waters.

Although the adults are deepwater fish, the eggs and young are found in the surface waters of the ocean, usually over the continental shelf. Their eggs are 3–3.5mm (0.1in) in diameter. They are slow-growing fish, and long lived, one estimate being that they may live for 20 years or more.

There are several genera, *Argentina* and *Glossanodon* being the typical argentinids, whereas *Microstoma* and *Nansenia* are greatly elongated species and *Xenophthalmichthys* is bizarre with tubular eyes that look forwards like a pair of automobile headlights.

The family Bathylagidae is divided into two superficially very different groups. The group that gives its name to the family consists of small, dark, large-eyed fishes found worldwide at depths of about 3,500m (11,500ft) or less. Nothing is known of their biology beyond the fact that some species eat crustaceans.

The other group, the opisthoproctids, is altogether more remarkable. There are six genera and probably about a dozen species in tropical and temperate seas down to about 1,000m (3,300ft). All of the species have tubular eyes. In the deep-bodied species (*Opisthoproctus*, *Macropinna* and *Winteria*) the eyes point directly upwards. In the fragile, slender-bodied forms (*Dolichopteryx* and *Bathylynchnops*) the eyes point forwards. The remaining species (*Rhynchohyalus natalensis*), which is known from only three examples, is alleged to be intermediate.

Opisthoproctus grimaldii grows to about 10cm (4in) long and lives in the North Atlantic. Its body is silvery, with a scattering of dark spots on the back. The sides of the body are covered with very deep scales. A swim bladder is present. The skull is so transparent that in live or freshly dead specimens the brain can be seen clearly behind the eyes. The spherical lenses in the tubular eyes are pale green. The ventral edge of the body is flattened and expanded into a shallow trough known as the sole. The base of the sole is silvery but covered with large thin scales and a dark pigment. The sole is believed to act as a reflector for the light produced by bacteria in a gland near the anus. The light from the gland passes through a lens and is then reflected downwards by a light-guide chamber just above the flattened part.

Opisthoproctus soleatus, a widespread species, has a different pigmentation pattern on the sole from its only congener, so it is thought that the sole enables species recognition in the areas where the two species live together. The upward pointing, tubular eyes, which afford excellent binocular vision, would easily be able to perceive the light directed downwards. The main food of the North Atlantic species (*O. grimaldii*) seems to be small, jellyfish-like organisms.

Dolichopteryx longipes is slender and very fragile. It is also very rare. The fins are elongated like a filament and there is no swim bladder. The muscles are very poorly developed; indeed, it has lost so much of its ventral musculature that the gut is enclosed only by transparent skin. It is, therefore, probably a very poor swimmer, and the tubular eyes may be advantageous in avoiding predators. Unlike *Opisthoproctus*, there is a light-producing organ associated with the eye. The species has been caught, infrequently, in all tropical and subtropical oceans between 350 and 2,700m (1,150–8,860ft) deep.

The family Bathyprionidae is based on fewer than six examples of its only species *Bathyprion danae*. It is a pike-like fish, living at depths of some 2,500m (8,200ft) in the South Indian Ocean and the North and Southeast Atlantic. It probably hovers waiting for very small fishes or crustaceans to come close when, so its fin positions would suggest, it surges forward to grab the prey.

Neither they, nor their close relatives the Alepocephalidae, have an adipose fin. The latter family are called **slickheads** because the head is covered with a smooth skin whereas the body has large scales. Most species are dark brown, violet or black in color. Light organs are rare among slickheads, but one genus, *Xenodermichthys*, is distinguished by the presence of tiny, raised light organs on the underside of the head and body.

The members of the remaining family in this assemblage, the Searsiidae, are deep-sea fishes with large, extremely light-sensitive eyes. The lateral-line canals on the head are greatly enlarged and expanded. They are found in all except polar waters and the family is characterized by a unique light organ on the shoulder above the pectoral fin. Light-producing cells are contained in a dark sac which opens to the outside by a backwards-pointing pore. When the fish is alarmed, a bright cloud is squirted out which lasts a few seconds and enables it to escape into the darkness.

A living example of the genus *Searsia* "was seen to discharge a bright luminous cloud into the water on being handled. The light appeared as multitudinous bright points, blue-green in color." There are also series of stripelike or rounded luminous organs underneath the body.   RMcD/KEB

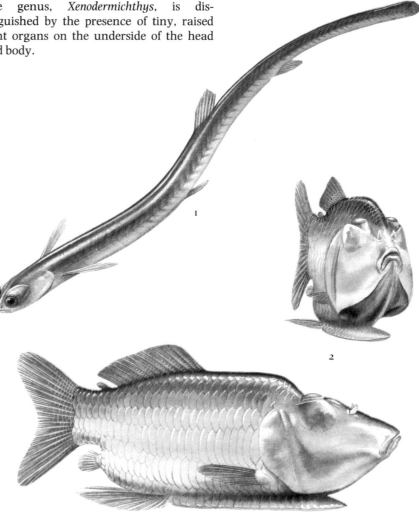

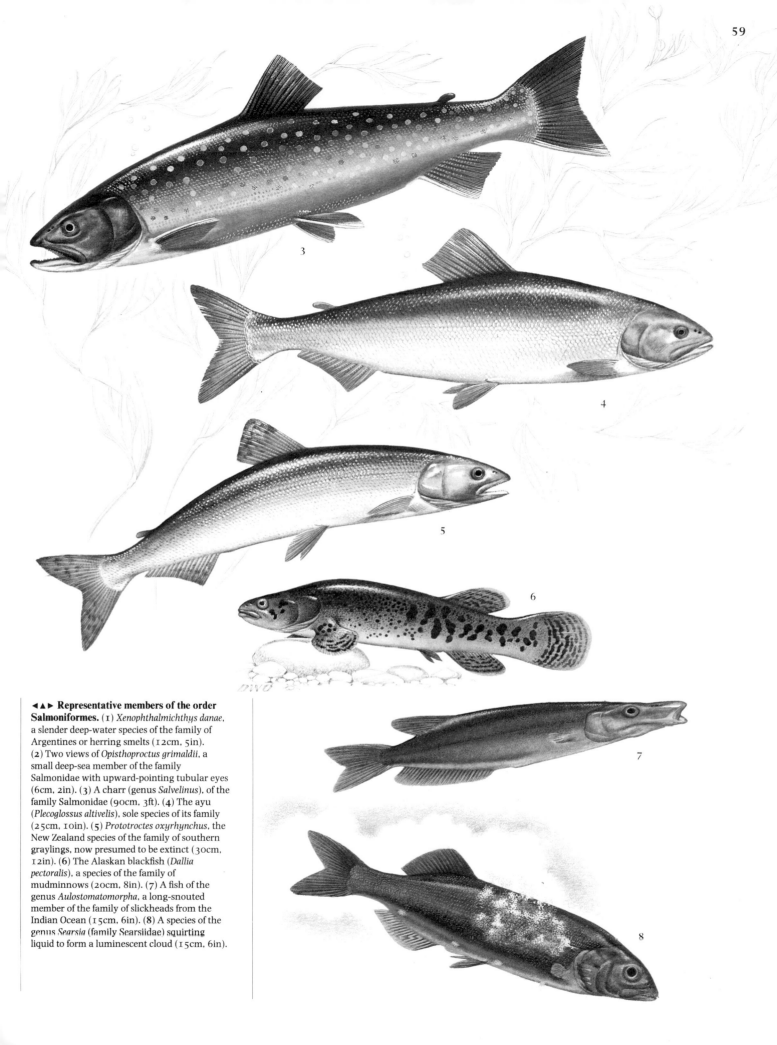

◄▲► **Representative members of the order Salmoniformes.** (1) *Xenophthalmichthys danae*, a slender deep-water species of the family of Argentines or herring smelts (12cm, 5in). (2) Two views of *Opisthoproctus grimaldii*, a small deep-sea member of the family Salmonidae with upward-pointing tubular eyes (6cm, 2in). (3) A charr (genus *Salvelinus*), of the family Salmonidae (90cm, 3ft). (4) The ayu (*Plecoglossus altivelis*), sole species of its family (25cm, 10in). (5) *Prototroctes oxyrhynchus*, the New Zealand species of the family of southern graylings, now presumed to be extinct (30cm, 12in). (6) The Alaskan blackfish (*Dallia pectoralis*), a species of the family of mudminnows (20cm, 8in). (7) A fish of the genus *Aulostomatomorpha*, a long-snouted member of the family of slickheads from the Indian Ocean (15cm, 6in). (8) A species of the genus *Searsia* (family Searsiidae) squirting liquid to form a luminescent cloud (15cm, 6in).

BRISTLE MOUTHS AND ALLIES

Order: Stomiiformes
About two hundred and fifty species in over 50 genera and about 9 families.
Distribution: all oceans within the temperate zones but discontinuous within these limits.

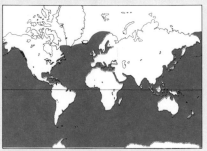

Size: maximum adult length 35cm (14in); most are much smaller.

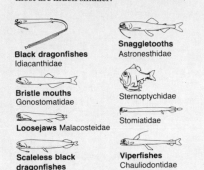

Black dragonfishes
Idiacanthidae

Bristle mouths
Gonostomatidae

Loosejaws Malacosteidae

Scaleless black dragonfishes
Melanostomiidae

Snaggletooths
Astronesthidae

Sternoptychidae

Stomiatidae

Viperfishes
Chauliodontidae

Luminous organs, bristle-like teeth, eyes on stalks—these are some of the characteristics of the order Stomiiformes, a worldwide group of deep-sea fishes contained in nine families. The classification of the order is very uncertain and the subject of active research, hence some of the families included here may well be transferred to other orders.

Practically all stomiiform species have luminous organs and many also have a luminous chin barbel, thought to act as a lure. They are flesh-eating fishes with a large, wide-gaping, toothed mouth. Most are scaleless, black or dark brown in color, but one family in particular—the Gonostomatidae—are mostly silvery, midwater fishes. Typically an adipose fin is present, but this and the pectoral and dorsal fins have been lost in some lineages.

The Gonostomatidae are known as bristle mouths because of their fine, bristle-like teeth. The genus *Cyclothone*, with over 20 species, occurs in all oceans. It is probably the commonest genus in the world regarding numbers of individuals: trawls can haul up of tens of thousands of these small fish at a time. They feed on small crustaceans and other small invertebrates, and in turn are a most important source of food for larger fishes, including some of their relatives.

The maximum size of a particular stomiiform species depends upon the richness of the environment. In areas of abundance, such as the Bay of Bengal and the Arabian Sea, the species may reach 6cm (2.5in) long, but in a polluted area, like the depths of the Mediterranean, or in an area poor in food they are much smaller. The Mediterranean species *Cyclothone pygmaea* grows to only 2.5cm (1in).

Some species occur worldwide whereas others have a very limited distribution. As well as a two-dimensional distribution that can be shown on a map there is also a three-dimensional distribution as species are separated vertically. Generally, silvery or transparent species live nearer the surface than the dark-colored species. It is also generally true that the species living deeper have weaker and fewer light organs than those living above them. The swim bladder is poorly developed, which may explain why they do not undertake the extensive daily vertical migrations common in many deep-sea fishes.

Cyclothone species show differences between the sexes not with different light-organ patterns, as in the lantern fish, but in the nature of the nasal complex. In males, as they mature, the olfactory organs grow out through the nostrils. In some species, eg *Cyclothone microdon*, the hydrodynamic disadvantage of the protruding nasal plates (lamellae) is thought to be compensated for by the development of an elongate rostrum.

The Idiacanthidae is a family of elongated, scaleless fishes also lacking a swim bladder. Worldwide in distribution, the number of species is uncertain; there may be less than half a dozen or just one variable species.

The North Atlantic species *Idiacanthus fasciola* is remarkable for extreme differences between the sexes and its most peculiar larvae. The sex of the larvae cannot be determined until they are about 4cm (1.5in) long. They are, however, so peculiar that they were assigned to a separate genus (*Stylophthalmus*) until it was realized that they were the young of *Idiacanthus*. They are stalk-eyed, ie the eyes are at the end of cartilaginous rods which extend up to one-third of the body length. The body is transparent, the intestine extends beyond the tail and the pelvic fins are not developed. However, the pectoral fins (lost in the adult) are well developed. The larva lacks luminous organs.

During metamorphosis the eye stalks gradually shorten until the eyes rest in an orthodox position in the orbit of the skull. The pectoral fins are lost and only the female

develops pelvic fins. She also grows a luminous chin barbel, develops rows of small luminous organs on the body and strong jaws with thin, hooked teeth. The male never has a barbel nor teeth but develops a large luminous organ just below the eye. The female is black, the male brown. The male does not grow after metamorphosis so remains less than 5cm (2in) long, whereas the female feeds actively on prey of suitable size and can grow to over 30cm (12in) long.

The general biology of *Idiacanthus* is poorly known. The smallest larvae are caught at the greatest depths and metamorphosing larvae at about 300m (1,000ft), so they probably spawn at considerable depths. As the catches of larvae are sporadic it has been thought that they may shoal or otherwise agglomerate. The adults undergo a daily vertical migration from 1,800m (6,000ft) deep during the day to reach the surface at night.

The family Chauliodontidae contains about six species in the one genus *Chauliodus*. These are mid-water fishes distributed worldwide in oceans between 60 degrees N and 40 degrees S. *Chauliodus sloani* has been recorded from all oceans, but distinct and discrete populations have been recognized by some authors. "From all oceans" does not necessarily mean that the species is equally and universally distributed within the stated range. Oceans consist of distinct water masses varying in temperature, salinity, current, food supplies, etc. The way that fish are distributed within these water masses is exemplified by the distribution of *Chauliodus* species.

Two small species of *Chauliodus*, *C. danae* and *C. minimus*, live respectively in the central water masses of the North and South Atlantic. A larger species, *C. sloani*, lives in the richer waters that flow around the poorer central water masses. *Chauliodus sloani* grows to over 30cm (1ft) long, more than twice the length of the central-water species. Even the oxygen content of water can limit a distribution: *Chauliodus pammelas* lives only in the deep waters off Arabia which have a low oxygen content. To cope with these conditions it has gill filaments much longer than those of its congeners.

Chauliodus species are highly specialized predators. The second ray of their dorsal fin is elongated, highly mobile, and has a luminous lure at the end. Their teeth vary in shape within the mouth but this variation is remarkably consistent throughout each species. The front teeth on the upper jaw have four sharp ridges near the tip and are used for stabbing. The longest teeth (which

▲ **Denizen of the depths,** the head of a deep-sea fish (genus *Chauliodus*).

◄ **The fine teeth** clearly seen on the jaws of this *Gonostoma atlantica* are the feature of the family Gonostomatidae from which their common name bristle mouths is derived.

▼ ► **Loosejaws** have remarkably distensible jaws. (The species BELOW is *Malacosteus niger*.)

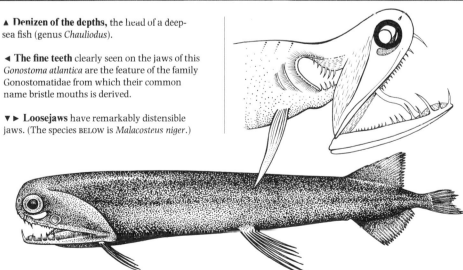

imply a remarkable gape) are the front two on the lower jaw. Normally when the mouth is closed they lie outside the upper jaw, but when they impale the prey their natural curvature tends to push the prey into the roof of the mouth. At the base of both the second and third upper jaw teeth there is a small tooth sticking out sideways which is thought to protect the large luminous organ below the eye.

These specialized, predatory modifications do not end with the teeth. The heart, ventral aorta and gill filaments are all much further forwards than is usual—in fact they lie between the sides of the lower jaw, the gill filaments extending almost to the front of it. Bearing in mind the fragility and importance of the gill filaments, how does large prey pass through the mouth without damaging them? The answer lies in the backbone. In almost all fish this consists of a series of firmly articulated bones which allow normal flexibility. In *Chauliodus* the front vertebrae generally do not develop and the spinal column remains a flexible cartilaginous rod. Although this is normal in embryonic and juvenile states (the notochord) bony vertebrae generally replace it in adults. However, its retention in *Chauliodus* enables the head to enjoy a remarkable freedom of movement. As the back muscles pull the head upwards, the hinge between the upper and lower jaws is pushed forwards. At the same time the mouth is opened, and the shoulder girdle, to which the heart is attached, is pulled backwards

and downwards. Special muscles pull the gill arches and their filaments downwards, away from the path of the prey. Movable teeth in the throat then clutch the prey and slowly transfer it to the elastic stomach. With the prey stowed away, these organs return to normal.

Chauliodus has been seen alive from a deep-sea submersible. It hangs still in the water, head lower than tail, with the long dorsal fin ray curved forwards to lie just in front of the mouth. The body is covered with a thick, watery sheath enclosed by a thin epidermis. This gelatinous layer is thickest dorsally and ventrally and thinnest laterally. It contains nerves, blood vessels and many small luminous organs.

Ignoring the lure, *Chauliodus* species have various kinds of luminescent organs (photophores). Along the ventral part of the fish are complex organs with two kinds of secretory cells, a pigment layer and a reflector. This type of organ is specialized below the eye, protected by teeth and transparent bones, with pigment layers and reflectors so arranged that the light shines into the eye. It is believed that this makes the eye more sensitive to light. Small light organs above and in front of the eye are thought to illuminate possible prey, suggesting that sight is an important factor in feeding. Scattered throughout the gelatinous sheath and inside the mouth are small, simple photophores which in life emit a bluish light. These are spherical and their bioluminous product is secreted into the hollow center of

▲ **Obscured by light.** Deep sea hatchet fishes have ventrally directed luminous organs which help to prevent them being seen by a predator below.

▼ **Fashioned by depth.** Species of the deep-sea fish family Stomiatidae live in all tropical oceans down to 1,400m (4,600ft). Larger species normally live at greater depths than smaller ones.

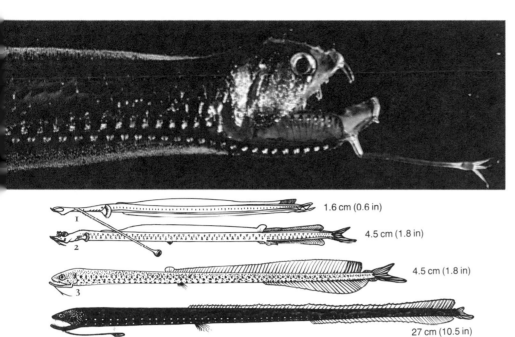

1.6 cm (0.6 in)

4.5 cm (1.8 in)

4.5 cm (1.8 in)

27 cm (10.5 in)

3.8 cm (1.5 in)

▲ **An awesome orifice.** TOP With its large teeth and luminous chin barbels *Stomias boa* displays the typical characteristics of the family of scaled dragon fishes.

▲ **The extraordinary development** of *Idiacanthus fasciola*. (1) A sexually indeterminate larva. (2) A post-larval female. (3) An adolescent female. (4) An adult female. (5) An adult male. Note the eye-stalk of the juvenile and the very different appearance of the adult male and female.

the organ. They are controlled by nerves, unlike the ventral luminous organs, and their function is unknown, but some interesting observations have been made.

When a *Chauliodus* is tranquil its ventral organs produce a bluish light. When touched, pulses of light illuminate the whole body. On top of this it has been demonstrated in an experiment that the intensity of the ventral photophores can be adjusted to match the amount of light received by the fish from above. In the clearest parts of the oceans all traces of sunlight have been absorbed at about 900m (3,000ft). Fishes below that depth lack ventral photophores. *Chauliodus* lives higher so is therefore affected by low light levels. The photophore by the eye varies its intensity with that of the ventral photophores and presumably, by balancing internal and external light levels, *Chauliodus* can produce the correct level of light from its ventral photophores to match the background illumination, thus making itself less liable to predation from below.

Species of *Malacosteus*, a black-skinned genus of the family Malacosteidae, are known as loosejaws. They have no floor to the mouth and the stark jaws with their long teeth are reminiscent of the cruel efficiency of gin traps. Some species in this family are unusual in that they have a cheek light organ producing a red light. Most photophores produce a blue-green light.

The species of the family Melanostomidae are called scaleless black dragon fish. They

live in all oceans and are predators. Some species are elongated whilst others are squatter. All have dorsal and anal fins set far back and large, fang-like teeth and rows of small, ventral, luminous organs. Almost all species have chin barbels which can vary from the very small, through multibranched versions, to ones six times the length of the body.

In the family Sternoptychidae species are laterally compressed and deep-chested, so are known as deep-sea hatchet fishes. Some have tubular eyes and in many the mouth is vertical. Species in the genus *Argyropelecus* have upwardly directed tubular eyes with a yellow, spherical lens. They feed on very small crustaceans. Some species undergo a small daily vertical migration. The genus *Polyipnus* has about 20 species, mostly in the western Pacific. All stay close to land at depths of 45–450m (150–1,500ft). Like all members of this family they have large, elaborate, downwardly pointing light organs. The photophores of species in the genus *Sternoptyx* have been studied intensively. They have two elliptical patches on the roof of the mouth which lack pigment, reflectors, or colored filters. They luminesce independently of the ventral organs and can glow for about half an hour before gently fading. Apart from, presumably, attracting prey there is a sort of light guide that allows some of this light to be guided close to the eye, where it may be used to relate the light production from the ventral organs to match the background daylight. The sternoptychids are exceedingly beautiful silvery fishes, though what advantage their hatchet shape gives them is unknown. *Maurolicus muelleri*, a small, sprat-like fish often found at night at the surface of the North Atlantic, is thought to be a primitive relative of the deep-sea hatchet fishes. It resembles them in many internal details but has a more normal fish shape.

Members of the family that has given its name to the order—the Stomiidae—are known as the scaled dragon fishes. There are only two genera, *Stomias* with eight species and *Macrostomias* with one. Elongated predators, lacking an adipose fin, they are found in the Indo-Pacific and Atlantic Oceans. They have large, easily shed, hexagonal scales that produce a pleasing honeycomb pattern on their dark bodies. *Stomias* species lack a swim bladder, so they can easily undertake extensive daily vertical migrations. There are usually two rows of small light organs along the ventral margin of the body. At their longest the scaled dragon fishes rarely exceed 30cm (12in).   KEB

Light from Living Organisms

How fish produce and use artificial illumination

As a terrestrial creature who in his waking hours is accustomed to light, man has little appreciation of life-styles that must cope with continual darkness. Yet that is precisely what life would be like in the world's oceans were it not for bioluminescence, the production of light by living organisms. Not to be confused with phosphorescence or fluorescence, light that results from the excited state of inanimate crystals, bioluminescence is the result of the chemical reaction of a substance, usually luciferin, and an enzyme, a luciferase, which is controlled by a living organism. Bioluminescence also occurs on land—it is best known from fireflies in an evening sky or glowing fungi on a forest floor; it does not occur in freshwater fishes, but its greatest display in variety and intensity occurs in the sea.

Of the nearly 20,000 living species of fishes, perhaps 1,000–1,500 bioluminesce. None of the lampreys and hagfishes or lungfishes are known to, but six genera of midwater and benthic sharks and species representing nearly 190 genera of marine bony fishes are luminescent. This is best seen in the lantern fishes, the gonostomatids, the families of dragonfishes (Melanostomiatidae, Malacosteidae, Chauliodontidae, etc), the slickheads (Alepocephalidae and Searsiidae) and the angler fishes (various families). Several shallow-water and bottom-living fish families also contain luminescent species, and these are better understood behaviorally and physiologically in that they are amenable to capture and study. Included are the ponyfishes (Leiognathidae), flashlight fishes (Anomalopidae), pinecone fishes (Monocentridae), midshipman fishes (Batrachoididae), and several of the cardinal fishes (Apogonidae).

The origin of the light and the associated chemistry may be conveniently divided into two categories. The first are those with self-luminous photophores—specialized structures usually arranged in rows and consisting of highly complex lenses, reflectors and pigmented screens. The skin photophores produce light by the photogenic cells and reflect it through the lens and cornea-like epidermis. The more than 840 photophores in the belly of the California midshipman (*Porichthys notatus*) produce a gentle, even glow whose intensity can be slowly modulated to match the downwelling moonlight upon the sandy bottom. The second category involves luminous bacterial symbionts, maintained in complex organs and nurtured by the host fish in exchange for a more brilliant level of light. This extrinsic form of light production does not allow the fish to control the bacterial light which continues to effuse, therefore the hosts have evolved fascinating mechanisms to turn out their lights when they would be a hindrance.

▶ **A complex luminous organ** (photophore) ABOVE from a deep-sea hatchet fish (genus *Argyropelecus*). A common pattern is shown here. The light-producing organs are partly screened by pigment cells and backed by a reflective layer which directs the light to the lens, in some cases through filters to change the color of the light emission.

▶ **Light in the shallows.** The Flashlight fish (belonging to the order Beryciformes of spiny-finned fishes) is one of the few shallow-water luminous fishes. It has a large light organ below the eye which flashes on and off. Only recently have public aquaria learned how to keep these fish so that all can see the beauty of their natural light, though they have not yet been bred in captivity. The light is produced by bacteria housed in cells well provided with blood vessels. The function of the organ is uncertain, but in parts of the eastern Indian Ocean it is used by native fishermen as a bait.

▶ **The strings of beads** FAR RIGHT on the underside of this fish (*Valenciennellus tripunctatus*, of the superorder of bristle mouths and allies) are its luminescent organs. Each organ is a highly complex structure.

▼ **Photophores in profusion,** a deepsea viperfish photographed under ultraviolet light (*Chauliodus sloani*, of the superorder of bristle mouths and allies).

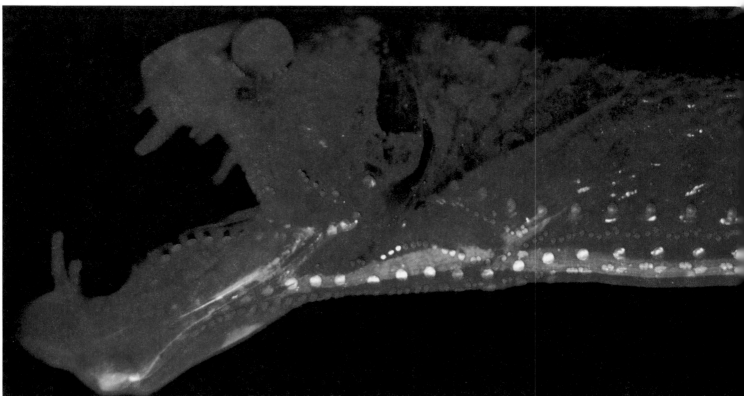

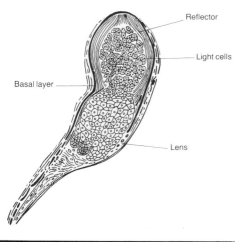

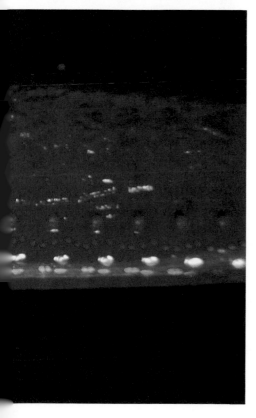

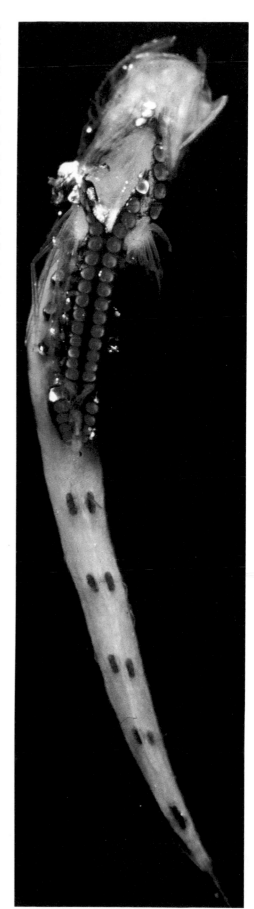

What then is the function of bioluminescence for fishes? In the case of most deepsea fishes it is one of camouflage through counter-illumination. Even at 1,000m (3,300ft) in clear tropical waters, downwelling light would silhouette a lanternfish to an upward-searching predator were it not for the weak glow of the rows of photophores on its underside which cancel its apparent presence. Other photophores on its body can be used as well to advertise its species and its sex. The large suborbital light organs of *Aristostomias*, *Pachystomias* and *Malacosteus* emit red light which, when coupled with its red-sensitive retina, might act like a "snooperscope" to hunt prey that can only see blue-green hues. Other uses in the deepsea include luring, as has been achieved by many melanostomiatid dragon fishes with elongate luminous chin barbels or angler fishes with luminous escas at the end of their modified first dorsal spine. Concealment is also a function whereby slickheads and certain macrourids presumably behave like squids and octopuses, leaving a predator snapping at a luminous ink cloud.

The behavior and bioluminescent function of shallow-water fishes is becoming better understood as a result of foolhardy nocturnal observers with scuba gear and the improvement in aquarium collecting and husbandry. The pine-cone fishes (*Monocentris japonicus* and *Cleidopus gloriamaris*) live in shallow water and apparently lure nocturnal crustacean prey to their jaws with the light organs located in their mouth and on their jaws. Bioluminescence is best perfected in the anomalopid flashlight fishes. These small, black reef associates possess an immense subocular light organ, capable of emitting enough light to be seen from 30m (100ft) away. In evening twilight they migrate up from deep water to feed along the reef edge and return before daylight to the recesses of the deep reef. They use the light for a multiplicity of purposes, including finding food, attracting food, communication and avoiding predators. Flashlight fishes continually blink, either by raising a black eyelid-like structure or by rotating the entire organ into a dark pocket. The living light produced by the flashlight fish *Photoblepharon palpebratus* is the most intense yet discovered.

This brief summary of fish bioluminescence reflects the meager knowledge man has of life in the nocturnal sea. As our ability to descend into the deep ocean at night improves, many more forms of extraordinary behavior associated with bioluminescence will surely come to light.   JEM

LIZARD FISHES, LANTERN FISHES

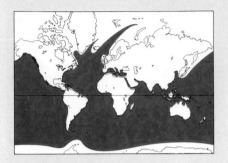

Lizard fishes and allies or aulopoids
Order: Aulopiformes
About one hundred and ninety species in about
40 genera and 11 or 12 families.
Distribution: all oceans.
Size: maximum length 2m (6.5ft).
Families, genera and species include:
Aulopidae, including **Sergeant Baker** (*Aulopus purpurissimus*); **barracudinas** (family
Paralepididae); Bathypteroidae, including
tripod fishes (genus *Bathypterois*); **Bombay
ducks** (family Harpadontidae);
chlorophthalmids (family Chlorophthalmidae);
giganturoids (family Giganturidae); **lancet fish**
(family Alepisauridae); **lizard fishes** (family
Synodontidae).

Lantern fishes
Order: Myctophiformes
About two hundred and fifty to three hundred
species in 35 genera and 2 families.
Distribution: deeper water of all oceans.
Size: maximum length 30cm (12in).

Lizard fishes

Aulopidae

Giganturoids Giganturidae

Lizard fishes Synodontidae

Ipnopidae

Lantern fishes

Barracudinas Paralepididae

Lancet fishes Alepisauridae

Bathypteroidae

Lantern fishes Myctophidae

Neoscopelidae

Bombay ducks Harpadontidae

Omosudidae

Chlorophthalmids Chlorophthalmidae

Pearleyes Scopelarchidae

Daggertooth Anotopteridae

Sabertooth fishes Evermannellidae

VARIOUS extraordinary fishes belong to the two orders of lizard fishes and lantern fishes: tripod fishes, which sit on their stiffened pelvic fins and lower tail lobe; species in the genus *Ipnops*, which have greatly flattened eyes covering the top of the skull; the lantern fishes, speckled with luminous organs; that famous delicacy the Bombay duck; and others.

The 14 or so families in these two closely related orders were formerly considered by some authorities to be more realistically included in one order. The arrangement here follows the latest classification in which just two families, Myctophidae and Neoscopelidae, comprise the Myctophiformes; all the others are Aulopiformes.

Within the order of **lizard fishes** and allies the number of families is uncertain, and one of the reasons for this stems from the deep-sea tripod fishes. They have been photographed on the deep-sea floor resting on their stiffened pelvic fins and the lower lobe of the tail, facing into the current with their bat-like pectoral fins raised over the head in the manner of forward-pointing elk horns. The 18 or so species of tripod fish are found worldwide in deep water. However, they are only found in a particular type of oceanic water mass called central oceanic water. Formerly they were grouped in the genera *Bathysauropsis*, *Bathytyphlops*, *Bathymicrops* and *Bathypterois*, all placed in the family Bathypteroidae. Recent research has shown that they should all be included in one genus, *Bathypterois*. It was known that their family was related to two others, the Chlorophthalmidae and the Ipnopidae (which contains just one genus of most peculiar deep-sea fishes). Further research placed this assemblage in perspective with the rest of the aulopoids and the conclusion was reached that they are all more closely related to each other than to the rest of the aulopoids, hence they are now just one family—the Chlorophthalmidae—not three. As not all workers agree with this the result is vague statements on the number of families present in the order.

Each species of tripod fish lives only in a particular area, defined by subtle parameters of temperature and salinity. *Bathypterois atricolor* may occur in waters as shallow as 300m (1,000ft) whereas *B. longicauda* lives as deep as 6,000m (19,700ft); *B. filiferus* lives only off South Africa whereas *B. longipes* is circumglobal. All, however, can only live where the sea floor is composed of ooze or very fine sand which permits a firm "foothold" for their fins. One puzzle is that there are large areas of deep sea, eg the North Pacific, with ideal conditions (as far as can be seen) in which they do not live. For deep-sea fish they are common. Off the Bahamas intensive surveys have shown there can be almost 90 fish per sq km (233 per sq mi). The bat-like pectoral fins have an elaborate nerve supply, so are sensory, but for what use is not known. The eyes are extremely small. The fish feed on small crustaceans which we can but presume are detected by the fins as they drift past in the current. There is so much that remains unknown, including the life history of these fish.

Fish in the genus *Ipnops*, with 3 or 4 species, look a bit like a flattened tripod fish without the tripod. They appear to be eyeless. They have achieved fame because of the puzzle presented, since they were first discovered, by the two large, flat, pale yellow plates on the top of the head. For about half a century it was thought that they were luminous organs that directed light upwards. The advent of deep-sea photography revealed that these plates were highly reflective. The flash from a deep-sea camera that photographed one was clearly reflected from the plates.

In the last few years deep-sea collecting has made more specimens available to scientists and the puzzle of the plates has been solved: they are a mixture of modified eyes and skull bones. Each plate, which covers half of the head width, is a transparent skull bone. The reflective layer below this is a very much changed eye retina. Ordinary eye structures, like the lens, have been lost; only the light-sensitive retina remains and this has been spread out over the top of the head and is protected by the skull. Although this is remarkable, it is difficult to understand what it means to the fish. It can

▼ **Wide-eyed repose.** This species of lizard fish (*Saurida gracilis*) from the Great Barrier Reef shows the typical observant resting position.

detect light coming down from above, yet it cannot focus on an object. Furthermore, it eats marine worms which live below it.

The chlorophthalmids are the orthodox members of the family. They are laterally compressed, silvery fishes with large eyes. They grow up to 30cm (12in) long, have dorsal and ventral fins well forward and an adipose fin directly above the anal fin. *Chlorophthalmus* species are widespread and fairly abundant in the North Atlantic, where they shoal at depths of 200 to 750m (660–2,460ft). In all species the lateral line system is well developed and greatly expanded into special organs on the snout, head and gill covers. These organs enable the fish to detect the approach of small prey.

The dim light from the perianal organ (an organ around the anus) is produced by bacteria. The presence of this light is believed to enable a fish to maintain contact with its fellows and perhaps to facilitate mating opportunities among those in breeding condition. The scientific name for the family comes from the green light reflected by the tapetum (a reflective layer behind the eye) in some species. Yellow eye lenses are common; the yellow coloring is believed to act as a selective filter enabling the fish to "see through" the downwardly directed luminous camouflage emanating from the ventral light organs of the small crustaceans that form its prey.

Only three families in this group are shallow- or moderately shallow-water fishes, the Aulopidae, the lizard fishes and the Bombay ducks. The last family is famous in Indian restaurants as the crispy, salty delicacy (the aforementioned Bombay duck) eaten before the meal. What you are eating is the sun-dried, salted fillet of a large-mouthed, large-toothed fish from the Ganges estuary. Its slender, cylindrical body with a soft dorsal fin and an adipose fin is typical of many aulopoids. Almost all are predatory, using their curved needle-like teeth to seize fish and invertebrates. The members of the other major shallow-sea family, the lizard fishes, are bottom dwellers, spending time propped up on their pectoral fins waiting for dinner to swim by. They get their common name not just from the very lizard-like shape of the head but also from their rapid feeding movements. Both these families live in warm waters.

The lancet fish are large, voracious, mid-water predators. None have luminous organs but all have large stabbing teeth and distensible stomachs. Species in the genus *Alepisaurus* can reach 1.8m (6ft) in length but their bodies are so slender that a fish of this size will only weigh 1.8–2.3kg (4–5lb). They have a small anal fin and a very long, very high dorsal fin that can be folded down into a deep groove along the back and become invisible. The function of this large dorsal fin is unknown but it has been suggested that it might be used like the similarly large dorsal fin of the sail fishes in helping to "round up" shoals of small fishes. The diet of the Atlantic lancet fish *Alepisaurus ferox* consists of deep-sea hatchet fish, barracudinas, squids, octopi and almost anything else available. Most of the fish eaten do not undertake daytime migrations. Lantern fish, although abundant, are rarely eaten and it is thought that a result of their daily migration is to avoid predation by lancet fish. Lancet fish are, however, eagerly eaten by tunas and other surface predators when the opportunity arises.

The scaleless body of the Pacific lancet fish is dark gray to greenish dorsally, silvery on the sides, somewhat iridescent with pale spots. The large dorsal fin is black or dark blue/gray with steely blue reflections. The pectoral, caudal and adipose fins are dark, the others paler.

Anotopterus pharao lacks the large dorsal fin of *Alepisaurus* but has a similarly shaped body. This species, which lives in cool polar waters, is sometimes placed in its own family, the Anotopteridae.

A 75cm (30in) long, intact specimen of *Anotopterus pharao*, taken from the stomach of a whale in the Antarctic, was found to have two barracudinas, 27 and 18cm (11 and 7in) long, both engorged with krill, in its distensible stomach.

The barracudinas derive their common name from their superficial resemblance to the predatory barracudas of coral reefs to which they are not related. Barracudinas are slender fishes with large jaws, many small, sharp, pointed teeth on the upper jaw and a mixture of larger stabbing teeth on the lower. A small adipose fin is present and the single, soft-rayed dorsal fin is in the rear half of the body. Scales, when present, are fragile and easily shed. This is thought to be an adaptation towards the swallowing of large prey by permitting easier body expansion. The anus is often much closer to the pelvic fins than to the anal fin. In many species there is a fleshy keel between the anus and the anal fin.

The family is widespread but the range of individual species is more limited. *Notolepis coatsi*, a species growing to more than 10cm (4in) long, is confined to Antarctic waters. *Notolepis coruscans* comes from the eastern North Pacific. The barracudinas are

▲ **Fish of the warm seas,** TOP a lizard fish (genus *Synodus*) from Fiji.

▲ **Lizard in the heather,** ABOVE a lizard fish (genus *Synodus*) from the Great Barrier Reef.

▶ **Miniatures of the deep seas.** Most lantern fishes are small, between 2.5cm and 15cm (1–6in). The light organs (photophores) are visible here as the small round dots.

unusual among the aulopoids in that they have light organs. In the genus *Lestidium* these consist of ducts that extend from the head to the ventral fins. The bacteria therein produce a bright, pale-yellow light. The function of the luminosity is unknown. It is probably not for camouflage because the species with luminous organs have an iridescent skin as well as being translucent. Perhaps the organs function as lighthouses enabling individuals of the same species to recognize one another.

The eyes of barracudinas are designed in such a way that their best field of binocular vision is directly downwards. However, barracudinas seen alive from submersibles oriented themselves head-up, or nearly so, in the water so that they were looking along a horizontal plane (ie straight ahead). It is also conjectured that the head-up posture presents a smaller silhouette for predators beneath them to spot.

The giganturoids are cylindrical, silvery fishes with forward-pointing tubular eyes set back on the snout, giving the fish the appearance of an old-fashioned racing car. Scales and luminous organs are absent. The pectoral fin is set high up on the body; the caudal fin has an elongated lower lobe; adipose and pelvic fins are missing. Also missing are a large number of bones. Indeed, the adult fish has been subjected to a loss of many features typically present in its relatives. Many of the remaining bones are still cartilaginous. The teeth are large, sharp and depressible to ease the passage of large prey. The inside of the mouth and stomach are lined with dense black pigment which, it has been suggested, blacks out the luminous organs of its last meal. The abdominal region is elastic so giganturoids can swallow food much larger than themselves. The pectoral fins are inserted above the level of the gills and are thought to help ventilate them during the slow passage of a large fish down the throat. Certainly some such device would have to be present as the normal water currents cannot flow through the blocked mouth to the gills.

Giganturoids are widespread in tropical and semitropical oceans at depths down to about 3,350m (11,000ft). All are small

fishes, rarely exceeding 15cm (6in).

Members of the family Aulopidae, which has given its name to the order, are the most primitive aulopoids. They live in warm shallow waters in the Atlantic, parts of the Pacific and around southern coasts. They are scaled, bottom-living fishes with slender bodies and large heads. The second ray of the dorsal fin is characteristically elongate. An adipose fin is present. The teeth are small and lie in closely packed rows. Small, bottom-living invertebrates are the main food.

They are surprisingly colorful for fishes that live as far down as 915m (3,000ft); browns, reds and pinks are well represented among the dozen or so species. *Aulopus purpurissimus* is a highly colored, edible species from Australia. The edges of the scales are crimson against a purple or scarlet background. The fins are yellow with rows of red spots. Its common name of Sergeant Baker apocryphally alludes to the name of the soldier who first caught this fish in New South Wales. There is no evidence for this, but many red-colored fishes have a military common name in allusion to the red jackets of British troops.

The 300 or so species in the order Myctophiformes are commonly called **lantern fishes** because of the appearance created by their extensive speckling of luminous organs. These carnivorous fish live in the middle depths of all oceans and rarely exceed 30cm (1ft) long. The overall body shape is very similar in all species, but the pattern of light organs differs from species to species and is the basis for species definition. As well as the small photophores there are larger organs—upper and lower glands—mostly near the tail, which indicate the sex of the fish. Usually the female lacks the upper glands and the lower ones are less conspicuous or even absent. The fish react strongly to light signals. One animal being studied in an aquarium appeared to be most interested in the investigator's luminous wristwatch. Brighter light sources have little effect.

Lantern fishes live 300–700m (1,000–2,300ft) down. There are both silvery-bodied and dark-bodied forms. Many display an upward migration at night, occasionally to as little as 50m (165ft). Those with swim bladders have less fat than those without. Fat, being lighter than water, helps the latter maintain neutral buoyancy. Again to emphasize the theme that generalizations on fish are almost impossible, *Tarltonbeania crenularis* has no gas in its swim bladder, almost no fat and is negatively buoyant (ie is less dense than water). **KEB**

CHARACINS, CATFISHES, CARPS...

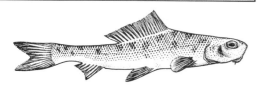

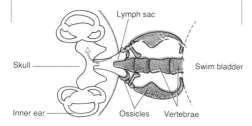

▲ **In want of lineage.** *Ellopostoma megalomycter* is known only from a handful of specimens. It has not yet been associated with any of the major cyprinoid families.

▶ **Dull fish in Africa.** There are fewer species of characins in Africa than in South America, and none has achieved the brilliant colors of some of the small Amazonian tetras. Africa's nearest approach to these popular aquarium fish is probably the Congo tetra (*Phenacogrammus interruptus*), shown here.

▼ **A sleek, silvery fish** BELOW RIGHT from the Indo-Pacific Ocean—is the milk fish or bandang. It enters fresh water to spawn. A single fish can lay up to six million eggs.

▼ **Diagram of the Weberian mechanism,** seen from above, a characteristic feature of many members of the superorder Ostariophysi. It transmits vibrations from the swim bladder to the inner ear.

CARPS, catfishes, characins, suckers, loaches and their allies are the dominant freshwater fishes in Eurasia and North America and arguably so in Africa and South America. (Only the catfishes are native to Australia.) The approximately 6,000 species are predominantly freshwater fish. Just two families of catfishes and one species of cyprinid are found at sea, although several genera may spend time in brackish water.

The major groups listed opposite are well defined (although their relationships are the subject of much controversy). However, one species, *Ellopostoma megalomycter* from Borneo, is a puzzle because it does not quite fit in anywhere.

The ostariophysi as a whole are split into two series, the otophysi and the anotophysi, the former having two hundred times more species than the latter. The otophysi have two main unifying characters. Firstly, the presence of an "alarm substance" secreted from glands in the skin when a fish is threatened, which causes a fright reaction in other otophysans. Perhaps understandably, the alarm substance is not present in families of heavily armored catfishes, but less comprehensible is its absence in some of the subterranean characins and cyprinids.

The second significant character is the Weberian mechanism. This is an elaborate modification of the first few vertebrae into a series of levers that transmit compression (ie sound) waves, received by the swim bladder, to the inner ear. Consequently, they have acute hearing. No one is certain how this complex "hearing aid" developed, but a clue may come from the anotophysi where there are "head ribs" that may be a primitive trial for such a mechanism.

The anotophysi are a diverse and somewhat incongruent group. The milk fishes are a genus of food fish from the region of Southeast Asia. They look rather like large herrings with small, silvery scales but lacking ventral scutes. They are intensively cultured in fish ponds in many areas and can grow to well over 1m (3.3ft) in length. They can tolerate a wide range of salinity.

Gonorhynchus gonorhynchus is the only member of its family (Gonorhynchidae). A shallow-water species from the temperate and tropical Indo-Pacific, it has an elongate body, long snout and a ventral mouth. There is no swim bladder. What could be fossil relatives have been found in Alberta, Canada.

Unlike *Gonorhynchus*, *Phractolaemus* from West African fresh waters has a dorsal mouth that extends like a short periscope. Its swim bladder is divided into small units and can be used for breathing atmospheric air. Growing to a little more than 15cm (6in) long, it is confined to the Niger and parts of the Zaïre basin.

The remaining anotophysan family is the Kneriidae, consisting of small freshwater fishes that feed on algae in tropical and nilotic African fresh waters. They exhibit marked sexual dimorphism whereby the male has a peculiar rosette, of unknown function, on its operculum. The genera *Cromeria* and *Grasseichthys* are neotenic, that is, they become sexually mature while still having a larval body form. By some authorities they are considered to be kneriids, by others they are placed in a separate family. These small, transparent fishes lack scales and a lateral line. They live in West African rivers.

The characins, catfishes and carps are a highly successful group and display a remarkable mixture of evolutionary conservatism and extreme radicalism which, when coupled with the plasticity or variability at the species level, makes their taxonomy very difficult. For such common fish, they are an enigmatic group.   KEB

Lymph sac

Skull

Swim bladder

Inner ear   Ossicles   Vertebrae

The 4 Orders of Characins, Catfishes, Carps and Milk Fishes

Series: Otophysi

Characins

Order: Characiformes
Over one thousand two hundred species in over 250 genera and probably 10 families.

Distribution: fresh waters of Africa, South and Central America, and Southern North America.
Size: maximum length 1.5m (5ft).

Families, genera and species include: Characidae, including **tiger fishes** (genus *Hydrocynus*); **freshwater hatchet fishes** (family Gasteropelecidae); **glandulocaudines** (subfamily Glandulocaudinae); Lebiasinidae, including **Splashing tetra** (*Copella arnoldi*); Serrasalminae, including **piranhas** (genus *Serrasalmus*), **silver dollars** (genera *Metynis, Myleus*), **South American salmons** (genera *Catabasis, Salminus*), **wimple-piranhas** (genus *Catoprion*).

Catfishes

Order: Siluriformes
Over two thousand species in probably over 400 genera and 29 or 31 families.

Distribution: most habitable fresh waters; members of two families (Ariidae, Plotosidae) can inhabit tropical and subtropical seas.
Size: maximum length 3m (10ft).

Families, genera and species include: **amphiliids** (family Amphiliidae); **callichthyids** (family Callichthyidae); Clariidae, including **Walking catfish** (*Clarias batrachus*); **crucifix fish** (family Ariidae); Ictaluridae, including **blue** and **channel catfishes** and **bullheads** (genus *Ictalurus*), **flatheads** (genus *Pylodictis*), **madtoms** (genus *Noturus*); **loricariids** (family Loricariidae); **Parasitic catfishes** (family Trichomycteridae), including **candirus** (genera *Vandellia, Branchioca*); **pimelodids** (family Pimelodidae); **schilbeids** (family Schilbeidae); Siluridae, including **European wels** (*Silurus glanis*); **sisorids** (family Sisoridae).

Carps and allies or cyprinoids

Order: Cypriniformes
About two thousand five hundred species in over 250 genera and 6 or 7 families.

Distribution: North America, Europe, Africa, Asia almost exclusively in fresh water.
Size: maximum length 3m (10ft).

Families, genera and species include: cyprinids (family Cyprinidae), including **bighead carps** (genus *Hypophthalmichthys*), **bream** (genus *Abramis*), **carp** (*Cyprinus carpio*), **grass-carp** (genus *Ctenopharyngodon*), **mahseers** (genus *Barbus*), **roach** (genus *Rutilus*), **snow-trouts** (genus *Schizothorax*), **squawfish** (genus *Ptychocheilus*), **tench** (*Tinca tinca*), **yellow-cheek** (*Elopichthys bambusa*); **gyrinocheilids** (family Gyrinocheilidae); **Homalopterids** (family Homalopteridae); **loaches** (family Cobitidae), including **weather fish** (*Misgurnus fossilis, M. anguillicaudatus*); **psilorhynchids** (family Psilorhynchidae); **suckers** or **catostomids** (family Catostomidae).

Series: Anotophysi

Milk fishes and allies

Order: Gonorhynchiformes
About 30 species in 7 genera and 4 families.

Distribution: Indo-Pacific Ocean; fresh waters in tropical Africa.
Size: maximum length 2m (6.5ft).

Genera include: **milk fishes** (genus *Chanos*).

There are about 1,200 living species of **characins**, about 200 of which are found in Africa and the remainder in Central and South America. This discontinuous distribution implies that some 100 million years ago, characins were widespread in the area of the land mass Gondwanaland, which later split forming Africa and South America, Antarctica and Australia.

Superficially the characins resemble members of the carp family (cyprinids) but they usually have a fleshy adipose fin between the caudal and dorsal fins and have teeth on the jaws but not in the pharynx. In addition to their complete functional set of teeth characins also have a replacement set behind those in current use. In some species all the "old" teeth on one side of the upper and lower jaws drop out and the replacement ones take their place. Once these are firmly in position the teeth on the other side of the jaws are replaced. In predatory characins all the "old" teeth drop out and are rapidly replaced in one go. As soon as the replacement teeth are functional, new replacement teeth begin to grow in the tooth-replacement trenches.

There are five families of characin in Africa. They are both carnivorous and omnivorous, but are less varied and less interesting than the Neotropical species. The most primitive African characin is *Hepsetus odoe*, the sole member of the family Hepsetidae. It is a fish-eater of the lurking-predator type, with a large mouth provided with strong, conical teeth to prevent the prey's escape. Unusually for characiforms it lays eggs in a floating foam nest, which incidentally is considered a great gastronomic delicacy. On hatching, the larvae hang from the water surface by special adhesive organs on their heads.

Members of the family Ichthyboridae are elongate fishes that earn a living by snatching mouthfuls of scales and nipping notches out of the fins of other fishes. Their close relatives are a group of relatively harmless fishes belonging to the families Distichodontidae and Citharinidae. Their claim to fame is that they have scales with a serrated edge (ctenoid scales) unlike all other characins which have smooth (cycloid) scales. A few of these species are of local commercial importance.

The last family of African characins, Characidae (the characids), is shared with the Neotropics. There are only a few genera in Africa, but enormous numbers in Central and South America. The genus of African "tetras" (*Alestes*) is one of the best known groups of characids because of their popularity with aquarists. They are fairly colorful fishes, often with a single lateral stripe and red, orange or yellow fins. The body can be short and deep or elongate (fusiform). All show sexual dimorphism in anal fin shape; in females the margin is completely straight but in males it is convex. These fish also show a curious sexual dimorphism in the caudal vertebrae. How they recognize it and what benefit it confers on them is unknown. The African tetras have very strong, multicusped teeth, for crushing and grinding their food of insects, fish and insect larvae, plankton and assorted vegetation.

The tiger fishes, a genus of close relatives of the African "tetras," have gained notoriety and their common name from their long, conical teeth which overlap the outside of the jaws when the mouth is shut and from their black body stripes (though these are horizontal rather than vertical). One Zaïrean species grows to over 1.5m (5ft) and weighs more than 45kg (100lb). There are a few unsubstantiated reports of tiger fish attacking people. They are, however, excellent sport fish, and generate an African equivalent of the excitement of catching a large salmon. Oddly, tiger fish lose all their "old" teeth at once so sometimes toothless individuals have been caught, but it only takes a matter of days for the replacement teeth to become functional.

The diversity in Neotropical characins ranges from voracious predators through tiny vegetarians to blind subterranean species. The piranhas are the most fabled

▲ **A change of style.** In Africa a characoid (*Hepsetus odoe*) has evolved the characteristics of a "lurking predator," that is the dorsal and anal fins are set far back near the tail to produce quick and powerful acceleration. In South America at least two genera have evolved in parallel to achieve the same end, including *Boulengerella maculata* shown here.

► **The much maligned piranha** has short, triangular teeth with sharp edges. It normally feeds on fishes and carrion in the wild. Observations in captivity suggest that the alleged "feeding frenzy" does not occur unless about 20 fish are gathered together.

► **One of the less spectacular characins** of South America BELOW is the Bleeding heart tetra (*Hyphessobrycon rubrostigma*). The contrasting colors on the dorsal fin are thought to be a "presence indicator" to keep the members of shoals together in dark waters.

▼ **Outlines of representative species** of characins, catfishes, carps and milk fishes.

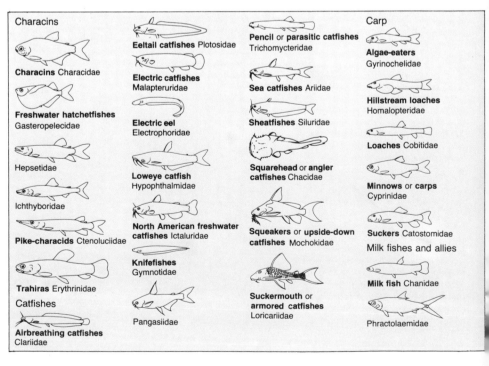

Characins

Characins Characidae

Freshwater hatchetfishes Gasteropelecidae

Hepsetidae

Ichthyboridae

Pike-characids Ctenoluciidae

Trahiras Erythrinidae

Catfishes

Airbreathing catfishes Clariidae

Eeltail catfishes Plotosidae

Electric catfishes Malapteruridae

Electric eel Electrophoridae

Loweye catfish Hypophthalmidae

North American freshwater catfishes Ictaluridae

Knifefishes Gymnotidae

Pangasiidae

Pencil or **parasitic catfishes** Trichomycteridae

Sea catfishes Ariidae

Sheatfishes Siluridae

Squarehead or **angler catfishes** Chacidae

Squeakers or **upside-down catfishes** Mochokidae

Suckermouth or **armored catfishes** Loricariidae

Carp

Algae-eaters Gyrinochelidae

Hillstream loaches Homalopteridae

Loaches Cobitidae

Minnows or **carps** Cyprinidae

Suckers Catostomidae

Milk fishes and allies

Milk fish Chanidae

Phractolaemidae

predatory fishes. They are stocky and rather ugly, having a deep head, short powerful jaws with triangular, interlocking, razor-sharp teeth. They are shoaling animals, and feed communally on smaller fish or allegedly on larger injured or supposedly healthy prey.

Of all the areas of fish lore the piranha legend is the least open to substantiation. Each piranha makes a clean bite of about 16 cubic cm (1 cu in) of flesh. Shoal size then dictates the speed with which the victim is despatched. In some areas the feeding frenzy is triggered by blood in the water and it takes only a few minutes for a victim to be reduced to little more than a skeleton. People wading or bathing in rivers have also been attacked. Legend has it that a man and horse fell into the water and were later found with all the flesh picked off, yet the man's clothes were found to be undamaged.

Piranhas are fairly small fishes, rarely exceeding 60cm (2ft), but are excellent sport fish. Anglers use a stout wire trace to catch them and they apparently make excellent eating. The piranhas' close relatives are the two vegetarian or omnivorous genera known to aquarists as silver dollars. Although they are similar in appearance, their nature is quite unlike that of the piranhas.

The wimple-piranhas eat scales of other fish. Their lower jaw is longer than the upper and their teeth are everted, which enables them to scrape the scales of their prey in a single upward swipe. In the presence of wary potential prey, they live on insects and other small invertebrates.

The genera of South American "salmons" have many features that suggest they are a primitive group. *Catabasis* is known only from a single preserved specimen. These fish are presumed to be extant but none have been seen since this fish was caught in 1900. *Salminus* is called a *dourados* by the Brazilians and is apparently the most primitive characiform known. Despite its importance as a genus of food and sport fish the classification of its four species is uncertain.

Acestrorhynchus is the Neotropical genus equivalent to the African genus *Hepsetus* but to exemplify the plethora of species and paucity of knowledge of South American characins, virtually nothing is known about it.

The family of freshwater hatchet fish is an exception to the above. It consists of deep-bodied fish, with long pectoral fins, that rarely exceed 10cm (4in) and are capable of powered flight. The deep chest houses powerful muscles which are necessary for turning the pectoral fins into wings. Prior to flight, hatchet fish may "taxi" for distances up to 12m (40ft), for most of which the tail and chest trail in the water. The flight phase is marked by a buzzing caused by the very rapid flapping of the fins. The flight distance rarely exceeds 1.5m (5ft) but they have been seen at a height of 90cm (3ft) above the water. The energy cost of flight to the fish is unknown, but it is thought that flight is used when frightened by predators. Normally hatchet fish are found near the water surface feeding on insects. Species of *Triportheus* also have their chest developed into a sort of keel and have large wing-like pectoral fins, similar to those of the hatchet fish. These fish are able to use the breast muscles and pectoral fins to jump about 1m (3.3ft) above the water surface. This is to escape predators but cannot be considered real flight.

Most characins are generalized egg-scatterers but in some there is specialized breeding behavior.

The male glandulocaudines have scales modified into special glands, known as "caudal glands," at the base of the caudal fin. To attract a female these glands secrete a pheromone—a chemical the opposite sex are supposed to find irresistible. If the pheromone alone is insufficient to attract a mate, some males are also equipped with a worm-like lure to signal to the female of his choice. It is unknown whether this is intended to be a prey mimic or whether it is a visual cue to induce the female to approach.

In the family Lebiasinidae is the Splashing tetra. To avoid the high predation on eggs in water the Splashing tetra lays its eggs on overhanging leaves or rocks. The male courts a gravid female until the point of egg-laying is reached. Leading her to the overhang of his choice about 3cm (1in) above the water, he makes trial jumps up to this spawning site. The female then follows, adheres briefly to its surface, using water surface tension, and lays a few eggs. The male then leaps and fertilizes them. This process continues until about 200 eggs are laid and fertilized. After spawning the female goes on her way but the male remains nearby to splash water over the eggs until they hatch. The eggs take about three days to hatch and once the fry fall into the water the male's "nursemaid" activities are over.

One of the South American characins, *Brycon petrosus*, indulges in terrestrial group spawning. Males are distinguished by a convex anal fin and short, bony spicules on the fin rays. About 50 adults move by lateral undulations or tail flips onto the banks and lay eggs on the damp ground. There is no parental care and the eggs take about 2 days to hatch.

Tetras, fishes beloved of aquarists, have received their common name from a contraction of their scientific group name, Tetragonopterini. Many of these highly successful fishes are brilliantly colored which, apart from rendering them commercially important in the aquarist world, holds the members of shoals together in the wild. Tetras are omnivorous to the extent that they will eat anything they can fit into their mouths. Within this group is the Mexican blind cave characin, discussed on p.84.

BB

There are about 2,400 species of **catfishes**, assigned to some 30 families. Most are tropical freshwater fishes, some inhabit temperate regions (Ictaluridae, Siluridae, Diplomystidae and Bagridae) and two families (Plotosidae and Ariidae) are marine.

Catfishes are so named because of their long barbels giving them a bewhiskered appearance (though barbels are not present in all species and are not a defining character of the group). Characters defining catfishes include the fusion of the first 4–8 vertebrae, often with modification of the chain of bones connecting the swim bladder to the inner ear (Weberian apparatus); lack of parietal bones; unique arrangement of blood vessels in the head; absence of typical body scales and strong dorsal and pectoral fin spines.

Although catfishes lack typical scales, the bodies of many are not entirely naked. Doradids, sisorids and amphiliids have bony scutes around the sensory pores of the lateral line and sometimes along the back. Loricariids and callichthyids may be completely encased in these scutes. Strong serrated pectoral and dorsal fin spines are widespread in catfishes. Locking mechanisms keep the spines erect and these coupled with the bony armor must deter potential predators. Catfish swim bladders may be partially or completely enclosed in a bony capsule. Why this should be is not clear, since the most obvious correlation, a reduction of the swim bladder with benthic habits, does not hold. For example, reduced and encapsulated swim bladders occur in active, fast-swimming species (of Hypophthalmidae and Ageniosidae) while the bottom-dwelling electric catfishes (Malapteruridae) have the largest swim bladders of any catfish.

There are more catfish species in South

▲ **A multitude laid bare.** Being nocturnal creatures and bottom-dwellers catfish are unfamiliar fish: they are, however, far more abundant than is generally realized. During the rainy season many move into flood plains to feed or breed. At the onset of the dry season those trapped in hollows will die unless, like some of the members of the family Clariidae, they can move over land back to the main river. This sad spectacle was seen in Lake Katavi, Tanzania.

▶ **Sublime symbolism.** The members of the family Ariidae are one of the two groups of catfish that have marine as well as freshwater representatives. In the Caribbean ariids are called crucifix fish because of the fancied likeness to Christ on the Cross presented by bones on the underside of the skull. The skull is often decorated and sold as a curio. Ariids lay a few large eggs which are protected in the mouth.

America than in the rest of their area of distribution. Both the world's smallest and largest catfishes occur here; *Scoloplax* (family Scoloplacidae) of Bolivia is a minute, partially armored fish whose total adult length is less than 13mm (0.5in) whereas *Paulicea* (family Pimelodidae) from Amazonia grows to more than 3m (about 10ft). Of the 16 families in South America most live in the Amazon Basin, but four are endemic to the Andes.

The largest of all catfish families is the Loricariidae with over 600 species. The mouth of these armored forms is sucker-like with thin, often comb-like teeth adapted to scraping algal mats. Most are active at night and hide during the day in crevices and logs. The males of some species bear long spines on their opercular apparatus (ie the bones forming the gill cover) and use them in cheek-to-cheek territoriality fights with other males. *Farlowella* species are long and thin and resemble dead twigs.

Closely related to the loricariids are the Astroblepidae (30 species) inhabiting Andean torrents. These are able to climb smooth, almost vertical rock faces by utilizing the muscles of the ventral surface and pelvic fins to induce suction.

The Callichthyidae have mail-like plates thought to resist desiccation when ponds dry up. Two genera (*Callichthys* and *Hoplosternum*) can withstand marked temperature changes and can move over dry land, using their pectoral spines for locomotion. They also build floating bubble nests. *Corydoras* is well-known to aquarists, who have recorded the breeding habits and development of many species.

The Doradidae vary greatly in size from *Physopyxis* (5cm, 2in) to *Megalodoras* (1m, 3.3ft). Most species are bottom dwellers and, surprisingly, play an important part in seed dispersal by their fruit-eating habits. The seeds are not digested and pass through the fish unharmed.

The Pimelodidae with 300 species is a family of great morphological diversity and includes some of the largest known catfishes. Most species are omnivorous, but the larger are fish-eaters or carnivores. One species has even been recorded as having eaten monkeys that fell in the river. Pimelodids form an important catfishery in parts of the Amazon Basin.

A strange family is the Trichomycteridae (280 species). Some are parasites, inhabiting and laying their eggs in the gill cavities of the larger pimelodid catfishes. The "candiru" is a notorious representative: mammals (including humans) have had their

urethra penetrated by this slender fish while urinating in the water. The candiru probably mistakes the flow of urine for water being expelled from the gill chamber of a large catfish. As well as several eyeless species there are also two peculiar nonparasitic genera which have large fat-filled organs above the pectoral fins. One of these, *Sarcoglanis*, known only from a single specimen 4cm (1.5in) long, was collected in 1925 by Dr Carl Ternetz from the San Gabriel Rapids on the Rio Negro. Forty years later Dr George Myers, visiting the same locality, caught another similar fish but belonging to a new genus. Nothing is known about these puzzling forms.

The Cetopsidae (12 species) is also a poorly known group. Some species prey on other catfishes by biting out circular pieces of flesh with their saw-like teeth.

The Auchenipteridae (60 species) contains species that have conspicuous spots and stripes. They range from Panama to La Plata and are nocturnal, detrital feeders, inhabiting hollow logs by day where they line up in ranks.

The Hypophthalmidae are most unusual. Unlike most other catfishes they feed on plankton which they sieve from the water through fine gill rakers. Long barbels help to funnel the plankton into the mouth. As it is a surface feeder *Hypophthalmus* has high fat content and paper-thin bones to increase buoyancy.

Of the eight families of catfishes in Africa, only three are endemic. Four are shared with Asia and one, the Ariidae, also occurs in the coastal waters of Asia, Australia and North and South America. Although there is less diversity in form among African catfishes than there is among South American ones, there are many unusual forms, some displaying remarkable parallels with those of South America. The richest diversity of species in Africa occurs in the principal equatorial river basin, the Zaïre, which may be regarded as Africa's equivalent of the Amazon.

A major difference between the continents, however, is the series of rift valley great lakes in Africa; some of these harbor endemic groups of catfishes.

The most widespread family is the Bagridae (over 100 species). Some species of *Bagrus*, a Nilotic catfish genus, weigh over 5kg (11lb). Small bagrids, Zaïrean endemics, live in torrents, while a "flock" of several *Chrysichthys* species inhabits Lake Tanganyika. This generalized family also has members in Asia.

The Clariidae (about 30 species and 10

genera) are long-bodied with long dorsal and anal fins and broad flat heads. Some species have an organ at the top of the gill chamber that enables them to breathe atmospheric air and so survive when water is deoxygenated. Clariids, like the South American callichthyids, can travel overland from one water body to another. During a drought some rivers appear to be alive with the heaving, wriggling bodies of *Clarias*. The largest clariid is *Heterobranchus*, exceeding 50kg (110lb). Other genera are small, anguilliform fishes with burrowing habits; *Uegitglanis* from Somalia is subterranean and eyeless; *Clarias cavernicola* from Namibia is also eyeless. The family also occurs in Asia, but there are fewer species.

The endemic family Malapteruridae with its two species is the only family of electric catfishes. All catfishes appear capable of *detecting* electrical activity, but only *Malapterurus* is capable of producing it. The dense, fatty electric organ covers the flanks of the fish and gives it a cylindrical, sausage-shaped appearance. The strong electric impulses (up to 450 volts) are used for both defense and stunning prey. Some specimens exceed 1m (3.3ft) in length and weigh up to 20kg (44lb).

The family Mochokidae is exclusively African. Over 100 of its 150 species belong to the genus *Synodontis*. A few species of that genus swim upside down to utilize food on the surface as well as feeding normally on

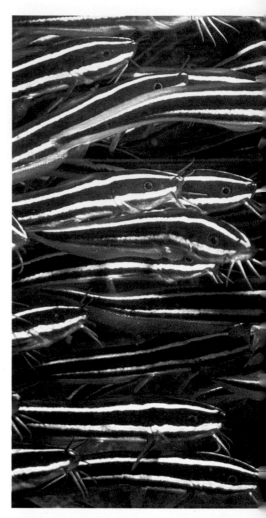

▲ **Shades of youth.** The family of eeltail catfishes (Plotosidae) has many marine representatives. The bright and conspicuous markings of *Plotosus lineatus* (sometimes called the Barber eel) are interpreted as a warning that this species has poisonous glands in its dorsal and pectoral fins. However, these bright colors are only present in the young. The adults are a dull brown.

◄ **Invisible camouflage.** Not all catfishes are bottom-dwellers. Some, for example those of the genus *Schilbe* in Africa, of *Pangasius* in Southeast Asia and of *Helogenes* in South America, are mid-water shoaling species. Among these at least two, *Physailia pellucida* and *Kryptopterus bicirrhus* (seen here) in Southeast Asia, have become transparent as a means of camouflage. The physiology of transparency is an enigma. It does not persist for long after death.

the bottom. This habit has led to the common name of the family being "upside-down catfishes."

The endemic family Amphiliidae contains about 45 small species. Some (eg *Phractura* and *Andersonia*) display remarkable parallelism with the South American *Farlowella* (Loricariidae) in their elongate, plated bodies. Amphiliids live largely in the faster, cooler upper reaches of rivers, clinging to the cobbled substrate.

The Schilbeidae, a family shared with Asia, has about 20 species in Africa. They have a short dorsal fin and compressed, rather deep bodies. They are fast-swimming shoalers which, in their large numbers, are both important predators on other fishes and in their turn form a large food source for fish-eating perches. Schilbeids are ubiquitous fishes that can change their diets easily (euryphagous) and have become quickly adapted to artificial situations such as dammed lakes and reservoirs.

After carps and their allies catfishes are the dominant element of Asia's fish fauna. Compared with African catfishes, Asian catfishes are not well-known, scattered as they are throughout the Indonesian islands and isolated rivers and lakes of China and high Asia. Twelve families occur in and around the continent, seven of which are endemic.

The Bagridae are widely distributed with the species-rich genus *Mystus* being typical. *Bagrichthys* from Borneo and Sumatra is unusual in that the dorsal fin spine of the adult is nearly the length of the fish's body.

The Siluridae are a significant family with species in Eurasia, Japan, and offshore islands. The family contains the glass catfish, *Kryptopterus*, a favorite of aquarists, as well as the 2m (6.5ft) long voracious predator *Wallago*. This giant follows shoals of carps in their upstream migrations and can leap clear of the water during a feeding frenzy.

Species of the endemic families Sisoridae (100 species) and Amblycipitidae (30 species) are small inhabitants of mountain streams, clinging to the substrate by the partial vacuum caused by corrugations of their undersides. The two species of the small family Chacidae from Borneo are well-camouflaged, flattened fishes with cavernous mouths and large heads. They resemble some angler fishes.

The Pangasiidae family, containing about 10 species, is possibly the most economically important of the Southeast Asian catfishes. In Thailand *Pangasius* species have been pond-reared on fruit and vegetables for over a century. One of the world's largest fresh-

water fishes, *Pangasianodon* occurs in the Mekong river, growing to 2m (6.5ft). Despite the evidence from travelers' tales, it too is vegetarian.

The family Plotosidae contains about 40 species, some of which live in the Indo-Pacific Ocean. Two genera, *Tandanus* and *Neosilurus* (the eel-tailed catfishes), live in the fresh waters of Australia and New Guinea. Specimens of *Tandanus* can weigh up to 7kg (15lb). The fishes build circular nests, about 2m (6.5ft) in diameter, from pebbles, gravel and sticks. The several thousand eggs are guarded by the male until they hatch after seven days. The young of *Plotosus* collect into moving "feeding-balls" which make them appear like sea urchins. Plotosids are notorious for the painful and dangerous wounds they inflict with their pectoral spines which have venomous glands at their base.

Of the Clariidae the widespread species, *Clarias batrachus*, has been introduced into Florida, USA, where it is known as the Walking catfish. Closely related to the Clariidae is the family Heteropneustidae, species of which possess long air sacs that extend backwards from each gill cavity.

The sea cats (the family Ariidae) have a circumtropical, largely marine distribution. Their most distinctive feature is that the males carry the eggs in their mouth. Up to 50 large, fertilized eggs will be carried for as long as two months, during which time the male starves.

Apart from a few ariid species reaching coastal regions, only one family, the Ictaluridae, is present in North America, with its 7 genera comprising some 50 species. The flatheads (*Pylodictis* species) are the largest, growing to some 40kg (88lb). The blue and channel catfishes (*Ictalurus* species) form the basis of a large catfishery in the Great Lakes and the Mississippi valley. The madtoms (*Noturus* species) are so named because they have venomous glands at the bases of their pectoral fins. There are three cave-dwelling (troglobitic), eyeless ictalurids which appear to have evolved independently. The genus *Prietella* is known only from a well in Coahuila, Mexico; *Satan* and *Trogloglanis* species have only appeared in 300m (1,000ft) deep artesian wells near San Antonio in Texas. It is thought they live in deep water-bearing strata.

Apart from introduced ictalurids only the family Siluridae is present in Europe. The European wels occurs in central and eastern Europe, growing to 5m (16ft) and over 300kg (660lb); adults of 15 years have been recorded.   GJH

The **cyprinoids** form a major lineage of largely freshwater, egg-laying fishes. All lack jaw teeth but most have a pair of enlarged bones in the pharynx, the teeth of which work against the partner bone and a pad on the base of the skull. A less conspicuous unifying feature is a small bone (the kinethmoid) which enables the upper jaw to be protruded. Cyprinoids are indigenous to Eurasia, Africa, and North America. Unlike the other major otophysan lineages they are not native to South America and Australia.

Of the six families, with over 2,500 species, by far the largest is the family Cyprinidae with over 2,000. The Cyprinidae—the chubs, minnows, mahseers, carps, barbs, ghillimentjes etc—reflect the distribution of the cyprinoids and are well known to both freshwater anglers and aquarists. Even in areas where they never lived naturally, many are familiar to fishermen and pond-keepers as well as to gourmets.

The carp epitomizes the family. Probably native to Central Europe and Asia, it has been introduced to all continents capable of supporting fish life. The carp is extremely tolerant of a wide range of conditions. In Central Africa, where it was introduced to provide food for expatriates, it has colonized areas so successfully that it is now the commonest cyprinid. In the United Kingdom, to which it was probably introduced in Roman times, it is loved by anglers, for whom catching an 18kg (40lb) specimen is a lifetime's goal. In South Africa, where the carp was introduced in the early part of this century, a 38kg (83lb) specimen was caught. The Japanese in particular have cultivated carps as objects of beauty: several hundred years of intensive breeding has released the potential colors and many carps (*koi*) are sold for high prices to beautify ponds and aquaria. In its native Eurasia it is bred commercially for food. Even there modifications have been made. Careful selection firstly produced carp with a few scales, then with none. Further selective breeding has resulted in strains lacking the hair-like intermuscular bones that create problems of decorum for diners.

The majority of cyprinids are smallish fishes, ie 10–15cm (4–6in) long when adult. But, as ever with fish, there are exceptions. The mahseer of the Himalayan and Indian rivers reach over 1.8m (6ft) long and can weigh 54kg (120lb). In North America the squawfish (*Ptychocheilus*) of the Colorado and Sacramento rivers, which were an important food source to the American Indians of those areas (hence the common name), used to grow to over 1.8m (6ft) long. Now they are practically extinct in the Colorado (because of damming) and much smaller and rarer in the Sacramento (because of overfishing). Of similar length is the yellow-cheek from the Amur River in northern China. *Elopichthys* and *Ptychocheilus* are unusual among cyprinids (which are toothless) in that they are specialized fish-eaters. Other, but smaller, examples of fish-eaters are *Barbus mariae*, from a few rivers in East Africa; *Luciobrama macrocephalus* from southern China—a fish with a disproportionately elongate head; and *Macrochirichthys macrochirus* from the Mekong River. The last is a strongly compressed fish with a large, angled mouth and a hook and notch on the lower jaw. It is capable of raising its head to increase the gape when lunging at its prey.

Most cyprinids, including the mahseer, will eat almost anything: detritus, algae, mollusks, insects, arthropods, sausages, cheese sandwiches and luncheon meat. The bighead carps (*Hypophthalmichthys*) of China are specialized plankton feeders with gill rakers modified into an elaborate filtering organ. The grass-carp subsists on plant food and has been introduced to many countries to clear weeds from canals, rivers and lakes.

The shape and distribution of the pharyngeal teeth often indicate the diet. Mollusk-eaters have crushing molar-like teeth, closely packed; fish-eaters have thin, hooked teeth; vegetarians have thin, knife-like teeth for shredding. Omnivores come somewhere in the middle. But even within one species the teeth can vary. In Africa fish of the species *Barbus altianalis* living in a lake with no snails have "middle-of-the-road" type teeth, while those in a lake a few kilometers away rich in snails have thicker, lower, more rounded teeth. The young all start off with the same type of teeth which

▲ **The carp** TOP (foreground) has been cultivated for centuries in Europe. Various forms, showing degrees of reduction in scales, are now known. The Crucian carp (*Carassius carassius*) from Asia is a much smaller species and a close relative of the goldfish.

▲**Push fish—a pair of Apollo sharks** (*Luciosoma setigerum*) engaged in an open-mouth, sideways-pushing display. This species is distributed in Southeast Asia.

◄**Most loaches** are eel like in shape but many members of the southern Asian genus *Botia* are strongly laterally compressed. Seen here is *Botia hymenophysa* from Sumatra, Indonesia.

light and neon tetras (order Characiformes).

Totally without color, however, are the subterranean cyprinids. Members of the genera *Barbus*, *Garra* and *Caecocypris* have lost all pigment—and their eyes—as a result of living in caves. They are discussed in detail in "Fish Underground" (pp84–85). The epigean *Garra* species are found in Africa, India and Southeast Asia. They are bottom-dwelling fishes with a suctorial and sensory disk on the underside of the head. The related genus *Labeo*, with a similar distribution, has an elaborate suctorial mouth and grazes on algae. One African species often feeds on the flanks of submerged hippopotamuses.

In the cold, mountain streams of India and Tibet live the poorly known snow-trouts (Schizothoracines), cyprinids that imitate the life-style of trout and salmon. They are elongate fish, up to 30cm (1ft) long with very small scales (or none at all) except for a row of tile-like scales along the base of the anal fin. Nepalese fishermen capture snow-trouts by fashioning a worm-shaped wire surrounded by a loop which is tightened when the fish strikes at the "worm."

Although earlier it was stressed that cyprinids are essentially freshwater fishes some can tolerate considerable degrees of salinity. In Japan species of *Tribolodon* have been found up to 5km (3mi) out at sea. In Europe the familiar freshwater roach and bream live in the Baltic Sea at about 50 percent salinity. These are, however, among the few exceptions.

As if cyprinids are not already full of eccentric habits, at least two species are known to get drunk! In Southeast Asia both *Leptobarbus* and *Hampala* gorge themselves on the fermented fruit of the chaulmoogra tree when it drops into the water. They even congregate before "opening time." When intoxicated they float helpless in the water, but are relatively safe as their flesh is now unpalatable. Perhaps as a compensation for such antisocial cyprinid behavior the Eurasian tench has gained a reputation from countrymen (whose observation is often acute) of being a doctor fish. It is reported that injured fish seek out tench and rub their wounds in its slime. It is true that the tench has a copious coating of slime, but as for its healing properties . . . ?

The suckers (Catostomidae) have many species in North America and a handful in north Asia. For a long time they were thought to be the most primitive cyprinoids. This was because of the shape of their pharyngeal bones and teeth. The pharyngeal bone is formed from one of the

makes cyprinid classification very difficult.

The widespread and species-rich genus *Barbus* derives its name from the (usually) four barbels around the mouth which are provided with taste buds so that the fish can taste the substrate before eating. A particular specialization of some African species is to vary the shape and thickness of the mouth and lips according to diet. A broad-mouthed form with a wide, sharp-edged lower jaw will feed on epilithic algae. A narrow-mouthed form with thick "rubber" lips feeds by sucking up stones and their associated fauna. That these two extremes are found in the same species explains why what is now regarded as one species (*B. intermedius*) formerly had over 50 scientific names.

Most cyprinids do not exhibit sexual dimorphism but some of the small *Barbus* species do. In Central Zaïre, living among submerged tree roots, are small "butterfly barbs," 4cm (1.5in) long, including *Barbus hulstaerti* and *B. papilio*. In these strikingly marked species the males and females have conspicuously different color patterns.

Although generally the cyprinids are not as brilliantly colored as the characins, some of the Southeast Asian *Rasbora* species have come close. The small Malaysian species *Rasbora brittani* and *R. axelrodi* are the cyprinid equivalents of the popular glow-

plain, cylindrical bones in the fifth gill arch. In catostomids, however, it is less modified and its homology is more evident. Suckers have also developed highly complex sacs from the upper gill-arch bones and it now seems likely, taking into account their distribution, that they are a highly specialized group of cyprinoids. They are innocuous, unspectacular fish, apart from two genera, one in China (*Myxocyprinus*) and one in the Colorado River (*Xyrauchen*). Both are deep-bodied, with a triangular profile. The function of this peculiar shape (the others are just "fish"-shaped) is to force them close to the bottom of the river during flash floods. Both live in rivers susceptible to flash floods. These suckers are therefore a good example of parallel evolution.

The loaches are a family of eel-like fishes with minute, embedded scales and a plethora of barbels around the mouth. Bony processes enclose most, or all, of the swim bladder, making its normal volume changes rather awkward. All the cyprinoids have a tube connecting the swim bladder to the pharynx, allowing the fish to swallow or expel air. But with the constricted swim bladder loaches use this more often than many. One species, known as the Weather fish, from eastern Europe has been kept for centuries by peasants as a living barometer. Its agitation in expressing air as the atmospheric pressure increases with, say, the approach of thunderstorms made it one of the earliest weather-forecasters.

Most loaches have small dorsal and anal fins symmetrically placed near the rear of the body. The poorly known *Vaillantella* has a dorsal fin the length of the body. Loaches can be divided into two subgroups: those that have an erectile spine below the eye and those that do not. Many are secretive fish, liking to hide under stones during the day. It was an overdue discovery when the first cave-living species was found in Iran in 1976. Since then two more have been found in southwest China.

Loaches live in Eurasia, not in Africa (apart from an arguably introduced species in European North Africa), except for the enigmatic *Noemacheilus abyssinicus*. A Mr Degen was collecting for the British Museum in Ethiopia in 1900. His collection from the mouth of a feeder stream entering Lake Tsana purportedly contained the specimen described under that name some years later. No more have ever been found, but again, no one has revisited the site and used his collecting techniques. Or perhaps one jar could have been misplaced on a museum shelf.

Gyrinocheilids, mistakenly known in the aquarium trade as Siamese algal eaters, sift detritus. There are only three or four species, all from Southeast Asia. They have a ventral, protrusile mouth, like the hose of a vacuum cleaner. With this they suck in the fine substrate and filter out the edible material. *Gyrinocheilus* lacks a pharyngeal tooth apparatus; whether this is lost or has never been developed is unknown. The gyrinocheilids are unique among cyprinoids in that the gill cover is sealed to the body for most of its length leaving top and bottom openings. The top opening is covered by a valve and takes *in* water to oxygenate the gills and expels it through the bottom opening. Thus breathing is through the gill cover rather than the mouth.

Almost nothing is known of the Indo-Chinese psilorhynchids. They are small, camouflaged bottom-living fishes.

▼▶ **Representative members** of the orders of milk fishes, characins, catfishes and cyprinoids. (1) *Gobiobotia longibarba*, a poorly known species of gudgeon or loach-like cyprinoid from China (15cm, 6in). (2) Two examples of *Gastromyzon*, a hillstream cyprinoid adapted for life in the torrential waters of South and Southeast Asia by having paired fins forming sucking disks (8cm, 3in). (3) *Xyrauchen texanus*, a rare species of cyprinoid from the Colorado River, USA (60cm, 24in). (4) A milk fish (*Chanos chanos*), from shallow waters of the Indian Ocean and adjacent fresh waters (2m, 6.5ft). (5) *Luciobrama*, a pike-like cyprinoid predator which however lacks teeth on the jaws (1m, 3.3ft). (6) *Rhaphiodon*, a highly specialized characoid piscivore from South America (40cm, 16in). (7) A tiger fish (genus *Hydrocynus*), an African characoid occurring in both rivers and lakes (maximum length about 2m, 6.5ft). (8) A species of sea cat, from the catfish family Ariidae. (9) Two examples of *Vandellia cirrhosa*, a species of candiru (6cm, 2.5in), investigating the gills of a catfish.

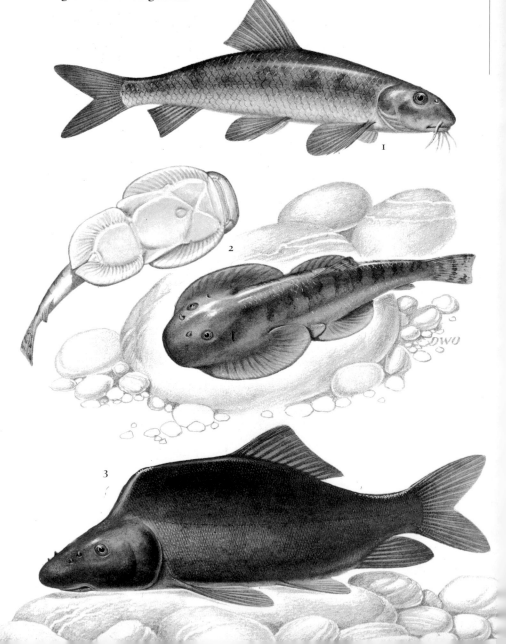

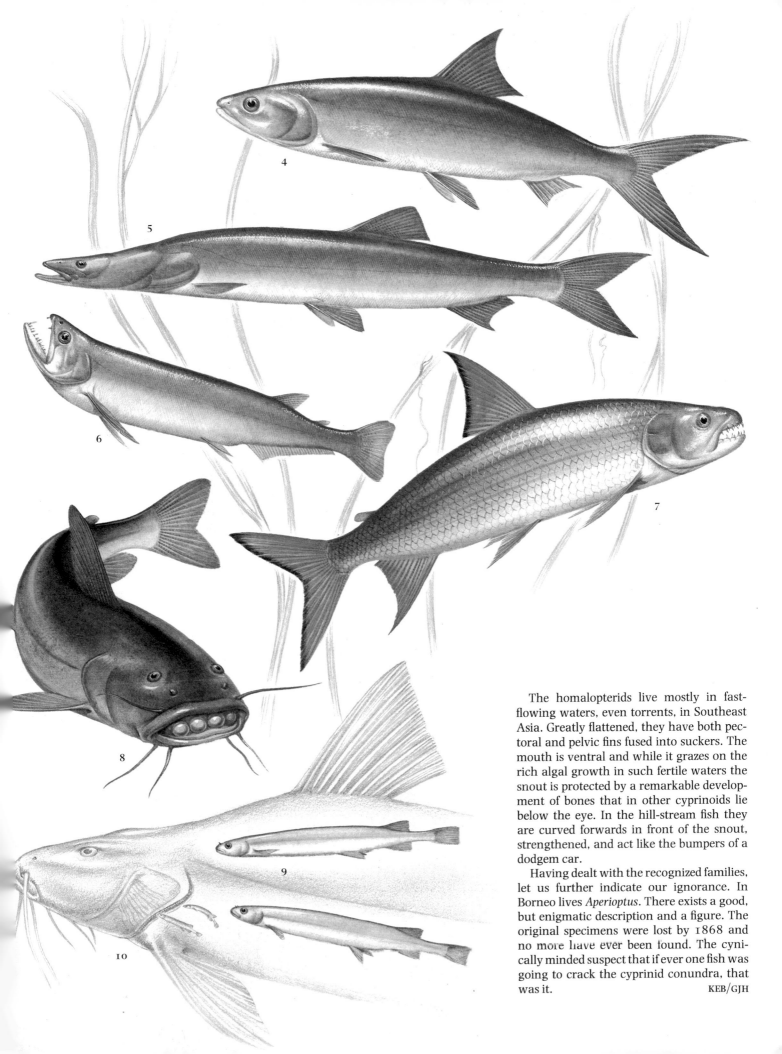

The homalopterids live mostly in fast-flowing waters, even torrents, in Southeast Asia. Greatly flattened, they have both pectoral and pelvic fins fused into suckers. The mouth is ventral and while it grazes on the rich algal growth in such fertile waters the snout is protected by a remarkable development of bones that in other cyprinoids lie below the eye. In the hill-stream fish they are curved forwards in front of the snout, strengthened, and act like the bumpers of a dodgem car.

Having dealt with the recognized families, let us further indicate our ignorance. In Borneo lives *Aperioptus*. There exists a good, but enigmatic description and a figure. The original specimens were lost by 1868 and no more have ever been found. The cynically minded suspect that if ever one fish was going to crack the cyprinid conundra, that was it.

KEB/GJH

Fish underground

The locations, forms and lives of cave fishes

Chologaster cornutus

There are now known to be about 38 species of fishes, belonging to 13 different families, that spend their lives in lightless, underground waters. In some cases it is uncertain whether an underground population represents a separate species or merely a highly modified population of a surface-living (epigean) species.

Cave fishes are colorless and have either minute eyes or no eyes. The scales are reduced by varying degrees in different species. Some fishes living under rocks in rapids may also have the same characteristics. Not all "cave fishes" live in caves. Some, for example, live in water-bearing strata (aquifers) where the rock is honeycombed with water-filled channels.

Subterranean fishes occur in tropical and warm temperate countries that have not been affected by recent glaciation. In Australia two species, an eleotrid and a synbranchid, live in the Yardee Creek wells on the North West Cape. Madagascar has two eleotrids. Africa has a barb in Zaïre, a synbranchid in Mauritania and a clariid catfish in Namibia; three species, a catfish and two cyprinids, live in Somalia. Oman has two species of the cyprinid genus *Garra*—neither yet described scientifically. Iran has a subterranean loach and another garrine cyprinid; Iraq, yet another garrine (*Typhlogarra widdowsoni*) and *Caecocypris*, a cyprinid of unknown affinities. Wells in Kerala, southern India, hold *Horaglanis*, a small catfish that subsequent investigation has suggested does not belong to the family in which it was originally placed (Clariidae). Three species (two loaches and yet another garrine) have recently been found in China. Cuba has two ophidioids, Mexico has one and also has characoids and a synbranchid. The subterranean fauna of Brazil consists of catfishes and characoids. The USA has a rich cave fauna in many sites from the Ozark and Cumberland plateaux to Texas comprising catfish and ambylopsids. There may also be cave fish in New Guinea and Thailand. Europe has no cave fish but there is a cave-adapted amphibian (*Proteus*) in Yugoslavia.

The eleotrids of Madagascar and Australia are related to the gobies and are fundamentally marine fishes, as are the ophidioids of Cuba and Mexico. Fishes of the family Ophidiidae show how difficult it is to categorize cave fishes neatly. They are mostly fishes living in deepish water of somewhat reclusive habit. Shallow-water representatives are likely to be found in salt-water-bearing strata vertically inshore in islands. When, as in the Bahamas, an artesian well is sunk, their appearance should not be unexpected. In the geological past fishes living in such conditions could have been trapped in such underground waters as land masses moved and then slowly adapted themselves to brackish or fresh water over a long period of time. Such speculations have been proposed to explain the presence of *Stygichthys* and *Lucifuga* in Cuba. How other cave fish came to live where they do is uncertain. There is some evidence from *Chologaster*, a close relative of the American ambylopsids, of preadaptation. *Chologaster* has small eyes and actively shuns light. In desert areas the fish may well have, where possible, followed the falling water-table underground.

So far as we know cave fish live longer than their surface-living relatives. This may be a response to their irregular and sparse food supply, which comes into the cave during floods as detritus or is provided by other cave animals all of which are ultimately supported by "the outside."

Reproduction in cave fishes has never been observed in the wild, except in the Cuban ophidioids, which give birth to fully formed young. Other cave-fish groups are probably egg-layers. In captivity the only form bred (and commercially available to aquarists) is the so-called *Anoptichthys jordani* which recent research has shown to be subterranean populations of the widespread *Astyanax fasciatus*. The reproductive strategies of its surface-living and subterranean populations are similar.

It is important to be able to breed cave fishes because, first, wild populations are low and, second, we cannot answer questions about them until we have done so. At least two species in captivity have been reared until sexually ripe, but all the individuals died before spawning. It seems that a trigger to induce spawning is missing. A possible clue to a trigger comes from caves in Zaïre. Here the young are found after the rains—the force of the rain water flooding into the caves is enough to flatten a gasoline can to the contours of the rock against which it lodges. It is not known what effect changes of temperature or other factors may have, but experiments are now being made using current-machines. KEB

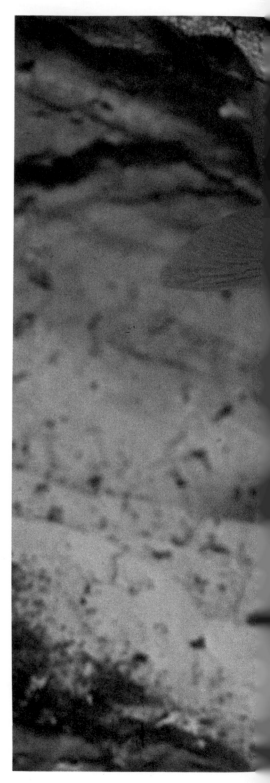

◄► **The relinquishing of sight.** These North American species are superficially alike. *Chologaster* shuns light and hides under stones. A continuation of this proclivity could have led to the occupation of dark caves and the nondevelopment of eyes and pigment.

▼ **The only cave fish bred in captivity,** "*Anoptichthys jordani,*" is actually a blind form of the widespread species *Astyanax fasciatus.*

Amblyopsis spelaea

Superorder Paracanthopterygii

About one thousand one hundred species in about 200 genera, about 30 families and 6 orders.

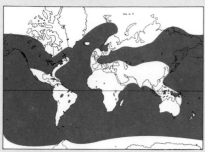

Position uncertain (*Incertae sedis*)

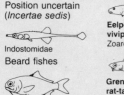

Indostomidae
Beard fishes

Beard fishes Polymixiidae
Trout-perches and allies

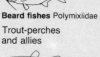

Amblyopsidae

Pirate perch Aphredoderidae

Trout-perches Percopsidae

Cods, hakes, grenadiers . . .

Carapids or **pearl fishes** Carapidae

Bregmacerotidae

Cods Gadidae

Eelpouts or **viviparous blennies** Zoarcidae

Grenadiers or **rat-tails** Macrouridae

Hakes Merlucciidae

Moridae

Frog fishes and angler fishes

Caulophrynidae

Ceratiidae

Linophrynidae

Oneirodidae

▲ **Outlines of representative species** of cods and angler fishes.

▶ **The Poor cod** is a common inshore fish of the Northeast Atlantic Ocean and Mediterranean. It is of little commercial importance but large numbers are caught by anglers.

For many years the number of cod caught per year in the North Atlantic region was about 400 million. This is some indication of the familiarity achieved by fish from this superorder, though more often on the plate than in the aquarium. The superorder holds a bewildering variety of forms, from plain fishes like whiting and haddock to fishes with peculiar shapes and colors, fishes flattened from top to bottom, camouflaged fishes, fishes with lures and the bizarre parasitic angler fishes (see p96).

The superorder, containing six orders, was erected in 1966 to remove from the major grouping of spiny-finned fishes (Acanthopterygii) all the forms that were of a similar evolutionary grade but lacked the characteristics of the spiny-finned fishes. Although this act enabled the acanthopterygians to be somewhat more succinctly defined, the coherence of the new superorder of cods and angler fishes was still rather tenuous. Since that time there have been additions and rearrangements and some workers still have doubts about whether the paracanthopterygians constitute a natural group.

Most members of this superorder are marine, inhabiting both deep and shallow waters worldwide. Exceptional are individual species of some otherwise marine families which live in fresh waters and the handful of species in the order of trout-perches which are confined to the fresh waters of northern America.

The family of **beard fishes** with probably just one genus and a few species is believed to be the most primitive member of the superorder. Prior to its inclusion here it was thought to be one of the most primitive members of the Acanthopterygii.

Polymixia nobilis, the type species of the genus, lives at depth between 150 and 365m (500–1,200ft) in the North Pacific, Atlantic and Indian Oceans. It has a deep, compressed body, a large eye and a single dorsal fin about half the length of the body. There are two long, fleshy barbels under the chin. The jaws and palate have fine teeth. The body color varies from greenish to reddish-brown and the dorsal and caudal fins have black tips. Almost nothing is known of its biology. It grows to about 30cm (1ft) long and has no commercial value. *Polymixia japonica* is very similar—indeed some workers think it should not be classified as a separate species.

The **trout-perches** and their allies are small freshwater fishes from northern America. Fossils clearly belonging to this group are known from the Lower Cretaceous era (100–65 million years ago).

The trout-perches derive their common name from their vaguely "perch-like" first dorsal fin, which has spines at the front, and the presence of a "trout-like" adipose fin. There are only two species. *Percopsis transmontanus* comes from slow-flowing, weedy parts of the Columbian river system. It has scales with a comb-like free margin (ctenoid scales) and a cryptic greenish color with dark spots. Growing to about 10cm (4in) long it reaches only half the size of its widespread congener *Percopsis omiscomaycus* which occurs from the west coast of Canada to the Great lakes and the Mississippi-Missouri river system. Although two rows of spots are present, the body is translucent and the lining of the abdominal cavity can be seen through the sides. Both species feed on bottom-living invertebrates and are themselves eaten by predatory fishes.

The Pirate perch, from still and slow-flowing waters of the eastern USA, is the only member of the family Aphredoderidae. A sluggish, dark-hued fish, it grows to about 15cm (6in) long and lacks the adipose fin of the trout-perches. It is a predator on invertebrates and small fish. Its most unusual feature is its mobile vent. On

An Anomalous Fish

An interesting example of the difficulties involved in classifying fish is the case of *Indostomus paradoxus*. It is a peculiar little fish, scarcely 2.5cm (1in) long and native to Lake Indawgyi in Upper Burma. It is very slender, covered with bony plates, yet so prehensile or flexible that it can touch its nose with its tail. It has the unusual attribute for a fish of being able to move its head up and down. When frightened it tends to leap out of the water.

Because it has scutes and five isolated spines in front of the dorsal fin it was originally placed with the sticklebacks and pipefishes. Later research showed that not only was it quite unrelated to either group but that it did not also appear to be closely related to anything else either. It has a series of unique anatomical features that do not help to establish its relationships. These include a lower jaw 20 times longer than the upper, and hardly any muscles in the rear half of its body; the tail fin is worked by a series of long tendons.

Out of desperation, and on the basis of a few very tenuous features, it was placed in an order of its own in the Paracanthopterygii. Later workers have disagreed with this association, but have not been able to find a satisfactory home for it. So here it still sits, *incertae sedis*, "position uncertain."

The 6 Orders of Cods and Angler fishes

Position uncertain (*incertae sedis*)
Indostomus paradoxus
Distribution: Lake Indawgyi in Upper Burma.
Size: maximum length 2.5cm (1in).

Division (a): Polymixiomorpha

Beard fishes
Order: Polymixiiformes
Probably five species of the genus *Polymixia* and the family Polymixiidae.

Distribution: N Pacific, Atlantic and Indian oceans.
Size: maximum length 40cm (16in).

Division (b): Salmopercomorpha

Trout-perches and their allies
Order: Percopsiformes
Seven species in 7 genera and 3 families.

Distribution: N America.
Habitat: fresh water.
Size: length 15cm (6in).
Families and species include: Amblyopsidae, including **Southern cavefish** (*Typhlichthys subterraneus*); **Pirate perch** (Aphredoderidae, sole species *Aphredoderus sayanus*); **trout-perches** (family Percopsidae).

Cling fishes
Order: Gobiesociformes
About thirty five genera of the family Gobiesocidae (number of species unknown).

Distribution: Atlantic, Indian and Pacific oceans in shallow waters; some species in fresh water in surrounding areas.
Size: most species under 10cm (4in); two reach over 30cm (12in).

Cods, hakes, grenadiers and their allies
Order: Gadiformes
Between three hundred and fifty and six hundred species in 70–100 genera and 9 families.

Distribution: northern hemisphere in cool marine coastal and moderately deep waters and cool fresh water.
Size: 4cm–2m (1.5–6.5ft).
Families, genera and species include: **brotulids and cusk eels** (family Ophidiidae), including **kingklip** (*Genypterus capensis*); **carapids** or **pearl fishes** (family Carapidae); **eelpouts** or **viviparous blennies** (family Zoarcidae); Gadidae, including **Atlantic cod** (*Gadus morhua*), **bib** (*Trisopterus luscus*), **burbot** (*Lota lota*), **haddock** (*Melanogrammus aeglefinus*), **lings** (genus *Molva*), **North Pacific cod** (*Gadus macrocephalus*), **pollack** (*Pollachius pollachius*), **Poor cod** (*Trisopterus minutus*), **poutassou** (*Micromesistius poutassou*), **saithe** (*Pollachius virens*), **Wall eye pollack** or **whiting** (*Theragra chalcogrammus*), **whiting** (*Merlangius merlangus*); **hakes** (family Merlucciidae), including **North Atlantic hake** (*Merluccius merluccius*), **Pacific hake** (*M. productus*), **Stock fish** (*M. capensis*); Moridae, including **Red cod** (*Physiculus bachus*); **rat-tails** or **grenadiers** (family Macrouridae).

Frog fishes and angler fishes
Order: Lophiiformes
About two hundred and seventy species in about 60 genera and 15 or 16 families.

Distribution: oceans worldwide, mostly in deep water.
Size: maximum length 2m (6.5ft) but most are much smaller.
Species and families include: **bat fishes** (family Ogcocephalidae); **ceratiid angler fishes** (family Ceratiidae); **frog fishes** or **antenariids** (family Antenariidae), including **Sargassum fish** (*Histrio histrio*); **linophrynids** (family Linophrynidae); **lophiid angler** or **goose fishes** (family Lophiidae); **oneirodids** (family Oneirodidae).

Toad fishes
Order: Batrachoidiformes
Over sixty species in 19 genera of the family Batrachoididae.

Distribution: almost worldwide with species in coastal and deep sea waters and fresh water.
Size: maximum length 60cm (24in).
Genera include: **midshipmen** (genus *Porichthys*).

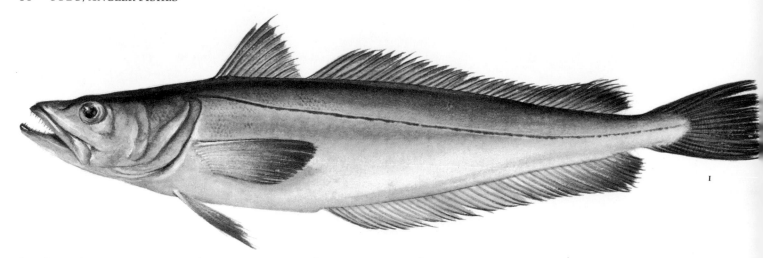

hatching, the anus is in a normal position, ie just in front of the anal fin. As the fish grows the anus moves forward until adulthood when it lies under the throat (see p89).

The family Amblyopsidae contains four genera, three of which live only in caves in the limestone regions of Kentucky and adjacent states. *Chologaster cornutus* is an eyed and pigmented fish found in sluggish and still waters from West Virginia to Georgia. Despite having functional eyes it shuns light and hides away under stones and logs during the day. *Chologaster agassizii* lives in subterranean waters in Kentucky and Tennessee. It lacks the dark stripe of its only congener but still has functional eyes. In both species, but especially the latter, there are series of raised sense organs on the skin.

The Southern cavefish lacks visible eyes, pigment and pelvic fins. Rows of papillae, sensitive to vibrations, are present on the body and on the tail fin. It is thought to have achieved its wide distribution, from Oklahoma to Tennessee and northern Alabama, by traveling through underground waterways. *Ambylopsis spelaea* is white, has minute eyes covered by skin, and tiny pelvic fins. Like its relatives it has vertical rows of sensory papillae on the body. Its reproductive strategy is unusual. The female lays a few relatively large eggs which, once fertilized, are carried in the gill chamber of the mother for up to ten weeks until they hatch.

The remaining genus, *Speoplatyrhinus*, is exceedingly rare and discussed in "Endangered Fishes" (see p9).

The order of **cods, hakes, grenadiers and their relatives** contains some of the commercially most important fishes, such as lings and whiting, as well as many smaller and deep-water species of biological but not financial importance.

They are mostly elongate fishes, having ventral fins well forward, often further forward than the pectoral fins. The fins lack spiny rays. The dorsal fin is very long and may be divided up into two or three separate units and may or may not be contiguous with the caudal fin, which in turn may or may not be contiguous with the long anal fin, which may be divided into two.

Almost all are marine fishes found in the cool waters of both hemispheres or in the deep sea. In some northern regions cooler currents enable them to live farther south than would be expected. The Pacific hake and the Pacific tomcod, for example, occur along the Pacific coast of America from Alaska to California.

Many of the commercially important species form shoals and the number of individuals they contained was, until severe overfishing had an effect, enormous. It was estimated that for many years 400 million cod were caught each year in the North Atlantic region. Bearing in mind the doubt about whether the North Pacific cod is the same species as the Atlantic cod, at any one time the number of individuals of that species was most impressive. As a female cod may produce over 6 million eggs it should not be too long before the fish stocks can regain their former abundance.

Some of the families in this order are of little importance and only brief reference will be made to them.

The Moridae is a worldwide family of deep-sea cods, found in all oceans almost from pole to pole. There are some 70 species often grouped into 17 genera. The configuration of dorsal and anal fins is variable but many species have a forked tail. Species in this family rarely exceed 90cm (3ft) in length. The biology of many species is unknown.

Antimora rostrata has been found in the North Pacific, North and South Atlantic and Indian Oceans at depths from 500 to 1,200m (1,650–3,900ft). The first ray of the first short-based dorsal fin is very long. The body color is dark violet to blackish brown. The Red cod—named from its death color—is common off New Zealand where it is used as a food fish. It occurs in much shallower waters than many of its relatives,

▲ ► ▼ **Some important species** belonging to the superorder Paracanthopterygii. (1) A North Atlantic hake (*Merluccius merluccius*; family Merluccidae; order Gadiformes) (1m, 3.3ft). (2) A haddock (*Melanogrammus aeglefinus*; family Gadidae; order Gadiformes) (50cm, 20in). (3) The ventral view of a cling fish (*Chorisochismus dentex*; family Gobiesocidae; order Gobiesociformes) (12cm, 4.7in). (4) A trout-perch (*Percopsis omiscomaycus*; family Percopsidae; order Percopsiformes) (20cm, 8in).

4

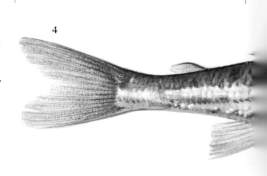

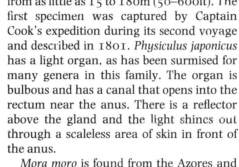

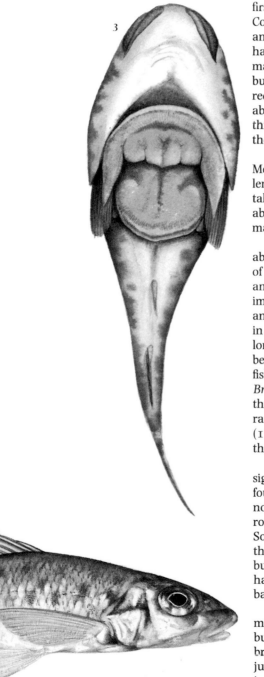

from as little as 15 to 180m (50–600ft). The first specimen was captured by Captain Cook's expedition during its second voyage and described in 1801. *Physiculus japonicus* has a light organ, as has been surmised for many genera in this family. The organ is bulbous and has a canal that opens into the rectum near the anus. There is a reflector above the gland and the light shines out through a scaleless area of skin in front of the anus.

Mora moro is found from the Azores and Mediterranean to the Faroe Islands. Its eye lens has been studied and photographs taken through it have shown a complete absence of the optical aberration that normally occurs in transparent spheres.

The Bregmacerotidae is a family with about six species living in the surface waters of tropical and subtropical oceans. Their anal and second dorsal fins are mirror images of each other; both have long bases and are higher at the front and back than in the middle. The first dorsal fin is just a long single ray with its base on the head just behind the level of the eyes. All are small fishes, growing to a few centimeters long. *Bregmaceros macclellandi* is as widespread as the rest of the family but has a greater depth range, from the surface to 3,660m (12,000ft). Next to nothing is known of their biology.

The family Gadidae is of most commercial significance in this order. Most species are found on the continental shelves of the northern hemisphere, but one genus of rocklings (*Gaidropsarus*) also occurs on South African shores. They have two or three dorsal fins and one or two anal fins, but none of the fins has spines. Many species have a chin barbel and some have additional barbels on the snout.

Practically all species are marine, but one most interesting exception is the burbot. The burbot is an eel-like fish with a mottled brown body. The first dorsal fin is short and just contacts the long, second dorsal fin. It is widespread throughout sluggish or still waters of northern Eurasia and northern America. During the last century it was common in east coast rivers of England, from East Anglia to Yorkshire. It is now probably extinct there. Certainly no reliable reports of its presence have been forthcoming in recent years. A major factor contributing to its demise has been the dredging of drains and canals. This has not only removed the weeds that gave the young protection but also increased the current speed, which sluggish fish find antipathetic.

The burbot is a winter spawner; a large female (about 90cm, 3ft long) may lay up to three million small eggs. Burbot are largely nocturnal and feed on invertebrates and bottom-living fish. Although their flesh is nutritious and tasty and the liver contains a lot of vitamin A it is hardly ever eaten nowadays. Only in northern Russia, where the fish fauna is sparse, does it have any commercial importance.

The lings look rather like large marine burbot. *Molva molva* is mostly a species of the eastern North Atlantic, but a few have been caught off the Canadian coast. It is commonest at depths of about 300m (1,000ft). A large female can live for 15 years and reach a weight of 22kg (50lb). It is a fish of some commercial importance, but its main claim to fame is one of statistical importance. A female ling 1.5m (5ft) long and weighing 24kg (54lb) was found to have 28,361,000 eggs in her ovaries, the largest number of eggs ever noted in a vertebrate.

The Atlantic cod and the haddock are highly prized, shoaling, prime-food fish from the North Atlantic. The North Pacific cod is regarded by some workers as merely a subspecies of the Atlantic cod. During the 19th century when the fishing grounds in the far North Atlantic were first fished, cod weighing up to 90kg (200lb) were recorded. These days, because of intensive fishing, an 18kg (40lb) cod is regarded as a giant and most of those caught commercially average about 4.5kg (10lb). Throughout its range the Atlantic cod exists in fairly discrete populations, but the migration of individuals from one population to another has precluded the accordance of subspecific status to these groups.

Cod spawn in late winter and early spring in water less than 180m (600ft) deep. The eggs are pelagic and scattered widely by the currents. The young hatch at the surface and feed on small plankton, but when they are about two months old, and about 2.5cm (1in) long, they descend to live close to the bottom.

Adult cod feed on large invertebrates—fish form a larger part of the diet as they grow. During daytime cod form dense shoals off the bottom, but at night they separate. The northern populations migrate south to spawn. The life style of the haddock is similar to that of the cod, but they are much smaller fishes hardly reaching 3.5kg (8lb) in weight.

There are many smaller and unimportant genera in both the North Atlantic and North Pacific. Common in the Atlantic are the Poor cod, bib, poutassou, whiting, pollack and saithe. The last three have some small commercial value; they are frequently sold by fishmongers for pet food. The "whiting" of the north Pacific, which is officially called Wall eye pollack, is caught to feed farmed mink.

The rocklings are a group of small, shore-dwelling, eel-like cods, particularly characterized by a chin barbel and up to four barbels on the snout. Their common name alludes to their abundance in rock pools and under rocks on the shore.

The species in the hake family have elongated bodies, a short first dorsal fin and a much longer second dorsal fin. Both the second dorsal and the anal fin are separate from the tail. In the Pacific hake the anal and second dorsal fins possess a deep notch which nearly, but not quite, divides these fins in half.

The North Atlantic hake, the range of which extends into the Mediterranean, can grow to over 90cm (3ft) long and weigh more than 4.5kg (10lb). It is a nocturnal feeder, its large mouth eager for squids and small fishes, even for its own species, which are caught in midwater. By day they live close to the bottom. They begin spawning in spring in water deeper than 180m (600ft), but as the season moves along they migrate and continue to spawn in shallower water. The eggs float at the surface and the future hake stocks are very dependent on the weather. If the wind blows the eggs away from the rich inshore feeding grounds

very few young survive, thereby occasioning the failure of the fishery a few years later.

In European waters there is a substantial hake fishery, especially off the Iberian peninsula. By contrast, the Pacific hake is regarded as a nuisance because its flesh is not esteemed and it raids salmon nets.

The Stock fish is found around South Africa, especially off the west coast where the water is richest. It is very similar to the North Atlantic hake in appearance but grows slightly longer and is a shoaling fish. The shoals appear to undergo random migrations which are not yet fully understood. The Stock fish is fished for both by trawling and by long lines. When fresh the flesh is delicious and well textured, but loses its flavor and texture with keeping.

Hakes also occur off New Zealand (*Merluccius australis*), and off the Pacific and Atlantic coasts of South America (*Merluccius gayi* and *Merluccius hubbsi* respectively). These, and the other species, are commercially important and when not used as prime food for humans are used as pet food and fertilizer.

It should be noted, however, that despite the different scientific names given to the hakes of various regions, there is much overlap in characters and some "species" may just be isolated populations of other species.

The rat-tails or grenadiers are a family of deep-water fishes with long, tapering bodies. The mouth is on the underside of the large head and is often overshadowed by a pointed snout. In many species the males have drumming muscles attached to the swim bladder which can produce a surprisingly loud noise. They possess a luminous gland, similar to that of *Physiculus*, and are distributed worldwide.

The family Ophidiidae contains unfamiliar but interesting fishes commonly called brotulids and cusk eels. Both brotulids and cusk eels are elongate fishes with long dorsal and anal fins which may fuse with the tail fin. Brotulids are deeper-bodied and broader-headed than cusk eels, and have thread-like pelvic fins below the rear of the head. The similar pelvic fins in cusk eels are under the throat. In some species the male has a penis-like intromittent organ with which he transfers parcels of sperm into the genital duct of the female. Some species are egg layers whereas in others the eggs hatch inside the mother and emerge as fully formed individuals.

Most of the cusk eels are small, secretive and burrowing fishes from warm seas. *Otophidium taylori* from the eastern Pacific

▶ **An exit on the move.** An extraordinary feature of the Pirate perch is the movement of the relative position of its anus while it is growing. Seen here are: (1) the position of the anus when the fish is 0.8cm (0.3in) long; (2) at 1cm (0.4in) (3) at 1.4cm (0.5in); (4) at 2.5cm (1in); (5) at 6.5cm (2.5in).

▶ **Fish with a flourish.** This is one of the rat-tails or grenadiers. Distributed worldwide, they characteristically have a large head and a very long tapering body. Maximum length is about 80cm (31in). This species is *Odontomacrurus murrayi*.

▼ **Sluggish, bottom-living fishes,** the toad fishes are predators that rely on their camouflage in catching their prey. Seen here is an Atlantic species, *Opsanus tau*.

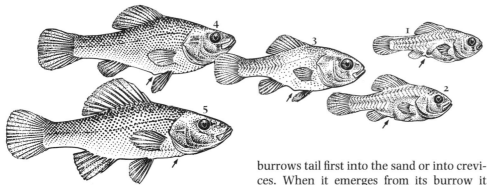

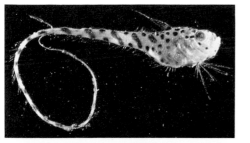

burrows tail first into the sand or into crevices. When it emerges from its burrow it aligns itself vertically, with only the last part of its body in the substrate. The kingklip is found only off the coast of South Africa from Walvis Bay to Algoa Bay. Growing to 1.5m (5ft) long, it is a giant among the cusk eels. Its flesh is excellent and the liver apparently has a quality much sought after by gourmets. It is not commercially important, however, as catches are irregular and sparse.

The brotulids are found worldwide, mostly in deep water. The few species that live in shallow waters hide away among rocks or in corals. This light-shunning habit may have led to the evolution of cave-living forms in the Yucatan Peninsula of Mexico and in Cuba. The caves are all close to the sea and the water is brackish but variable in salinity. *Typhlisiana pearsi* from Mexico is known by a very few specimens. Its eyes are minute and covered by skin. *Lucifuga subterranea* and *Stygicola dentata* (both placed in *Lucifuga* by some authors) have the best claim to being freshwater members of their family. Their coloration is remarkably variable, from off white to deep violet or dark brown. They give birth to fully formed young. Also placed in *Lucifuga* is *L. spaeleotes*, recently discovered in a sink hole in the Bahamas. The future of this species is in doubt as the area around Mermaids Pool, where it was found, is threatened with commercial redevelopment.

The carapids or pearl fishes are commonest in tropical and warm seas. All are slender fish with long, pointed tails. They have a complicated life history in which they pass through two dissimilar larval stages known as the vexillifer and tenuis stages. For a long time these two larval forms were thought to belong to two separate groups of fish.

All carapids are secretive and live inside the bodies of sea cucumbers, clams, tunicates, sea urchins or any other animal with a suitable body cavity. Some small species living inside oysters have been found embedded in the shell wall, hence "pearl fish." The Mediterranean species *Carapus apus* grows to about 20cm (8in) long. It lives in large sea cucumbers which, like many of its relatives, it enters tail first. Rather surprisingly it has been known to feed on the internal organs of its host. The larvae are free-living and this semiparasitic habit is only taken up by the adult. Small invertebrates are the usual diet for carapids.

The eelpouts or viviparous blennies live in cold seas from the shore to considerable depths. There are many species of these small fishes in both the Arctic and the Antarctic. Not all species give birth to fully formed young ("viviparous") but the name was generally adopted from one of the best known species that does, the North Atlantic *Zoarces viviparus*. All the species are similar, being moderately elongated with a single confluent median fin.

Toad fishes constitute an order of sluggish, bottom-living fishes, mostly from shallow warm seas. The body is stocky, the head

▲ **Roosting in the weed,** a Sargassum fish. This is a pelagic species that occurs almost worldwide along with Sargassum weed. It crawls around in floating clumps of weed.

▼ **Flashing in the depths,** a Plainfin midshipman (*Porichthys notatus*). By day this Pacific species burrows into mud. When fully grown it measures between 30 and 40cm (12–16in). The luminous organs along its scaleless body, clearly visible here, are used in courtship.

wide and the eyes lie on top of the head. A large mouth is well provided with teeth: toad fishes are predatory. Many species are colored to match their background. Some species, eg *Opsanus tau* of the eastern American seaboard, show some parental care. The large eggs are laid in a protected spot and guarded by the male until they hatch. *Opsanus tau* is one of the species that make grunting noises.

Thalassophryne maculosa occurs on both sides of Central America southwards and is an extremely poisonous species. Spines in the dorsal fin and on the operculum are hollow and linked to venom sacs. When trodden on the spines act like hypodermic needles and inject poison into the offending foot.

The midshipmen, *Porichthys notatus* from the Pacific and *P. porosissimus* from the Atlantic, burrow during daytime. Each species has a pattern of light organs which used in courtship. A wide range of sounds, from whistles to grunts and growls, also occurs during courtship.

Some of the species of the genus *Halophryne*, from the Indo-Pacific, can live in fresh water and are sometimes sold in aquarists' shops. As they have poisonous spines on the gill covers the aquarist should think twice before buying these cryptic, predatory fishes.

The scaleless **cling fishes** are mostly small (less than 30cm, 1ft long), and occur worldwide, but are commonest in shallow tropical seas. Their name is derived from their pelvic fins which are fused to form a strong sucker. Many species live in the intertidal zone where the sucker, enabling them to cling to rocks, is a great advantage in combating wave action. Almost all species are squat, stocky and bottom-dwellers. *Diademichthys lineatus*, from the Indo-Pacific, is an exception. It is long, thin and conspicuously colored (black with a horizontal white stripe). The snout is extremely long, even more so in the female than in the male. The peculiar shape and coloring are adaptations to life among the spines of the long-spined sea urchin *Diadema savigny*. The fish lives head down among the slender black spines. *Diademichthys* is not particularly grateful for its host's protection, as it supplements its diet by eating the urchin's tube feet.

The order of **frog fishes** and allies divides conveniently into two groups: first the frog and bat fishes, second the angler fishes. They are characterized by their first few dorsal rays being separated from the rest of the fin, and the first in particular is placed on the snout and modified into a lure. Although dorso-ventral (ie top–bottom) flattening is rare in bony fishes, several genera in this taxon have become so shaped as an adaptation to bottom dwelling.

The frog fishes rarely grow to more than 30cm (1ft) long. They are either flattened from side to side or more nearly globular. Highly cryptic forms, they live camouflaged either on the bottom or among weeds. The Sargassum fish lives among the floating clumps of Sargassum weed. Like those of the other members of the family, its pectoral and ventral fins are muscular, have a few rays and resemble little hands. Although the Sargassum fish lives at the surface in the Sargasso Sea, in other parts of the world frog fishes are bottom-dwellers, but their color always resembles that of their background. The lure of *Histrio* is little more than a stout tentacle, but in some species of *Antennarius* it is a short thin whip with a flap at the end. When this lure is rapidly wiggled around in a figure of eight the little flap follows behind looking very like a small fish. The thinness of the whip hopefully dissociates the conspicuous, but artificial, "fish" from the inconspicuous, but real, fish waiting below for its meal.

The bat fishes (family Ogcocephalidae) are depressed fishes found in most tropical seas. Viewed from above they look either like a triangle with a tail or a near circle with a tail. The snout is at the apex and the pectoral fins are at the bottom angles in the triangular species and equivalently placed in the nearly circular ones. The pelvic fins

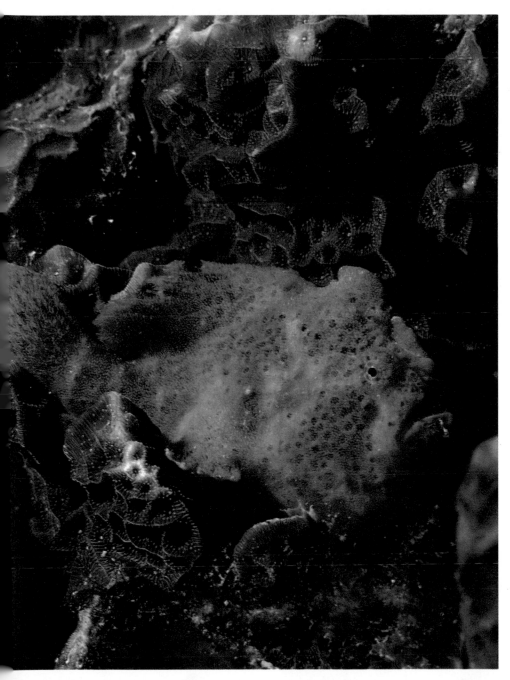

lie under the throat and they, and the muscular pectoral fins, are used for "walking" across the sea floor. The gill openings are near the hind end of the body, just before the pectoral fins. The eyes are placed dorsally and the mouth is below the snout.

Most species are inhabitants of deep water and all spend the day on the bottom waiting to eat prey tempted by their lures. Some kept in aquaria and some shallow-water species come to the surface at night, which suggests they might not be the inefficient swimmers they appear to be. The southern African species *Halieutea fitzsimonsi* has been recorded as entering brackish water. The Pacific deep-water species *Dibranchus stellulatus*, which grows to only a few centimeters long, has an elaborate network of bony stars rather than nodules in the skin.

In the Far East bat fishes are called rattle fish. The inside of the fish is cleaned out and the skin left to dry in the wind. Some pebbles are placed inside it, the hole sewn up, the spines smoothed and the whole then used as a baby's rattle.

The family Chaunacidae is represented by one genus, probably by only one species, and is worldwide in distribution. Living at depths of 180–300m (600–1,000ft) it has a rough, pinkish skin, is compressed rather than depressed and has a large mouth. Its belly is highly inflatable.

The lophiid angler (or goose) fishes are large, depressed, bottom-living fishes. Their heads and mouths are enormous. *Lophius piscatorius* is found in both the North and South Atlantic, at depths down to 550m (1,800ft) outside the breeding season. It can reach 1.8m (6ft) in length and is a voracious predator. Normally an angler fish feeds on the bottom. Its outline is broken up by a series of irregular fleshy flaps and the prey is attracted by the lure. It is known that anglers occasionally come to the surface because one was found dead having choked

▲ **Lazing on the bottom,** an Atlantic species of frog fish (*Antennarius multiocellata*). The colors of frog fishes vary considerably.

▶ **Two species of batfishes** (frog fishes).
(1) *Halieutichthys aculeatus* (10cm, 4in).
(2) *Ogcocephalus parvus* (15cm, 6in).

▶ **Frog fish "skinscape"** OVERLEAF. The goosefish or Lophiid angler (*Lophius piscatorius*) is a shallow-water species with a flattened body and a large wide head. Its fishing lure is visible here in front of the eyes.

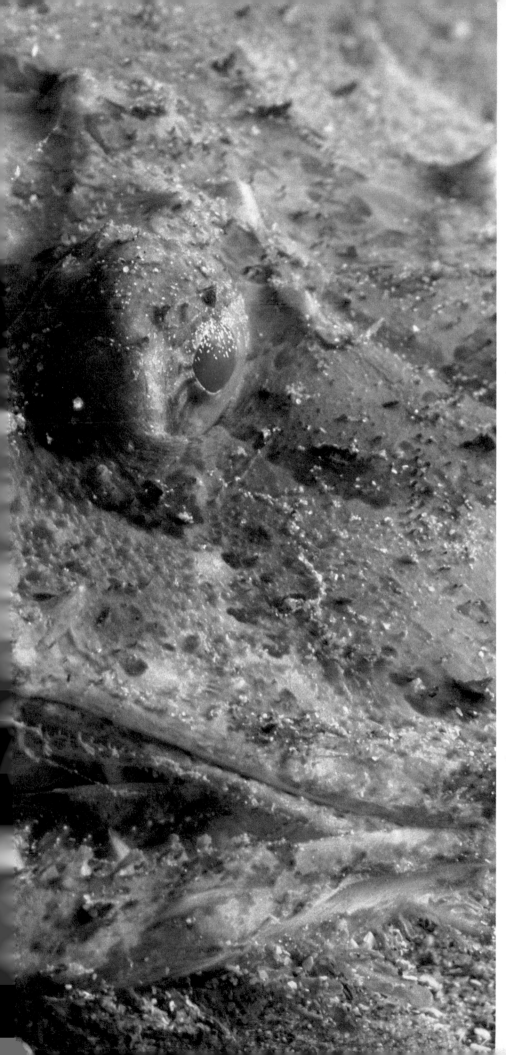

on a sea gull. Their voracity is further evidenced by many comments that *Lophius* usually comes up in a trawl with a full belly, having been unable to resist gorging itself on its fellow captives. *Lophius* is relatively uncritical about its diet and consumes any food item that comes close. As far as is known its only predator is the sperm whale in Icelandic waters.

In the North Atlantic angler fish move into deep water 1,800m (6,000ft) to spawn in spring. The egg mass forms a remarkable ribbon-like gelatinous structure up to 9m (30ft) long and 60cm (2ft) wide. The eggs usually form a single layer within this mass. The young hatch near the surface and feed on plankton, until they are 5–8cm (2–3in) long when they settle on the bottom.

The systematics of this group are not well known. Other genera, *Lophiodes*, *Lophiomus* and *Sladenia*, have been described from the Indo-Pacific region, but there is much dispute about the validity of some species included in them.

Although usually described as "flabby and revolting," the appearance of angler fish belies the nature of their flesh which is well-textured and delicately flavored, popular in Europe where it is sold as monk fish. It has even been known for unscrupulous fishmongers to substitute scoops of angler fish tail for scampi. When breaded and fried it is very difficult to tell the difference.

The most bizarre members of this super-order, and probably the most bizarre vertebrates, are the ceratioid angler fishes. It was noted above that in the frog fishes and *Lophius* the first dorsal fin ray was displaced forwards towards the top of the head, where it developed a fleshy flap and by muscular control is moved around like a fishing rod with the fleshy flap forming the bait. While this system works well in shallow, sunlit seas, it is useless in the lightless waters where the ceratioids dwell. But they have retained the system and improved it. The fleshy flap has been developed into a sometimes elaborate luminous organ called the esca, while the fishing pole has become very much longer, much more mobile, and is known as the illicium. Naturally, there are exceptions. The female *Neoceratias* lacks an illicium; in *Caulophryne* the esca is not luminous; *Gigantactis* has an illicium several times its body length whilst in *Melanocoetus* it is very short.

Many species also have a chin barbel, with several branches, the tips of which may be luminous. One species (*Himantolophus*),

with a much branched chin barbel, has been kept alive in an aquarium and the ends of the barbels, in motion, resemble fingers crawling through the water.

There are about 100 species of ceratioid angler fishes, the majority belonging to one family, the Oneirodidae. Although the ceratioids are found worldwide, they are patchily distributed, preferring highly productive waters. Generally they are found at greater depths in the tropics than at the poles, but all these generalizations can be affected by local factors. Ceratioids like cool waters with a particular salinity, so hotter water extending from the surface to a considerable depth will form a barrier. In the tropics they have been found below 2,750m (9,000ft), whereas off Greenland one of the largest species, *Ceratias holboelli*, occasionally comes in to very shallow waters. Some species appear to have rather restricted distributions, but *Oneirodes eschrichti* occurs all over the world in suitable waters. However, it must be emphasized that our knowledge

of the distribution of deep-sea fishes reflects little more than the distribution of the hauls of research vessels.

Angler fishes lack pelvic fins, but in some cases the pelvic fin bones are still present beneath the skin. In the other fins the number of rays is low, but the rays are thickened. The fan-shaped pectoral fin is, in some species, supported on a fleshy pedicel. The anal and dorsal fins are placed opposite one another.

In most species the skin is naked, but in *Himantolophus* and *Ceratias* bony nodules are present, and in the oneirodid *Spiniphryne* there are numerous close-set spines in the skin and along the bases of the fins.

In all species the skeleton is poorly

developed, both in terms of bone dimension and degree of calcification. Much cartilage persists and many bones are reduced to load-bearing struts. The operculum—a fan-shaped bone in most fishes—is reduced to a V in the oneirodids. To bear the large teeth the jaw bones are well developed by ceratioid standards, but fragile when compared with most other fishes. The lightening of the skeleton helps to achieve neutral buoyancy in the absence of a swim bladder.

Most ceratioids are dark brown or black but a few are gray and at least one is pallid. As far as we know most ceratioids are small fishes, rarely growing more than 15cm (6in) long. *Ceratias holboelli* is a comparative giant, reaching 90cm (3ft) in length. Ceratioids may be larger than we think, because deep-sea trawling is a cumbersome and inefficient process. To trawl at 1.6km (1mi) deep may mean that the boat is towing 4.8km (3mi) of cable so the net is only moving very slowly and creating a great disturbance in the water. Larger and more agile individuals (if there are any) may well be able to avoid capture. Some of the globular species have their lateral-line organs set out from the body on stalks, to increase the fishes' awareness of pressure changes, and they may well be able to detect the approaching net without difficulty.

It is presumed that the lateral line of adult ceratioids is sensitive because their eyesight is poor. The eyes are small, as are the olfactory organs, thereby enhancing the need for an efficient lateral-line system. In juveniles, where the lateral-line system is not yet well developed, the eyes and olfactory organs are relatively much larger. Free-living male *Linophryne* have developed tubular eyes.

Angler fishes will eat fish, squids, crustaceans, arrow worms or any other life form they can attract. The stomach is remarkably elastic. A specimen of *Linophryne quinqueramus* 8.9cm (3.5in) long contained: a deep sea eel (*Serrivomer*) 3.3cm (1.3in) long, a deep sea Hatchet fish (*Sternoptyx*) 2.5cm (1in) long, two bristle mouths (*Cyclothone*) 2.6–6cm (1–2.5in) long and five shrimps 1.2–3cm (0.5–1.2in) long.

To help the prey into the stomach, and keep it there, the long, curved teeth of the globular female ceratioids are depressible. They give way to allow the food to pass down and then spring back to prevent its escape. In *Neoceratias spinifer* the longest teeth originate on the outside of the jaws. *Lasiognathus* has the upper jaws expanded as a flap on each side with the teeth pointing outwards until the flaps are closed down over the lower jaw.

It is thought that in species in which the illicium is short, and only moves from the vertical to the horizontal, ie is scarcely retractable, once the prey is near either a sudden lunge or buccal suction is used to secure the meal.

In some of the ceratioids and oneirodids the illicium is so long that the lure can be placed far in front of the mouth and then slowly retracted as the unsuspecting prey appears.

Gigantactis has a very thin, whip-like illicium several times the body length emanating almost from the tip of the snout. In what manner and for what purpose this small-mouthed species uses it are unknown.

A logical development of the retractable illicium is shown in *Thaumatichthys*. This is a somewhat flattened species of angler known only from a very few specimens. Its illicium is short and its luminous two-lobed esca hangs down from the roof of the mouth. It swims around slowly or, as it is rather flattened, may lie on the bottom with its mouth open. Any prey that approaches the esca is already in the mouth and the jaws merely have to close gently around it. **KEB**

Representative genera of angler fishes.
(1) *Edriolynchus* (8cm, 3in). (2) *Melanocoetus* (14cm, 5.5in). (3) *Gigantactis* (15cm, 6in). (4) *Lasiognathus* (7cm, 3in). (5) *Thaumatichthys* (8cm, 3in). (6) *Linophryne* (7cm, 3in).

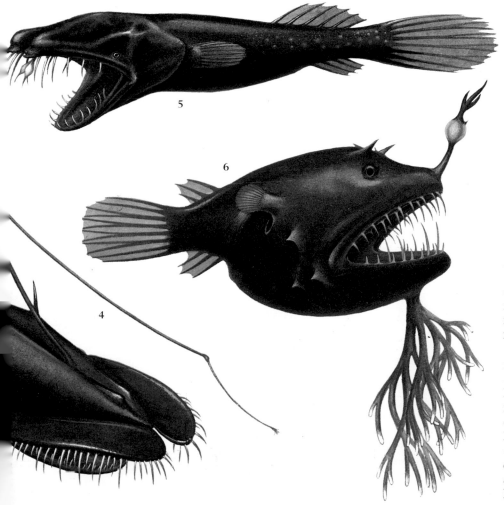

A Self-fertilizing Hermaphrodite
Reproduction strategies in angler fishes

Some of the deep-sea angler fishes reproduce in a manner unique among vertebrates. Deep-sea ceratioid angler fish lay and fertilize their eggs in deep water, after which the eggs float up to the surface. In some species the eggs are very numerous. A 6.3cm (2.5in) long *Edriolynchus* had over 9,000 eggs and a 65cm (26in) long *Ceratias holboelli* carried nearly 5 million immature eggs. (It has been suggested that the absence of a swim bladder makes more room for eggs.)

When the eggs hatch the larvae are, in the case of *Himantolophus groenlandicus*, about 2mm (0.1in) long. Even at this stage the sex of the larva is apparent because on the snout of the females is a small bud that will develop into the illicium. The larvae feed and grow in the surface waters. In the North Atlantic most species spawn in summer. The linophrynids, however, spawn in spring and their larvae live in deeper water (100–200m, 330–660ft). The larval phase lasts about two months and as the fishes grow they go deeper. A rapid metamorphosis then occurs; they change into subadults, looking more like their parents as sexually different characters become apparent.

By the time of metamorphosis the non-parasitic male has already developed large testes. Although his jaws are rather feeble pincer-like structures, he carries on feeding and grows very slowly. He is sexually mature at this small size, but the female is not mature for years after metamorphosis. Then she may be up to 20 times the length of her mate (according to species). When the breeding season arrives the male nips hold of the female's skin with his pincer-like jaws. Fertilization of the eggs is effected and the cycle starts again. *Himantolophus* and *Melanocoetus* are two genera believed to use this system.

The reproductive strategy of parasitic males is the most remarkable among vertebrates and, indeed, is the only known example of complete vertebrate parasitism.

After metamorphosis only the females feed. The male again has pincer-like teeth but is unable to eat even small food items. As a result of larval feeding he has a thick gelatinous layer which is believed to act as a food reserve until he finds a female. If he fails to find a mate before the reserve runs out he dies. For the good of the species, many more males hatch than females. How the male finds the female is not certain, but as he still has well-developed eyes and olfactory organs there must be clues for him to follow. How he finds a female of the same species will be discussed below. In contrast to the previous system, the male's testes are not developed. On meeting a female the male bites her skin—anywhere—and the two bodies fuse. The male loses most of his organ systems and becomes little more than a sac containing testes which do not develop until the female so dictates. He is fed by the nourishment in the female's blood *via* a placenta-like arrangement. Discharge of his sperm is presumably under the control of hormone levels in the female's blood. More than one male may attach to a female. By incorporating the male as part of her body, the female has developed into a self-fertilizing hermaphrodite. The mechanism of this fusion has aroused some interest in those working on problems of tissue rejection in humans.

In at least two genera, the oneirodid *Leptacanthichthys* and *Caulophryne*, both free-living males with developing testes and parasitic males have been found. Similarly, females with maturing ova but without attached males have been caught. It seems that these two taxa display "facultative parasitism" (ie some individuals have opted to be parasitic).

How do the males identify their own species and do they ever make mistakes? The answer to the last part of the question is "rarely." Hundreds of ceratioid anglers have been caught and there is only one known mistake, where a male *Melanocoetes* was latched on to, but not fused with, a female *Caulophryne*. It was mentioned above that the eyes and nostrils of the males are probably involved. Doubtless a female will exude a particular scent to which the male is sensitive. Correlated with the eye development is the fact that the structure and shape of the esca (the luminous organ at the end of the illicium) are specific to each species, so it is likely that they act as a lighthouse for the male.

► **At her command,** two parasitic male angler fish fused to the body of a female (*Edriolynchnus schmidti*).

▼ **The metamorphosis of a deep-sea angler fish** (*Ceratias holboelli*). (1) An unmetamorphosed male with a thick gelatinous skin. (2) An unmetamorphosed female. The beginning of the illicium can be seen above the eye. Unmetamorphosed angler fish are normally found near the surface of the ocean. Metamorphosis occurs at a depth of about 1,000m (3,300ft). (3) A metamorphosed male. Note the pincer-like teeth and the absence of an illicium. From this time on males do not eat. (4) A young metamorphosed female with a well-developed illicium. (5) An adult female with an attached parasitic male. The largest known male was about 15cm (6in) long, the largest female nearly 2m (6.5ft).

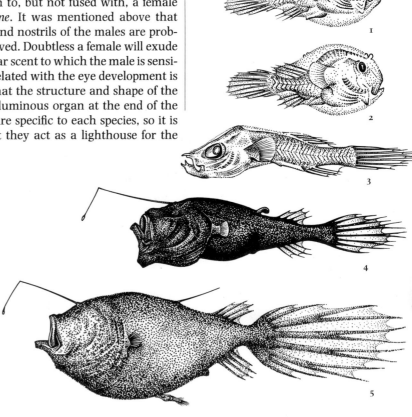

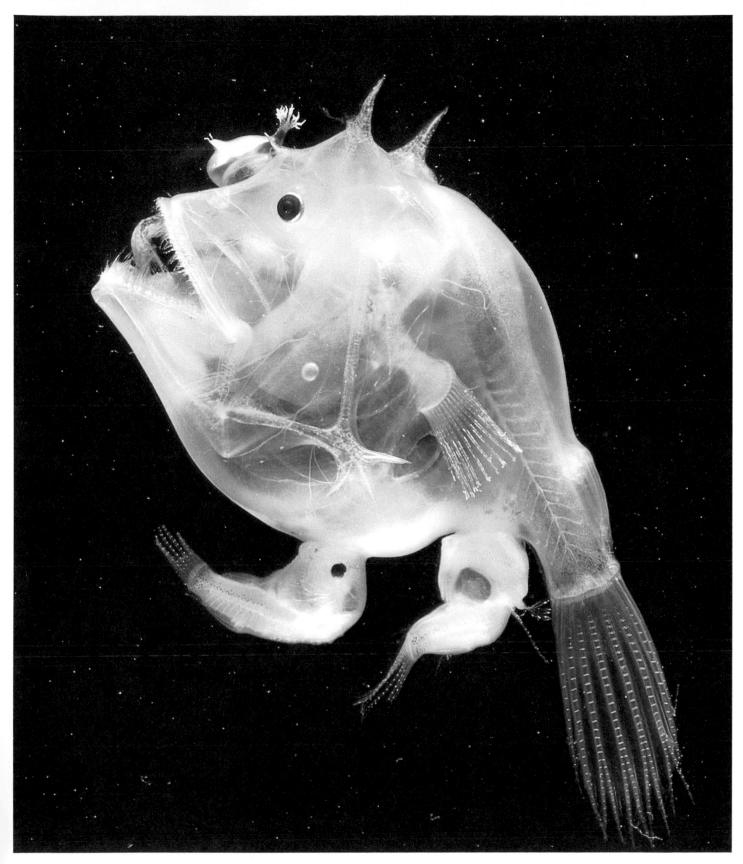

The light is produced by bacteria in the central bulb of the esca. Beneath the thin skin is a thick layer of connective tissue with nerve fibers. Then there is a dark layer and inside that a highly reflective layer. The central part is hollow and has glandular cells (to feed the bacteria?), blood vessels, and at the middle are the luminous bacteria.

The presence of nerves has led to speculation that the esca may be sensitive to touch. The light in ceratioids is yellowish-green or bluish and has been found to shine in a series of flashes.

The esca is a lure for prey as well as a mate. The light from the chin barbel of *Linophryne* is produced by chemical means.

Each organ is like a tubercle with a lens surrounded by blood vessels. Inside are the light-producing cells (photocytes) and loops of capillaries. Light production is controlled by oxygen in the blood stream. In the laboratory the addition of hydrogen peroxide to the organs will produce light.

KEB

SILVERSIDES, KILLIFISHES...

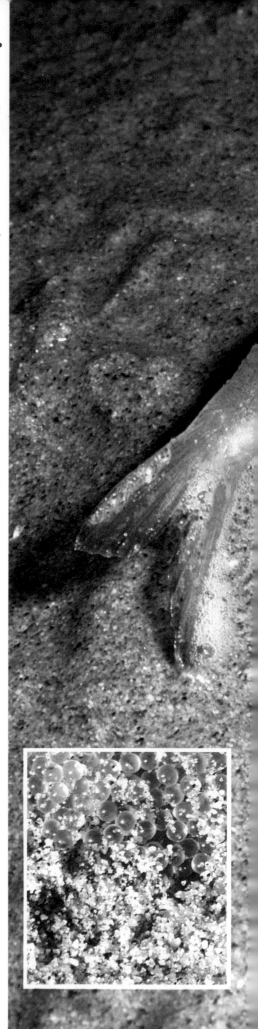

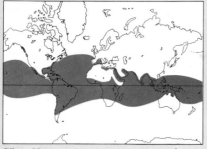

Series: Atherinomorpha
Superorder: Acanthopterygii.
About one thousand species in about 165
genera, 21 families, 2 orders and 1 division.

Silversides
Division: Atherinoidei
Two hundred and forty species in 50 genera
and 7 families.
Distribution: worldwide, fresh and sea waters.
Size: maximum length 60cm (24in).
Families: Atherinidae, including the **grunion**
(*Leuresthes tenuis*); Bedotiidae;
Dentatherinidae; Isonidae; Melanotaeniidae;
Telmatherinidae; Phallostethidae.

Killifishes
Order: Cyprinodontiformes
Six hundred species in 80 genera and
9 families.
Distribution: pantropical and north temperate
regions in fresh and brackish waters.
Size: maximum length 30cm (12in).
Families: Anablepidae, including **quatro ojos**
(genus *Anableps*); Aplocheilidae;
Cyprinodontidae; Fundulidae; Goodeidae;
Poeciliidae, including **guppy** (*Poecilia reticulata*),
mosquito fish (*Gambusia affinis*); Profundulidae;
Rivulidae; Valenciidae.

Ricefishes and their allies
Order: Beloniformes
One hundred and sixty species in 35 genera
and 5 families.
Distribution: worldwide, fresh and sea waters.
Size: maximum length 1m (3.3ft).
Families: **flying fishes** (family Exocoetidae),
halfbeaks (family Hemiphamphidae),
needlefishes (family Belonidae), **ricefishes**
(family Adrianichthyidae), **sauries** (family
Scombersocidae).

▶ **Moonlighting grunion.** The best-known
example of a fish that synchronizes its
spawning behavior with lunar cycles is that of
the grunion of the North American Pacific
coast. This is a marine species that moves
inshore and spawns in the sand of beaches at
night, following the highest of the spring tides
after both the new and full moons. As the
waves roll in the fish come ashore and deposit
and fertilize their eggs in the wet sand near the
top of the high-tide mark. The eggs are
normally hidden about 8cm (3in) below the
surface INSET LEFT. The eggs develop in the
sand, awaiting the next set of spring tides, some
12–14 days later. Then the eggs hatch INSET
RIGHT and the young are washed into the sea.

FISH that fly, fish that give birth to live
young, fish used to control mosquitoes
and fish that spawn to the cycles of the
moon: these are some of the unusual life-
styles found in the series Atherinomorpha.
Some of its members are extremely well
known, thanks to their wide use in experimental studies of embryo development and
their adaptability to aquarium conditions.
The series comprises some 1,000 species of
minute to medium-sized fishes distributed
worldwide in temperate and tropical
regions, inhabiting fresh, brackish and sea
water. Here it will be described as consisting
of three groups: silversides, killifishes, and
ricefishes and their allies.

The silversides are not placed in a separate named order because at present no
uniquely derived characteristics are
recognized that set apart these seven
families from all other teleosts. Silversides
are characterized by having a silvery lateral
band at midbody, hence their common
name, but this character is found in many
other groups of fishes. Most silversides are
narrow-bodied and elongate, though some
are relatively deep-bodied, such as species of
the genus *Glossolepis* of New Guinea, which
are used as food fish.

Many atherinomorphs have a prolonged
development time with fertilized eggs taking
one week or more to hatch, as opposed to
the more usual time in teleosts of from one
to two days. The grunion, a species of silver-
side found on the West Coast of North
America from southern to Baja California,
is well-known because of its spawning
behavior which is correlated with the lunar
cycle. Grunion spawn during spring tide;
the fertilized eggs are stranded during low
tide, and hatching is stimulated by the
waters of the returning high tide two weeks
later.

Silversides, such as species of the North
American genus *Menidia* found in coastal
and gulf drainages, are used as bait fish.
Another common name for silversides is
smelts, though they are not related to the
true smelts of the salmoniform family
Osmeridae.

The killifishes are probably best known to
the public at large by fishes in the family
Poeciliidae. Included here are the guppy,
certainly one of the most common domestic
animals, and the mosquitofish, which consumes mosquito larvae and pupae and so is
used throughout the world as a mosquito-
control agent. Poeciliids are also of great
interest to biologists because populations of
some species occur composed entirely, or
almost entirely, of females. One such species,

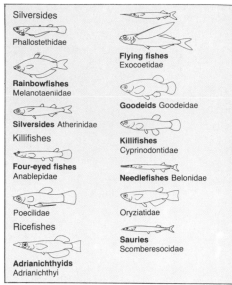

Silversides
Phallostethidae

Rainbowfishes
Melanotaeniidae

Silversides Atherinidae

Killifishes

Four-eyed fishes
Anablepidae

Poeciliidae

Ricefishes

Adrianichthyids
Adrianichthyi

Flying fishes
Exocoetidae

Goodeids Goodeidae

Killifishes
Cyprinodontidae

Needlefishes Belonidae

Oryziatidae

Sauries
Scomberesocidae

▲ **Outlines of representative families** of silversides, killifishes, ricefishes and their allies.

◄ **A popular aquarium fish,** the Red rainbow fish (*Glossolepis incisus*) from Lake Sentani in Irian Jaya in Indonesia. (Male in foreground.)

► **A Goldie River rainbow fish** (*Melanotaenia goldiei*) from Papua New Guinea.

▼ **The Mosquito fish** from Mexico has been introduced elsewhere to control mosquitoes.

Poecilia formosa, reproduces by spawning with males of another poeciliid species which do not contribute genetic material to the offspring but simply stimulate development of the egg.

Other killifishes or killies in the Old and New World tropical families Aplocheilidae and Rivulidae, respectively, are also popular aquarium fishes, as well as pest-control agents. Their popularity as aquarium fishes is no doubt in part due to their bright and beautiful coloration, as well as to their hardy nature, which is so well known that hobbyists often exchange fishes around the world by mailing them, wrapped only in a plastic bag with a little water and air and shipped in an insulated container. Included in these two tropical families are the annual killifishes, which are so named because adults rarely live longer than one rainy season, at the end of which time they spawn, leaving fertilized eggs in the drying muddy substrate. The eggs spend the dry season buried in the mud, lying quiescent until the rains return the following season. When the rains begin again the eggs are stimulated and hatch and the cycle is repeated.

North American killifishes are less brightly colored, as are most temperate fishes, but no less well-known, at least to biologists. The mummichog, a species found in brackish water, from Canada to the southern USA, is widely used in experimental embryological studies. Its biology is probably better known than that of any other species of bony fish.

Some of the most spectacular killifishes are those of the genus *Anableps*, found from southern Mexico to northern South America. These are the largest killifishes, some reaching to over 30cm (12in). They are certainly most well-known by the characteristic from which their common name of quatro ojos or "four eyes" is derived. Each eye is divided horizontally into two sections: there are separate upper and lower corneas and retinas. *Anableps* species are usually found just below the surface of the water, and seen from above only by the tops of their eyes which protrude above the surface. The upper eyes are used for vision above the water, whereas the lower eyes are used for vision below.

Killifishes of the families Poeciliidae, Anablepidae, and the Goodeidae of the Mexican plateau and western North America, all have species that are viviparous (ie there is internal fertilization of females by males, and females subsequently give birth to live young). The males of viviparous species within these families have anal fins that are modified for sperm transfer—the first few anal rays are generally more elongate and elaborate than those in the rest of the fin and are modified into what is called a gonopodium. At one time it was believed that all viviparous killifishes formed a natural group, ie they were more closely related to each other than any was to a group of oviparous or egg-laying killifishes. This has been found not to be the case, with egg-laying killifishes often judged to be more closely related to a particular viviparous group. This knowledge of killifish relationships has led us to understand that viviparity, although characterized by many

complex anatomical and behavioral modifications, is a way of life that has arisen several times within the evolution of killifishes.

The order Beloniformes comprises two groups, the ricefishes (Adrianichthyoidei) and the halfbeaks, flying fishes, needlefishes and sauries (Exocoetoidei).

All ricefishes are contained in one family. They are so-called because they were discovered in Oriental rice paddies. The scientific name of the common ricefish genus *Oryzias* is in fact derived directly from the name of the rice plant *Oryza*. Ricefishes are common in fresh and brackish waters from the Indian subcontinent throughout coastal Southeast Asia into China, Japan and along the Indo-Australian archipelago into the Celebes (Sulawesi).

Members of the halfbeak family are freshwater and marine fishes characterized by a very elongate lower jaw and a short upper jaw, hence "half a beak." Most halfbeaks are oviparous, but some, such as the Indo-Australian *Dermogenys pusillus*, have internal fertilization and are viviparous.

Species in the family of flying fishes do not exhibit true flight as the name implies. They have expanded pectoral (and sometimes pelvic) fin rays that allow them to glide for several seconds after they propel themselves above the water surface.

In the needlefish family species have an elongate upper as well as lower jaw—they are more or less fully beaked. The common name is a reference to the extremely sharp, needlelike teeth in the jaws. Most of the cosmopolitan temperate and tropical needlefishes are marine whereas some, such as *Potomorhaphis guianensis* of the Amazon, live in fresh water. Needlefishes are characterized by having bones, and also often muscle tissue, that are rather greenish in color. This does not, however, prevent them from being used as a food fish.

Sauries are commercially among the most important beloniform fishes. *Cololabis saira*, found in both the eastern and western Pacific, is an important species in fisheries in Japan. The scientific name *Scomberesox*, the type genus in the family, is a composite of *Scomber*, a name for mackerels, and *Esox*, the name for pikes and pickerels. Apparently sauries appeared to early workers as having characteristics of those two distantly related groups—five to seven finlets behind the dorsal and anal fins being reminiscent of the mackerels and the moderate-sized jaws with strong teeth being much like those of pikes and pickerels.   LP

SPINY-FINNED FISHES

Series: Percomorpha
Superorder: Acanthopterygii.
About ten thousand five hundred species in
2,000 genera, 230 families and 11 orders.

▶ **Zebrafishes** ABOVE live in coastal waters of
the Indo-Pacific Ocean. Their graceful
elongated fin rays bear extremely poisonous
spines. Their genus (*Pterois*) contains four or
five similar species (family Scorpaenidae, order
Scorpaeniformes).

▶ **Outlines of representative species** BELOW of
spiny-finned fishes.

▼ **This mosaic-like fish** is *Cleidopus glorimaris*,
an Australian pineapple fish (order
Beryciformes) with a multitude of common
names. Like the other members of the family
Monocentridae it has a covering of large spiny
scales on its deep rounded body. It has two open
pits beneath the chin which contain luminous
bacteria.

THE superorder Acanthopterygii contains
the latest flowering of bony fish evolution, some 60 percent of all living fishes
divided into two series. The earliest forms
referable to the group are found in the Cretaceous period (135–65 million years ago).
The common name alludes to the presence
of spines either in front of the soft dorsal fin
(although this also occurs independently in
other taxa, for example notacanths, see
p36), or incorporated into the anterior part
of the dorsal fin, or as a separate first dorsal
fin. Anal fin spines occur similarly and
spines in the ventral fins are widespread.
However, with such a successful, dominant
and species-rich group any or all of these
spines have been secondarily lost or modified in many species. The presence of rough-
edged (ctenoid) scales is another superficial
character of spiny-finned fishes but, again,
these have been modified into a wide array
of scutes or plates, or have even been lost.
There is also a tendency for the pelvic fins
to move forward on the body compared with
other superorders. The difficulty of dis-
tinguishing between basal acanthopterygians and paracanthopterygians has been
exemplified with regard to the status of
Polymixia on p86.

By far the major group of the spiny-finned
fishes are the Perciformes—the classical
"perch-like" fishes, whose classification is
very debatable (see p110). However, within
the rest of the spiny-finned fish there are a
number of discrete, well-defined orders.

The Berciformes (with some nine families) are large-headed, deep-water marine
fishes found in all temperate and subtropical
oceans. Most families have rough scales, but
the family Monocentridae, with only two
species, has been given the graphic common
name pine-cone fishes. These rounded little
fish (of which *Monocentrus japonicus* is the
best known) live in small schools in the
Indo-Pacific Ocean. The body is covered
with irregular bony plates and the soft-
rayed dorsal fin is preceded with a few large,
alternately angled spines. The pelvic spines
are massive and erectile. Although pine-
cone fishes do not grow to more than 23cm
(9in) long, they are commercially viable in
Japan, where they are eaten. They have two
small luminous organs under the lower jaw
with colonies of bacteria providing the light.
Their novel appearance coupled with
luminosity has made them popular aquarium fishes in the Far East.

Luminous organs are also present under
the eye in members of the family Anomalopidae. Each organ is a peculiar flat white in
daylight but at night glows with a blue-
green light. The luminous organs blink on
and off when functioning and are controlled
by rotating the entire gland. These are
shallow-water fishes living among the coral
reefs of both the East and West Indies. They
are of no commercial importance although,
if caught, the luminous gland is removed
and used as bait in subsistence fisheries.

The family Berycidae contains mid-water,
large-eyed species with compressed bodies.
Most species are red or pink. The flesh is
excellent and, in the North Atlantic at least,
commands a good price when offered at
markets.

Species belonging to the other families are
poorly known. Many have fine spines on the
head and anterior part of the body, especially when young. This once led to the
adults of *Anoplogaster cornuta* being known
as *Caulolepis longidens*. Despite the difference
in appearance between juveniles and adults
the mistake was realized when intermedi-
ates were found. However, as the young
were described first the former name
remains in use.

The order Zeiformes comprises deep-
bodied, extremely compressed fishes. The
John Dories (the European *Zeus faber*, the
American *Zeus ocellata*) are North Atlantic
fishes with a lugubrious appearance. They
have extremely protrusile jaws and feed on
small fishes and crustaceans. The origin of

Beryciformes

Pine-cone fishes
Monocentridae

Zeiformes

Dories Zeidae

Lampridiformes

Opahs or **moon-fishes**
Lampridae

Ribbonfishes Trachipteridae

Mail-cheeked fish

Pataecidae

Synancejidae

Sea moths

Sea moths Pegasidae

Sticklebacks and pipefishes

Seahorses Syngnathidae

Shrimp fishes Centriscidae

Perch-like fishes

Archer fishes Toxotidae

Barracudas Sphyraenidae

Butterfly fishes
Chaetodontidae

Cichlids Cichlidae

Cod icefishes Nototheniidae

Combtooth blennies
Blenniidae

Dolphins Coryphaenidae

Giant gourami
Osphronemidae

Goatfishes Mullidae

Jacks and pompanos
Carangidae

Louvars Luvaridae

Mackerels and tunas
Scombridae

Mullets Mugilidae

Nursery fishes Kurtidae

Parrot fishes Scaridae

Perch Percidae

Pikehead Luciocephalidae

Quill fishes Ptilichthyidae

Remoras or **shark-suckers**
Echeneidae

Sailfishes Istiophoridae

Sea basses Serranidae

Spiny eels Mastacembelidae

Stargazers Dactyloscopidae

Swordfish Xiphiidae

Tetraodontiformes

Leatherjackets Balistidae

Molas Molidae

Porcupine fishes Diodontidae

Three-toothed puffer
Triodontidae

Flatfish or pleuronectids

Psettodids Psettodidae

THE 11 ORDERS OF SPINY-FINNED FISHES

Order: Beryciformes

Between one hundred and one hundred and eighty species in 30–40 genera and 9–14 families.

Distribution: temperate and subtropical oceans.

Size: maximum length 1m (3.3ft).

Families include: **pine-cone fish** (Monocentridae).

Order: Zeiformes

About thirty-five species in at least 20 genera and 6 families.

Distribution: oceans in mid to deep water.

Size: maximum length about 1m (3.3ft).

Genera include: **John Dories** (*Zeus*).

Order: Lampridiformes

About forty species in 20 genera and 11 families.

Distribution: oceans worldwide.

Size: maximum length 9 or 10m (30 or 33ft).

Families and species include: **crest fish** (family Lophotidae); **deal fish** (family Trachipteridae); **opahs** or **moon-fish** (family Lampridae); **Oar fish** (*Regalecus glesne*, sole species of the family Regalecidae).

Swamp eels or synbranchids

Order: Synbranchiformes

Fifteen species in 4 genera and the family Synbranchidae.

Distribution: fresh or brackish waters of Mexico, C and S America, West Africa, Asia, Indo-Australian archipelago, NW Australia.

Species include: **Rice fish** (*Monopterus albus*).

Mail-cheeked fishes

Order: Scorpaeniformes

Over one thousand species in about 300 genera and 20 families.

Distribution: worldwide in sea and fresh waters (though distribution more disjunct in fresh waters).

Size: maximum length 2m (6.5ft).

Families and species include: **armored sea robins and pogges** (family Agonidae); Cottidae, including **bullhead** or **Miller's thumb** or **sculpin** (*Cottus gobio*); **gurnards** (family Triglidae); **sable fish** (family Anoplopomidae); **stone fishes** (family Synaceidae).

Flying gurnards

Order: Dactylopteriformes

Five species in 4 genera and the family Dactylopteridae.

Distribution: tropical oceans.

Size: maximum length 50cm (20in).

Sea moths

Order: Pegasiformes

Five species in 1 or 2 genera and the family Pegasidae.

Distribution: Indian Ocean and West Pacific.

Size: Maximum length 13cm (5in).

Sticklebacks and pipefishes

Order: Gasterosteiformes

About two hundred and fifty species in about 70 genera and 8 families.

Distribution: worldwide in tropical and temperate seas; fresh water in northern hemisphere

Size: maximum length 1.8m (5.9ft).

Families, genera and species include: **flute mouths** (families Aulostomatidae, Fistulariidae); **pipefishes and seahorses** (family Syngnathidae); **shrimp fishes** (family Centriscidae, genus *Centriscus*); **sticklebacks** (family Gasterosteidae), including **Brook stickleback** (*Culea inconstans*), **Fifteen-spined stickleback** (*Spinachia spinachia*), **Four-spined stickleback** (*Apeltes quadracus*), **Nine-** or **Ten-spined stickleback** (*Pungitius pungitius*), **Three-spined stickleback** (*Gasterosteus aculeatus*); **snipe fishes** (family Macrorhamphosidae); **tube snouts** (family Aulorhynchidae).

Perch-like fishes or perciforms

Order: Perciformes

About eight thousand species in about 1,400 genera and about 150 families.

Distribution: worldwide in both marine and fresh waters.

Size: maximum length 5m (16ft).

Families, genera and species include: **anabantids** (family Anabantidae), including **climbing perch** (genus *Anabas*); **archer fish** (family Toxotidae, genus *Toxotes*); **barracudas** (family Sphyraenidae); **butterfly fishes** (family Chaetodontidae), including **Blue angelfish** (*Pomacanthus semicirculatus*), **Forceps fish** (*Forcipiger longirostris*); **cardinal fishes** (family Apogonidae), including **Cardinal fish** (*Apogon endekataenia*); **cichlids** (family Cichlidae); **clownfishes** (family Pomacentridae); **drums** (family Sciaenidae); **groupers** (family Serranidae), including **Queensland grouper** (*Epinephalus lanceolata*); **Kissing gourami** (*Helostoma temmincki*, sole species of the family Helostomatidae); **leaffishes** (family Nandidae); **louvar** (family Luvaridae, sole species *Luvaris imperialis*); **mackerels, tunas and bonitos** or **scombrids** (family Scombridae), including **Common mackerel** (*Scomber scombrus*), **Frigate mackerel** (*Auxis thazard*), **Horse mackerel** (*Trachurus trachurus*), **Skipjack tuna** (*Euthynnus* or *Katsuwonus pelamis*); **Man-o'-war fishes** (family Nomeidae); **mudskippers** (family Gobiidae); **parrot fishes** (family Scaridae); **perch** (family Percidae), including **North American darters** (genera *Ammocrypta*, *Etheostoma*, *Percina*), **Eastern sand-darter** (*Ammocrypta pellucida*), **North American zander** (*Stizostedion vitreum*), **perch** (*Perca fluviatilis*), **ruffe** or **pope** (*Gymnocephalus cernua*), **zander** (*Stizostedion lucioperca*); **pilot-fish** (family Carangidae, genus *Naucrates*); **ragfishes** (family Icosteidae), including **Brown ragfish** (*Acrotus willoughbyi*), **Fantail ragfish** (*Icosteus aenigmaticus*); **remoras** or **shark-suckers** (family Echeneidae); **sailfishes, spearfishes and marlins** or **istiophorids** (family Istiophoridae), including **Black marlin** (*Makaira indica*), **sailfishes** (genus *Istiophorus*); **scats** (family Scatophagidae); **snakeheads** (family Channidae); **stargazers** (family Dactyloscopidae), including **European stargazer** (*Uranoscopus scaber*); **surgeonfishes** (family Acanthuridae), including **Common surgeonfish** (*Acanthurus bahianus*), **Unicorn fish** (*Naso unicornis*), **Yellow surgeon** (*Zebrasoma flavescens*); **swordfish** (*Xiphius gladius*, sole species of the family Xiphiidae); **wrasses** (family Labridae).

Flatfish or pleuronectids

Order: Pleuronectiformes

About five hundred species in about 120 genera and 7 families.

Distribution: worldwide in both marine and fresh waters.

Families, genera and species include: Citharidae; **left-eye flounders** (family Bothidae); **pleuronectids** (family Pleuronectidae), **European flounder** (*Platichthys flesus*), **halibut** (*Hippoglossus hippoglossus*), **plaice** (genus *Pleuronectes*); **psettodids** (family Psettodidae); Soleidae; **tongue soles** (family Cynoglossidae); **turbot** (family Scophthalmidae), including **European turbot** (*Scophthalmus maximus*), **scald-fish** (genus *Arnoglossus*).

Order: Tetraodontiformes

About three hundred and thirty species in over 90 genera and 8 families.

Distribution: worldwide in tropical and temperate waters.

Size: maximum length 2m (6.5ft), maximum weight 1,000kg (2,200lb).

Families include: **aracanids** (Aracanidae); **box** or **cow fishes** (Ostraciontidae); **file fishes** (Monacanthidae); **oceanic sun fishes** (Molidae); **porcupine** or **puffer fishes** or **diodontids** (Diodontidae); **puffer fishes** or **tetraodontids** (Tetraodontidae); **triacanthoids** (Triacanthidae); **trigger fishes** or **balistids** (Balistidae).

▶ **A fish full of puzzles.** The John Dories (order Zeiformes) are greatly compressed fishes with extremely protrusile jaws (that is jaws capable of being pushed forward). They live in temperate seas in both hemispheres. All have various fin rays elongated; in some the arrangement of scales (squamation) is reduced. The function of the spot on the side (the St Peter mark) is unknown, as is the origin of the John Dories' common name.

▼ **Night watchmen of the tropics.** Soldier or squirrel fishes (family Holocentridae, order Beryciformes) are widespread in coral reef or rock regions in tropical and subtropical seas. Despite a wide distribution the family contains few genera. Many are nocturnal (hence the large eyes) and hide in suitable holes during the day. All species make noises, especially when courting.

occurs so often in them. As the real creature grows to over 9m (30ft) long the legends are well founded.

The Oar fish, the single species in the family Regalecidae, is distributed worldwide and probably lives in moderately deep water when healthy (those at the surface are usually moribund and those cast up definitely dead). Almost nothing is known of its biology. A few caught with a full stomach reveal a diet of small crustaceans. Largely on theoretical grounds, it has been argued (with good reason) that in life it swims at an angle of 45 degrees with the long red mane streaming out horizontally. The blades of the pelvic fins bear a large number of chemoreceptor cells. The pelvic fins are now regarded as chemical probes that are held out in front of the animal so that the fish can detect its prey before it reaches it, and organize its respiratory cycle for maximum efficiency in sucking in the small crustaceans. Locomotion is probably by undulations of the dorsal fin which would produce a haughty and dignified passage through the water.

With regard to body shape, there are two distinct groups of Lampridiformes. The Lophotidae, Stylophoridae, Regalecidae and Trachipteridae have ribbon-shaped bodies; propulsion is effected largely by the dorsal fin and the skin has prickles or tubercles whose function is to reduce water drag. The other group contains the family Veliferidae and the opahs or moon-fish (Lampridae). The body is deep and propulsion is by enlarged, wing-like pectoral fins powered by powerful red muscles attached to an enlarged shoulder girdle.

The coloration of the opahs (family Lampridae) is extraordinary. Their back is azure, which merges into silver on the belly. The sides have white spots and the whole is overlain with a salmon-pink iridescence that fades quickly after death. The fins are bright scarlet. Despite being toothless and of a seemingly cumbersome shape, they feed on midwater fish and squids, which is a testimony to the swimming efficiency of the wing-like pectoral fins. There is one worldwide species and it can grow to over 1.5m (5ft) in length and over 90kg (200lb) in weight.

The ribbon-like lophotids are called crest fish because the bones of the top of the head project forward over the eyes and in front of the mouth like a dorsal keel. The expansion of the bones is used as a base to attach some of the muscles that work the locomotory dorsal fin. Apart from their surrealistic shape they also possess an ink sac that lies

the common name is a subject of much controversy. Some argue that it derives from the French *jaune d'orée* (with a yellow edge) in allusion to the yellowish color of the body. In some parts its common name is the vernacular for St Peter Fish. This name alludes to the single dark blotch on each side which is romantically thought to represent the thumb and forefinger prints of St Peter, who allegedly took the tribute money from its mouth. The same honor has, however, been accorded to the haddock in other countries.

Lampridiform fish (order Lampridiformes) are scarce, spectacular in color, weirdly shaped and cause great excitement when washed ashore. Some species have found a place in folklore as harbingers of abundance or paucity of food fish. The Oar fish is also known as King of the Herrings by northern Europeans while the Pacific Northwest Indians call the related *Trachipterus trachypterus* the King of the Salmon.

The Oar fish is a silvery, ribbon-like fish, with a crimson or scarlet dorsal fin the length of its body. The head profile resembles that of a horse. The anterior rays of the dorsal fin are elongated like a mane and the scarlet pelvic fins are very long with blade-like expansions at the end (hence the common name). The caudal fin is reduced to a few streamers. This most spectacular animal is thought to have been the origin of many sea-serpent stories as a red mane theme

close to the intestine and discharges via the cloaca. Very few undamaged specimens have been found, but they grow to over 1.2m (4ft) long. Their distribution is uncertain but they have been found off Japan and South Africa.

The deal fishes (family Trachipteridae) have the silvery body and fins of their relatives as a color pattern. Some add dark spots and bars to this pattern. As with many of the long-bodied lampridiformes the swim bladder is reduced and the skeleton correspondingly lightened to achieve neutral buoyancy. When adult, the caudal fin consists of a few rays of the upper lobe, elongated and turned upwards at right angles to the body. The life-style of the deal fishes is thought to be like that of other members of the order.

The **swamp eels** (order Synbranchiformes) are freshwater fishes from South and Central America, West Africa, southern Asia and Australia. They are given their common name on account of their shape, and because they live in poorly oxygenated waters. Instead of having a gill opening on each side of the head there is a single, common slit on the underside. In some species the gill chamber is divided internally into two chambers by a tissue dividing-wall (septum). Often the gill chamber is distensible and, by filling it with water, the fish can "breathe" while it travels overland. Species in stagnant water absorb atmospheric air either through a modified section of the gut, well provided with blood vessels, or via lung-like chambers extending from the branchial cavity. In both cases the air is taken in through the mouth.

At least one species, the widespread South American *Synbranchus marmoratus*, can burrow in the mud and aestivate to avoid droughts, much like the lungfish. Pectoral and pelvic fins are absent, and the dorsal and anal fins are reduced to mere ridges of skin without fin rays.

The Rice fish (*Monopterus albus*) from Asia has colonized the irrigation ditches in paddy fields. They grow to 90cm (3ft) long and are an important source of food, not least because they can stay alive, hence fresh, for a long time if just kept moist. The male makes a bubble nest into which the female lays the eggs. They and the newly hatched young are guarded by their father.

Most swamp eels are nocturnal, hiding away in mud during the day and coming out to feed at night. There are two trends in swamp eel sexuality. In some the sexes are fixed from birth, whereas in other species the young are all female on hatching and

only later do some change into functional males. At least three species, *Monopterus boueti* from West Africa, *Ophisternon infernale* from Yardee Creek Wells, NW Australia and *Ophisternon candidum* from the Yucatan, Mexico, live only in subterranean waters.

The relationships of this order are uncertain, but recently a case has been made for considering them to be the sister group of the family Mastacembelidae (spiny eels), usually regarded as perciform fishes.

Mail-cheeked fishes or Scorpaeniformes are an order of predominantly shallow-water, marine fish. Usually reddish in color, they have a general appearance similar to that of the Perciformes (ie spiny first dorsal

▲ **Extraordinary camouflage** is a feature of some species in the order Scorpaeniformes. This is a species of the genus *Scorpaeonopsis*, photographed off Florida, USA.

▶ **A reclusive fish.** The Miller's thumb is a freshwater member of the largely marine family Cottidae of the order Scorpaeniformes. It is a retiring species, hiding away under stones in clean streams and rivers in Europe. Other species of the genus occur in North America. All feed on bottom-living invertebrates, small fishes and fish eggs.

fin, ctenoid scales, spines on the head and pelvic fins well forward). The order contains over 20 families including the gurnards, the extremely poisonous stone fishes of Australasia and the armored sea robins and pogges.

Species of the family Cottidae are common in fresh water. The large head of *Cottus gobio* has given it the name bullhead in the United Kingdom. In North America they are called sculpins. A closely related family, the Cotto-comephoridae, is endemic to the deepest lake in the world, Baikal in Siberia. Here, because the lake is capable of supporting life down to the bottom (in other deep lakes the lower waters are deoxygenated), a whole

series of "deep-sea" freshwater fishes has evolved. Although luminosity has not been developed, many of the cottocomephorid characteristics parallel those of deep-sea fishes. *Cottocomephorus comephoroides* is a species of the open sea living down to depths of 1,000m (3,300ft) in summer. In winter they live closer to the bottom, but usually in shallower water.

The **flying gurnards** (order Dactylopteri-formes) are not gurnards and probably do not fly. They are heavily built, bottom-living fish with a heavy, bony skull. The pectoral fins are greatly expanded into colorful fans and may well serve to frighten away potential predators.

The Red Breast of the Three-spined Stickleback

It is probably for its breeding behavior that the Three-spined stickleback is best known. The male, in his breeding dress of red breast and bright blue body, is well known to children as a prized tiddler to be caught in spring. The male builds a roughly spherical nest from strands of water plants stuck together by secretions from his kidneys. The choice of site varies and, oddly, there is some evidence that males with more bony scutes prefer a sandy locality while those with fewer plates a muddier one. The males' bright colors serve both to advertise the nest site to the female and to warn other males to keep away. An attracted female is courted, shown the nest, and, if she approves, lays her eggs there. More than one female may be induced to lay. The

fertilized eggs are guarded by the male who fans them and removes diseased eggs. During the parental phase the male changes to an inconspicuous dark livery. As the eggs hatch, the male progressively destroys the nest and, a few days after hatching, the young are left on their own.   KEB

Another order with large pectoral fins is the **sea moths** (Pegasiformes). They are small, tropical marine fishes with a body encased in bony scutes and variously developed snouts. There are probably only six species. Because of their curious shape, and the fact that the shape is maintained when the fish is gutted and dried, they are often sold as curios in the Far East. Apart from the fact that they are egg-layers and eat very small bottom-living invertebrates, little is known about their biology.

Some of the problems with classification were mentioned in the introduction. Here we have a practical example. The families included in the group of **sticklebacks and pipefishes** (including seahorses, shrimp fish, flute mouths and some without common names), are now known not to constitute a natural assemblage. Despite superficial similarities, ie bony scutes, isolated spines before the dorsal fin, and some elongation of jaws, which led to the association, there are three quite distinct units. Therefore, ere too long, this order will cease to exist in its present form.

The "humble" stickleback is found in most fresh, brackish and sometimes coastal waters throughout Eurasia and northern America. It is highly variable in form. Although mostly "three-spined" two- and four-spined individuals occur. There are populations in Canada that never have pelvic fins. Freshwater forms usually have fewer bony scutes than brackish or marine forms. Generally, sticklebacks never grow to more than 7.5cm (3in) long but in lakes in the Queen Charlotte Islands, off the coast of British Columbia, there are dark pigmented forms that grow to 20cm (8in) long. Both the abundance and variability of this fish have led to extensive studies of its behavior

as well as to various distinctive forms having formerly been accorded status as individual species.

The Fifteen-spined stickleback is a solitary species most of the year but it also makes a nest of seaweeds. The Nine-spined stickleback (which usually has 10 spines) has at most only a few small scutes. It is nearly as widespread as the Three-spined but rarely enters brackish or saline waters. Two genera are confined to northern America: the Brook stickleback (usually with 5 spines) which is fairly widespread and may even be found in coastal waters of reduced salinity in the north, and the Four-spined stickleback which only occurs in the northeastern part of northern America.

Despite the presence of protective spines on the back and in the pelvic fins, sticklebacks form an important part of the food chain and are eaten by larger fish, birds and otters. Even so, there was a report of cleaning symbiosis between a three-spined stickleback and a pike.

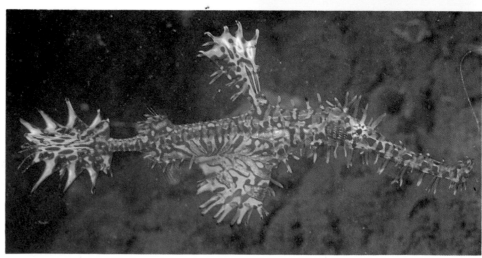

nently in fresh water. Important unifying characters of these fish are long snouts with a small terminal mouth, the elongation of the first few vertebrae (in the shrimp fish, the first six vertebrae form over three-quarters of the length of the vertebral column) and the peculiar structure of the first dorsal fin which, when present, consists not of fin rays but of prolonged processes associated with the vertebrae.

The shrimp fishes, have an extremely compressed body entirely enclosed in thin bony sheets. Only the downturned posterior part of the body is free, to allow tail-fin movement for locomotion. They live in shallow warm seas, sometimes among sea-urchin spines where they shelter for protection. The Deep-bodied snipe fish lives in deeper water and is covered with prickly denticles and a row of scutes on the chest. Apart from the lack of parental care, little is known of their reproductive behavior.

The pipefishes, however, show a remarkable series of reproductive adaptations. The simplest strategy, in the subfamily of nerophiine pipefishes, is for the eggs to be loosely attached to the abdomen of the male. A more elaborate condition is present in some syngnathines where the eggs are individually embedded in spongy tissue covering the male's ventral plates. Further protection in other groups is provided by the development of lateral plates partially enclosing the eggs. In all cases it is the male that carries the eggs.

The seahorses, which are merely pipefishes with the head at right angles to the body, a prehensile tail and a dorsal fin adapted for locomotion, exhibit the ultimate in egg protection. The trend seen in the development of protective plates is continued until a full pouch (or marsupium) is formed, with a single postanal opening. The female has an ovipositor by which the eggs are placed in the male's pouch until the pouch is full. Apparently this simple act is not always done without mishap.

Hatching time varies with temperature, the young leaving the pouch between 4 and 6 weeks after the eggs were deposited. In some larger species, the male helps the young to escape by rubbing his abdomen against a rock, in others there are vigorous muscular spasms which may expel the young with considerable velocity. After "birth" the male flushes out his brood sac by expansion and contraction to expel egg remains and general debris to prepare for the next breeding season. This may occur relatively soon as three broods a year are not unknown.   KEB

▲ **Hanging around,** a school of shrimp fish (*Aeoliscus strigatus*), members of the order of sticklebacks and pipefishes.

◄ **Highly variable coloration** ABOVE is a feature of *Hippocampus kuda*, a seahorse widespread in Indo-Pacific coastal regions. Although often offered for sale in aquarium shops they are hard to keep in captivity as it is difficult to provide their diet of very small live fish and invertebrates.

◄ **In the warm southern oceans** lives *Solenostomus paradoxus*, a relative of the seahorses and pipe fishes. It is about 8cm (3in) long and cryptic. The female (seen here) has cape-like pelvic fins which form a pouch. Within this the eggs are attached to short filaments. When they hatch the young become independent immediately.

The tube snouts are primitive relations of the stickleback found in the cooler waters either side of the North Pacific. The American species *Aulorhynchus flavidus*, which resembles the European Fifteen-spined stickleback, lives in huge shoals. No nest is built but the female lays sticky eggs along the stipe of the giant kelp (large seaweed) which is first bent over and then glued down. The male defends the eggs. Its Japanese relative *Aulichthys japonicus* is poorly known but is reputed to lay its eggs inside a seasquirt (*Cynthia*).

The seahorses, and their relatives the pipefishes, flute mouths, shrimp fish and snipe fish, are an almost entirely marine group; only a few pipefishes live perma-

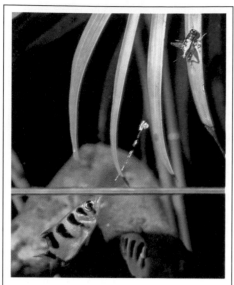

Shooting Down Insects

Some members of the perciforms have evolved elaborate techniques for catching prey. Six species of archer fish are popular with aquarists; they occur naturally in both fresh and salt water from India and Malaysia to northern Australia. The name archer fish is rather inappropriate as it really *spits* droplets of water to catch its prey. The archer fish has a groove in the roof of the mouth. The tongue is thin and free at the front but thick and muscular with a mid-line fleshy protuberance at the back. When the tongue and fleshy protuberance are pressed against the roof of the mouth, it converts the groove into a narrow tube. The thin, free end of the tongue acts as a valve. When the archer fish spots a "juicy" insect, the tongue is pressed against the roof of the mouth, the gill covers are jerked shut and the tip of the tongue is flicked shooting out the drops of water. Archer fish are able to compensate for refraction of light by placing the body vertically below the prey. A fully grown fish can shoot down an insect from up to 1.5m (5ft) whereas babies can only shoot a few centimeters without loss of accuracy. Results of experiments with adult archer fish suggest that their aim may be quite haphazard and the "shooting down" of an insect is due to massed firepower rather than sharp shooting. It seems that when the archer fish is only a few centimeters from the insect it will jump out of the water and snatch the insect with its jaws rather than shoot it down. Archer fish often frequent murky water where they are found cruising just below the surface.   BB

▲ **On target,** *Toxotes jaculator*, one of the species of archer fish, from Java.

▶ **Free food for the taking**—a grouper feeding on a shoal of small fish.

No other order of fishes approaches that of the **perch-like fishes** in number of species and variety of form, structure and ecology. The perciformes comprise the largest vertebrate order, containing over 150 families embracing about 8,000 species. Whether the perch-like fishes form a natural (monophyletic) assemblage is debatable. At present the perciforms are ill-defined and lack a single specialized character or combination of characters to define the group.

The most "typical" members of the perciforms are the species in the perch family. The perch body is typically deep and slender; the two dorsal fins are separate; the pelvics are near the "throat" and the operculum ends in a sharp, spine-like point. They are adapted to northern hemisphere temperatures; warm winters retard the maturing of sperm and eggs.

The European perch is a sedentary species, preferring lakes, canals and slow-flowing rivers. The ruffe or pope of Europe and southern England is a bottom-feeding species, frequenting canals, lakes and the lower reaches of rivers. Confusingly it has contiguous dorsal fins.

The zander is a native of eastern Europe that has been introduced to Britain on numerous occasions from the 19th century onwards. Its British distribution is patchy, notably the River Ouse and other slow-flowing Cambridgeshire rivers and some Bedfordshire lakes, but it is apparently extending its range. The North American zander (or walleye) occurs naturally in wide, shallow rivers and lakes.

The North American darters are the most speciose percids with nearly 100 species. The common name is derived from their habit of darting between stones, as they are bottom-dwelling fishes that lack swim bladders. While many species of darter are brightly colored, the Eastern sand-darter is an inconspicuous translucent species, usually found buried in sandy stream beds with only the eyes and snout protruding.

Like that of all other teleosts, the percid skeleton is basically bone, with some cartilage, though the skeletons of the perciform families of louvars and ragfishes are largely cartilaginous. The louvar, the only species in its family, is probably related to the mackerels and tunas. It lives in tropical seas, grows to about 1.8m (6ft) and has a tapering, pinkish colored body. Its pectoral fins are sickle shaped; the pelvics minute; its dorsal and anal fins are long, low and set far back on the body. The louvar feeds on jellyfish; consequently its intestine is very long, with numerous internal projections which increase the absorbent surface area of the gut. The significance of the cartilaginous skeleton remains a mystery.

The two species of ragfishes are so called because they appear to be boneless and look like a bundle of rags that has been dropped on the floor. Little is known of their biology. The Fantail ragfish grows to 45cm (18in) and has a deep, laterally compressed scaleless body with small spines on the lateral line and fins. The Brown ragfish grows to some 2.1m (7ft). Its body is elliptical and lacks scales, spines and pelvic fins. It is apparently eaten by Sperm whales.

The large spotted groupers are voracious predators. Sometimes they are found with black, irregular lumps, either lying in the body cavity or bound by tissue to the viscera: these are mummified sharp-tailed eels. Each eel is swallowed by the grouper and in its death throes punctures the gut; it gets squeezed into the body cavity where it becomes mummified.

The Queensland grouper, a native of Australian seas, may weigh up to half a tonne and is another sea bass with a hearty appetite! This fish has been known to stalk pearl and shell divers, much as a cat stalks a mouse, a habit that has led to stories of divers being swallowed by groupers.

The barracudas are another group of perciforms reported to attack divers. They are tropical marine fish which in some areas, especially the West Indies, are more feared than sharks. The body is elongate and powerful; the jaws are armed with sharp, dagger-like teeth. Barracudas eat other fish and seemingly herd shoals, making the food easier to catch. Large individuals tend to be solitary, but younger barracudas aggregate in shoals. The barracuda makes very good eating but is notorious for being sporadically poisonous due to the accumulation of toxins acquired from the fish on which it feeds.

The mackerels, tunas and bonitos or scombrids are also delicious perciforms. The scombrids are mainly schooling fishes of the open seas, cruising at speeds of up to 48km/h (30mph). Their bodies are highly streamlined, terminating in a large lunate caudal fin. Some scombrids have slots on the dorsal surface of the body in which the spiny dorsal fin fits to reduce water resistance. Behind the dorsal and anal fins are a series of finlets, the number of which varies according to the species. In all species the scales are either very reduced or absent.

The Common mackerel is found on both sides of the North Atlantic. On the European side it ranges from the Mediterranean to Ireland. It is a pelagic fish which in summer

The Mystery of the Swordfish

The swordfish is the only species in its family. It is also a solitary fish, and may weigh up to 675kg (1,500lb). The snout is produced into a powerful, flattened sword. Swordfish live in all tropical oceans but will enter temperate waters, occasionally straying as far north as Iceland.

The sword has a coat of small denticles similar to those found on sharks. Its function is unknown but suggestions include its use as a weapon (ie the swordfish strikes a shoal of fishes with lateral movements afterwards devouring the mutilated victims) and as extreme streamlining with the snout acting as a cutwater.

There are numerous accounts of large fish attacking boats, but often there is no attempt to discriminate between swordfish, spearfish and sailfish, all of which have similar habits. There is no doubt that a swordfish could pierce the bottom of a boat and have the sword snap off in its struggles to withdraw it. In the British Museum (Natural History) there is a sample of timber which a swordfish snout has penetrated to a depth of 56cm (22in). It is also reported that the ship HMS *Dreadnought* sprang a leak on voyage from Ceylon (Sri Lanka) to London. Examination of the hull revealed a 2.5cm (1in) hole punched through the copper sheathing which was reputed to have been made by a swordfish. Periodically swords are found in whale blubber. Whether these attacks on ships and whales are deliberate remains unresolved. The most likely explanation is that when a swordfish, which can travel at speeds up to 100km/h (60mph), encounters a boat or whale it finds it impossible to divert in time and a collision becomes inevitable.   BB

---

forms enormous shoals at the water's surface near coasts to feed on small crustaceans and other plankton. In winter the shoals disband and move to deeper water, where the fish approach a state of hibernation.

The skipjack tuna is a cosmopolitan marine species that owes its name to its habit of "skipping" over the surface of the water in pursuit of smaller fish.

The swordfish and the family of sailfishes, spearfishes and marlins (istiophorids) are all fast-swimming fishes closely related to the scombrids (see box). The istiophorids include some of the world's most popular marine sport fishes and are often referred to as billfishes, a term also used, confusingly, for the swordfish. In cross-section the bills of these fish are rounded and they have two ridges on each side of the caudal peduncle, as opposed to just one in the swordfish. Several billfish species are known to be migratory, possibly to follow food. Billfishes are fish-eaters; they erect the dorsal fin to prevent prey escaping. They can also use the bill as a club to maim their victims as they rush through a school of preferred fishes, such as the Frigate mackerel.

Marlins lack the pelvic fins present in the sailfish and spearfish. Their classification is unresolved. The only easily identifiable species and probably the largest is the Indo-Pacific Black marlin whose pectoral fins are permanently and rigidly extended.

Sailfishes undergo a remarkable change during their larval development. Larvae of about 9mm (0.3in) have both jaws equally produced, armed with conical teeth; the edge of the head above the eye has a series of short bristles; there are two long pointed spines at the back of the head; the dorsal fin is a long, low fringe and the pelvic fins are represented by short buds. At 6cm (2.4in) it begins to resemble the adult: the upper jaw elongates, the teeth disappear, the dorsal fin differentiates into two fins, the spines at the back of the head become reduced and the bristles disappear. Young swordfish also undergo a similar series of changes.

The scats are found in Southeast Asia, Australia and eastern Africa. At about 1.7cm (0.75in) they start to resemble the adult; before this they are very different. The larval phase is known as the "Tholichthys" stage because the famous ichthyologist Albert Günther failed to recognize the young and described them as a separate genus, *Tholichthys*.

Scats are primarily marine but range into brackish or fresh water, adjusting very rapidly to waters of different salt levels. It is still unknown how scats can regulate their body salts so rapidly, but however they manage it, they can cross sea water concentration gradients that are lethal to over 90 percent of other teleosts.

Closely related to and having a similar larval stage as the scats are the butterfly fishes. Butterfly fishes are exquisitely colored

How Tuna Fish Keep Warm

Tunas differ from other scombrids and all other teleosts in their ability to retain metabolic heat *via* a countercurrent heat-exchange system which operates in the muscles and gills. Red muscle occurs in large proportions in tunas. It is extremely vascular, so the muscle cells are supplied with oxygen- and carbohydrate-enriched blood, enabling them to utilize highly efficient, aerobic metabolism. Aerobic metabolism liberates energy to drive the muscle and as heat, which is retained in the body by the countercurrent heat-exchange system. White muscle is found in large proportions in all other fish. White muscle has a very poor blood supply and carbohydrate is metabolized anaerobically, which liberates just enough energy to drive the muscle.

Fishes normally lose heat, through their gills during respiration but the tunas' countercurrent heat-exchange system ensures that the metabolic heat is returned to the body. The advantage to the tunas is twofold, the muscles operate at a higher temperature, helping the fish to achieve high speeds and allowing it to range further north.   BB

▲ **Where impairment is best sustained** TOP.
Butterfly fishes have a dark stripe to disguise
their eyes and a false eye on the dorsal fin to
minimize the effects of predation. They live
around coral reefs. This is the Filament
butterfly fish (*Chaetodon auriga*).

▲ **Unexpected attractiveness.** Many of the
wrasses (family Labridae) are both large and
brightly colored. In some species there are
marked differences in color and pattern
between the sexes. They live in shallow areas
of the oceans worldwide except for colder
regions. This is a Purple queen (*Mirolabrichthys
tuka*).

and move with a flitting motion. They are
distributed worldwide in warm waters
around coral reefs. Despite their bright
coloration, the pattern camouflages the eye
and it is difficult for a potential predator to
distinguish the head from the tail. Many species have a dark vertical bar that runs
through the eye, further disguising it, and
to add to the confusion many species also
have an eye spot near the caudal fin. Butterfly fishes delude would-be predators by
swimming slowly backwards. Once the
predator lunges at the eye spot, the butterfly
fish darts forwards, leaving its attacker
rather confused. The Indo-Pacific butterfly
fish known as the forceps fish is so called
because its snout is extremely long and used
like a pair of forceps to reach deep into reef
crevices.

Included by some workers in the butterfly
fishes family are angelfishes, distinguished
from butterfly fishes by their larger, rather
rectangular bodies and heavy spines at the
base of the gill cover. The Blue angelfish has
a dark blue ground color broken by a series
of narrow, curved white stripes with a
similarly marked caudal fin. Some years ago
a specimen appeared in a Zanzibar market
with the caudal fin lines broken, resembling
Arabic characters reputed to read *Laillaha,
Illalahah* ("There is no God but Allah") and
the other side *Shani-Allah* ("A warning sent
from Allah"). Blue angelfish are normally
sold for a few cents; this fish eventually
fetched some 5,000 rupees.

The surgeonfishes are another family of
reef-dwelling fishes which, during development, have a larval or "acronurus" stage
where the larvae are totally dissimilar from
the adult. There are almost 1,000 species of
oval-bodied surgeonfishes. The name alludes to a razor-sharp, lancet-like spine on

either side of the caudal peduncle. In most
species the spines lie in a groove and are
erected when the fish is disturbed or excited.
The spines are a formidable weapon, inflict-
ing slash wounds into the victim, as the
surgeonfish lashes its tail from side to side.
The unicorn fish has a large lump on its
head and has two immovable spines on
either side of the caudal peduncle.

Surgeonfishes do not school but travel in
small groups using their small incisor-like
teeth to scrape plants and animals from reefs
and rocks. Frequently surgeonfishes show
variation in color—the Yellow surgeon has
yellow and gray-brown color phases. The
Common surgeonfish, found in the Atlantic,
changes from blue-gray to white anteriorly
and dark posteriorly when chasing another
member of the species in territorial disputes.

In the Indo-Pacific surgeonfishes are a
tasty food fish but the offending caudal peduncle is cut off prior to sale.

Members of the family of stargazers are
widely distributed in warm seas and earn
their name from their eyes which are set on
top of the head so that they appear to be star-
ing at the sky. Stargazers have electric
organs situated just behind the eyes which
deliver a shock sufficient to stun small fish—
which are then eaten. The European
stargazer is common in the Mediterranean;
it grows up to 30cm (1ft) and has flaps of
tissue in the mouth that resemble worms,
tempting potential prey to approach. Predators are deterred from eating the stargazer
by grooved spines situated above each pectoral fin. At the base of the spine is a poison
gland; as the spine inflicts a wound, poison
is trickled into it via the groove. Stargazers
are usually found buried in the sand, with
only the eyes and snout tip protruding.

Some perciforms sometimes form unusual

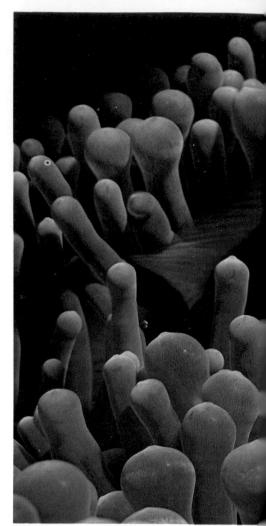

relationships with other vertebrates and invertebrates and even with floating objects! The remoras or shark-suckers are slim fishes usually associated with sharks, large fishes and occasionally turtles. Their dorsal fin is modified into a sucking disk, the rim of which is raised, and the platelike fin rays can be adjusted to create a strong vacuum between the disk and a remora's chosen partner. It is unknown what benefit such an association is to either the remoras or the animals to which they attach. It has been suggested that remoras are simply "hitching a ride," a phenomenon known as phorecy, or that they associate with sharks to feed on the scraps they can snatch from the shark's meal. Remoras have been observed entering the mouths of manta rays, large sharks and billfishes, hence it has been mooted that they may fulfill a role like that of cleaner fish. However, there is no documented evidence of remoras undertaking cleaning duties. Occasionally, fairly large remoras (*Echeneis naucrates*) have been found in the stomachs of Sand sharks (*Odontaspis taurus*).

Despite their usual attachment to sharks, remoras are competent swimmers and often leave the "host" to forage. When free-swimming in a group, remoras arrange themselves with the largest on top, smallest at the bottom, reminiscent of a stack of plates. The group swim in a circular fashion; it seems remoras do not like to swim unaccompanied.

Ancient legend has it that remoras can impede the progress of sailing vessels, even stop them. The remora is also reported to have magic powers and a potion including one was supposed to delay legal proceedings, arrest aging in women and slow down the course of love!

Pilot fishes in the family Carangidae also associate with sharks and rays. It was thought that the pilot-fishes guided the sharks to their prey and in return received protection from their enemies by their proximity to such a formidable companion. Sharks and rays are in fact seeking food, and although the pilot-fish gain from the hunting efforts of sharks they never lead the foray. Pilot-fish never stray too close to the shark's mouth but they have been seen to take refuge in the mouths of rays that do not eat fish.

The young of another carangid, the common Horse mackerel, shelter in the bell of the sombrero jellyfish (*Cotylrhiza*). Why these small fishes do not get stung is unexplained but possibly lack of glutathione (an amino acid that stimulates release of sting

cells) in their mucus coats protects them.

Members of two other families also associate with jellyfish. Young butterfishes are laterally compressed fish which lack pelvic fins and shelter under the protection of the Portuguese man-of-war (genus *Physalia*). The closely related Nomeidae are known as "man-o'-war" fishes and are distinguished from the preceding family by the presence of pelvic fins. Again, it is unknown how these fish gain immunity from the stinging cells of this jellyfish.

Among the family Gobiidae, fishes in the genus *Evermannichthys* habitually live inside sponges. The bodies of these little fishes are slender and nearly cylindrical, allowing them easy access to the larger orifices on the sponge's surface. Scales are either absent or poorly developed, but along the lower posterior line of the sides are two series of large, well-separated scales whose edges are produced into long spines. A further series of four spined scales is situated in the middle line, behind the anal fin. It is thought that these structures are used by the fish for climbing up the inner surfaces of the sponge cavities.

Another Indo-Pacific genus of goby, *Smilogobius*, is commensal with a snapping shrimp. The goby is usually found at the

▲ **Protection arranged.** The clown fishes (genus *Amphiprion*) live, with immunity, among the stinging tentacles of sea anemones. They may be protected by their mucus lacking the protein that normally triggers the anemones' stinging cells into action.

◄ **Protection afforded.** The brightly colored cleaner fishes are wrasses that feed on the ectoparasites of the teeth, mouth and gills of larger fish.

► **Protection taken.** Blennies (family Bleniidae) are found worldwide, mostly in shallow seas. Here a blenny takes refuge in a brain coral.

▼ **Protection desired,** remoras following a ray.

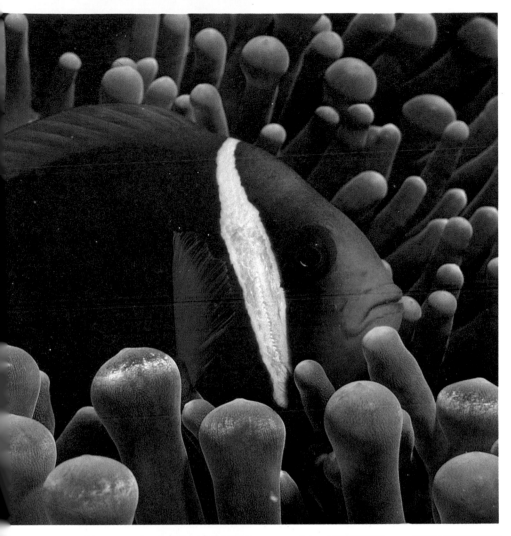

burrow entrance while the snapping shrimp busily excavates it. When danger threatens, the goby dives into the burrow; this also alerts the shrimp which follows the fish inside. The snapping shrimp will not emerge until the goby is once again on sentry duty at the burrow entrance.

The five species of the family of clownfishes are small, brightly colored fishes of warm shallow seas where they live in association with large sea anemones. The relationship is intimate: the clownfish remain inside the anemone when it withdraws its tentacles. The clownfish benefit by protection from predators so they never stray far from their anemones, but the latter can happily exist without the clownfish. Probably the clownfish feed on particles stolen from the tentacles of anemones. It is unknown whether the anemones' sting-cells are lethal to nonsymbiont fishes. If they are then the clownfishes' immunity to their sting is also unexplained. Current views are that the toxin of the sting-cells is mild and the clownfishes' mucus coat is sufficient protection, or that the mucus coat lacks glutathione which triggers the sting-cells.

The Cardinal fish lives among the spines of certain long-spined sea urchins. The urchins do not benefit from the relationship, but the Cardinal fish gains protection from its partner's spines. Another cardinal fish (*Astropogon stellatus*) is only 5cm (2in) long and common in the Caribbean and lower Florida coasts. This fish is found in either the mantle cavity of giant conch shells or the inner cavities of sponges. The host gains little benefit from the association but the cardinal fish gains protection from predators.

A centrolophid rudder fish (*Schedophilus*) has formed the curious habit of accompanying logs, planks, barrels, broken boxes and other flotsam, which has earned it the title "wreck fish." The rudder fish gains shelter and feeds on barnacles and other invertebrates which live on the flotsam.

A number of species of wrasse (*Labroides*) have an unusual "cleaning" relationship with other fishes: they remove ectoparasites and clean wounds or debris. There are some 600 species in the wrasse family, usually nonschooling, brilliantly colored and found on reefs in all tropical and temperate marine waters. The wrasses have well-developed "incisor" teeth which protrude like a pair of forceps from a protractile mouth and which in noncleaning species are used for removing fins and eyes of other fishes.

The cleaner fishes are small, brightly-colored reef dwellers that occupy a specific area, the "cleaner station." Their diet is mostly parasitic organisms on the bodies and gills of fishes. The association between cleaner and customer is not permanent. Fishes requiring "cleaning" congregate at the cleaning stations and follow a specific behavior pattern that invites the cleaner to get to work. The customer allows the cleaner to move all over the body, including such sensitive areas as the eyes and mouth, and even to enter the branchial cavity to remove parasites from the gills. The cleaners benefit by immunity from predation during the cleaning and presumably at other times, since many of the customers are predators on fishes the size of these wrasses.

A feature common to many wrasses is their ability to burrow and usually to sleep buried in sand. Nearly 50 species of Hawaiian labrids sleep buried in the sand at night. The exceptional labrid (*Labroides phthirophagous*) secretes a mucous nightshirt around itself.

The parrotfishes family is closely related to the wrasses and some species (eg *Pseudoscarus guacamia*) secrete a mucous nightshirt that surrounds the whole body. This mucous cocoon may take up to half an hour to secrete and as long for the fish to release itself. Interestingly this cocoon is not secreted every night but only under certain conditions, the causal factors of which are a mystery. It seems that the mucous cocoon may be a protective device, preventing odors from the parrotfishes reaching predatory fish, eg moray eels.

Species in the family of grunts are pretty, tropical marine fishes that have earned their name from the noise they produce by grinding together well-developed pharyngeal teeth. Grunts are deep-bodied fishes that usually travel in schools. Several reef-dwelling species indulge in a form of kissing. Two individuals approach with their mouths wide open and touch lips. The behavior is reminiscent of the "kissing gouramis" and is thought to play a role in courtship.

Members of the drums family are rather dull-colored fishes—they also derive their name from the noise they produce. The noise is caused by muscles vibrating the swim bladder, not always attached directly thereto but running from either side of the abdomen to a central tendon situated above the swim bladder. Rapid twitches of the muscles vibrate the swimbladder walls which have a complex structure and act as a resonator to amplify the drumming sound. A swim bladder is absent in the drum genus *Menticirrhus* so this fish produces only a weak noise by grinding its teeth.

Species in the leaffishes family rely on crypsis (pretending to be something else) to catch their food. They are found in tropical fresh waters in Africa, Southeast Asia and South America. The Southeast Asian leaf-fishes are very perchlike—none mimics leaves and the body is only slightly compressed. The most common leaffish of this area is *Badis badis* (sometimes placed in a family of its own, Badidae) which has a large number of different color forms and is found in streams of India and Indochina. The most spectacular leaffishes are those found in South America. These are deep-bodied fish with soft dorsal fin rays: they closely resemble floating leaves in both contours and marks—even a "leaf stalk" is present, emanating from the lower jaw. Leaffish usually hide beneath rocks or in crevices, where they look like a dead leaf that has become wedged, then dart out to capture prey. The most famous leaffish (*Monocirrhus polyacanthus*) is found in the Amazon and Rio Negro basins of South America. Its body is leaf-shaped and tapered towards the snout, with an anterior barbel mimicking a leaf stalk. The fish reaches about 10cm (4in) in length and is a mottled brown similar to dead leaves. It drifts with the current and on approaching a potential meal the leaffish bursts into action and assisted by its large protrusile mouth engulfs fishes up to half its size.

A perciform family with one of the largest numbers of species is that of the cichlids. They are also very popular aquarium fishes.

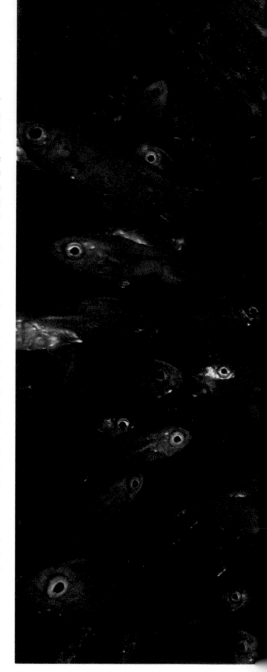

▲ **An assembly of cardinals.** There are many species of cardinal fishes—small, brightly colored shoaling fish of tropical and temperate seas. Most species are reef dwellers in the Indo-Pacific Ocean.

▶ **Extractive industry.** Parrot fishes (family Scaridae) have teeth fused into a "beak" with which they bite off lumps of living coral. Large grinding teeth in the throat then chew the coral to extract the food. The limestone residue is excreted.

◀ **A couple of cichlids** involved in premating behavior. Members of the family Cichlidae live in South America, Africa, parts of southwest Asia and India. Most species are territorial and display parental care of the eggs and young. The great rift-valley lakes of Africa have produced cichlids with extremely specialized feeding habits, such as scale-scraping, eye-eating and snaffling young from a parent's mouth. The South American species shown here (the Butterfly cichlid, *Papiliochromis ramirezi*) is a more generalized feeder.

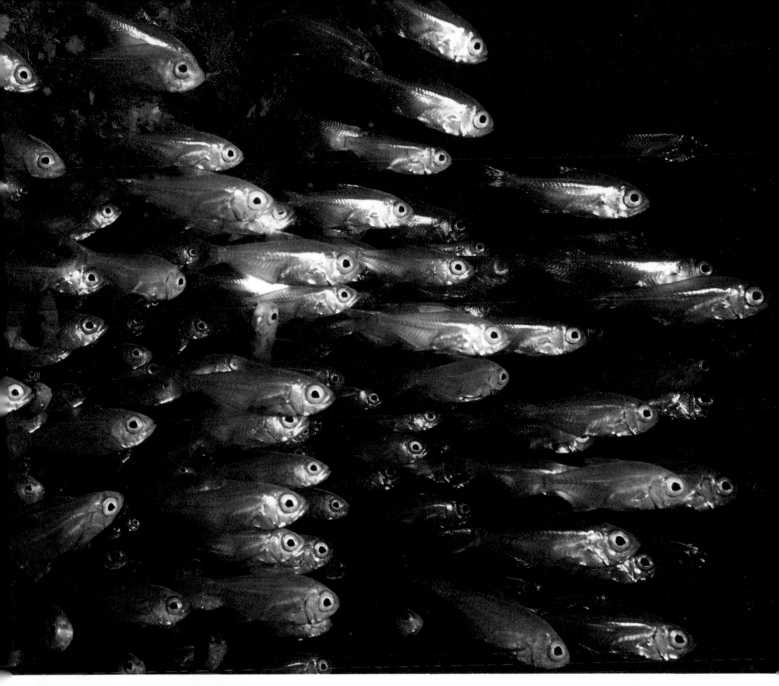

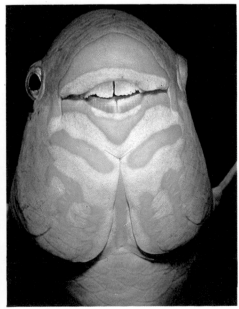

They are characterized by a single nostril on each side of the head and two lateral lines on each side of the body. The pharyngeal bone is triangular, lying on the floor of the "throat"; its function is to break up food against a hard pad at the base of the skull, and it is of diagnostic importance in the identification of species.

Cichlids are widely distributed in Central and South America, Africa, Syria, Madagascar, southern India, Sri Lanka and Iran. There are at least 1,000 species, over half of which are found in Africa, especially the Great Lakes, Victoria, Malawi and Tanganyika, each of which boasts 100 or more endemic species.

Cichlids have evolved all kinds of dentition to cope with their varied diet. Vegetarians have bands of small notched teeth in the jaws, sometimes with an outer chisel-like series for cutting weed or scraping algae off rocks. Fish-eating species have large mouths armed with strong pointed teeth for securing struggling fish. The mollusk-eating varieties have strong, blunt pharyngeal teeth to grind up mollusks, although in some species the lateral jaw teeth are modified, enabling the fish to remove the snail from its shell before swallowing it. Finally, in some species the dentition is greatly reduced and deeply embedded in the gums of a very distensible mouth. These species feed almost entirely on the eggs and young of mouth-brooding cichlids which they force the parent to "cough up." In Lake Barombi, Cameroon, there is a genus of cichlid (*Pungu*) whose diet consists entirely of freshwater sponges.

In the rapids of the lower Zaire river there lives an unusual cichlid, *Lamprologus lethops.* The head of this fish is rather flattened, its body nearly cylindrical and the eyes are

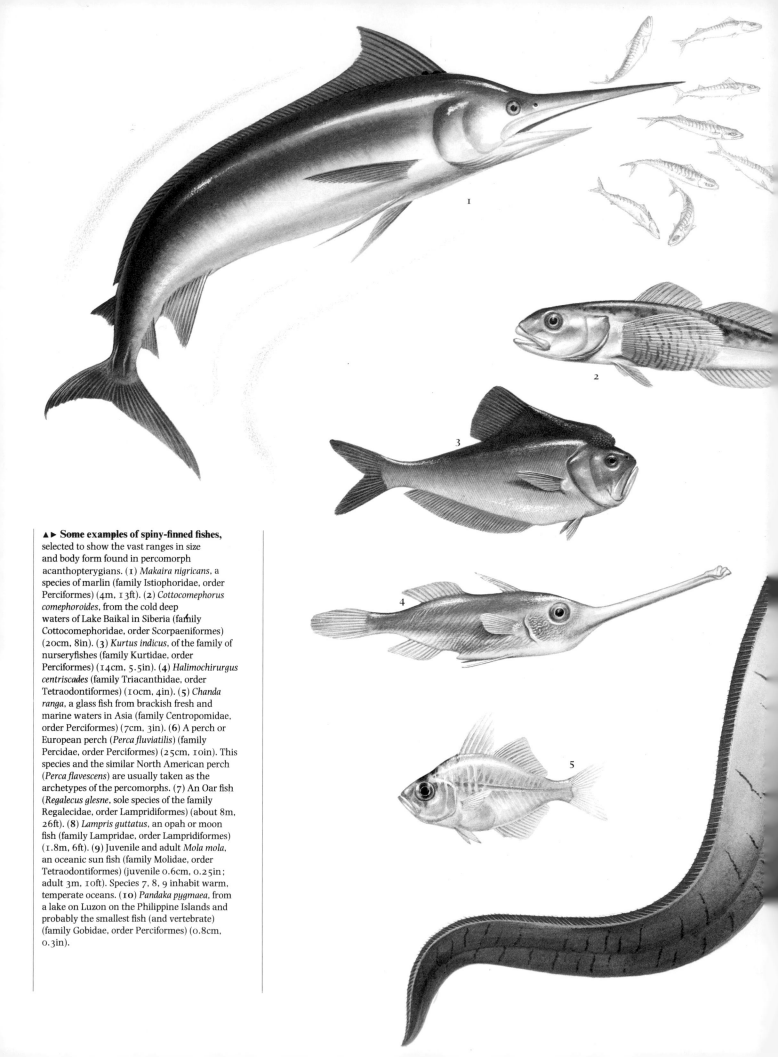

▲ ▶ **Some examples of spiny-finned fishes,** selected to show the vast ranges in size and body form found in percomorph acanthopterygians. (1) *Makaira nigricans*, a species of marlin (family Istiophoridae, order Perciformes) (4m, 13ft). (2) *Cottocomephorus comephoroides*, from the cold deep waters of Lake Baikal in Siberia (family Cottocomephoridae, order Scorpaeniformes) (20cm, 8in). (3) *Kurtus indicus*, of the family of nurseryfishes (family Kurtidae, order Perciformes) (14cm, 5.5in). (4) *Halimochirurgus centriscades* (family Triacanthidae, order Tetraodontiformes) (10cm, 4in). (5) *Chanda ranga*, a glass fish from brackish fresh and marine waters in Asia (family Centropomidae, order Perciformes) (7cm, 3in). (6) A perch or European perch (*Perca fluviatilis*) (family Percidae, order Perciformes) (25cm, 10in). This species and the similar North American perch (*Perca flavescens*) are usually taken as the archetypes of the percomorphs. (7) An Oar fish (*Regalecus glesne*, sole species of the family Regalecidae, order Lampridiformes) (about 8m, 26ft). (8) *Lampris guttatus*, an opah or moon fish (family Lampridae, order Lampridiformes) (1.8m, 6ft). (9) Juvenile and adult *Mola mola*, an oceanic sun fish (family Molidae, order Tetraodontiformes) (juvenile 0.6cm, 0.25in; adult 3m, 10ft). Species 7, 8, 9 inhabit warm, temperate oceans. (10) *Pandaka pygmaea*, from a lake on Luzon on the Philippine Islands and probably the smallest fish (and vertebrate) (family Gobidae, order Perciformes) (0.8cm, 0.3in).

completely covered by skin and tissues of the head. It is further distinguished from other species of *Lamprologus* by its small scales and lack of pigment. The eyes are thought only to be able to perceive light but do not form visual images. In such fast-flowing water eyes would probably be of little benefit and easily damaged and so they are reduced in this "blind" cichlid. The species is currently known from only a few dead specimens, so details of its biology remain a mystery.

Also found in the same Zaïre rapids is *Mastacembelus brichardi* (family Mastacembelidae), a spiny eel. The eyes of this species are minute, deeply embedded in the head and covered by expanded jaw muscles. Its body is more or less depigmented, so is a milky white or pinkish color. Once again little is known of the biology of this species.

"Like a fish out of water" is an ill-quoted expression as a number of fish are quite at home out of the water. Many of the spiny eels are air-breathers and utilize this ability to survive in poorly aerated water or mud. Like many species of spiny eel, *Macrognathus aculeatus* spends the daytime in a mud burrow, excavated by rocking and forward-wriggling movements that submerge it at a constant rate, leaving only the tips of the nostrils protruding.

The mudskippers, of the goby genus *Periophthalmus*, are found in tropical Africa and spend a great part of their time walking or "skipping" about mangrove roots at low tide. During these periods the branchial chamber is filled with water and oxygen exchange continues over the gills. When the oxygen in this water is exhausted the mudskippers replace it with oxygenated water from a nearby puddle. The mudskippers can also respire through the skin (cutaneously) and have a highly vascular mouth and pharynx through which gaseous exchange can take place; they are therefore often seen to sit with their mouths gaping.

Mudskippers are very competent amphibious fishes. Their pectoral fins are highly modified and the membraneous part is carried at the base of a highly muscular "stump" that can be moved backwards, forwards and sideways. Movement is achieved by "crutching." The mudskipper swings the pectorals forwards while the weight of its body is supported by the pelvic fins. Then, by pressing downwards and backwards with the pectorals, the body is both lifted and drawn forwards. At the end of the movement the body falls and rests again on the pelvic fins. Rapid movement is achieved by curling the body to one side and

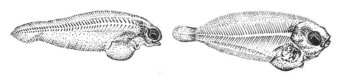
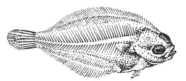
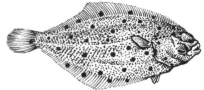

suddenly straightening it, "flipping" itself distances up to 60cm (2ft). Mudskippers can also skitter across the surface of the water in a series of jumps as well as swim normally when submerged.

When the tide returns most of the mudskippers retreat into elaborate burrows with enlarged living chambers in the mud, over which they are highly territorial. Those mudskippers that do not possess burrows, usually juveniles, spend most of their time out of the water, and are obliged to climb surrounding mangrove trees as the incoming tide brings with it numerous other predatory fish.

The Anabantidae and Ophiocephalidae are two more perciform families renowned for their ability to breathe atmospheric air. The anabantids include the climbing perch, the Kissing gourami and the genus of Siamese fighting fish. They are also called labyrinth fishes for their labyrinth-like accessory breathing organ. The accessory breathing organs are housed at the top of each gill chamber; they are hollow and formed from highly vascular skin lining the gill chambers. As the fish grows the organs become more convoluted, increasing the surface area available for respiration. The anabantids rely on atmospheric air for survival and quickly suffocate if denied access to the surface of the water.

The gouramis are tropical freshwater fish found from India to Malaya. The Kissing gourami is found in the Malay Peninsula and Thailand. The name is derived from the broad-lipped mouths by which two individuals approach each other and appear to be kissing. This apparent affectionate gesture is in reality a threat, issued in exercise of territorial rights or mate selection.

The Siamese fighting fish that are so brightly colored in aquaria are an unexciting brownish red color in the wild. The highly colored varieties are obtained by selective breeding, a process by which characters potentially present in wild forms are encouraged to develop in captivity by breeding those individuals which demonstrate the required character most strongly.

The snakehead family is closely related to the anabantids but contains long cylindrical fishes with flattened, rather reptilian-looking heads. These fish inhabit rivers,

ponds and stagnant marsh pools in Southeast Asia. The various species of snakehead differ in the extent to which each has developed the habit of breathing air. The accessory breathing organs are simpler than those found in anabantids, consisting of a pair of cavities lined with a vascular thickened and puckered membrane. These lung-like reservoirs are not derived from the branchial chambers but are pouches of the pharynx.

The snakeheads are also able to move overland but do so by a rowing motion of the pectoral fins. During prolonged drought snakeheads survive by burying themselves in mud and aestivating; in hot dry weather they become torpid.

The examples of perciforms given are a small representation of the amazing diversity of this group as it is currently recognized.

The origin of **flatfish** is obscure; fossils are known from the Eocene epoch (54–38 million years ago) but these are as specialized as the living species. There are about 500 extant species divided into seven families (Psettodidae, Citharidae, Scophthalmidae, Bothidae, Pleuronectidae, Soleidae and Cynoglossidae). The members of the most primitive family, Psettodidae, have rather perch-like pectoral and pelvic fins; only the eyes and long dorsal fin distinguish them from the sea-perch, suggesting that flatfish evolved from perch-like ancestors.

All adult flatfish are bottom-living fish but their eggs, which contain oil droplets, float at or near the sea surface. The larvae take a few days to hatch; the fish appear symmetrical, with an eye on each side of the head and a ventrally situated mouth, further suggestive of their perch-like ancestry. When about a centimeter (0.5in) long a metamorphosis occurs which has profound effects on the symmetry of the skull and the whole fish. The changes are initiated when one eye migrates across the head to lie alongside the other, its passage being assisted by resorption of the cartilaginous bar of skull separating them. The nostril simultaneously migrates to the eyed or colored side. Except for the psettodids, the mouth also twists into the same plane as the eyes. In the soles (genus *Solea*) the mouth twists unusually onto the blind side. The eye that migrates is often characteristic of

▲ ► ▼ **A change of direction.** Flatfish start life as normally shaped fish with an eye on each side and a horizontal mouth. As the larva grows ABOVE one eye migrates to the other side of the head and the mouth twists until the adult comes to lie permanently on one side. The consequent disformation in the adult can be seen in this photograph of a plaice RIGHT. Adult flatfish are adept at adapting the color of their pigmented side to match that of the surrounding seabed—see the flounder BELOW. Although they are normally bottom-dwellers some species, for example the large halibut (genus *Hippoglossus*) which grows to over 2m (6.5ft), can swim actively and catch fish in midwater.

particular families. Members of the family Bothidae are called left-eyed flounders because their right eye usually migrates, so the uppermost, colored side is the left one. Pleuronectidae are right-eyed flounders because ultimately the right side is uppermost. In the psettodids equal numbers are found lying on either side.

While these radical changes are taking place the little fish sinks to the sea bottom. Flatfish do not have a swim bladder, so they remain lying at or near the bottom, on their blind side. The body shape of adult flatfish is quite variable—the European turbot and its relatives are nearly as broad as long, whereas the tongue soles are long and narrow. Frequently, flatfish bury themselves, by flicking sand or by wriggling movements of the body, leaving just their eyes and upper operculum exposed. There is a special channel which connects the gill cavities. Water is pumped from the mouth over both sets of gills, but the expired water from the gills on the buried side is diverted through the channel and expired from the exposed side.

Many flatfish are predominantly brown on the colored side, although they often have spots and blotches of orange thus enabling them to blend with the substrate. The pleuronectids, however, are masters of disguise among fish as they can change their color to match the substrate. When placed on a chequered board some species can reproduce the squares with reasonable accuracy. The American flounders can even produce an effective camouflage on colored backgrounds, although their attempts at reproducing reds are inaccurate.

All flatfish are carnivorous but their methods of catching prey are quite diverse. Members of the family Bothidae are daytime hunters that feed on other fish. They swim actively after their prey and have very acute vision. Species belonging to the families Soleidae and Cynoglossidae hunt at night for mollusks and polychaete worms which they locate by smell. These families of flatfish both have innervated filamentous tubercles instead of scales on the blind side of the head which probably enhance their sense of smell. The pleuronectids are intermediate: some, eg the halibut, actively prey on fish, others, eg plaice, hunt polychaete worms and crustacea, relying on smell and visual acuity to locate their prey.

The majority of flatfishes are marine, but a few species can live in sea or fresh water. The European flounder frequently migrates up rivers to feed and is found up to 65km (40mi) inland in the summer, returning to spawn in the sea in the fall. The American flatfish *Achirus achirus* is a freshwater species, often kept by aquarists. It has a large surface area to weight ratio and can suspend itself by surface tension at the water surface. It can also "stick" itself to rocks or the sides of aquaria by creating a vacuum between the underside of its body and the substrate.

There is no obvious difference between the sexes in most species of flatfish, although in the scald-fish genus the male has some filamentous dorsal and pelvic fin rays. The males in the closely related genus *Bothus* have spines on the snout, their eyes are wider apart than the females' and the upper rays of the colored side pectoral fin are elongate.

Judging by their worldwide distribution the flatfish are apparently a very successful group of fish. They are found from the Antarctic to beyond the Arctic circle. In temperate regions flatfish are represented in enormous numbers and have become commercially exploited. In the North Atlantic, food fishes, eg plaice, have been so intensively fished that they are farmed to replenish natural stock.

The tetraodontiformes are an order of mostly marine fishes that have the teeth fused into a beak. Among their number are poisonous fishes, inflatable fishes, and one of the largest oceanic teleosts. None has scales; instead they are covered either with spines or with skin so thick that little can penetrate it.

The group of marine fishes called triacanthoids derives its name (meaning three-spined) from the presence of a conspicuous spine in the first dorsal fin and from the spine that forms each pelvic fin. Rarely, there is a small fin ray behind the pelvic spine. The group is widespread in the Indo-Pacific Ocean, largely in shallow waters, and a few species occur in the tropical West Atlantic.

Very little is known of the biology of triacanthoids. Some species have massive teeth and remains of hard shelled mollusks and crustaceans have been found in the stomachs. They are of no great significance to humans. A few species inhabit deeper waters and even less is known about them than about shallow-water species. This is a pity because they are so bizarre that their life style invites much speculation. *Halimochirurgus* has a long, slender, upcurved snout that may form half of the body length. The mouth is terminal, faces upwards and has just a few small, widely spaced teeth. The fish lives over mud flats and sand beds and is presumed to feed on worms. *Macrorhamphosodes* has a slightly

shorter but wider and straighter snout. The lower jaw has well-developed stocky teeth but those on the upper jaw are few and feeble. Most remarkable, however, is that as the fish grows, the mouth twists to the right or the left.

The trigger fishes are named for the interlocking "trigger-like" mechanism of their first and second dorsal fin spines; the small second spine must be released before the larger first spine can be depressed. Triggers, with their bony scales, have an easily recognized overall appearance, with their opposite and almost symmetrical dorsal and anal fins actively undulating as the major propulsion mechanism. Many have striking color patterns and are inhabitants of coral reefs. The file fishes are a group of elongated balistids with very small, rough scales. The dorsal spines are much further forward than in the trigger fishes. They have extremely small mouths and feed by picking up small invertebrates. Many have an expandable dewlap between the chin and the anal fin, a feature taken to extremes by *Triodon bursarius* (see below).

The box fishes or cow fishes have been described as bony cuboid boxes with holes for the mouth, eyes, fins and the vent. Some

The diodontids are exclusively marine and have larger spines than the tetraodontids. This latter family has the "beak" in each jaw in separate halves. Some tetraodontids live in fresh water. *Tetraodon mbu* is a striking black and yellow species widespread throughout the Zaïre system and some other West African rivers. This, and some of the other African freshwater species, are occasionally kept in aquaria, but they are aggressive inhabitants. Tetraodontids can inflate themselves with water and also by gulping air. It is difficult to evaluate the survival value of the latter technique because they float upside down at the surface where they are attacked by birds. All are, however, very poisonous fishes but, despite that, some are valued as a delicacy. Particularly in Japan, the more poisonous parts are eaten as *fugu*, and are prepared by specially trained cooks. Even so, death by poison from ill-prepared *fugu* is not unknown.

The family Triodontidae contains only one species, the rare and poorly known *Triodon bursarius*, which is found throughout the Indo-Pacific Ocean. The lower jaw has a single beak-like tooth but the upper jaw beak is in two separate halves. The extensive, thin dewlap is not expandable as in file fishes, but can be pulled forwards by the pelvic fin skeleton which forms its anterior edge. The function of this flap is unknown.

The oceanic sunfishes are the giants of the order. *Mola mola* is the largest species, probably weighing up to 1,000kg (2,200lb). Seen from the side this browny-blue fish is nearly circular, with the caudal fin reduced to a mere skin-covered fringe, but with the dorsal and anal fins produced into "oars" used for locomotion. It is most often seen lying on its side at the surface, allegedly basking but probably dying. A rare film of a young specimen alive shows that it swims rapidly in an upright position by vigorous sculls of the expanded fins. The diet consists of jellyfish and other small, soft items. Below the scaleless skin is a very thick layer of tough gristle. Although not common, *Mola mola* lives worldwide in tropical and subtropical waters. In common with the related genus *Masturus*, the Sharp-tailed sunfish, the young are spherical and have long spines on the body which become reduced and lost as the adult shape is developed. The last genus in the family, *Ranzania*, is more elongated. It is also a more colorful species—deep blue on the back paling ventrally, with diagonal black-edged silver stripes on the sides. Like its few close relatives, it is found worldwide in warmish waters.   BB/KEB

▲ **Set in a monochrome world** this is a species of file fish (*Oxymonacanthus longirostris*), sometimes known as the Beaked leatherjacket. It has erectile dorsal spines (not visible here) and lacks scales. It lives in the West Pacific and East Indian Ocean on the deep side of coral reefs and belongs to a group whose members grow to only 8–10cm (3–4in). Many file fishes are brightly colored. (Order Tetraodontiformes.)

◄ **Barbed fish.** The aracanids or cowfish come from the South Pacific. They have rounded bodies and a remarkable series of scythe-like spines along the sides and back of the carapace (that is the shield along the back). In some species the male has a hump on the snout and his color may differ from that of the female. (Order Tetraodontiformes.)

▼ **A prickly precaution.** Porcupine and puffer fishes can inflate themselves as a means of defending themselves against or deterring predators. When so inflated, either with water or air, the lateral spikes of the scutes stick out.

species also have two small, horn-like processes over the eyes, hence cow fish. The rigid outside skeleton (exoskeleton) is formed by fused bony scutes. Box fishes are slow-swimming, brightly colored fishes of shallow tropical seas. In case their armor should be thought inadequate against predators, box fish can also secrete a virulent toxin if molested. There are about 50 species, none of which grows to more than 60cm (2ft); most are shorter than 30cm (1ft).

Aracanids are commonest in the South Pacific and look like ovoid versions of box fishes armed with a series of curved or scythe-like spines. They are brightly colored and, like the box fishes, the coloring may differ between sexes.

The puffer fishes (diodontids) derive their main common name from having the body covered in spines which rise from their normal flat position when the body is inflated as a defense tactic.

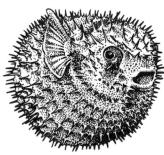

COELACANTH, LUNGFISHES, BICHIRS

Orders: Polypteriformes, Coelacanthiformes; superorder: Ceratodontimorpha
Probably seventeen species in 6 genera, 4 or 5 families and 4 orders.

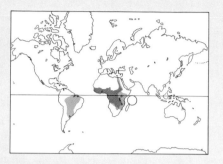

Bichirs or polypterids

Order: Polypteriformes
Probably ten species in 2 genera of the family Polypteridae.
Distribution: fresh waters in Africa.
Size: maximum length about 1m (3.3ft) but most are smaller.
Genera: **bichirs** (*Polypterus*), **Reed fish** (*Erpetoichthys calabaricus*).

Coelacanth

Order: Coelacanthiformes
Family: Latimeriidae.
Sole species *Latimeria chalumnae*.
Distribution: around the Comoro Islands.
Size: maximum length 1.8m (5.9ft).

Lungfishes

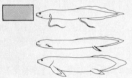

Superorder: Ceratodontimorpha
Six species in 3 genera, 2 or 3 families and 2 orders.
Distribution: fresh waters of Brazil, Paraguay, Africa and Queensland, Australia.
Size: maximum length 2m (6.5ft).

▶ **Strange fish of uncertain affinity** ABOVE. Polypterids live only in the fresh waters of Africa. There is no certainty as to which other group of fish they should be related. They are active predators and also have the ability to survive a long while out of water.

▶ **The last of a famous line.** The anachronistic coelacanth only became known to science in 1938. Between the Devonian and Cretaceous periods (400–65 million years ago) they lived worldwide. Today they are found only off the Comoro Islands in the western Indian Ocean.

THREE disparate groups of fishes are linked together here solely on the grounds that all are anachronistic. Although the term "living fossils" has been applied to them, such a description, apart from being a contradiction in terms, does not help our understanding of their relationships. "Anachronistic" is a much more useful term and also reflects the truth.

The **bichirs** are endemic to the fresh waters of Africa. The family contains just two genera, the true bichirs with nine species and the Reed fish. To which other major group of fishes the bichirs are related is a source of much debate. Almost every group has been suggested at one time or another during the last century.

Bichirs are primitive-looking fishes. They have thick, diamond-shaped scales that articulate by a "peg and socket" joint, gular plates, and the upper jaw fixed to the skull. As a group they retain a surprising number of primitive features. Their fossil record is scanty, the oldest known remains coming from deposits of the Cretaceous period (135–65 million years ago). All the fossils are in Africa, largely within the present area of distribution.

True bichir species owe their generic name (*Polypterus*, "many fins") to the row of small finlets on the back, each finlet consisting of a stout spine supporting a series of rays. The pectoral fin has a stout, scale-covered base. Although the tail is symmetrical in appearance, its internal structure retains the primitive, upturned, heterocercal condition.

Bichirs live in sluggish fresh waters. Their swim bladders have a highly vascularized lining that is used as a lung, enabling them to live in deoxygenated waters. The young have external gills. They are largely nocturnal fishes, not known for an active lifestyle. At night they feed on smaller fishes, amphibians or large aquatic invertebrates. Within Africa they are confined to the tropical regions in drainages emptying into the Atlantic Ocean or the Mediterranean. Living bichirs rarely grow to more than 75cm (2.5ft) long but, to judge from the size of the scales, some fossil species might have been twice as long.

The Reed fish is a slender, eel-like version of the true bichirs. It lacks pelvic fins and the subsidiary rays on the isolated spines. A much smaller species, it is found only in reedy areas in coastal regions of West Africa near the Gulf of Guinea. It appears to eat mostly aquatic invertebrates.

Fossil **coelacanths** first appeared in rocks of the Devonian period (400–350 million years ago) and continued to occur until about 70 million years ago when, along with the dinosaurs, they disappeared from the fossil record. Since the coelacanth was discovered in 1938 and again in 1952 (see box) over a hundred have been transferred to scientific institutions but, although its anatomy has been detailed, little is known of its biology and its relationships are still an unresolved controversy. Indeed, in a symposium volume published by the California Academy of Sciences in 1979 there are several contradictory papers each advocating different groups of fishes as the closest relatives of the coelacanth.

Much of the biology of the coelacanth has to be inferred from its anatomy and catching records. No specimens have been kept alive for long.

Apart from the mysterious first catch, all

coelacanths have been caught off the islands of Grand Comore and Anjouan at depths between 70 and 400m (230–1,300ft). Most are caught during the first few months of the year in the monsoon season. The two islands are made of highly absorbent volcanic rock and it has been argued, with some possible corroboration from its kidney structure, that the coelacanth lives in areas where the fresh rainwater leaks out into the sea. The coelacanth has the build of a lurking predator and fish remains have been found in the stomach. It has very large eggs; 20 about the size of a tennis ball were found in a 1.6m (5.3ft) long female. The only known embryos, which still had yolk sacs so were probably not close to birth, were more than 32cm (12.5in) long and it has been estimated that the gestation period is well over a year. The females are larger than the males and can live for at least 11 years.

Although the coelacanth is superficially similar to the last known fossilized coelacanths there are some interesting anatomical differences. In the present-day coelacanth, unlike its Cretaceous forebears, the swim bladder is nonfunctional and filled with fat. It retains the peculiar triple tail, the lobed paired fins, a skull with a hinge in it and the rough cosmoid scales of its fossil forebears.

"Anachronisms" do not, as one might have hoped, solve problems—they tend to create more. Why do coelacanths, formerly worldwide, now only live around two small islands? Why have some of the anatomical changes occurred? Why, if they are purportedly freshwater fishes, do they live in the sea? Above all else, how did just one species survive? We will probably never know.

The three genera and six species of **lungfishes** are now confined to the fresh waters of the Amazon, West and Central Africa and the Mary and Burnett Rivers of northern Queensland. Their fossils, by contrast, are found in rocks from Greenland to Antarctica and Australia. As with some other archaic fishes, we know that they represent an important organizational advance somewhere on the march from aquatic vertebrates to land-living vertebrates, but exactly how they fit in is still a matter of some uncertainty.

Living lungfishes are grouped into two families: the South American *Lepidosiren paradoxa* and the African *Protopterus* (4 species) are in the Lepidosirenidae whereas the Australian *Neoceratodus forsteri* is the sole occupant of its family (Ceratodontidae).

Neoceratodus retains more of the primitive features of its Devonian ancestors than do its relatives. It has large scales and lobed, paired fins fringed with rays retaining the primitive characteristic of being more numerous than the supporting bones. The dorsal, caudal and anal fins are confluent. Unlike *Protopterus* (see below) it cannot survive drought and will die if it is out of water for any length of time. The lungs open by a ventral slit in the oesophagus; the air tube then runs dorsally so that the partially two-lobed lungs lie dorsal in a similar position as the swim bladder in bony fishes but not like the lungs in other vertebrates which lie ventral to the alimentary tract. The Australian lungfish normally uses its gills for respiration but also breathes atmospheric air in adverse conditions.

Lungfishes are carnivores, and when fully grown (about 1.5m, 5ft) will eat frogs and small fish. The teeth consist of a pair of sharp plates in each jaw which work in a shearing action.

Reproduction has been observed in shallow water in August when, following a rudimentary courtship consisting of the male nudging the female, the eggs are scattered over a small area of dense weed and fertilized (hopefully). There is no indication of any further parental concern. Although native to two small river systems, the interest in this rare species is such that stocks have been transplanted into other Australian rivers. At least three of the introductions have been successful and the fish have bred.

The biology of the African species is better known. At least two, *Protopterus annectens* from west and southern Africa and *P. dolloi* from the Zaïre basin, are known to survive drought by a type of summer hibernation (aestivation) in cocoons. The widespread East African species *P. aethiopicus* is thought to be capable of aestivating but in the wild rarely does so as its waters are less likely to dry up.

African lungfish are more elongate than their Australian relatives. Small scales cover the body and the paired fins are long and threadlike. They are aggressive predators and can grow to more than 2m (6.5ft) long. The lungs are paired and lie in a ventral position as in terrestrial vertebrates. The use of lungs in breathing here necessitates a four-chambered heart, so unlike bony fish the lungfish's auricle and ventricle are functionally divided by a partition so that blood is circulated to the lungs as a bypass from the normal body and gill circulation.

All the African lungfish make nests. In *Protopterus aethiopicus* it is often a deep hole made by the male who will then guard the newly hatched young for about two

months. As well as driving away would-be predators he also aerates the water in the nest. The nest of *P. dolloi* is much more elaborate and has an underwater entrance. The terminal brood chamber may be in swampy ground and may be open at the top. Aeration vents are also built into it by the male.

The larvae of African (and South American) lungfish have external gills, the degree of development varying with the amount of oxygen in the water. At metamorphosis the external gills are usually resorbed and the lung and gill respiration takes over. Occasionally vestigial external gills remain throughout life.

Protopterus annectens lives in swamps and rivulets that are prone to drying out, often for months. To survive the fish burrows into the soft mud as the water level falls. Using its mouth and general body pressure the lungfish widens out the bottom of the tube until it can turn round. As the water falls below the opening of the tube the fish closes the mouth of the tube with a block of porous mud, curls up in the lower chamber and then secretes much special mucus which hardens to form an encasing cocoon with only an opening for the mouth. The cocoon conserves moisture and the porous mud plug allows breathing. During aestivation, as in hibernation, the metabolic rate is

▲ **An aborigine.** The Australian or Queensland lungfish (*Neoceratodus forsteri*) is native only to rivers in Queensland. Less adapted than its relatives to air breathing, it relies on gill respiration. Its diet includes fish, as here, which it chews with bony plates in the mouth. The Australian authorities have been trying to introduce this exciting fish to other rivers.

▲ **Survival posture** ABOVE RIGHT. At least two of the four species of African lungfishes (genus *Protopterus*) can survive drought by burrowing into mud, sealing themselves off and entering summer hibernation (aestivation). This can last for at least three years, during which time they feed off their own muscle tissue.

▶ **From familiarity to fame.** Before the coelacanth became known to the scientific world it was familiar to the inhabitants of the Comoro Islands. They called the fish *kombessa*. To them it was valueless. Now the fish has scientific value it is a highly prized catch.

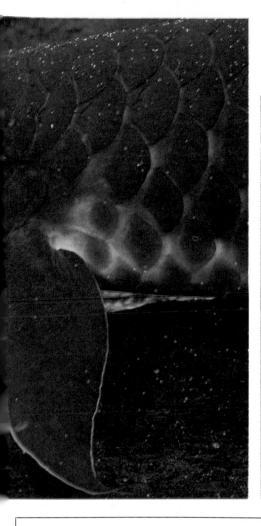

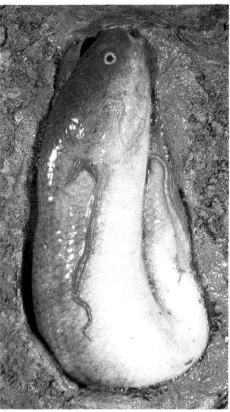

greatly reduced and the basic energy needed for survival comes from the breakdown of muscle tissue. In this state they have been known to survive four years of drought although normally an incarceration of only a few months would be necessary. When the river floods again the waters dissolve the cocoon and the fish emerges.

The South American lungfish (*Lepidosiren*) looks similar to the African with an eel-like body and feeler-like fins. It also possesses the general characteristics of a cartilaginous vertebral column and a common opening for the rectal, excretory and genital products.

Lepidosiren aestivates but its refuge is much simpler than that of its African relatives and no cocoon is produced. An elaborate nest is made in the breeding season during which time the pelvic fins of the male bear a large number of blood-rich filaments. The function of these is unknown, but suggestions have included the release of oxygen into the water of the nest or alternatively acting as supplementary gills to reduce the number of visits to the surface while guarding the young.

KEB

The Finding of the Coelacanth

It is an axiom in biology that absence of records does not imply certainty of extinction.

On a hot summer's day in 1938 Captain Goosen's boat *Nerine* docked at East London in South Africa. At that time Mary Courtnay-Latimer was the curator at the East London Museum and local skippers were accustomed to her frequent visits to the port to obtain fish specimens for the museum. At 10.30am that morning (22 December) she was telephoned and told that the *Nerine* had returned with some specimens for her. Among the catch was a large blue fish with flipper-like fins and a triple tail which she had never seen before. After several attempts she found a taxi driver willing to take her and her 1.5m (5ft) long, oily, smelly prize back to the museum. There, after searching through reference books, the nearest she could come to identifying the fish was as "a lung fish gone barmy."

Realizing that her find was important she tried to contact Dr J. L. B. Smith (whose name will be forever linked with the fish), the ichthyologist at Rhodes University, Grahamstown. It was then mid summer in South Africa; temperatures were high; how was an important oily fish to be preserved? The local mortuary refused to have this corpse in its cold store. Finally a local taxidermist, Mr R. Centre, although admitting inexperience in fish-stuffing, agreed to help. He wrapped the body in cloth, soaked it in formalin and placed it in a makeshift ichthyosarcophagus.

On 26 December there was still no reply from J. L. B. Smith. An examination of the body showed that the formalin had not penetrated and that the internal organs were rotting. Pragmatism dictated that the decaying parts should be thrown away and what could be preserved should be preserved.

On 3 January 1939 a telegram arrived from J. L. B. Smith. It read "MOST IMPORTANT PRESERVE SKELETON AND GILLS = FISH DESCRIBED." The ensuing search of local rubbish heaps failed to find the missing organs. Parallel disasters now revealed themselves. The early photographs taken of the fresh fish had been spoiled. The museum trustees, not thinking the fish important, had ordered the skin to be mounted before, on 16 February, J. L. B. Smith finally arrived. He stared at the mounted skin and said: "I always knew, somewhere, or somehow, a primitive fish of this nature would appear."

He described the fish as *Latimeria chalumnae*. The origin of the generic name needs no explanation and it is followed by an allusion to the Chalumna River, off which the fish was purportedly caught.

Why had this conspicuous, spiny-scaled fish remained unnoticed for so long? Its large eye and lurking predator shape suggested it did not normally live off East London. Unless the only specimen had been caught, there must be more, but where?

The chase was long and extremely exciting.

It took 14 years, involved a great deal of work and almost the commandeering of the South African Prime Minister's private aircraft. For all the details the reader is recommended to read J. L. B. Smith's book *Old Fourlegs*.

Just before Christmas 1952 (note the date again), J. L. B. Smith received a telegram from Eric Hunt from the Comoro Islands. It read: "REPEAT CABLE JUST RECEIVED HAVE FIVE FOOT SPECIMEN COELACANTH INJECTED FORMALIN HERE KILLED 20TH ADVISE REPLY HUNT DZAOUDZI." The search had finally come to ground. Dzaoudzi is in the Comoros and there lives the coelacanth.

Although this discovery was very exciting to the scientific community, the natives on the Comoro Islands were unmoved. They knew this fish, had given it the name of Kombessa, and considered it to be a worthless catch although the rough scales could be used in roughing a bicycle tyre tube prior to mending a puncture.

SHARKS

Subclass: Selachii
About three hundred and seventy species in 74 genera, 21 families and 12 orders.
Distribution: worldwide in tropical, temperate and polar oceans at all depths.

Size: adult lengths 25cm–20m (10in–65ft), adult weights 2–12,000kg (1–26,500lb).

▲ **A potent image of the modern world,** the teeth of a Great white shark. With cinema's need for gripping visual subjects the Great white shark has, perhaps inevitably, become one of the great stars of the wide screen. Celluloid has magnified and popularized this shark's infamy in a manner only possible in the contemporary world. Serving a need for spectacle has however brought distortion and misrepresentation. The number of deaths attributable to sharks is minimal, and among sharks there is a great variety of forms, which tends to be overlooked. Even teeth vary considerably: in some sharks teeth for grinding are predominant; others are filter-feeders.

▶ **Beauty in shape, movement and color,** a requiem shark (genus *Carcharinus*), photographed off the Maldive Islands in the Indian Ocean. Most of the hundred or so species in the order of requiem sharks are restricted to the tropics.

Most people think that sharks are large, fast-swimming, elegant savage predators. This is true of some species, but only of a minority. The group should be of general interest on account of the intriguing aspects of biology found within it: all sharks have an exceptional sense of smell; some can detect minute electrical discharges; some sharks give birth to live young.

One of the most notable features of sharks is their teeth. In the large highly predaceous sharks these are large and razor sharp, used for cutting and shredding their prey into bite-sized pieces. However, this is not the case in all species. Bottom-feeding species, which eat mollusks and crustaceans, have teeth flattened for crushing the shells of their prey. Fish-eaters have very long, thin teeth to help them catch and hold on to a struggling and slippery fish. At any one time a shark may have up to 3,000 teeth in its mouth, arranged in 6 to 20 rows, according to species. In most sharks only the first row or two are actively used for feeding. The remaining rows are replacement teeth, in varying stages of formation with the newest teeth being at the back. As a tooth in the functioning row breaks or is worn down it falls out and a replacement tooth moves forward in a sort of conveyor-belt system. A shark can replace its teeth every few days. Thus the functional row, or rows, of teeth are always kept sharp. It has been estimated that a shark may use over 20,000 teeth in its lifetime. The strong jaws of a shark can exert a biting strength of 3,000kg per sq cm (44,000lb per sq in) on the teeth. Contrast this with only 10kg per sq cm (150lb per sq in) in humans.

Sharks find their prey through a number of sensory systems. Many have poor eyesight, but a few, especially the isuroids (eg thresher sharks) and requiem sharks have good eyesight (as good as man's). Some species have sensory barbels around their mouth to taste the sea bed for prey. All sharks have a very keen sense of smell. Their nostrils are used only for smelling, not for breathing. The part of a shark's brain that deals with smelling is twice as large as the rest—evidence of the importance of smell. Sharks can detect one part of blood in a million parts of water—about 1 drop of blood in 115 liters (25 gallons) of water!

Sharks (along with skates and rays and chimaeras) have a lateral line system: a series of canals on the entire body and head which are filled with a jelly-like substance, sensory receptors which pick up pressure waves caused by the movements of another animal or even by the shark itself approaching a stationary object.

On the snout are the so-called ampullae of Lorenzini, a series of electro-receptive pits (named for Stefano Lorenzini, who first studied them in 1678). They are the most sensitive electro-receptive devices found in any animal. They are capable of picking up one-millionth of a volt, which is less than the electric charge produced by the nerves in an animal's body! Thus sharks can find their prey from the prey's natural electric output.

The most primitive living shark is the **Frilled shark**. It has broad-based trident-like teeth which are otherwise found only in fossil sharks. Its common name derives from its long, floppy gill flaps, forming a frill around the head. Its primitive and unique characters place it in its own group. First discovered off Japan on Sagami Bay in the 1880s it was long thought to occur nowhere else. In fact for unknown reasons it only enters water as shallow as 30m (100ft) in Sagami Bay. Recent deep trawling has shown that it lives at depths of 300–600m (1,000–2,000ft) over a wide area off the coasts of Australia, South Africa, Chile, California and Europe. It grows to 2m (6ft) long and has a thin eel-like body. It feeds on small fish swallowed whole. The female develops eggs in her body (ovoviviparous), producing 6–12 young per litter.

The **six- and seven-gilled sharks** are so named because they have developed one or two extra sets of gill slits. They prefer cold water, and in the tropics live in deep water. Individuals have been photographed more than 1,800m (6,000ft) down. Only in cold polar waters do they come into shallow water. They reach 4.5m (15ft) in length and feed on other fishes. Upper-jaw teeth are long and tapered; lower-jaw teeth are short and wide with unique, strong, multiple serrations. They too develop eggs internally and produce up to 40 young.

The horn sharks live in the Indian and Pacific oceans. These sluggish, bottom dwellers, 1m (3ft) long, feed on shallow-water mollusks and crustaceans. The genus name (*Heterodontus*—"different teeth") reflects the fact that, unlike other sharks, they have pointed front teeth and rear teeth adapted into crushing plates. They are rather stout, stocky sharks and have prominent brow ridges above their eyes which give them the appearance of having horns—hence horn sharks. All species are spotted or patterned, making them rather handsome. This, coupled with their small size and hardiness, has resulted in them being desired by aquarists. They are easy to care for and will live for

THE 12 ORDERS OF SHARKS

Frilled shark

Order: Chlamydoselachiformes
Family: Chlamydoselachidae.
Sole species *Chlamydoselachus anguineus*.

Distribution: off coasts of California, Chile, Europe, S Africa, Japan, Australia.

Six- and seven-gilled sharks or hexanchoids

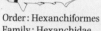

Order: Hexanchiformes
Family: Hexanchidae.
Five species in 3 genera.

Distribution: worldwide in cold marine waters at shallow to moderate depths.

Horn or Port Jackson sharks or heterodontoids

Order: Heterodontiformes
Family: Heterodontidae.
Six species of the genus *Heterodontus*.

Distribution: tropical Pacific Ocean.

Orectoloboids

Order: Orectolobiformes
About thirty species in 12 genera and 5 families.

Distribution: Indo-Pacific Ocean; two species in Atlantic Ocean.
Families, genera and species include: **banded catsharks and epaulette sharks** (family Hemiscyllidae), including **epaulette sharks** (genus *Hemiscyllium*); **carpet sharks** (family Orectolobidae); **catsharks** (family Parascyllidae); **nurse sharks** (family Ginglymostomatidae); **Whale shark** (*Rhincodon typus*, sole species of the family Rhincodontidae).

Catsharks and false catsharks or scylloids

Order: Scyliorhiniformes
Eighty-seven species in 18 genera.
Distribution: worldwide in cold or deep marine water.
Families, genera and species include: **catsharks** (family Scyliorhinidae), including **Lesser spotted dogfish** (*Scyliorhinus caniculus*), **swell sharks** (genus *Cephaloscyllium*); **False catshark** (*Pseudotriakis microdon*, sole species of the family Pseudotriakidae).

Smooth dogfish sharks or triakoids

Order: Triakiformes
Family: Triakidae.
About thirty species in 5 genera.
Distribution: worldwide in tropical, subtropical and temperate seas.
Species include: **Leopard shark** (*Triakis semifasciata*), **Soupfin shark** (*Galeorhinus zyopterus*).

Sand tiger, false sand tiger and goblin sharks or odontaspoids

Order: Odontaspidiformes
Seven species in 3 genera and 3 families.
Distribution: worldwide in tropical and temperate seas.
Families, genera and species include: **False sand tiger shark** (*Pseudocarcharias kamoharrai*, sole species of the family Pseudocarchariidae): **Goblin shark** (*Scapanorhynchus owstoni*, sole species of the family Scapanorhynchidae); **sand tiger sharks** (genus *Odontaspis*, family Odontaspididae).

Thresher, mackerel and megamouth sharks or isuroids

Order: Isuriformes
Ten species in 6 genera and 3 families.
Distribution: worldwide in tropical and temperate seas.
Families, genera and species include: **mackerel sharks** (family Isuridae), including **Basking shark** (*Cetorhinus maximus*), **Great white shark** or **maneater** or **White shark** (*Carcharodon carcharias*), **Mako shark** (*Isurus oxyrinchus*), **Porbeagle shark** (*Lamna nasus*), **Salmon shark** (*L. ditropis*); **megamouth sharks** (family Megachasmidae); **thresher sharks** (family Alopiidae).

Requiem sharks or carcharhinoids

Order: Carcharhinoidiformes
Family: Carcharhinidae.
One hundred species in 10 genera.
Distribution: worldwide in tropical and temperate seas.
Species and genera include: **Black-tipped shark** (*Carcharhinus melanopterus*), **Bull shark** (*C. leucas*), **Great hammerhead shark** (*Sphyrna mokarran*), **hammerhead sharks** (genus *Sphyrna*), **Tiger shark** (*Galeocerdo cuvier*), **White-tipped shark** (*Carcharhinus albimarginatus*).

Spiny dogfish and allies or squaloids

Order: Squaliformes
Family: Squalidae.
About seventy species in about 12 genera.
Distribution: worldwide in cold or deep seas.
Species and genera include: **bramble sharks** (genus *Echinorhinus*), **cigar sharks** (genera *Isistius*, *Squaliolus*), **sleeper sharks** (genus *Somniosus*), **Spiny dogfish** (*Squalus acanthias*).

Angelsharks or squatinoids

Order: Squatiniformes
Family: Squatinidae.
About ten species of the genus *Squatina*.
Distribution: worldwide in tropical and temperate seas.

Sawsharks or pristiophoroids

Order: Pristiophoriformes
Family: Pristiophoridae (sawsharks).
Five species in 2 genera.
Distribution: Bahamas, off coast of S Africa, western Pacific Ocean from Japan to Australia.

▶ **Port Jackson shark** ABOVE (*Heterodontus portjacksoni*), a species in the order of horn sharks. These are as distinctive as the more familiar sharks but in their own ways. Their specific features include the horn-like ridges above the eyes, pointed front teeth and crushing plates, and bottom-dwelling habits.

▶ **Among the smallest of the sharks** are those species belonging to the order of catsharks and false catsharks. Most are no larger than 1–1.5m (3–4.5ft). Consonant with their size is their diet, which includes non-fish prey and bottom-dwelling fishes. The swell sharks have developed techniques for defense against predators. This is the Gulf catshark (*Halaelurus vincenti*),

years in captivity. They lay eggs which are unique in shape, having a spiral form. These are forced into crevices between rocks or pieces of coral by the female. The screw shape helps hold the egg in place.

The **orectoloboids** are an order of five closely related families of sharks. They are tropical to subtropical fishes. Only two species are found in the Atlantic Ocean, the rest are in the Indo-Pacific.

In size the orectoloboids range from the epaulette sharks (genus *Hemiscyllium*)—1m (3ft) long—to the Whale shark, which has been reliably reported as reaching over 15m (50ft) in length and is the world's largest fish. In most species the young are born with beautiful spotted or banded patterns, which are lost when they mature.

Most species lay eggs (ie are oviparous), but a few develop them internally (ie are ovoviviparous). Usually less than 12 young are produced per litter.

All species except the Whale shark are bottom dwellers, spending most of their time just sitting on the ocean floor. The skeletons of the pectoral fins are modified so that they can actually use the fins for walking on the ocean floor. Many species, even when disturbed, will crawl away rather than swim.

Most orectolobiform species feed on mollusks and crustaceans and have teeth for crushing and grinding. All have sensory barbels around the mouth. The carpet sharks eat mainly fish, so they have long, thin teeth. The Whale shark is a filter feeder, having its gill arches specially modified to act like a sieve to filter out the planktonic organisms upon which it feeds. As its teeth are redundant they are minute. Its vast bulk needs constant fuel, so the Whale shark swims incessantly, filtering the ocean for its food. A slow swimmer, it has been involved in collisions with boats as it often swims at the surface where there is most food. (Boat and passengers are usually more damaged than the shark.) It is found worldwide, in all tropical and subtropical oceans. Many orectolobiforms are commonly kept by aquarists on account of their small size, hardiness and beautiful patterns.

The **catsharks and false catsharks** comprise about 18 genera and 87 species. Most live in cold or deep waters. They are found worldwide. Many are spotted or patterned and do not lose this coloration as they mature. These stocky forms live on the ocean floor. They feed on mollusks, crustaceans and bottom-dwelling fishes, using their short, stout teeth. Some species have sensory barbels to help them locate their prey.

Most are small, maturing at about 1–1.5m (3–4.5ft). However, some species are larger, eg the False catshark, which is reported to reach 3m (9ft). The swell sharks are aptly named. When frightened or attacked they swallow water or air into their stomachs, enlarging their girth to three or four times its normal size.

The tasteful flesh of catsharks is much appreciated. In England the Lesser spotted dogfish is commonly sold in fish and chip shops under the name of rock salmon or huss.

The **smooth dogfish sharks** (triakoids) live in shallow waters of tropical, subtropical and temperate seas. They are moderately large sharks reaching about 2m (6ft) in length. They are bottom dwelling, feeding upon mollusks, crustaceans and fish, but they do not sit or crawl. The majority of species have modified crushing and grinding teeth.

The Leopard shark from the Eastern Pacific has a beautiful color pattern with dark gray to black spots on a silvery background which makes it a favorite for public aquaria. The Soupfin shark is commercially fished in the Orient for its fins which are used to make shark fin soup.

Virtually all species are migratory, spending winter in the tropics and migrating to temperate waters in the summer. Evidence suggests these migrations are regulated by water temperature. Females of these species develop embryos in the uterus and deliver 10–20 young at a time. Overall the triakoids epitomize the general public's concept of typical sharks.

The **sand tiger**, **false sand tiger** and **goblin sharks** are all fairly large, reaching 3–3.5m (10–12ft) in length. The sand tiger sharks (5 species) are found worldwide in shallow, temperate and tropical waters. Largely fish-eaters, they have the appropriate long, thin teeth. The tooth shape produces an extremely ferocious appearance which, combined with their rather docile nature, has made them favorites in public aquaria, where they can live for over 20 years. (The term sandshark is often, but inaccurately, applied to any smallish shark that comes into shallow water, including sand tiger sharks, smooth dogfish, spiny dogfish, nurse sharks and requiem sharks.)

Reproduction in this order is viviparous, ie embryos develop in the uterus, and quite unusual. The female begins with 6–8 embryos per uterus. As they grow these embryos devour one another until one embryo per uterus is left. Hence only two young are born at a time. This unique prenatal feeding habit is called intra-uterine cannibalism.

The False sand tiger shark lives in deeper waters off China and East and West Africa. One specimen caught off West Africa by the German research vessel *Walther Herwig* gave birth to four young when it was brought on deck. Thus the False sand tiger shark does not seem to have intra-uterine cannibalism. Otherwise very little is known of the biology of this large-eyed fish-eater.

The Goblin shark is perhaps the most bizarre of all living sharks. Projecting from its "forehead" is an extremely long, thin, broad, shovel-shaped process. Its function is unknown. The Japanese fishermen that first caught it called it *tenguzame*, which means "goblin shark." Like the Frilled shark, it was originally caught in the 1890s in Japan's Sagami Bay. Since then it has been caught in the Indian Ocean, off South Africa and off the coast of Portugal at depths of 300m (1,000ft) and more. Again more deep-water collecting will probably extend its known range. In life Goblin sharks are a translucent white color, but become very dark brown soon after death. Judging by its teeth it is a fish-eater. Apart from its length, 4.3m (14ft), we know practically nothing about it.

The **thresher**, **mackerel** and **megamouth sharks** are among the largest sharks in the world. They are found in tropical and temperate seas.

The thresher sharks derive their name from the extremely long, thin upper lobe of their caudal fin, which may be as long as the rest of their body. Swimming into a school of small fishes they use their tail like a whip, thrashing it among the school, killing or stunning the fishes, which are then eaten. They reach a length of about 6m (20ft) and give birth to only a few young, but those of the largest species are about 1.5m (5ft) long! These pelagic sharks rarely come into shallow water.

The most recent exciting shark discovery was that of the aptly named megamouth, which was first described only in 1983. This 4.5m (15ft) long fish—only two are known —was caught accidentally off Hawaii in 1977 in 150m (500ft) of water. It is a filter feeder, but unlike other known filter feeders its mouth has many organs that are thought to be luminous. A second specimen was caught in 1984 off the California coast. How such a large animal has evaded capture is a great mystery.

The family of mackerel sharks contains some of the best known sharks, including the unjustifiably infamous Great white

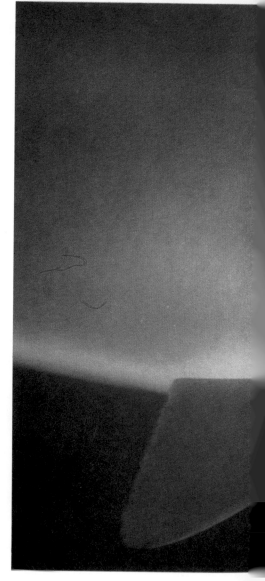

◄ **Poised for emergence,** a dogfish in its egg capsule. In sharks there are three modes of egg development. Most sharks are ovoviviparous, that is eggs develop and hatch in the womb. Several species are viviparous: the eggs develop in the womb connected to the wall by a yolk "placenta." Following the gestation period, the females of ovoviviparous and viviparous shark species give birth to fully developed young. A few species are oviparous: the female lays fertilized eggs. A female dogfish normally lays 18–20 eggs, each measuring about 5 × 11cm (2 × 4in). She deposits them among vegetation or sponges, and each capsule fastens itself to a support with a filament. From laying to hatching normally takes 8–9 months.

▼ **Poised for attack,** a sand tiger shark. The six species of the genus *Odontaspis* live in the tropical and subtropical seas. They grow to 6m (20ft) in length, and are voracious predators. Usually they are sluggish: invigoration appears when the time comes for hunting, and they head for shallow waters.

shark. Also here are the Porbeagle shark, the Mako shark and the Basking shark. The mackerel sharks are large (the Basking shark reaches 9m, 30ft), and live in all tropical and temperate seas.

They all have an unusual caudal fin with the lobes being nearly equal in length, caudal keels on either side of the tail, and are relatively fast swimmers (in part due to the previous two characters). Most species are fish-eaters. Some make spectacular leaps into the air—one of the reasons they are popular game fish. It is not known why they make these leaps, but it has been suggested that it is an attempt to dislodge skin parasites. The Basking shark is known to have collided with boats during such leaps, which have caused numerous human deaths. Most, if not all, species are homeothermic, ie they maintain their body temperature above that of their surroundings.

The Mako shark is probably the fastest fish in the world. It has been recorded to swim at speeds over 95km/h (60mph), and is known to have out-swum and eaten swordfish. Reaching over 6m (20ft) in length, it is a prized catch because its delicious flesh is highly sought after and valuable.

The most famous shark is probably the Great white shark, also called the White pointer, Man eater, White death, or just White shark. It is the species most often cited in references to shark attacks on humans. Mainly feeding on marine mammals (the only shark to do so), its broad, serrated teeth are designed for biting large chunks of flesh from whales and seals. Although cited as reaching about 12m (36ft) in length, authentic records show that it rarely grows over 7m (23ft) in length. It is known to be viviparous, ie embryos develop in the

Shark Attacks

The most famous characteristic of sharks is their alleged propensity to kill and eat humans. But less than one tenth of all shark species have been implicated in unprovoked shark attacks.

The idea that there is a shark off every beach waiting for people to go into the water so that it can attack them is completely untrue. To put it in perspective, every year thousands more people are killed by other people than by sharks; thousands more people are killed in automobile accidents than by sharks, and more people are killed by lightning than by sharks. Each year humans kill tens of thousands of sharks yet there are fewer than 100 shark attacks upon humans.

The Great white shark, often called the man-eater, has the greatest reputation for attacking humans. Recent studies of Great white sharks show that they mainly feed on sea mammals—seals and porpoises. Seen from below a person swimming looks much like a seal, with arms and legs sticking out. The shark usually surprises its victim, be it man or seal, with one massive bite, and then retreats in order to allow its victim to die before eating it. For this reason many humans survive the attack of a Great white shark if they are saved before being eaten. Death, however, may result from this one massive bite, from blood loss or damage to organs.

Attacks by other species of sharks, mostly requiem sharks, are for feeding. The most dangerous are the Tiger shark and various species of hammerhead sharks. Large individuals of these species can eat a whole human, whereas smaller ones would only take bites. The only other species that is a confirmed man-eater is the Australian sand tiger shark. Sharks, unlike tigers, do not acquire a taste for human flesh. (Sharks have small brains and cannot "learn." Feeding is instinctive.) Simply, if one of these sharks is hungry and a person is in the water the shark will eat the person just as readily as it will eat anything.

Much research has gone into finding out how to prevent shark attacks. Various chemicals, air bubbles and electric fields have been tried to deter attacks. Some are useless; others will deter some species of sharks but actually attract others! Dr Eugenie Clark has experimented with a skin secretion produced by the Moses sole, found in the Red Sea. This chemical deters attacks by certain sharks, but is not a universal repellent.

uterus, but a pregnant female has never been caught, so there is no information on litter size or the size of young at birth. The smallest free-swimming specimen known was about 1m (3ft) long. Only the belly is white, the dorsal side usually being grayish.

The Porbeagle shark and the Salmon shark are the smallest members of the mackerel sharks family, 2.7m (9ft) long. They live in the Atlantic and Pacific oceans respectively. They are fish-eaters and have an unusual mode of embryo development. Only one embryo develops in each uterus, but instead of eating other embryos, the female produces more infertile eggs during gestation which the two embryos eat for their nourishment.

The Basking shark is second in size to the Whale shark. Commonly over 10m (30ft) long and recorded up to 15m (45ft) long, it is a filter feeder. Its teeth are minute and virtually useless, and modified gill rakers are used for sieving the plankton. Its liver yields vast amounts of oil and the fish has been the subject of local fisheries in the North Atlantic. The Basking shark's name comes from its behavior of often swimming and resting, ie basking, at the surface.

The **requiem sharks** are probably the largest group of living sharks with about 100 species in 10 genera. In body shape and behavior they are the "typical" shark people think of. They reach 3m (10ft) in length and occur in all tropical and temperate seas.

The Bull shark, found worldwide, commonly enters fresh water for lengthy periods and has been found more than a thousand kilometers from the ocean in the Zambesi River and the Mississippi River. Originally sharks from these fresh waters were thought never to enter the ocean and thus to be distinct species and were appropriately named as "*Carcharhinus nicaraguensis*" and "*C. zambeziensis.*"

All species of *Carcharhinus* are widespread and in summer will migrate long distances into temperate waters. They have a metallic gray or brown dorsal coloration. Some, however, will have the edges of the fins tipped in white or black, hence the names White-tipped and Black-tipped sharks. Usually feeding and swimming close to the surface, their large triangular first dorsal fin will often be sticking out of the water, a sign that people have closely associated with shark's proximity.

The largest requiem shark, the Tiger shark, reaches over 6m (18ft) in length and is unquestionably one of the most dangerous sharks. As it swims through the water it will swallow anything that it can

▲ **A threat to marine mammals.** The Great white shark is the only shark known to feed regularly on dolphins, sea lions and other seals. Humans are normally attacked when it mistakes a swimming man or woman for its usual prey.

◄ **Possible protection for divers,** a "shark-proof" suit being tested.

get down its throat—literally anything, including rolls of tar paper, shoes, gasoline cans, automobile license plates, and cans of paint, not to mention human parts! One of the most unusual cases of this occurred in April 1935 when a Tiger shark 4.4m (14.5ft) long caught off Maroubra Point, New South Wales, Australia, was brought alive to the Coogee Aquarium. Several days later it regurgitated its stomach contents, which included a human arm. It was found that the arm had not been bitten off by the shark but had been cut off at the shoulder with a sharp instrument, probably a knife. From its tattoos and fingerprints the arm was identified as coming from a man who had been reported missing a week earlier. Police theorized that the man had been murdered, dismembered and the pieces scattered at sea. Although a suspect was arrested he was never proven guilty because the arm regurgitated by the shark was the only piece of evidence that the crime had been committed! Young Tiger sharks have dark bands on a silvery gray background, hence their name. The bands fade with age.

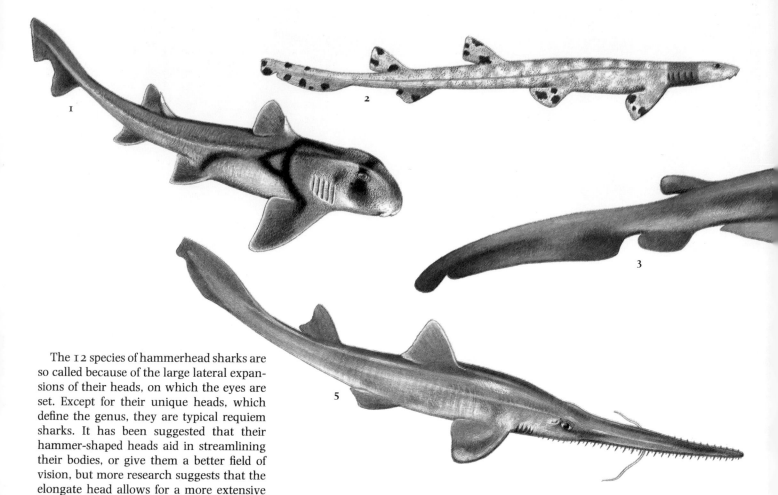

The 12 species of hammerhead sharks are so called because of the large lateral expansions of their heads, on which the eyes are set. Except for their unique heads, which define the genus, they are typical requiem sharks. It has been suggested that their hammer-shaped heads aid in streamlining their bodies, or give them a better field of vision, but more research suggests that the elongate head allows for a more extensive electro-detecting system, the ampullae of Lorenzini. The Great hammerhead shark is the largest, growing to more than 5m (15ft) long. Like other requiem sharks it is highly predaceous and has been involved in several attacks on humans.

The **spiny dogfish sharks** are cold-water forms, worldwide in distribution. All develop eggs internally (ovoviviparous), producing about 12 young per litter. In size they range from less than 30cm (1ft) to over 6m (20ft). Many, especially the deepwater species, feed on squid and octopuses.

In the North Atlantic the Spiny dogfish is probably the most abundant shark. Tens of millions of kilograms are caught every year; it is an important food fish. Spiny dogfish rarely exceed 1m (3ft) in length, travel in schools and migrate long distances, moving into Arctic waters each summer. Each dorsal fin is preceded by a spine, which has venom-producing tissue at its base. To humans the venom is painful, but not fatal.

Many deepwater species, especially in the genus *Etmopterus*, have light-producing organs along the sides of their body. Speculation is that these organs may attract their prey, deepwater squids, as well as providing camouflage through "counter-illumination." Their eyes are very large and sensitive at low light levels.

The small but very thin cigar sharks (especially genus *Isistius*) have their lower-jaw teeth greatly elongated. They swim up to a larger animal (a fish, squid, or even a cetacean), bite it, and then with a twist of the body cut out a perfectly circular piece of flesh from the prey.

Sleeper sharks, the giants among spiny dogfish sharks, are the only sharks permanently inhabiting the Arctic water. They have been seen under the polar ice cap. They feed upon seals and fishes, and are probably the only sharks to have flesh that is poisonous to both humans and dogs.

The bramble sharks are unusual in having very large, flat dermal denticles widely scattered on the skin, giving them a "brambly" appearance. There are probably two species, one in the Atlantic (*E. brucus*) and one in the Pacific (*E. cookei*). Although large, over 2.7m (9ft) long, their skeleton is uncalcified and so is extremely soft.

The **angelsharks** are unusual, being very flat, and are considered to be more closely related to the skates and rays than to the more "typical" sharks. They grow to more than 1.8m (6ft) in length and there are about 10 species in the genus *Squatina*, found in all tropical to temperate seas. An anterior lobe of the pectoral fins extends in front of their gill slits. They have long, thin

▲ ► **Representative species and genera of sharks.** (1) An adult horn or Port Jackson shark (*Heterodontus portusjacksoni*, order Heterodontiformes) (about 3m, 10ft).
(2) A variolated catshark (genus *Parascyllium*, order Orectolobiformes) (about 60cm, 24in).
(3) A Goblin shark (*Scapanorhynchus owstoni*, order Odontaspidiformes) (3.3m, 11ft).
(4) A Frilled shark (*Chlamydoselachus anguineus*, order Chlamydoselachiformes) (2m, 6.5ft).
(5) A sawshark (genus *Pristiophorus*, order Pristiophoriformes) (about 2.5m, 8ft). (6) A six-gilled shark (*Hexanchus griseus*, order Hexanchiformes) (about 6m, 20ft). (7) A luminous deepwater dogfish (*Etmopterus hillianus*, order Squaliformes) (30cm, 12in). (8) A False catshark (*Pseudotriakis microdon*, order Scyliorhiniformes) (3m, 10ft). (9) A humantin (*Oxynotus centrina*, order Squaliformes) (1m, 3.3ft).

teeth, and sit camouflaged in fairly shallow water, waiting for prey to swim by, when they quickly lunge out and capture it. Usually lethargic, they move very rapidly when catching prey or when they are hooked. If landed on a boat they will snap viciously at anything that comes near, and anything siezed they will hold with great tenacity. They are ovoviviparous, with about 10 young per litter.

Having a long rostrum with lateral rostral teeth very much like a sawfish, the **sawsharks** are true sharks, with lateral gill openings. They are quite rare and grow to a length of about 1.8m (6ft). One species (*Pliotrema warreni*), from the coast of South Africa, has an extra set of gills. The genus *Pristiophorus* has four species, three in the western Pacific, from Japan to Australia in shallow waters, and one Atlantic species found only off the Bahamas in very deep water. They have a pair of long thin barbels under the rostrum which help them find their prey—mollusks and crustaceans—on the sea floor. Their teeth are flat and broad for crushing and their "saws" appear to be used only for defense. Sawsharks are ovoviviparous, producing about 12 young per litter.   GDi

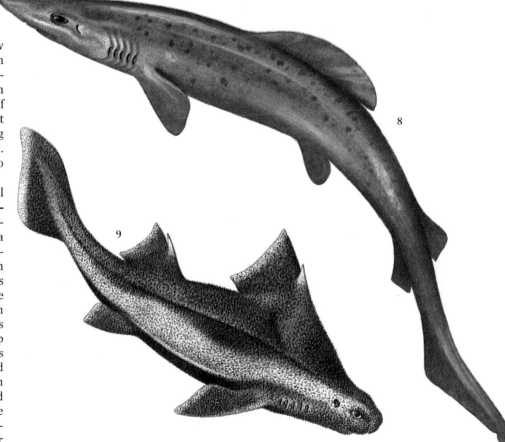

SKATES, RAYS, CHIMAERAS

Order: Batiformes
Skates and rays
About three hundred and eighteen species in
50 genera and 7 families.
Distribution: worldwide in tropical, subtropical
and temperate waters; two genera in fresh
water in S America.

Subclass: Holocephali
Order: Chimaeriformes
Chimaeras
About twenty-three species in 6 genera and
3 families.
Distribution: subarctic, subantarctic, temperate
and tropical waters.

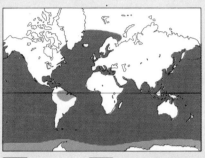

| Chimeras | Skates and rays |

▲ **Swan of the oceans,** a Blue-spotted stingray
(*Taeniura lymma*). Found in the Pacific and
Indian Oceans, this species often frequents
shallow waters. Its normal length is about 2m
(6.5ft).

▶ **Shark-like sawfish.** ABOVE The distinctive
feature of sawfish is their saw-like snout, which
is similar to that of the hammerhead sharks. It
is a general-purpose adaptation, used for
digging, killing and defense.

▶ **Spread like a butterfly,** a Spotted eagle ray
(*Aetobatus narinari*), a species of eagle ray,
photographed near Christmas Island in the
Pacific. It inhabits tropical and subtropical
shallow waters.

THE mere mention of rayfish to a mariner
conjures up images of gigantic jumping
devilfishes, potently venomous stingarees,
toothy snouted sawfishes and the electric
ray which can stun the unwary: all these
reputations reflect truth and are based on
anatomical adaptations, but human fear
can be excessive. These flattened relatives of
sharks respond defensively to our attempts
to capture and subdue them. Their real contribution to our lives and to the seashore
environment is the role they play as benthic
scavengers, cleaning the bottoms of our
bays and coastal shelves of detritus, and by
their participation in maintaining balance
in the food chain. Although they are bizarre
in appearance, **skates and rays** are edible
and consumed by most fishing cultures.

The order is most closely related to the
angelsharks and sawsharks. All are flattened (one can think of them as "pancake"
sharks), have their pectoral fins extending
well in front of the gill arches and fused to
the sides of the head, and gill slits underneath the body. They are mostly sedentary
and feed on mollusks, crustaceans and
fishes. They have very large openings
(spiracles) which take in the water which
is pumped over the gills and then out
through the ventral gill slits.

Members of the sawfish family are easily
recognized by the large "saw" that protrudes from the front, which is similar to
that of sawsharks. It is used both to capture
food and as a defensive weapon. Swimming
into a school of fishes, sawfish rapidly slash
their saw back and forth, stunning or killing
fish in the school. Then they return and eat
immobile fishes. The razor-sharp teeth on
the saw can cause lethal wounds to an
attacker many times the size of the sawfish.

The number of teeth on the saw varies
with the species and ranges from 12 to 30
pairs. The jaw teeth are short and flattened
for crushing the shells of crustaceans and
mollusks, on which sawfish also feed.
Including the saw, sawfish can reach a
length of about 7m (24ft). The worldwide
Common sawfish is found so often in rivers
and lakes that it has been argued that it is
actually a freshwater species that only
occasionally enters the sea. It even gives
birth in fresh waters. Female sawfish
develop their eggs internally (ovoviviparous), producing up to 12 young per
litter.

Of all the skates and rays, sawfish look
most like sharks, because of their stout
bodies, large dorsal fins and relatively small
pectoral fins.

Species in the guitarfish family look much
like sawfishes without their saws, but have
larger pectoral fins. They range in length
from about 60cm (2ft) for the Atlantic
guitarfish to about 3m (10ft) in the Giant
guitarfish from the Indo-Pacific. They feed
in shallow water on mollusks and crustaceans and have flattened crushing teeth. All
species develop their eggs internally.

The guitarfish genus *Rhinobatos* contains
the largest number of species. Members of
both *Rhinobatos* and *Rhynchobatus* have a
fairly long front extension (rostrum) thus
giving the front part of the body (also called
the disk) a heart-shaped appearance. The
remaining genera (*Rhina, Platyrhina, Zap-
teryx* and *Platyrhinoides*) have much shorter
rostra and the disk looks round. These four
genera are Indo-Pacific—only *Rhinobatos*
has species in the Atlantic. Many species

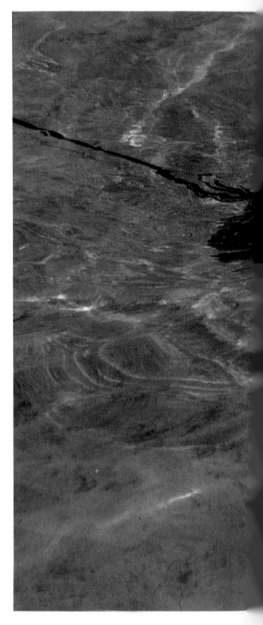

have enlarged dermal denticles, often called thorns, on their dorsal surface.

All members of the electric ray family produce electricity. There are over 30 species in 6 genera, found in all tropical and subtropical seas. Most live in shallow water though a few species are found at great depths. They are slow swimmers, spending most of their time on the ocean bed, and they feed on fishes which they capture by stunning them with electric shocks. Their large electric organs are located at the bases of their pectoral fins and can discharge over 300 volts.

The Lesser electric ray grows to about 30cm (1ft) whereas the Atlantic torpedo ray grows to over 2m (6ft) long and weighs over 90kg (200lb). Their disks are round, the tail short and stubby in most species and the

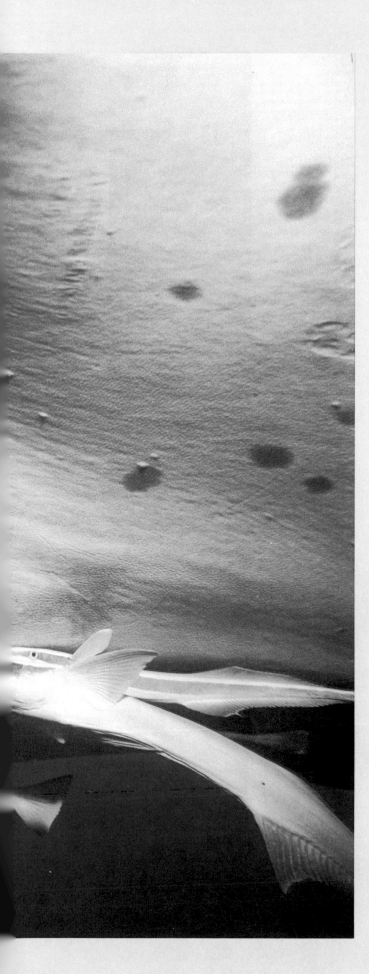

The 7 Families of Skates and Rays

Size: length: 60cm–7m (2–24ft), weight 0.5–1,600kg (1–3,500lb).

Eagle rays

Family: Myliobatidae
About twenty species in 3 genera.
Distribution: worldwide in tropical and subtropical oceans.

Electric rays

Family: Torpedinidae
About thirty species in 6 genera.
Distribution: tropical and subtropical oceans, mainly in shallow water. Species include: **Atlantic torpedo ray** (*Torpedo nobiliana*), **Lesser electric ray** (*Narcine brasiliensis*).

Guitarfish

Family: Rhinobatidae
About thirty-six species in 6 genera.
Distribution: tropical and subtropical oceans.
Species include: **Atlantic guitarfish** (*Rhinobatos lentiginosus*), **Giant guitarfish** (*Rhynchobatus djidensis*).

Mantas

Family: Mobulidae
About eight species in 2 genera.
Distribution: tropical and subtropical oceans.
Species include: **Atlantic devil ray** (*Mobula hypostoma*), **Pacific manta** (*Manta hamiltoni*).

◀ **Fast food.** Giant manta rays (here *Manta alfredi*) take in food — small fish and crustaceans — while on the move. A special adaptation enables them to sieve food from the water.

Sawfish

Family: Pristidae
Four species in 2 genera.
Distribution: tropical and subtropical oceans.
Species include: **sawfish** (*Pristis pristis*).

Skates

Family: Rajidae
About one hundred and twenty species in 12 genera.
Distribution: worldwide in cool or deep-sea waters.
Species include: **Big skate** (*Raja binoculata*), **Little skate** (*R. erinacea*).

Stingrays

Family: Dasyatidae
About one hundred species in about 19 genera.
Distribution: tropical and subtropical oceans; two genera in S America inhabit fresh water. Species include: **Atlantic stingray** (*Dasyatis sabina*), **Indo-Pacific stingray** (*D. brevicaudata*).

The 3 Families of Chimaeras

Size: length 60–90cm (2–3ft) in most species to 1.8–2.2m (6–8ft) in *Chimaera monstrosa* and *Hydrolagus purpurescens*; maximum weight 25kg (55lb).

Blunt-nosed chimaeras

Family: Chimaeridae
About thirteen species in 2 genera.
Atlantic, Pacific and Indian oceans.

Long-nosed chimaeras

Family: Rhinochimaeridae
Six species in 3 genera.
Atlantic and Pacific oceans, Caribbean Sea.

Plow-nosed chimaeras

Family: Callorhinchidae
Four species of the genus *Callorhinchus*.
Southern hemisphere.

eyes usually small. In some other species (eg *Typhonarke aysoni* from deep water off New Zealand) there are no functional eyes. The skin is naked in all species and is often beautifully marked. In reproduction eggs develop internally until hatching.

The largest group of skates and rays is the Rajidae family, with about 120 species. They are found worldwide in cool waters—even in the tropics they are common in deep, cold water, at depths greater than 2,100m (7,000ft). Always closely associated with the sea bed, they feed mostly upon mollusks and crustaceans, although they occasionally catch fish. The smallest species is the Little skate, which is common off the Atlantic coast of North America, only reaching about 50cm (20in) long and weighing less than 450g (1lb). The Big skate from the Pacific coast of North America has been recorded at over 2.5m (8ft) and 90kg (200lb).

The enlarged pectoral fins of species in this family and their fairly long snout give the disk a diamond shape. The enlarged pectoral fins of stingrays, eagle rays and manta rays as well as skates are often called wings; and from their graceful up-and-down movements in swimming it is easy to see why. In some skates the pelvic fins have been greatly enlarged and elongated and they can use these to "walk" over the sea bed. The tails of skates are long and thin, and covered with very strong, sharp thorns. The tail, so armed, is used as a defense weapon. Similar thorns or "bucklers" are on the back. The tails have weak electric organs, the four-volt impulses of which may play a part in courtship.

They lay leathery rectangular egg cases with long tendrils at each corner for anchoring to seaweed or rocks. Often washed up, the egg cases are popularly known as "mermaids' purses." The young take 6–9 months to hatch.

Stingrays derive their name from the one or more spines on the dorsal side of their tail. They are found worldwide in warm tropical and subtropical waters, some migrating into temperate waters in the summer. They spend a lot of time camouflaged on the seabed, often partially covered by sand. They can swim rapidly when disturbed, or in pursuit of fish. They also eat mollusks and crustaceans, and have flat, crushing teeth. The disk can be diamond shaped or almost round, causing a distinction between the round stingrays and the square stingrays. Two genera are called butterfly rays (*Gymnura* and *Aetoplatea*) with reference to their wide wings and short, stubby tails.

The spines on stingrays' tails are associated with venom sacs. The venom is very painful but rarely fatal to humans. The needle-sharp spines are used only for defense; each has angled barbs, which allow easy penetration, but make it very difficult to remove.

The Atlantic stingray measures only about 30cm (1ft) across its disk. The largest species is the Indo-Pacific Smooth stingray, which weighs over 340kg (750lb) with a total length of over 4.5m (15ft) and a disk width of over 2m (7ft). Several deaths have been attributed to this species in Australia. Even very large specimens will lie in water less than about 90cm (3ft) deep. Bathers swimming over them have been impaled in the chest or abdomen by the stingray's large spine, up to 30cm (1ft) long, and have subsequently died.

Although most species are marine, there are two South American genera that live only in fresh water (*Potamotrygon* and *Disceus*). Their internal physiology has so adapted to fresh water that if they are placed in salt water they rapidly die. They have an almost perfectly circular disk and are all beautifully marked with spots and bars. Lying in shallow water, covered by mud, they are greatly feared by the native indians because of their poisonous spines. Like all other stingrays, the females develop their eggs internally, producing up to twelve young per litter.

Although they have firm and tasty flesh, stingrays are very rarely eaten. However, their tail spines are commonly sold as curios, especially as letter openers.

The eagle rays are so called because of their very large pectoral wings. When seen from the front they look like eagles flying through water. There are three genera (*Myliobatus*, *Aetobatus* and *Rhinoptera*), found worldwide in tropical and subtropical seas, with about 20 species.

They have no frontal protrusion (rostrum), giving their head a very pug-nosed appearance. Their whip-like tail may be more than twice the length of their disk, with one or more spines at the base of the tail. The wings taper and are pointed at the tips, and both the eyes and vents (spiracles) are very large. The teeth are fused into large crushing tooth plates. They eat shellfish, which are found by squirting water from the mouth and blowing away the sand. Such excavations can be 30cm (1ft) deep. Only the soft flesh is swallowed; the crushed shells are spat out. Eagle rays can swim fast enough to take off and glide through the air. Why they do this is unknown. A large shoal can contain several hundred eagle rays.

▲ **Yellow stingrays** (*Urolophus jamaicensis*) shoaling in the Galapagos Islands.

► **This leathery disk** is a Thorny ray (*Urogymnus asperrimus*), one of the hundred species of stingray. Stingrays are found in all oceans, usually in waters less than 100m (330ft) deep, and are most abundant close to shore.

▼ **Blunt-nosed chimaeras** or ratfishes live in cold waters in the North Atlantic, off South Africa, and in the Pacific. The first dorsal fin is linked to a poison gland. This is *Hydrolagus colliei.*

The manta rays are the giants of the skates and rays. The Pacific manta reaches a wingspan of over 6m (20ft) and a weight of over 1,600kg (3,500lb), whereas the wingspan of the Atlantic devil ray is only 1.5m (5ft). Worldwide in tropical and subtropical seas, there are two genera (*Manta* and *Mobula*) and about 8 species.

They have two fins that project forwards from the head, which look like horns—hence the common name of devil-fish. They are truly inhabitants of the open sea, where they swim by flapping their strong pectoral wings. Occasionally they will float at the surface, apparently "basking." Like eagle rays they can swim at great speed and can make spectacular leaps into the air. It has been suggested that they leap either to knock off parasites or to stun schools of small fish just under the water's surface. There have also been several reports of females giving birth to their young during such leaps. (The young develop by feeding upon uterine "milk," like the eagle rays.) The tail is long and whip-like, sometimes with spines at its base. The back is a uniform dark gray to black, while the underside is pure white. Like some of the other giants they are filter feeders, sifting plankton and small fish out of the water by means of special modifications of the gill arches as they swim along. They travel in small groups. Although they are not commercially fished, one individual would provide a large quantity of meat.

A large, blunt head, incisor-like teeth at the front of the mouth, a spine with a venom-producing sac at its base and an elongate tail are some characteristics of the group of fish known as **chimaeras**. Their form recalls the she-monster of Greek mythology (which had a lion's head, a goat's body and a serpent's tail) from which the name chimaera is derived. These relations of the sharks are also known as ratfish or rabbit fish; indeed one of the generic names (*Hydrolagus*) literally means water (*hydros*) rabbit (*lagus*).

All chimaeras occur in cold water, living in subarctic waters or subantarctic waters, often at great depths—some have been recorded as deep as 2,400m (8,000ft). They are poor, slow swimmers. Instead of using powerful side-to-side body movements like most fishes, especially sharks, they swim by flapping their pectoral fins, which makes them bob up and down in an awkward and clumsy fashion. Chimaeras usually keep close to the bottom. They have been observed motionless on the bottom, sitting up on the tips of their fins.

Like all other cartilaginous fishes (sharks, rays etc), males have pelvic claspers to introduce sperm into the female. However, male chimaeras also have a unique clasper projecting from the middle of the forehead which, it is believed, helps to hold the female during mating. Long-nosed and plow-nosed chimaeras have very elongated, fleshy, flexible and unusually shaped head projections (rostra) which are heavily covered with electrical and chemical receptors, possibly used to help locate prey or breeding partners.

All the teeth are fused together to form a solid beak, like that of a bird or turtle, which is used to crush the shells of their food, bottom-dwelling mollusks and crustaceans.

Young chimaeras are covered by short, stout dermal denticles (minute teeth) which are lost as they mature, except in long-nosed chimaeras which keep some of them for life.

Chimaeras have very large, iridescent, metallic blue-green eyes, whose form relates to the low light levels at the depths where they live.

All chimaeras lay fairly large eggs (15–25cm, 6–10in) with a hard leathery shell. They hatch in about 3 months. It is thought they lay only two eggs at a time, one from each uterus.

Sharks take water in through the mouth to pass over the the gills and absorb the oxygen they require. Chimaeras do not take in water through their mouths but through large nostrils connected to special channels which direct the water to the gills (see also p143). Unlike sharks, chimaeras have their gills in a common chamber protected by a flap (operculum).

The first dorsal fin of the chimaeras is preceded by a freely movable spine, which is linked to a poison gland that produces venom reported to be painful to humans. The first dorsal fin is short and high; the second is long and low, and gently merges into the long, tapering caudal fin forming, in many species, a filamentous "whip" as long as, or longer than, the rest of the body. Only the plow-nosed chimaeras have an anal fin. In members of the genus *Chimaera* what appears to be a long, low anal fin is merely a fold of skin.

The flesh of chimaeras is very solid and tasty, but they are rarely eaten except in New Zealand and China, though in some parts of the Atlantic seaboards the flesh is appreciated for its strong laxative effect. In Scandinavia they are caught mainly for their liver oil, which is used for medical purposes and as a lubricant. GDi

AQUATIC INVERTEBRATES

AQUATIC INVERTEBRATES

The sequence of phyla is as they appear in this book. For evolutionary relationships, see p147 and also text on the individual phyla.

KINGDOM PROTISTA—SINGLE-CELLED ANIMALS

Subkingdom Protozoa

Flagellates, amoebae, opalinids
Phylum: Sarcomastigophora

Ciliates
Phylum: Ciliophora

Phylum: Apicomplexa—includes malaria, coccidiosis

Phylum: Labyrinthomorpha

Phylum: Microspora

Phylum: Ascetospora

Phylum: Myxospora

KINGDOM: ANIMALIA

Subkingdom Parazoa

Sponges
Phylum: Porifera

Subkingdom Mesozoa

Mesozoans
Phylum: Mesozoa

Subkingdom Metazoa—Multi-cellular Animals

Sea anemones, jellyfishes, corals and allies
Phylum: Cnidaria

Comb jellies or sea gooseberries
Phylum: Ctenophora

Endoprocts
Phylum: Entoprocta (Kamptozoa)

Rotifers or wheel animalcules
Phylum: Rotifera

Kinorhynchs
Phylum: Kinorhyncha

Gastrotrichs
Phylum: Gastrotricha

Lampshells
Phylum: Brachiopoda

Moss animals or sea mats
Phylum: Bryozoa (Ectoprocta)

Horseshoe worms
Phylum: Phoronida

Flatworms (flukes, tapeworms)
Phylum: Platyhelminthes

Ribbon worms
Phylum: Nemertea

Roundworms
Pylum: Nematoda

Spiny-headed or thorny-headed worms
Phylum: Acanthocephala

Horsehair worms or threadworms
Phylum: Nematomorpha

Segmented worms
Phylum: Annelida

Echiurans
Phylum: Echiura

Priapulans
Phylum: Priapula

Sipunculans
Phylum: Sipuncula

Beard worms
Phylum: Pogonophora

Arrow worms
Phylum: Chaetognatha

Acorn worms and allies
Phylum: Hemichordata

Crustaceans
Phylum: Crustacea

Chelicerates
Phylum: Chelicerata

Uniramians
Phylum: Uniramia
Terrestrial, including insects, millipedes,

Water bears
Phylum: Tardigrada

Tongue worms
Phylum: Pentastomida

Velvet or walking worms
Phylum: Onychophora
Terrestrial.

Mollusks
Phylum: Mollusca

Spiny-skinned animals
Phylum: Echinodermata

Chordates
Phylum: Chordata
Includes the aquatic sea squirts and lancelets, and also all vertebrates.

WHAT do crabs, sea urchins, earthworms, malaria parasites and corals have in common? In fact, these very diverse groups share very little, apart from the fact that they all lack a backbone. Of the 1,071,000 or so known species of animal about 1,029,300 are invertebrate, that is, over 95 percent are animals without backbones. Invertebrates make up the bulk of animals, measured both in terms of numbers of species recognized and numbers of individuals. Some invertebrates, like garden snails and earthworms, are conspicuous and familiar animals, while others, although abundant, pass unnoticed by most people.

Invertebrate body forms range in size from the lowly microscopic *Amoeba*, which may be just one micrometer in diameter, to the Giant squid 18m (59ft) in length, a ratio of 1:18,000,000. They include life forms as diverse as the Desert locust and the sea anemones. They inhabit all regions of the globe, and all habitats, from the ocean abyss to the air. Life almost certainly originated in the seas, and virtually all the major invertebrate groups (phyla) have marine representatives. Somewhat fewer (almost 14 phyla) have conquered freshwater. Fewer still (about 5 phyla) live on land and of these only the jointed-limbed groups (arthropods) have mastered their air and really dry places. Most numerous among arthropods are the uniramians (eg insects, millipedes, centipedes) and the chelicerates (eg scorpions, spiders, ticks, mites).

Many invertebrates, like slugs (whether of the garden or sea), are free living; others, such as barnacles, are attached to the substrate throughout their adult life; yet others live as parasites in or on the bodies of plants or other animals. Some inverte-

brates are of great commercial significance, either as direct food for man (eg prawns and oysters), or as food for man's exploitable reserves (eg the planktonic copepods on which herring feed). Others (eg earthworms) are much appreciated because they improve the soil for agriculture. Many invertebrates live as parasites, either in man himself or in the bodies of domestic animals and plants, where great damage may be wrought, and so are of great medical or agricultural importance.

This great diversity of form and life-style in animals has led zoologists to sort out, or classify, animals according to type and evolutionary connections. To be certain that they are speaking of the same animals, they give each species a unique scientific name (eg *Lumbricus terrestris*, the Common earthworm). Every species is classified into one of the major groups, or phyla. A phylum comprises all those animals which are thought to have a common evolutionary origin. According to their understanding of the probable evolutionary processes involved, zoologists may recognize some 39 animal phyla. With one exception, they are made up exclusively of invertebrate animals. The phylum Chordata includes all animals with a hollow dorsal nerve cord. Nearly all—including fishes, amphibians, reptiles, birds—have a backbone, but some are invertebrate, such as the sea squirts.

The technical classification that zoologists employ leads from the most primitive and simple animal types to the most complex and advanced. In order to achieve some form of system, various levels of organization are recognized which give clear distinctions between phyla. The most fundamental of these lies in the number of cells in the body. A cell is the smallest functional unit

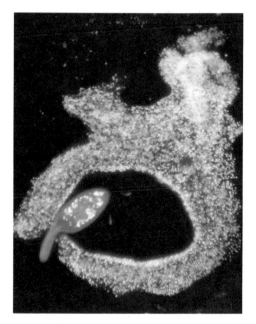

▲ **Cell eats cell.** A microscopic *Amoeba* engulfs a ciliate *Paramecium*, both single-celled animals (× 100).

◄ **Sophisticated cephalopod.** The octopus is the most intelligent of all invertebrate animals. It has a memory and can learn.

Classification of a species, *Asterias rubens*, the European Common starfish.

Kingdom Animalia
Subkingdom Metazoa
Phylum Echinodermata
Subphylum Asterozoa
Class Asteroidea
Subclass Euasteroidea
Order Forcipulatida
Family Asteriidae
Genus *Asterias*
Species *rubens*

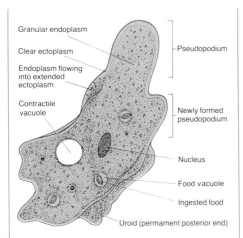

Granular endoplasm

Clear ectoplasm

Endoplasm flowing
into extended
ectoplasm

Contractile
vacuole

Pseudopodium

Newly formed
pseudopodium

Nucleus

Food vacuole

Ingested food

Uroid (permanent posterior end)

▲ **The single protozoan cell** may be quite a complex structure, with specialized parts called organelles responsible for different functions including feeding and locomotion (× 125).

▼ **Number and arrangement of cell layers** distinguish degrees of complexity in many-celled animals, from the single layer of monoblastic sponges and mesozoans (1), through the diploblastic jellyfishes and comb jellies (2), to the three layers of most animals. The bulky middle layer (mesoderm) of triploblastic animals may be solid, as in flatworms and ribbon worms (3), or divided into inner and outer parts separated by a cavity or coelom (4). Some groups (eg nematode worms) have a body cavity that is not formed within the mesoderm and is called a pseudocoel (5).

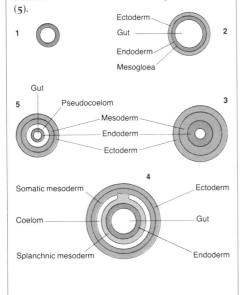

1

2

Ectoderm
Gut
Endoderm
Mesogloea

5

3

Gut
Pseudocoelom
Mesoderm
Endoderm
Ectoderm

4

Somatic mesoderm
Coelom
Splanchnic mesoderm
Ectoderm
Gut
Endoderm

► **Relationships between animal phyla,** showing grade of organization, numbers of cell layers, presence or absence of a coelom, and site of development of the mouth in the embryo—the protostome/deuterostome distinction.

of an animal, governed by its own nucleus which contains genetic material, known as DNA.

Protozoans—the "simplest animals"

The animals with bodies made up of a single cell represent a separate level of organization from all the rest, for their body processes are performed by the one cell, which therefore itself cannot be specialized. In multi-cellular animals the responsibility for different life processes is shared out, different cells being specialized for different tasks, eg receiving stimuli (sensory cells), communication (nerve cells), movement (muscle cells) etc.

The single-celled animals or protozoans (sometimes referred to as acellular, ie without distinct cells) are assumed to be low on the scale of evolutionary sophistication. In reality their single cell is often large and complex. The essential life processes are carried out by special regions (organelles) within the cell—nucleus for government, bubble-like food vacuoles for energy acquisition, vacuoles that contract and expand to regulate water levels within the cell and so on.

Another fundamental question may be asked about single-celled animals. Are they in fact all animals? True animals derive their energy from relatively complex organic material that is plant or animal in origin—they are heterotrophic; they obtain their nourishment by a process of breaking down organic materials (are catabolic). However, some protozoans (eg green flagellates) have developed structures characteristic of plants, called chloroplasts, which are incorporated into their cells. The chloroplast enables the cell to synthesize organic materials, such as sugars, from mineral salts, carbon dioxide and water in the presence of sunlight (photosynthesis). Like green plants, these protozoans can subsist on inorganic food (are autotrophic) from which they synthesize organic material (are anabolic). Botanists consider such protozoans to be single-celled plants. The difficulty is that while some protozoans are clearly plants and others clearly animals, some, like *Euglena gracilis*, can feed by both autotrophy *and* heterotrophy. For reasons such as this some authorities believe that most protozoans warrant classification as an animal subkingdom, with obviously

			PROTOSTOME	Protostome with some deuterostome tendencies	DEUTEROSTOME		
Organ grade	with coelom and blood vascular system	**Metameric** (many segments)	Arthropoda (7 phyla) Annelida		Chordata	Subkingdom **Metazoa**	KINGDOM ANIMALIA
		Oligomeric (few segments)	Pogonophora Mollusca	Brachiopoda Ectoprocta Phoronida Chaetognatha	Hemichordata Echinodermata		
	lacking true coelom	**Americ** (lacking segments)	Echiura Sipuncula Priapula				
		Pseudocoelomate vascular	Nemertea				
		Pseudocoelomate non-vascular		Endoprocta Acanthocephala Nematoda Nematomorpha Gastrotricha Rotifera Kinorhyncha			
		Acoelomate (no coelom)	Platyhelminthes				
		Triploblastic (3 layers)					
Tissue grade		**Diploblastic** (2 layers)	Ctenophora Cnidaria				
Cellular grade		**Monoblastic** (1 layer)	Porifera Mesozoa			Subkingdom **Parazoa** Subkingdom **Mesozoa**	
		Multi-cellular					
		Single-celled	Protozoa (7 phyla)				KINGDOM PROTISTA

plant-like protozoans treated as members of the plant kingdom, outside it.

Recently scientists have proposed a new classification of the living world which overcomes the problem of whether protozoans are plants or animals. This reorganization proposes five kingdoms: Monera (bacteria and blue-green algae); Protista; Plantae (plants); Fungi and Animalia. The protozoans are included in the Protista as the subkingdom Protozoa, along with some other groups previously classified as algae by botanists.

Origins of life

Protozoans are most important because, as the "simplest animals" they are likely to provide important keys to two fundamental questions. These concern the origin of life, and the origin of multi-cellular metazoan animals. The origin of life is shrouded in scientific speculation. The biblical account of the origin of life as presented in *Genesis* is an historic attempt to answer one of man's most fundamental questions. Belief in the idea of a Divine creation is a matter of faith which cannot be tested by science. Nor can science yet tell us how life first began, although it has also been shown that primitive ideas such as spontaneous generation of life are completely wrong.

It is thought that the earth is not quite 5,000 million years old, and realistic estimates suggest that life began in its simplest form 4,000 million years ago. The first sedimentary rocks, not quite 4,000 million years old, contain fossils of simple cells which resemble those of present-day bacteria, that is, they lacked a distinct nucleus (ie were prokaryotes). These lived in a primeval atmosphere devoid of oxygen. The appearance of oxygen on the earth, 1,800 million years ago, brought with it many new evolutionary possibilities. The protozoans, green algae, higher plants and animals appeared a lot later, the earliest known fossils having been taken from rocks 1,000 million years old. All these organisms are made up of cells with nuclei (ie are eukaryotes). Thus the startling fact emerges that for three-quarters of the period for which life has existed on earth the only cells were prokaryotes, ie resembling bacteria. It was not until the Cambrian period 600–500 million years ago that invertebrates such as mollusks, trilobites, lampshells and echinoderms became established. Invertebrates with soft delicate bodies, such as flatworms and sea squirts, have not left any fossil record.

For life to have appeared many conditions

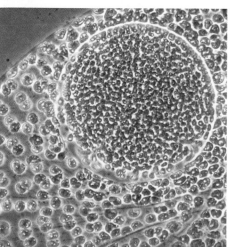

▲ **Colony of polyps.** Cnidarian colonies like this gorgonian arise from a many-celled floating planula larva which settles, develops into a polyp, and buds off other polyps asexually. The form of the colony varies with the species and situation.

◄ **Sharing common ancestry** with the higher animals. A colony of single-celled flagellates (*Volvox* species), with a daughter colony in formation (× 200). Ancestors of these colonies may have developed into permanent two-layered organisms somewhat resembling the planula larvae of sea anemones, corals and jellyfishes.

had to be fulfilled. It seems quite possible that the physical conditions prevailing on the surface of the early earth could have generated simple organic molecules such as amino acids, and then proteins, from inorganic molecules. The big unanswered question is how such substances could form themselves into organized living systems capable of reproducing their own kind.

Multi-cellular animals

The origins of multi-cellular animals are also speculative, but rather more can usefully be said about their possible early history. Because so many of the early animals had soft bodies they left very little fossil record. All theories about the early evolution of animals therefore rely mainly on the study of similarities between developing embryos and adults of animals in different groups. This allows inferences to be drawn about common ancestry. Two chief theories have been put forward.

According to one theory, protozoan animals gave rise to multi-cellular ones by colony formation. A number of types of colonial protozoan are known to exist, such as *Volvox*.

The famous 19th-century German biologist Haeckel proposed that a hollow *Volvox*-like ancestor could have developed into a two-layered organism. Views differ as to whether or not this was a planktonic or a bottom-dwelling organism, but it may have somewhat resembled the planula larvae of the sea anemones and jelly fishes (p149) and could have given rise to bottom-dwelling animals such as adult hydroids and sea anemones. Although most animals are composed of three layers of cells, often in a highly modified form, there is evidence that the evolution from two-layered animals did occur in evolutionary history.

A different theory proposes that multicellular animals arose from single-celled animals containing many nuclei by the growth of cell walls between each nucleus. A number of protozoans, for example *Opalina*, are like this (see p158). According to this theory, the primitive multi-cellular animal would lack a gut, as in present-day gutless flatworms (see p198). However, the latter have three layers of cells, which raises the question—what is the origin of the two-layered animals? Because of this and other criticisms, this theory is now generally discarded in favor of the colonial one.

While it is not certain how the multicellular animals evolved from protozoan ancestors, it is possible to distinguish groups of metazoans on the basis of relative sim-

plicity or complexity of structure. (Some biologists divide invertebrates into protostomes and deuterostomes, see below.)

In metazoans there are three categories that can be used to determine level of complexity: how the cells are organized; how many layers of cells are to be found within the body; and whether or not a body cavity is present.

There are few types of cell in the most lowly metazoans and in sponges and these cells are never arranged into groups of similar cells (tissues). Such animals are said to have a *cellular grade of organization*. In the next step, as found in jellyfishes and allies and comb jellies, cells with similar functions are arranged together into tissues, each tissue having its own function or series of functions—these animals have a *tissue grade of organization*. In all animals apart from those just mentioned, requirements for functional specialization increase such that specific organs (often comprising a series of tissues) have evolved. Thus all animals from flatworms to man are said to have an *organ grade of organization*.

The second way to divide up multicellular animals is to look at the number of layers of cells that make up the animal's body. Mesozoans and sponges consist of just one layer of cells, but in the jellyfishes and comb jellies two layers appear (ectoderm outside and endoderm inside). This "diploblastic" condition contrasts markedly with the single layer of cells seen in the protozoan colony *Volvox* (p158), which is described as monoblastic. The two layers develop from the egg and remain throughout adult life, separated from each other by a sheet of jelly-like mesogloea.

▲ **Drifting predator,** the jellyfish *Rhizostoma octopus* of northern waters. In jellyfishes the two cell layers of the body are separated by an extensive jelly-like mesogloea. Jellyfishes lack a brain and their movement is under control of a primitive nerve net. Most jellyfishes live in warm waters.

▶ **Crustacean in a mollusk shell.** A hermit crab (*Dardanus lagopodes*) takes shelter in the former home of a shellfish. Eyes, antennae, palps, pincers, external gills and legs are paired appendages on each body segment, evolved to fulfill a wide range of functions, including sense perception, feeding, locomotion and copulation.

▼ **Crowns of fanworms** (here *Protula maxima*) act as feeding organs and gills. Giant nerve fibers running the length of these polychaete worms enable them to retract the gaily colored crown with startling rapidity.

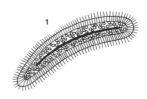

All animals "above" the jellyfishes and comb jellies are equipped with three layers of cells and are known as triploblastic. Here the ectoderm and endoderm are separated by a third cell layer, the mesoderm. The mesoderm forms the most bulky part of many animals, contributing the musculature of the body wall as well as that of the gut. In some of these triloblastic animals, eg flatworms (Platyhelminthes) and ribbon worms (Nemertea), the mesoderm is solid and not itself divided into two layers by a body cavity (coelom), so they are described as acoelomate. Without a body cavity these animals are at a disadvantage because the body movements affect the movements of the gut and vice versa. To be able to move the gut independently of the body wall is a great advantage, as it allows for sophisticated digestive activities. Such a condition is reached only in coelomate animals, in which the mesoderm is divided into an outer section forming the body wall and an inner section forming the muscles of the gut, and the two are largely separated by a body cavity (coelom) within the mesoderm. The possession of a fluid-filled body cavity is the hallmark of all the more advanced invertebrates and the major phyla, including mollusks, annelids, the different arthropods, echinoderms and chordates, all have a coelom (are coelomates).

In some groups there is a body cavity between the body wall and the gut, but this is a pseudocoelom, not a true coelom, for it is not formed inside the mesoderm. Generally it persists from an early stage in development and it contains fluid. In animals such as the nematode worms it performs an important function as a fluid skeleton.

Some evolutionary biologists suggest a different way of grouping invertebrates, believing that two main evolutionary lines have emerged in the animal world, the protostomes (first mouth) and deuterostomes (second mouth). In the early embryos of protostomes such as annelid worms and insects, the mouth is formed at or near the site of the blastopore. In deuterostomes such as the echinoderms (starfishes and sea urchins) and the chordates, the anus forms at the blastopore. Other distinctions can be seen by comparing the development of the two types. In the annelids the nerve cord is on the underside, a double, solid structure reminiscent of the nerve cord of insects. In the chordates it is a single hollow structure along the upper side of the animal. The annelid coelom is formed by the splitting of the mesoderm, but in the echinoderms and the chordates it develops from two pouches of the primitive gut. The sea-dwelling representatives of these two "lines" have characteristically different larvae.

These distinctions are not definite evidence of links between phyla, but indications of possible evolutionary affinities. Certainly the chordates are likely to have arisen from a deuterostome type of ancestor, and the form of development shown by annelids and mollusks places them a long way from the echinoderms and chordates in any phylogeny or "family tree."

Symmetry

One of the most obvious differences separating the phyla of the animal kingdom is the overall appearance of the animals. The majority of animals, including the vertebrates, worms and jointed-limbed forms, are bilaterally symmetrical; complementary right and left halves are mirror images.

▲▶ Larval types. In most aquatic invertebrates a larva hatches from the egg, floats free, and eventually undergoes a dramatic metamorphosis into adult form. In some an adult-like juvenile emerges. Generally, adults with simple body architecture have larvae that are simple in form (eg the planula of sponges and jellyfishes). The tadpole-like sea squirt larva may have provided the evolutionary springboard from which vertebrates developed. (1) Planula (sponge, jellyfish and other cnidarians); (2) trochophore (many worms); (3) cyphonautes (sea mat); (4) tornaria (hemichordates including acorn worm); (5) nauplius (first stage of many crustaceans); (6) zoea (crabs and other decapod crustaceans); (7) veliger (mollusk); (8) pluteus (brittle star, sea urchin); (9) auricularia (starfish, sea cucumber); (10) "tadpole" larva of sea squirt.

◀▼ Most animals are symmetrical. The majority are bilaterally symmetrical with the left mirroring the right side, and with front and rear ends. In many worms a head is barely distinguishable, but jointed-limbed animals such as crustaceans, and most mollusks, have clearly developed heads.

The straightforward symmetry of the Banana slug (*Ariolimax californicus*) BELOW with its paired tentacles conceals the internal twisting of body organs common to most gastropod mollusks.

Two major groups show a different, radial symmetry without a head—the jellyfishes and their relatives, and echinoderms such as this starfish *Marthasterias glacialis* ABOVE LEFT with its five-rayed symmetry. One arm is in the process of regeneration, its predecessor having been broken or torn off by a predator.

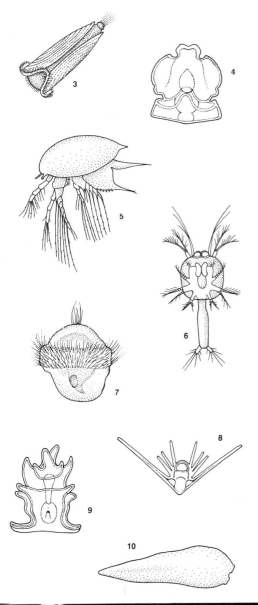

Their front and hind ends are however dissimilar. This is because in these animals there has been some specialization at the head end. In the lowly flatworms the head is only feebly developed, but it is identifiable. In the jointed-limbed animals, such as the insects, it is clearly defined. The evolution of a head end (cephalization) has come about in direct response to the development of forward movement. Clearly it is an advantage to have specialized sensory equipment (eg eyes, smelling and tasting receptors at the front of the body) to deal efficiently with environmental stimuli as they occur in the direction of travel. In this way prey and predators can be quickly identified.

The position of sense receptors is often associated with the mouth opening. These factors have led to the development of aggregations of nervous tissue to integrate the messages coming from the receptors and to initiate a coordinated response (eg attack or flight) in the muscle systems of the body. This brain, be it ever so simple, came to lie at the front end of the body, often near the mouth, and the typical head arrangement was formed. In addition to a distinct head, other features of bilaterally symmetrical animals are thoracic regions (modified for respiration) and abdominal regions (modified for digestion, absorption and reproduction). The function of locomotion may be undertaken by either or both of the regions. In the mollusks the principles of bilateral symmetry are somewhat disguised by the unique style of body architecture, but if these animals are reduced to their simplest form, as in the aplacophorans and chitons, a basic bilateral symmetry is visible.

Some animals have, however, managed without a head. Despite their relatively high position in the table of phyla, the echinoderms (eg starfishes, sea urchins) are headless. This is all the more surprising when one realizes that the earliest echinoderms and their likely ancestors were bilateral animals. Still, they managed without a head, and in the present-day echinoderms the nervous tissue is spread fairly evenly throughout the body and all sensory structures are very simple and widely distributed. Modern adult echinoderms are fundamentally radially symmetrical; the body parts are equally arranged around a median vertical axis which passes through the mouth. There is no clear front end, nor left and right sides, in most of the species. Radial symmetry is best for a stationary lifestyle, in which food is collected by nets or fans of tentacles. This was almost certainly the life-style of the earliest echinoderms, as it is of most of present-day Cnidaria (jellyfishes and their allies), the other phylum to display a clear radial symmetry. The radial symmetry in the Cnidaria is evident, whereas in the echinoderms a unique five-sided body form has been imposed on it, as can be seen in present-day starfishes and brittle stars.

Finally, there are those animals which are essentially without regular form or are asymmetrical. These are the sponges, which grow in a variety of fashions, including encrusting, upright or plant-like, or even boring into rocks. While each species has a characteristic development of the canal system inside the body, its precise external appearance depends very much on prevailing local conditions, such as exposure, currents and form of substrate. The simple body form of the sponges of course rules out the development of nerves and muscles, so a coordinating nerve system has not evolved.

The bilaterally symmetrical bodies of coelomate metazoans may be unsegmented (americ), divided into a few segments (oligomeric), or divided into many structural units (metameric). The bodies of many types (eg ectoprocts, or moss animals, and echinoderms) show division into a few segments during their development, but these may be masked in the adult. In the annelids and arthropods the repetition of structural units along the body (metamerism) allows for the modification of the segmental appendages to fulfil various functions. In the arthropods these appendages or limbs, often jointed, carry out a wide range of activities, from locomotion to copulation.

AC

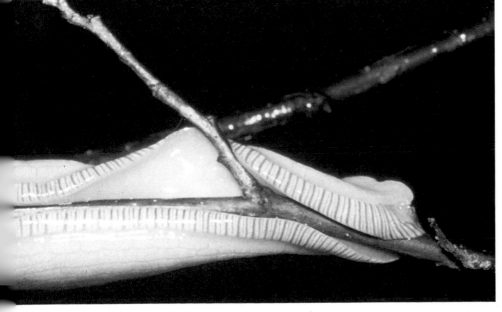

Drifters and Wanderers
The ecology of zooplankton

The word "plankton" means drifter or wanderer. It refers to those plants and animals that are swept along by water currents rather than by their own swimming ability. (Animals that swim and determine their own direction are called nekton.) The greatest diversity of plankton exists in the world's seas and oceans, but lakes and some rivers have their own plankton communities. Here examples from the sea will be used. Plants of the plankton are known as phytoplankton and animals of the plankton as zooplankton. While many zooplankton are small animals, less than 5mm (0.2in) long, a few are large, for example some jellyfishes, whose tentacles may reach 15m (49ft) in length or more. A number of zooplankton can actually swim, but not sufficiently well to prevent them from being swept along by currents in the water. However their swimming ability may be sufficient to allow them to regulate their vertical position in the water which can be very important as the position or depth of their food can vary around the daytime/nighttime cycle (for example phytoplankton rises by day and sinks by night whereas zooplankton does the reverse).

Seawater contains many nutrients important for plant growth, notably nitrogen, phosphorus and potassium. Their presence means that phytoplankton can photosynthesize and grow while they drift in the illuminated layers of the sea. Two forms of phytoplankton, dinoflagellates and diatoms, are particularly important as founders in the planktonic food webs, for upon them most of the animal life of the oceans and shallow seas ultimately depends. By their photosynthetic activity the dinoflagellates and diatoms harness the sun's energy and lock it into organic compounds such as sugars and starch which provide an energy source for the grazers that feed on the phytoplankton.

Zooplankton comprises a wide range of animals. Virtually every known phylum is represented in the sea, and many examples of marine animals have planktonic larvae. Such organisms may be referred to as meroplankton or temporary plankton. Good examples are the developing larvae of bottom-dwellers such as mussels, clams, whelks, polychaete worms, crabs, lobsters and starfish. These larvae ascend into the surface waters and live and feed in a way totally different from that of their adults. Thus the offspring do not compete with the adults for food or living space and the important task of dispersal is achieved by ocean currents. At the end of their plank-

tonic lives the temporary plankton have to settle on the seabed and change into adult forms. If the correct substrate is missing then they fail to mature. Often involved physiological and behavioral processes take place before satisfactory settlement can be achieved, and many settling larvae have elaborate mechanisms for detecting textures and chemicals in substrate surfaces.

In addition to the temporary plankton there is a holoplankton: organisms whose entire lives are spent drifting in the sea. Of these the most conspicuous element (around 70 percent) are crustaceans. The most abundant class of planktonic crustaceans are the copepods, efficient grazers of phytoplankton especially in temperate seas. The euphausids make up another very important group of crustacea, and in some regions, eg the southern oceans, they can occur in enormous numbers as "krill," providing the staple diet of the great whales. All these crustaceans have mechanisms for straining the seawater to extract the fine plant cells from it. Other noncrustacean holoplanktonic forms that sieve water for food are the planktonic relatives of the sea squirts. Some rotifers live as herbivores in the surface waters of the sea, but they are a much more important component of the plankton of lakes and rivers. Along with many invertebrate larvae, these herbivorous holoplanktonic forms are important in harvesting the energy contained in the planktonic and pass it on to the carnivorous zooplankton by way of the food webs of the sea's surface.

There are many types of carnivorous zooplankton in the seas of the world and

▲ **Drifters**—centric diatoms protected inside their strong cell walls. In some species these are ornamented or extended to give lift in the water column, functioning like miniature wings.

▼ **Cycle of nutrients and energy** in the sea. This diagram is considerably simplified since many animals obtain food from several levels, forming an intricate "food web."

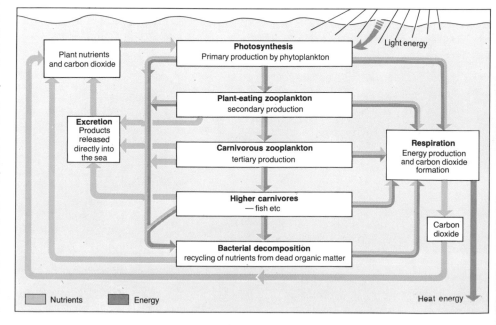

▼ **Myriads of animals** under the microscope: a mixture of medusae, copepods, crab larvae, crayfish larvae etc. Some of the larvae are being eaten by the medusae.

members of many phyla are involved. Protozoans feeding on bacteria or other protozoans occur. Some like the foraminiferans and radiolarians form conspicuous deposits on the seabed after they die because of the enduring nature of their mineralized shells or tests. Cnidarians provide a range of temporary and permanent planktonic carnivores. Many hydroid medusae spend only part of the life cycle of the hydroids in the plankton, while others like the Portuguese man-of-war are permanent plankton dwellers often taking food as large as fishes. The ctenophores, for example *Pleurobrachia* and *Beroë*, are efficient predators of copepods, often outcompeting fishes (for example her-

rings) for them. Thus they are of economic significance as competitors of commercial fish stocks. Other carnivores include pelagic gastropods, polychaetes and arrow-worms.

The occurrence of certain species in surface waters is taken by oceanographers as an indication of the origins of water currents. Thus in Northwest Europe plankton containing the arrow worm *Sagitta elegans* has been demonstrated to come from the clean open Atlantic whereas water containing *Sagitta setosa* is known to have a coastal origin. Different chaetognaths also appear at different depths in the ocean and are indicative of different animal communities.

AC

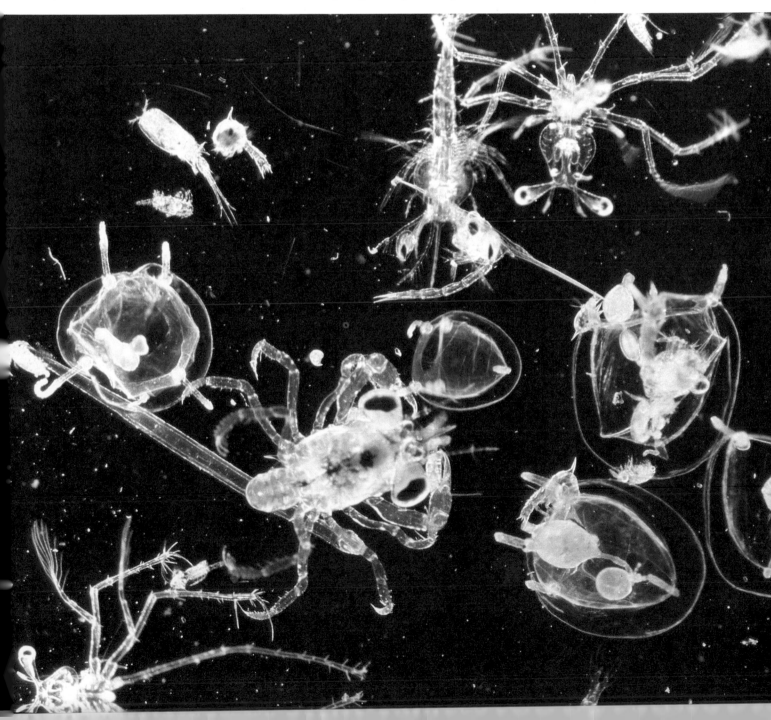

SIMPLE INVERTEBRATES

**KINGDOM PROTISTA—
SINGLE-CELLED ANIMALS**

Subkingdom Protozoa

Flagellates, amoebae, opalinids
Phylum: Sarcomastigophora

Ciliates

Phylum: Ciliophora

Phylum: Apicomplexa—includes malaria,
coccidiosis

Phylum: Labyrinthomorpha

Phylum: Microspora

Phylum: Ascetospora

Phylum: Myxospora

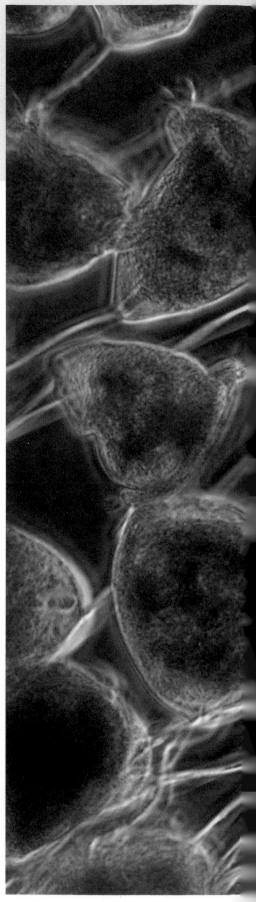

THE expression "Simple Invertebrates" implies an elementary level of organization and superficially this is the case. Here, protozoans, that is single-celled or non-cellular animals, and mesozoans (which have no common name) are simple when compared with other animal architecture. These animals are constructed from one cell or colonies of similar cells, so that there are no tissues and no organs as found in the more highly evolved invertebrates. In virtually every case one cell is capable of fulfilling all the requirements for life including energy acquisition, which may often involve food or prey detection and coordinated movement. This means that the single cell must have a means of detecting stimuli and in turn exciting and controlling a system of movement, for example by the contraction of fibrils or the beating action of cilia and flagella. In addition the cell must be able to control its water content and to have means of resisting drying out. It must be able to reproduce and the offspring must be able to grow in a coordinated fashion and to disperse through the environment. For all these operations to be carried out within one or a few similar cells means that the processes going on inside these cells have to be complex and involved. There is no division of labor in these animals. The result is that protozoan cells tend to be larger and to contain a diversity of parts (nucleus, vacuoles etc) called organelles. All processes necessary for life can be carried out in the one cell.

Because of the high surface-area-to-volume ratio, these animals lose water easily. They are therefore dependant on free water and as such live either in aquatic habitats or in damp places like soil. Many are successful parasites of other animals, flourishing in blood and guts.

In protozoans there are various symmetries, some animals having recognizable front and rear ends, for example *Paramecium*, others being attached by a stalk, for example *Vorticella*, and yet others being totally variable in shape as in amoebae. AC

▶ **Bell-shaped cells,** ringed with thread-like cilia are characteristic of the protozoan ciliate *Vorticella*. These animals are found commonly in freshwater attached by contractile stalks to water plants and larger animals. Minute food particles extracted from the water pass into the vortex created by the beating cilia.

▼ **Are they plants or animals?** This is a question that has often been asked about protozoans, which like true plants contain photosynthetic chloroplasts. Here two dinoflagellates (*Pyrocystis* species) are each in the process of dividing within their original cell walls.

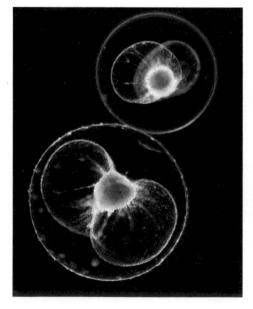

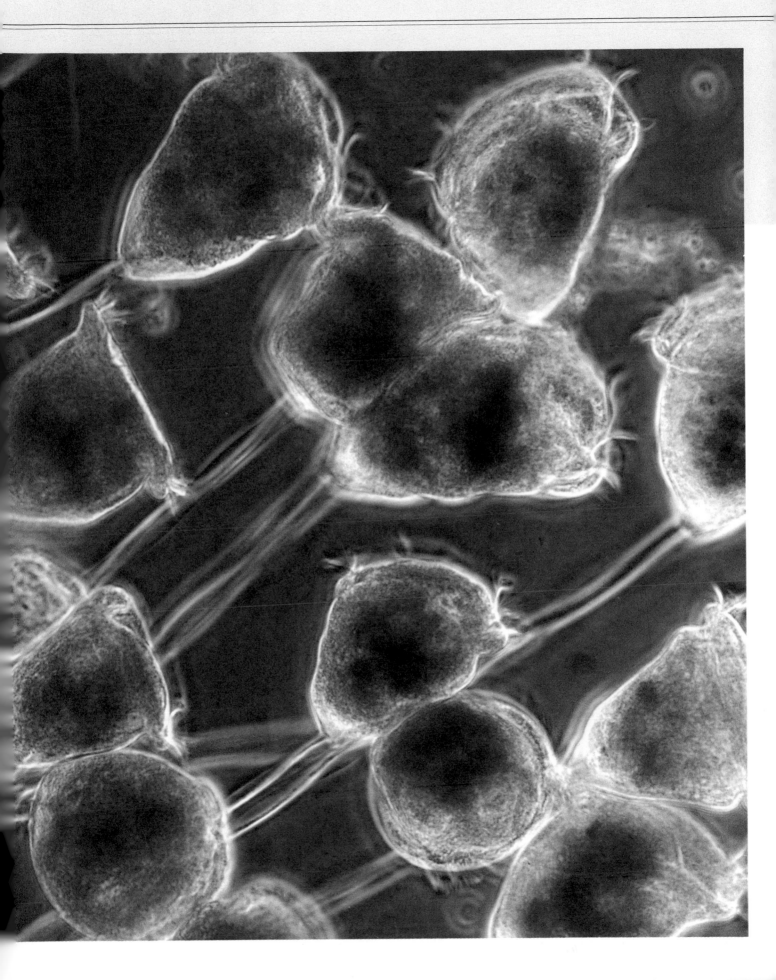

PROTOZOANS

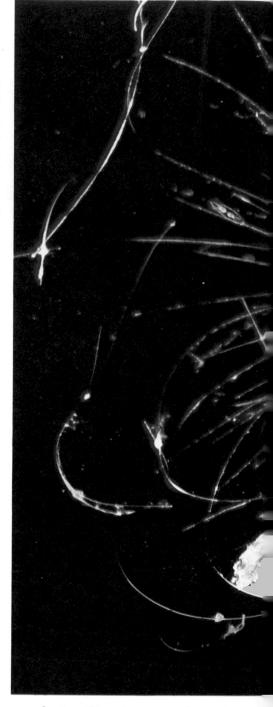

PROTOZOANS live in an unseen world. They are invisible to the naked eye, but occur all around us, beneath us and even within us—they are ubiquitous. Their over-riding requirement is for free water so they are mainly found in floating (planktonic) and bottom-dwelling communities of the sea, estuaries and freshwater environments. Some live in the water films around soil particles and in bogs, while others are parasites of other animals, notably causing malaria and sleeping sickness in humans.

Most protozoans are microscopic single-celled organisms living a solitary existence. Some, however, are colonial.

Colony structure varies; in the green flagellate *Volvox* numerous individuals are embedded in a mucilaginous spherical matrix, while in other flagellates, for example *Diplosiga*, a group of individuals occurs at the end of a stalk. Branched stalked colonies are characteristic of some ciliates, such as *Carchesium*.

Many protozoans have evolved a parasitic mode of life involving one or two hosts. Some species live in the guts and urogenital tracts of their hosts. Others have invaded the body fluids and cells of the host, for example the malarial parasite *Plasmodium* which lives for part of its life cycle in the blood and liver cells of mammals, birds and reptiles and the other part in mosquitoes (see pp164–165). Five of the seven protozoan phyla are exclusively parasitic, but among the mainly free-living flagellates, amoebae and ciliates there are some species which have opted for parasitism. Notable among the parasitic flagellates are the hemoflagellates causing various forms of trypanosomiasis, including sleeping sickness. The opalinid gut parasites of frogs and toads are another example. In most multicellular animals the essential life processes are carried out by specialized tissues and organs. In protozoans all life processes occur in the single cell and the building blocks for these specialized functions are small tubular structures termed microtubules. These have reached their most complex organization in the ciliates, the most advanced of all protozoans. Apart from a few flagellate groups, the ciliates are the only protozoans to possess a true cell mouth or cytostome. In addition they typically have two types of nuclei (dimorphism), each performing a different role. The macronucleus, which is often large and may be round, horseshoe shaped, elongated or resemble a string of beads, controls normal physiological functioning in the cell, while the micronucleus is concerned with the replication of genetic material during reproduction. It is quite common for a ciliate to possess several micronuclei. Other protozoans have nuclei of one type only, although some species may have several. The exceptions are the foraminiferans which show nuclear dimorphism at some stages in their life cycle. The most widely known and researched ciliates are species of the genera *Paramecium* and *Tetrahymena*, but these represent only a minute fraction of the 7,000 species so far described by scientists.

The majority of protozoans feed on bacteria, algae, other protozoans, microscopic animals and in the case of parasites on host tissue, fluids and gut contents. Their diet incorporates complex organic compounds of nitrogen and hydrogen, and they are said to be heterotrophic. Some flagellates, such

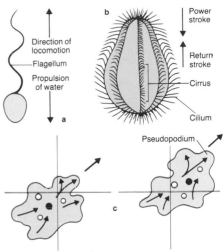

▲ **Movement in protozoans** is by three means. (1) By a single flagellum which, like a propeller, pulls the cell through the water. (2) By rows of cilia which are coordinated to act like the oars of a rowboat, having a power stroke and a recovery stroke. (3) By amoeboid movement whereby pseudopodia ("false feet") are extended and the rest of the body flows into them. Pseudopodia are also involved in prey capture.

◄ **Like exploding fireworks,** the slender filaments of a microscopic foraminiferan (*Globigerinoides* species) catch the light under the microscope. The filaments (pseudopodia) radiate out from the rest of the cell inside a hard shell or test. They are used to trap food and for movement.

▼ **Protozoan forms.** (1) *Actinophrys* (heliozoan). (2) *Opalina* (opalinid). (3) *Acineta* (suctorian). (4) *Elphidium* (foraminiferan). (5) *Euglena* (phytoflagellate). (6) *Trypanosoma* (hemoflagellate). (7) *Hexacontium* (radiolarian).

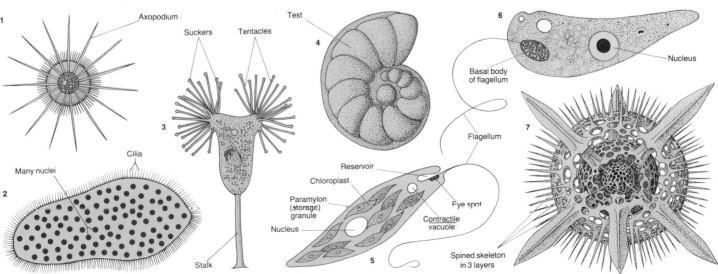

as *Euglena* and *Volvox*, however, possess photosynthetic pigments in chloroplasts. These protozoans are capable of harnessing the sun's radiant energy in the chemical process of photosynthesis to construct complex organic compounds from simple molecules—they are said to be autotrophic. A number of such flagellates must, however, combine autotrophy with heterotrophy, in varying degrees. Such organisms, sit on the boundary between an animal and plant-like nutrition (see p149).

Many protozoans are simply bound by the cell wall, but skeletal structures in the form of secreted shells or tests are common among the amoebae and usually have a single chamber. The exclusively marine foraminiferans, however, are exceptional in having shells with numerous chambers. Shells and tests may be formed of calcium carbonate or silica, or from organic substances such as cellulose or chitin.

Most free-living and some parasitic species need to move around their environment to feed, to move toward and away from favorable and unfavorable conditions, and in some cases special movement is required in reproductive processes. The various protozoan groups achieve movement using different structures.

Members of the subphylum Sarcodina (including the amoebae) produce so-called pseudopodia—flowing extensions of the cell. These may be extended only one at a time, as in *Naegleria* or several at a time, as in *Amoeba proteus, Arcella* and *Difflugia*. Heliozoan sarcodines, which resemble a stylized sun, possess long slender pseudopodia, called axopodia, which radiate from a central cell mass. Each axopodium is supported by a large number of microtubules arranged in a parallel fashion along the longitudinal axis. Heliozoans move slowly, rolling along by repeatedly shortening and lengthening the axopodia. A well-known example of these so-called sun organisms is *Actinosphaerium*. The foraminiferans, for example *Elphidium*, which bear complex chambered shells, have a complicated network of pseudopodial strands which branch and fuse with each other to produce a linking complex of what are termed reticulopodia. Like the axopodia of heliozoans the reticulopodia of foraminiferans are supported by microtubules.

The other means of movement is by the beating action of the filamentous cilia and flagella, which are permanent outgrowths of the cell rather than, like pseudopodia, its temporary pseudopodial extensions.

Cilia and flagella are structurally similar,

but cilia are shorter. Normally flagellates carry only one or two flagella, while in the ciliates the cilia are numerous and usually arranged in ordered rows each called kinety. The number of kinety is constant in each species and is used as an aid in identification. In some cases cilia may fuse to form cirri, which resemble short thick hairs, or structures which are sail-like. Each cilium and flagellum is about $0.15-0.3\mu$m in diameter and is supported by a core (axoneme) made up of two centrally positioned microtubules surrounded and joined by cross-bridges to nine double microtubules. This $9+2$ arrangement of microtubules is common in cilia and flagella throughout the living world—from amoebae to invaders of lung linings of humans. Movement in cilia and flagella involves the passage of waves along

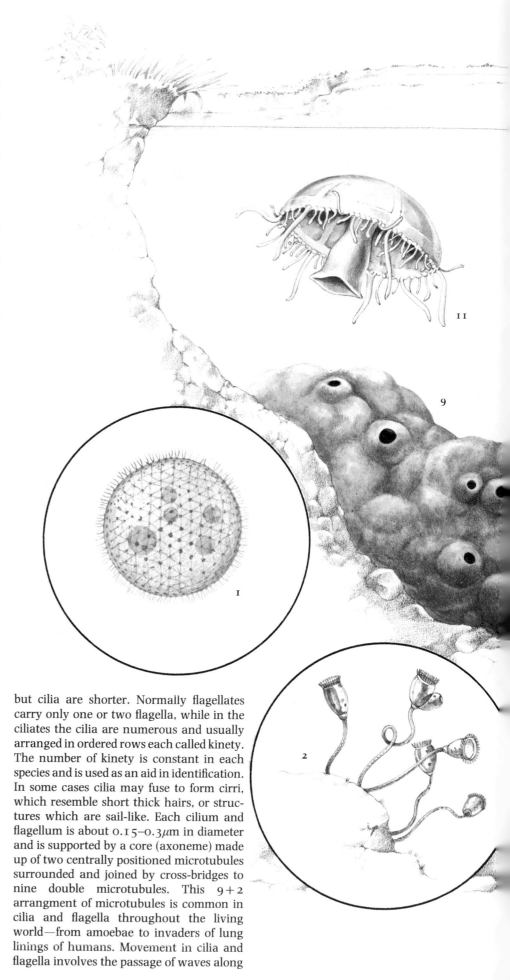

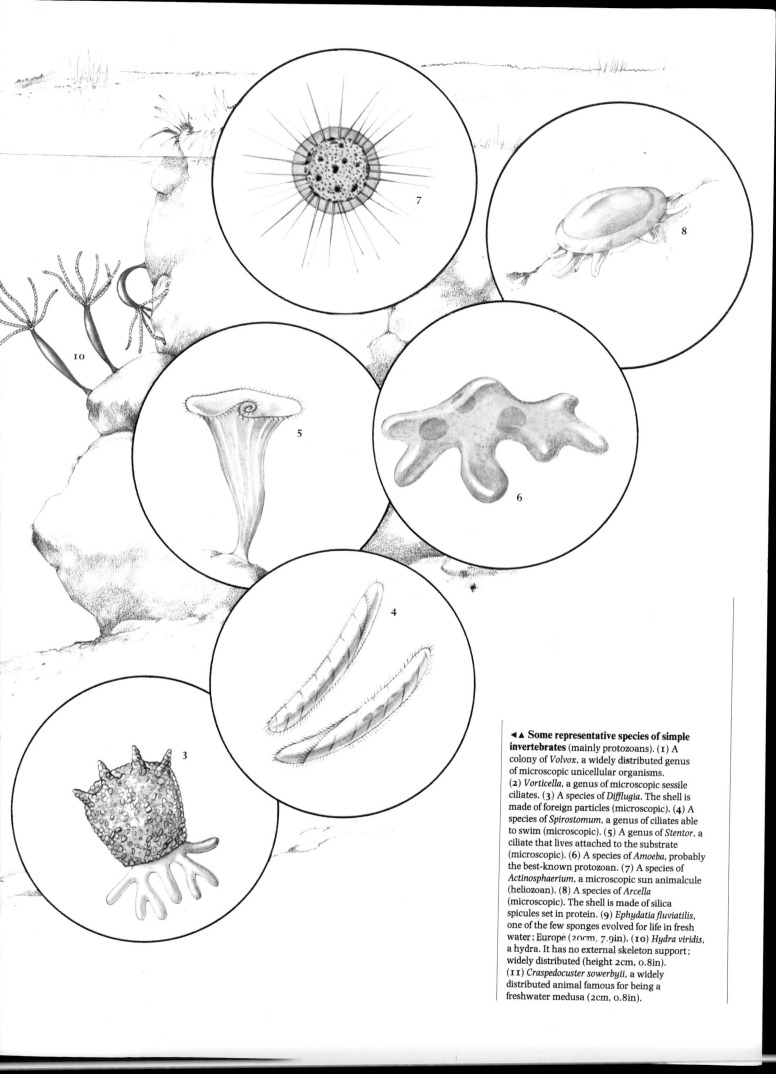

◄▲ **Some representative species of simple invertebrates** (mainly protozoans). (1) A colony of *Volvox*, a widely distributed genus of microscopic unicellular organisms. (2) *Vorticella*, a genus of microscopic sessile ciliates. (3) A species of *Difflugia*. The shell is made of foreign particles (microscopic). (4) A species of *Spirostomum*, a genus of ciliates able to swim (microscopic). (5) A genus of *Stentor*, a ciliate that lives attached to the substrate (microscopic). (6) A species of *Amoeba*, probably the best-known protozoan. (7) A species of *Actinosphaerium*, a microscopic sun animalcule (heliozoan). (8) A species of *Arcella* (microscopic). The shell is made of silica spicules set in protein. (9) *Ephydatia fluviatilis*, one of the few sponges evolved for life in fresh water; Europe (20cm, 7.9in). (10) *Hydra viridis*, a hydra. It has no external skeleton support; widely distributed (height 2cm, 0.8in). (11) *Craspedocuster sowerbyii*, a widely distributed animal famous for being a freshwater medusa (2cm, 0.8in).

them from one axis to the other. Most flagella move in two-dimensional waves, while cilia move in three-dimensional patterns coordinated into waves which result from fluid forces (hydrodynamic forces) acting on the automatic beating of each cilium.

Reproduction in the protozoans does not usually involve sex or sexual organelles—it is asexual. In most free-living species asexual reproduction occurs by a process called binary fission, whereby each reproductive effort results in two identical daughter cells by the division of a parent cell. In the flagellates, including the parasitic species, the plane of division is longitudinal, while in the ciliates it is normally transverse and prior to division of the cytoplasm the mouth is replicated. The amoebae do not normally have a fixed plane for division. In shelled and testate species the process is complicated by the need to replicate skeletal structures. In testate species of amoeba such as *Difflugia*, cytoplasm destined to become the daughter is extruded from the aperture of the parent test. Preformed scales in the cytoplasm then

Prey Capture in Carnivorous Protozoans

Carnivorous protozoans prey on other protozoans, rotifers, members of the Gastrotricha and small crustaceans. The mode of capture and ingestion is often spectacular and frequently the prey are larger than the predators.

Among the ciliates the sedentary Suctoria have lost their cilia, which have been replaced by tentacles, each of which functions as a mouth. When other ciliates, such as *Colpidium*, collide with a tentacle, they stick to it. Other tentacles move toward the prey and also attach. The cell wall of the prey is perforated at the sites of attachment and the prey cell contents are moved up the tentacle by microtubular elements within the tentacle. A single suctorian, for example *Podophrya*, can feed simultaneously on four or five prey. *Didinium nasutum* is a ciliate that feeds exclusively on *Paramecium*, which it apprehends using extrudable structures called pexicysts and toxicysts. The former hold the prey while the latter penetrate deeply into it releasing poisons. *Didinium* consumes the immobilized prey whole, its body becoming distended by the ingested *Paramecium*.

The heliozoan *Actinophrys* also feeds on ciliates which are captured, on contact, by the radiating axopodia. Once attached, the prey is progressively engulfed by a large funnel-shaped pseudopodium produced by the cell body. Occasionally when an individual *Actinophrys* has captured a large prey, other *Actinophrys* may fuse with the feeding individual to share the meal. In such instances, after digesting their ciliate victim, the heliozoan predators separate again.

The foraminiferan *Pilulina* has evolved into a living pit-fall trap. This bottom-dwelling species builds a bowl-shaped shell or test with mud, camouflaging the pseudopodia across the entrance. When copepod crustaceans blunder onto the pseudopodia they get stuck and are drawn down into the animal. The radiolarians, which possess a silica-rich internal skeleton, deal with copepod prey by extending the wave-flow along the axopodia the broad surfaces of the prey's exoskeleton, to and rupturing the prey by force. The axopodia then penetrate and prize off pieces of flesh which are directed down the axopodia to the main cell body for ingestion.

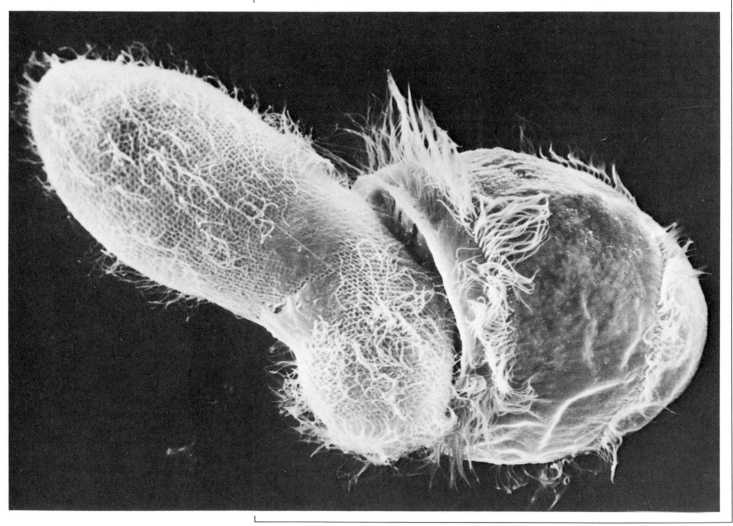

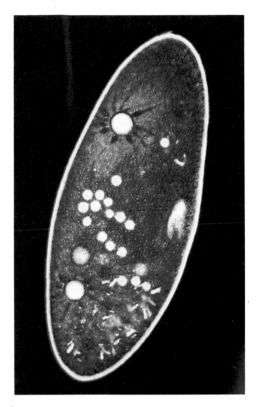

▲ **Slipper animalcules** (*Paramecium* species) mostly live in fresh and stagnant water, feeding on bacteria and particles of plant food. Here food in the oral groove shows stained orange and the star-shaped objects at top and bottom are contractile vacuoles which collect excess water and expel it from the body.

◄ **A struggle for life and death,** a *Paramecium* about to be devoured by a *Didinium*.

▼ **Divide and reproduce.** Asexual reproduction in protozoans is by simple division (binary fission). In flagellates (**1**) (eg *Euglena*) it is longitudinal. In ciliates (**2**) (eg *Tetrahymena*) division is complicated by the replication of the oral apparatus. In shelled amoebae (**3**) (eg *Euglypha*) the daughter is extruded from the parent. Amoebae (**4**) have no fixed plane of division.

form a test around the extruded cytoplasm. When the process is complete the two amoebae separate.

Most free-living species normally reproduce asexually providing conditions are favorable. Sexual reproduction is usually only resorted to in adversity, such as drying up of the aquatic medium when the normal cells would not survive. The ability to undergo a sexual phase is not widespread in the amoebae and flagellates and is restricted to a limited number of groups. Some species may never have reproduced sexually in their evolutionary history, but others may have lost sexual competence. Both isogamous (reproductive cells or gametes alike) and the more advanced anisogamous (reproductive cells or gametes dissimilar) forms of sexual reproduction occur.

The foraminiferans are unusual among free-living species in having alternation of asexual and sexual generations. Here each organism reproduces asexually to produce many amoeba-like organisms which secrete shells around themselves. When mature these produce many identical gametes which are usually liberated into the sea, where they fuse in pairs to produce individuals which in turn secrete a shell, grow to maturity and repeat the cycle.

Almost all of the ciliates are capable of sexual reproduction by a process called conjugation, which does not result in an immediate increase in numbers. The function of conjugation is to facilitate an exchange of genetic materials between individuals. During this process two ciliates come together side by side and are joined by a bridge of cell contents (cytoplasm). A complex series of divisions of the micronucleus occurs, including a halving of the pairs of chromosomes (or meiosis).

In the final stages a micronucleus passes from each individual into the other. Essentially the micronuclei are gametes. Each received micronucleus fuses with an existing micronucleus in the recipient. The ciliates separate and after further nuclear divisions eventually undergo binary fission.

All members of the parasitic phyla, except for some groups in the Apicomplexa, produce spores at some stages in their life cycles. The Apicomplexa contains a number of parasites of medical and veterinary importance, including the malarial parasites *Plasmodium* and the *Coccidia* responsible for coccidiosis in poultry. Some species, like *Plasmodium*, have complex life cycles involving two hosts with an alternation of sexual and asexual phases. In *Plasmodium* the sexual phase is initiated in humans and is completed in the mosquito; following this many thousands of motile spores (sporozoites) are reproduced which are infective to humans and are transmitted when the mosquito feeds. In humans repeated phases of multiple asexual division take place in the red blood cells and liver cells (see p164). The phyla Microspora, Ascetospora and Myxospora are parasites of a wide range of vertebrates and invertebrates, while the Labyrinthomorpha parasitize algae.

The ecology of protozoans is very complex, as one would expect in a group of ubiquitous organisms. They are found in the waters and soils of the world's polar regions. Some have adapted to warm springs and there are records of protozoans living in waters as warm as 68°C (154°F). Protozoans occur commonly in planktonic communities in marine, brackish and freshwater habitats, and also in the complex bottom-dwelling (benthic) communities of these environments. Little is known about protozoans in the marine deeps, but there is a record of foraminiferans living at 4,000 (13,000ft) in the Atlantic. Ciliates, flagellates and various types of amoebae are also common in soils and boggy habitats.

Since many protozoans exploit bacteria as a food source they form part of the decomposer food web in nature. Recent research suggests that protozoans may stimulate the rate of decomposition by bacteria and thus enhance the recycling of minerals such as phosphorus and nitrogen. The exact mechanism is not entirely clear, but protozoans grazing on bacteria may maintain the bacterial community in a state of physiological youth and hence at the optimum level of efficiency. There is also evidence that some protozoans secrete a substance that promotes the growth of bacteria.

JL-P

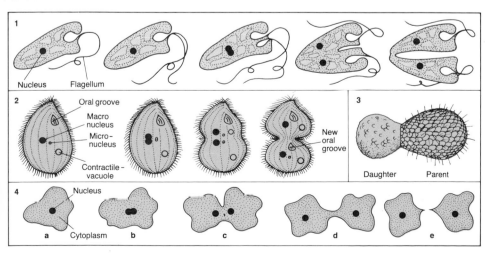

Malaria and Sleeping Sickness
Protozoan diseases of man

One million people die each year in Africa from malaria—this statistic exemplifies the virulence of protozoan diseases. The most serious pathogens are *Plasmodium*, which produces malaria, and various trypanosome species responsible for diseases broadly called trypanosomiasis or sleeping sickness.

Malaria is caused by four species of the genus *Plasmodium*. The life cycle is similar in each species but there are differences in disease pathology. *Plasmodium falciparum* causes malignant tertian malaria and accounts for about 50 per cent of all malarial cases. It attacks all red blood cells (erythrocytes) indiscriminately so that as many as 25 per cent of the erythrocytes may be infected. In this species stages not involving the erythrocytes do not persist in the liver, so that relapses do not occur. *Plasmodium vivax* produces benign tertian malaria, which invades only immature erythrocytes so that the level of cells infected is low. Here, however, other stages remain in the liver, causing relapses. Benign tertian malaria is responsible for approximately 45 per cent of malarial infections. The other two species are relatively rare. *Plasmodium malariae*, causing quartan malaria, attacks mature red blood cells and has persistent stages outside the blood cells. Little is known about *P. ovale* because of its rarity.

The diseases are named after the fevers which the parasites cause, tertian fevers occurring every three days or 48 hours and quartan fevers every four days or 72 hours. The naming practice is based on the Roman system of calling the first day one, whereas we would call the first day nought.

Once inside an erythrocyte, the parasite feeds on the red blood cell contents and grows. When mature it undergoes multiple asexual fission to produce many individuals called merozoites which by an unknown mechanism rupture the erythrocyte and escape into the blood plasma. Each released merozoite then infects another erythrocyte. The asexual division cycle in the red blood cells is well synchronized so that many erythrocytes rupture together—a phenomenon responsible for the characteristic fever which accompanies malaria. The exact mechanism producing the fever is not fully understood, but it is believed to be caused by substances (or a substance), possibly derived from the parasite, which induce the release of a fever-producing agent from the white blood cells, which fight the disease. When the parasite has undergone a series of asexual erythrocytic cycles, some individuals produce the male and female gametocytes which are the stages infective to the mosquito host. The stimulus for gametocyte production is unknown.

Malaria is still one of the greatest causes of death in humans. Tens of millions of cases are reported each year and many are fatal. Successful control measures are available, and in countries such as the USA, Israel and Cyprus the disease has been eradicated. In Third World countries, however, control measures have little impact on malaria. Broadly, eradication programs involve the use of drugs to treat the disease in humans, and a series of measures aimed at breaking the parasite's life cycle by destroying the intermediate mosquito host.

Like many insects, the mosquito has an aquatic larval stage. The draining of swamps and lakes deprives the mosquito of an environment for breeding and its larval development, but residual populations continue to breed in irrigation canals, ditches and paddy fields. Spraying of oil on the water surface asphixiates the larvae, which have to come to the water surface periodically to breathe. The poison Paris Green can effectively kill larvae when added to the water. Biological control using fish predators of mosquito larvae, such as the guppy, aid in reducing larval populations.

Adult mosquitoes can be killed by spraying houses with various insecticides such as hexachlorocyclohexane and dieldrin. In the past DDT was very successful but its toxicity

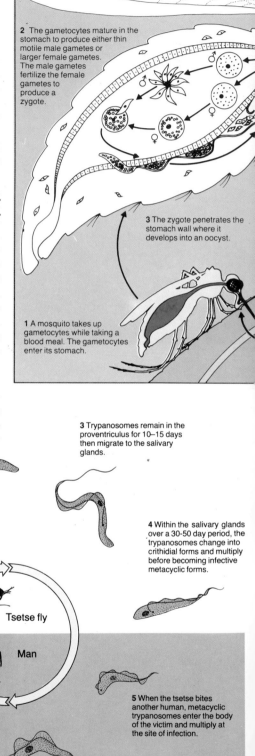

2 The gametocytes mature in the stomach to produce either thin motile male gametes or larger female gametes. The male gametes fertilize the female gametes to produce a zygote.

3 The zygote penetrates the stomach wall where it develops into an oocyst.

1 A mosquito takes up gametocytes while taking a blood meal. The gametocytes enter its stomach.

3 Trypanosomes remain in the proventriculus for 10–15 days then migrate to the salivary glands.

2 They enter midgut where division occurs before migrating forward after 48 hours to the foregut (proventriculus).

4 Within the salivary glands over a 30-50 day period, the trypanosomes change into crithidial forms and multiply before becoming infective metacyclic forms.

1 Trypanosomes in human blood taken up by tsetse fly while feeding.

Tsetse fly

Man

8 They are taken up by further tsetse flies to continue the cycle.

5 When the tsetse bites another human, metacyclic trypanosomes enter the body of the victim and multiply at the site of infection.

7 They enter the tissue space of various organs and lymph nodes and subsequently invade the central nervous system to cause typical sleeping sickness symptoms. The trypanosomes do not enter cells, but remain between them.

6 Trypanosomes then invade the blood stream and reproduce by binary fission. They they may follow two courses (7 or 8).

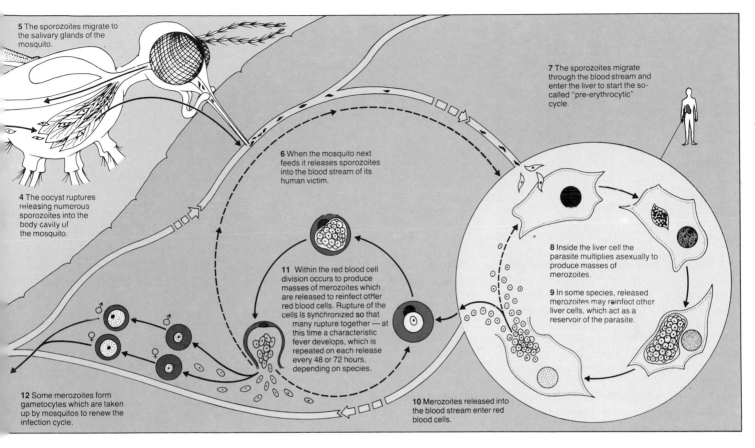

5 The sporozoites migrate to the salivary glands of the mosquito.

4 The oocyst ruptures releasing numerous sporozoites into the body cavity of the mosquito.

6 When the mosquito next feeds it releases sporozoites into the blood stream of its human victim.

7 The sporozoites migrate through the blood stream and enter the liver to start the so-called "pre-erythrocytic" cycle.

8 Inside the liver cell the parasite multiplies asexually to produce masses of merozoites.

9 In some species, released merozoites may reinfect other liver cells, which act as a reservoir of the parasite.

11 Within the red blood cell division occurs to produce masses of merozoites which are released to reinfect other red blood cells. Rupture of the cells is synchronized so that many rupture together — at this time a characteristic fever develops, which is repeated on each release every 48 or 72 hours, depending on species.

12 Some merozoites form gametocytes which are taken up by mosquitos to renew the infection cycle.

10 Merozoites released into the blood stream enter red blood cells.

▲ ◄ **Life cycles of malaria** ABOVE **and sleeping sickness** LEFT. A fifth of the world's four-billion population is probably threatened by malaria, while in Africa 50 million people are exposed to sleeping sickness or trypanosomiasis.

▼ **Hidden killer.** A mass of the pre-erythrocytic stage of *Plasmodium* (malaria) developing in the human liver.

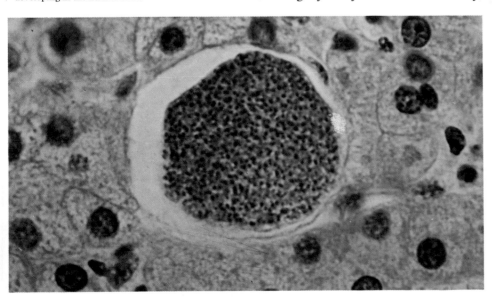

to higher animals now precludes its use. Biological control measures involve releasing sterile male mosquitos into the population, thereby decreasing reproduction rates, and the introduction of bacterial, fungal and protozoan pathogens of the mosquito.

Chemotherapy in humans involves four broad categories of treatment. Firstly, there are prophylactic drugs, such as Proguanil, which taken on a regular basis prevent recurring erythrocytic infections. Secondly,

there are drugs such as Chloroquine which destroy the blood stages of the parasite. Thirdly there are drugs which destroy gametocytes. Lastly there are drugs which, when taken up by the mosquito during feeding on humans, prevent further development of the parasite in the insect. Drug resistance by *Plasmodium* does occur; *P. falciparum* has become resistant to Chloroquine in some parts of Africa and South America, and has to be treated by a combination of quinine and sulfonamides.

The flagellates *Trypanosoma rhodesiense* and *T. gambiense* cause African trypanosomiasis or sleeping sickness. The two-host life cycle involves a tsetse fly (genus *Glossina*) and humans. In man, the trypanosomes live in the blood plasma and lymph glands, progressing later to the cerebrospinal fluid and brain. The disease is typified by mental and physical apathy and a desire to sleep. The disease is fatal if untreated, *T. rhodesiense* running a more acute course than *T. gambiense*. Control measures include insecticide use, introduction of sterile males, clearing vegetation in which *Glossina* spends the whole of its life cycle, and the use of drugs in humans. Control is complicated by the fact that *T. rhodesiense* also infects game animals, so that a reservoir population of the parasite persists.　　　JL-P

Subkingdom Parazoa
Sponges
Phylum: Porifera

Subkingdom Mesozoa
Mesozoans
Phylum: Mesozoa

Subkingdom Metazoa
Sea anemones, jellyfishes and their allies
Phylum: Cnidaria

Comb jellies
Phylum: Ctenophora

Endoprocts
Phylum: Endoprocta (Kamptozoa)

Rotifers
Phylum: Rotifera

Kinorhynchs
Phylum: Kinorhyncha

Gastrotrichs
Phylum: Gastrotricha

Lampshells
Phylum: Brachiopoda

Moss animals or sea mats
Phylum: Bryozoa (Ectoprocta)

Horseshoe worms
Phylum: Phoronida

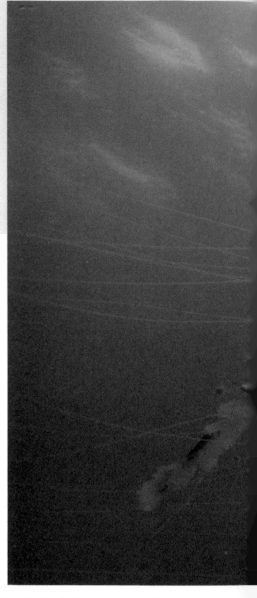

▲ **Deadly beauty.** Contact with the trailing tentacles of this Compass jellyfish (*Chrysaora hysoscella*) means certain death to small swimming organisms which are stunned by its stinging cells.

▶ **A sit-and-wait predator.** Sea anemones rest attached to the substrate waiting for prey to blunder into their outstretched tentacles. They are often well camouflaged against detection by eye.

LIFE appears to have originated in the oceans and virtually every group of animals known to man has representatives living in the sea. A few of these groups such as the sponges (phylum Porifera), hydroids (phylum Cnidaria) and the moss animals or sea mats (phylum Bryozoa) have invaded fresh water, but none has been very successful there.

Life in water offers all sorts of possibilities to animals. The drifting communities of plankton teeming in the surface waters of the open seas offer vast resources of food to those swimming animals that are able to utilize microscopic suspended food matter. The sea bed (and to a lesser extent the lake and river floor) provides a variety of habitats—hard stones, rocks, soft sands and muds—which can give support to animals if they live attached to the substrate.

Solitary invertebrates include many that need a firm base for attachment, for example sponges, hydroids and sea anemones, corals, endoprocts and sea mats. Invariably these animals adopt a plant-like growth form which confused the early naturalists. Many of them are colonial too; that is, a number of individuals live inside a common shared body mass. Being attached to a fixed object, rock, pebble or man-made structure has many problems, and these organisms frequently display either an asymmetry (as in the sponges) or radial symmetry (as in the hydroids, comb jellies, endoprocts etc). The organisms cannot move, either to gain food or to escape predation. This means that they have to exploit naturally occurring currents to bring them suspended particles of food. (Some other invertebrate groups like the bivalve mollusks (mussels, clams etc) can generate suitable currents by special pumping systems themselves.) Many sedentary organisms then resort to crowns of tentacles which will act as filters for food-gathering

and function simultaneously as gills in respiration. Radial symmetry is ideally suited to this type of function.

Defense against predation is usually achieved either by inedibility (for example sponges with a bad taste and hard spicules) or by specific weaponry (such as the stinging cells of hydroids and jellyfishes). In others the body is housed in a box-like external skeleton into which the delicate parts of the crown can be withdrawn (for example sea mats). The tendency of such organisms to form colonies is an asset since it allows some individuals to become specialized at certain roles, such as food-gathering, while others pursue functions in reproduction or defense.

Attached animals are often referred to as sessile, but sessile strictly means attached by a stalk, not directly encrusting the substrate. Not all sedentary animals are of course attached immovably by a holdfast (a grasping structure which resembles in appearance, but not in function, the roots of a plant). Sea anemones grip by means of a sole-like sucker and can creep around over rocks. Some can burrow in sand.

The free-swimming animals present a direct contrast, with their often elaborate systems of movement. These can range from pulsing muscular systems (jellyfishes), single-cilia systems (rotifers and gastrotrichs) and compound ciliary systems (comb jellies). These examples are drawn from those free-swimming animals that are closely or more distantly related to the sedentary bottom-living types also dealt with here. In the higher groups of invertebrates more elaborate systems will be shown. The links between the sedentary and free-swimming groups here treated are evidenced by the free-swimming medusae of the stalked hydroids which essentially act as reproductive and dispersal phases in the hydroids' life cycles.

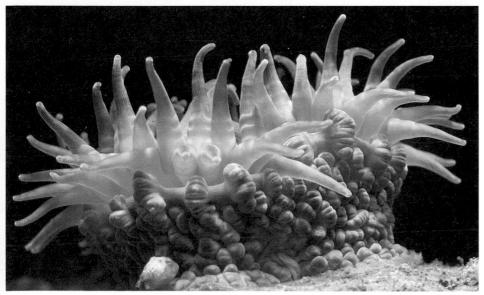

The comb jellies are the most remarkable form of swimming animals, occupying an interesting evolutionary position somewhat between the radial symmetry of the hydroids etc and the true bilateral body form of the flatworms. Comb jellies are virtually all true plankton-dwellers whose almost invisible transparent bodies drift in the oceans trailing tentacles like fishing lines. These are armed with unique lasoo cells that explode and ensnare their microscopic prey. The small and relatively insignificant rotifers, gastrotrichs and kinorhynchs have some importance in the food web but are otherwise of academic interest only. Many rotifers are sedentary, living in tubes or attached to other organisms like plants. In a few planktonic types their "wheel organs" serve to propel them through the water.

AC

SPONGES

Phylum: Porifera
Sole phylum of subkingdom Parazoa.
About 5,000 species in 790 genera and 80 families.
Distribution: worldwide, marine and freshwater, intertidal to deep sea.
Fossil record: originated in Cambrian 570–500 million years ago; 390 genera identified from Cretaceous (135–65 million years ago).
Size: from microscopic to 2m (6.6ft); the largest sponges occur in the Antarctic and the Caribbean.
Features: form variable; solitary or colonial; mostly porous, filter-feeding organisms mostly attached direct to substrate, without "stem"; lack organs and have little in way of definite tissues, but with complex array of cell types; skeleton lacking or of siliceous or calcareous spicules, or of organic spongin fibers; generally hermaphrodite; sexual and asexual reproduction.

Glass or siliceous sponges

Class Hexactinellida (Hyalospongiae)
About 600 species. Marine, below tidal levels but more common in deeper waters. Skeleton of complex silica spicules, with basic pattern of 6 rays. Includes *Aphrocallistes*, *Euplectella aspergillum* (**Venus' flower basket**), *Holascus*, *Pheronema*.

Calcareous sponges

Class Calcarea
About 400 species. Marine. Skeleton of calcareous spicules which are needle-like or 3- or 4-rayed. Includes *Acyssa*, *Clathrina*, *Leucilla*, *Leucosolenia*, *Scypha*.

Typical sponges

Class Demospongiae
About 4,000 species. Marine and freshwater. Skeleton lacking or of silica spicules, spongin fibers or both. Includes: *Aplysina*, *Cliona*, *Cribochalina vasculum* (**Caribbean sponge**), *Ephydatia*, *Haliclona*, *Hippospongia communis* (**bath sponge**), *Neofibularia nolitangere* (**Caribbean fire sponge**), *Siphonodictyon*, *Spongia officinalis* (**bath sponge**), *Spongilla*.

Coralline sponges

Class Sclerospongiae
About 15 species. Marine, in tropical, shallow, subtidal caves or underneath corals. Skeleton with calcareous base and entrapped silica spicules and organic fibers. Sponge forms thin layer over calcareous base. Includes: *Ceratoporella*, *Stromatospongia*.

▶ **Like a cluster of smokestacks,** a purple column or tube sponge (*Verengia lacunosa*) rises from the seabed. The large exhalent opening (osculum) can be seen at the top of the lower column.

The humble bath sponge has been used by people since earliest times, particularly in the Mediterranean region. Bath sponge species are the best known of a group of animals whose relationship to other organisms is a matter of debate. Until the early 19th century sponges were regarded as plants, but they are now generally considered to be a group (phylum Porifera) of animals placed within their own subkingdom, the Parazoa. They probably originated either from flagellate protozoans or from related primitive metazoans.

Sponges range in size from the microscopic up to 2m (6.6ft). They often form a thin incrustation on hard substrates to which they are attached, but others are massive, tubular, branching, amorphous or urn-, cup- or fan-shaped. They may be drab or brightly colored, the colors derived from mostly yellow to red carotenoid pigments.

All sponges are similar in structure. They have a simple body wall containing surface (epithelial) and linking (connective) tissues, and an array of cell types, including cells (amoebocytes) that move by means of the flow of protoplasm (amoeboid locomotion). Amoebocytes wander through the inner tissues, for example, secreting and enlarging the skeletal spicules and laying down spongin threads. Sponges are not totally immovable but the main body may show very limited movement through the action of cells called myocytes, but they often remain anchored to the same spot.

Although sponges are soft-bodied, many are firm to touch. This solidity is due to the internal skeleton comprised of hard rod- or star-shaped calcareous or siliceous spicules and/or of a meshwork of protein fibers called spongin, as in the bath sponge. Spicules may penetrate the sponge surface of some species and cause skin irritation when handled.

Sponges are filter-feeders straining off bacteria and fine detritus from the water. Oxygen and dissolved organic matter are also absorbed and waste materials carried away. Water enters canals in the sponge through minute pores in their surface and moves to chambers lined by flagellate cells called choanocytes or collar cells. The choanocytes ingest food particles, which are passed to the amoebocytes for passage to other cells. Eventually the water is expelled from the sponge surface, often through volcano-like oscules at the surface. Water is driven through the sponge mainly by the waving action of flagella borne by the choanocytes.

Sponges reproduce asexually by budding off new individuals, by fragmentation of

parts which grow into new sponges and, particularly in the case of freshwater sponges, by the production of special gemmules. These gemmules remain within the body of the sponge until it disintegrates, when they are released. In freshwater sponges, which die back in winter in colder latitudes, the gemmules are very resistant to adverse conditions, such as extreme cold. Indeed, they will not hatch unless they have undergone a period of cold.

In sexual reproduction, eggs originate from amoebocytes and sperms from amoebocytes or transformed choanocytes, usually at different times within the same individual. The sperms are shed into the water, the eggs often being retained within the parent, where they are fertilized. Either solid (parenchymula) or hollow (amphiblastula) larvae may be produced; many swim for up to several days, settle, and metamorphose into individuals or colonies that feed and grow. Others creep on the substrate before metamorphosis. Some mature Antarctic sponges have not grown over a period of 10 years.

▲ **A free run.** This encrusting red sponge spreads over the rock face particularly well in low light intensity because competition with algae for space is reduced when the illumination is too poor for the plants to flourish.

▼ **The three basic body forms of sponges.** (1) (2) (3) Sponges are filter feeders straining off bacteria and fine detritus from the water. Oxygen and dissolved organic matter are also absorbed and waste materials carried away. Water enters canals in the sponge through minute pores in their surface and moves to chambers lined by flagellate cells called choanocytes or collar cells (4). The choanocytes ingest food particles, which are passed to the amoebocytes for passage to other cells. Eventually the water is expelled from the sponge surface, often though volcano-like oscules at the surface. Water is driven through the sponge mainly by the beating action of the flagellae borne on the choanocytes. Choanocytes may line the body cavity (1), or the wall is folded so these cells line pouches (1, 2) connected to more complicated canal systems.

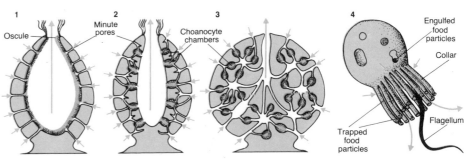

Sponges are found in large numbers in all the seas of the world. They occur in greatest abundance on firm substrates, relatively few being adapted to life on unstable sand or mud. Their vertical range includes the lowest part of the shore subject to tidal effects and extends downwards as far as the abyssal depths of 8,600m (27,000ft). One family of siliceous sponges, the Spongillidae, has invaded freshwater lakes and rivers throughout the world.

Sponges living between tide marks are typically confined to parts of the shore that are seldom exposed to the air for more than a very short period. Some occur a little higher up the shore, but these are only found in shaded situations or on rocks facing away from the sun.

Some sponges are killed by even a relatively short exposure to air, and it is in the shallow waters of the continental shelf that sponges achieve their greatest abundance in terms both of species and individuals.

Cavernous sponges are frequently inhabited by smaller animals, some of which cause no harm to the sponge, although others are parasites. Many sponges contain single-celled photosynthetic algae (zoochlorellae), blue-green algae, and symbiotic bacteria, which may provide nutrients for the sponge.

Sponges are eaten by sea slugs (nudibranchs), chitons, sea stars (especially in the Antarctic), turtles, and some tropical fishes.

Usually more than half of the species of tropical sponges living exposed rather than under rocks are toxic to fish. This is believed to be an evolutionary response to high-intensity fish predation, nature having selected for noxious and toxic compounds that prevent fish from consuming sponges. Some toxic sponges are very large, such as the gigantic Caribbean sponge (*Cribochalina vasculum*), while others, such as the Caribbean fire sponge (*Neofibularia nolitangere*), are dangerous to the touch—in humans they cause a severe burning sensation lasting for several hours. Toxins probably play an important role in keeping the surface of the sponge clean, by preventing animal larvae and plant spores from settling on them. Some sponge toxins may prevent neighboring invertebrates from overgrowing and smothering them.

Sponge toxins are becoming important in studies on the transmission of nerve impulses. They show considerable potential as biodegradeable antifouling agents and possibly as shark repellants.

Bath sponges have been fished in the Mediterranean since very early times. They

▲ **A blue theme.** These extensive beds of blue sponge encrust the coral limestone around Heron Island near the Great Barrier Reef. The brown lace-like growths are colonies of calcareous bryozoans.

▶ **A squadron of spicules.** Sponge spicules help to support the body of the sponge. They may be made from calcium carbonate or silica. Their various shapes may be characteristic of particular types of sponge and can be important in identifying them. A complex nomenclature has been developed. (**1**) Monaxon spicule with barbs (*Farrea beringiana*). (**2**) Monaxon spicule (*Mycale topsenti*). (**3**) Triaxon spicule (*Leucoria heathi*). (**4**) Hexaxon spicule (*Auloraccus fissuratus*). (**5**) Monaxon spicule with terminal processes (*Mycale topsenti*). (**6**) Monaxon spicule (*Raspaigella dendyi*). (**7**) Monaxon spicule with recurved ends (*Sigmaxinella massalis*). (**8**) Polyaxon spicule (genus *Streptaster*).

Mesozoans

Mesozoans are a taxonomic enigma. They comprise the phylum Mesozoa, which contains about 50 species parasitic on marine invertebrates and none bigger than 8mm (0.3in). They are multicellular animals constructed from two layers of cells, and are therefore distinct from the protozoa, but the layers do not resemble the endoderm and ectoderm of the metazoans (see p148). The features of the group render them unassignable to any other animal phylum. One view holds them to be degenerate flatworms, in other words they may have been previously more complex; the other more widely held opinion, however, is that they are simple multicellular organisms holding a position intermediate between Protozoa and Metazoa.

Mesozoans of the order Dicyemida are all parasites in the kidneys of cephalopods (for example octopus), while the Orthonectida infect echinoderms (starfish, sea urchins etc), mollusks (snails, slugs etc), Annelida (earthworms etc) and ribbon worms (see p198). Despite their simple morphology the Dicyemida have evolved complex life cycles

involving several generations. The first generation, called a nematogen, occurs in immature cephalopods. Repeated similar generations of nematogens are produced asexually by repeated divisions of special central (axial) cells which give rise to wormlike (vermiform) larvae (**1**). When the host attains maturity the parasite assumes the next generation or rhombogen, which looks superficially similar to the nematogen, but differs in its cellular makeup. The individuals are hermaphrodite and produce infusariiform larvae which look superficially like ciliate protozoans. The fate of the larvae is uncertain, but it is believed that another intermediate host is involved in the life cycle. Genera included in the order are *Dicyema*, *Dicyemmerea* and *Conocyema*.

The second of the two orders, the Orthonectida (**2**), live in the tissues and tissue spaces of their marine invertebrate hosts, for example nemerteans, polychaetes, ophiuroids and bivalves. The asexual phase looks like an amoeboid mass and is called a plasmodium because it resembles the protozoan *Plasmodium*.

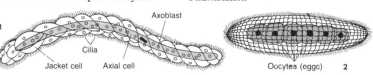

1   Axoblast
Cilia
Jacket cell   Axial cell   Oocytes (eggs)   2

owe their usefulness to the water-absorbing and retaining qualities of a complex lattice of spongin fibers; the fibers are also elastic enough to allow water to be squeezed out of the sponge. A number of species are harvested (mainly off Florida and Greece), the chief of which are *Spongia officinalis*, with a fine-meshed skeleton, and *Hippospongia equina*, with a coarser skeleton. These grow on rocky bottoms from low-tide level down to considerable depths and may be collected either by using a grappling hook from a boat, or by divers. The curing of sponges merely involves leaving them to dry in the sun, allowing the soft tissues to rot, pounding and washing them, leaving only the spongin skeleton.

Cultivation of sponges from cuttings has been successfully used although such projects are probably less economic than the making of synthetic products.

Large species, such as the Venus' flower basket have been prized as decorative objects, particularly in Japan.

Sponges contain a variety of antibiotic substances, pigments, unique chemicals such as sterols, toxins, and even anti-inflammatory, and antiarthritis compounds. Boring sponges of the family Clionidae (eg *Cliona* species) may cause economic losses by weakening oyster shells. Boring sponges are widespread within tropical stony corals and cause considerable damage by weakening them. These sponges excavate chambers by both chemical and mechanical methods.   GJB

SEA ANEMONES AND JELLYFISHES

Phylum: Cnidaria
About 9,400 species in 3 classes.
Distribution: worldwide, mainly marine; free swimming and bottom dwelling.
Fossil record: Precambrian (about 600 million years ago) to present.
Size: microscopic to several meters in width.

Features: radially symmetrical animals with cells arranged in tissues (tissue grade); possess tentacles and stinging cells (nematocysts); body wall of two cell layers (outer ectoderm and inner endoderm) cemented together by a primitively noncellular jelly-like mesogloea and enclosing a digestive (gastrovascular) cavity not having an anus; there are two distinct life-history phases: free-swimming medusa and sedentary polyp.

SEA ANEMONES, corals and jellyfishes are perhaps the most familiar members of the phylum Cnidaria. It contains a vast number of mainly marine animals. There are only a few freshwater species, of which the best-known are the hydras.

The cnidarians are multicellular animals and have a two-layered (diploblastic) construction in which both the differences between cells and organ development are limited. These restrictions have, however, been partially offset in colonial types by the specialization of individuals (polymorphism). There are two life-history phases: polyp and medusa. The polyp is the sedentary phase and consists of three regions: a basal disk or pedal disk which anchors it; a middle region or column within which is the tubular digestive chamber (gastrovascular cavity); and an oral region which is ringed by tentacles. In colonial types a tubular stolon links adjacent polyps. The medusa is the mobile phase and is effectively an inverted polyp. By virtue of the fluid (water) it contains the digestive cavity plays an important role in oxygen uptake and excretion. This fluid additionally acts as a hydrostatic skeleton through which body wall muscles can antagonize one another.

Since the medusa is the sexual phase, and it can be argued that it is the original life form, with the predominantly bottom-living (benthic) polyp acting as an intermediate, multiplicative asexual stage. However, in the class Hydrozoa the medusa is frequently reduced or even lost, and in the class Anthozoa totally absent. Emphasis in the class Scyphozoa lies, to the contrary, with the medusa stage, as the evolution of the highly mobile and graceful jellyfish testifies; the polyp phase in jellyfish is a relatively inconspicuous component in the life cycle.

The outer (ectodermal) and inner (endodermal) cell layers of the body are cemented

◄ ▼ **Some representative species of sea anemones and jellyfishes.** (**1**) *Cyanea lamarckii*, a jellyfish; N Atlantic (diameter 20cm, 7.9in). (**2**) *Aurelia aurita*, the Common jellyfish; medusa phase; Mediterranean and N Atlantic (diameter 25cm, 9.8in). (**3**) *Physalia physalis*, the Portuguese man-of-war; Atlantic (diameter of float 30cm, 12in). (**4**) *Actinia equina*, the Beadlet anemone; Mediterranean and N Atlantic in the intertidal zone (height 7cm, 2.8in). (**5**) *Metridium senile*, the Plumose anemone. (**6**) *Obelia geniculata*, showing a colony of polyps and close-up view; shallow rocky habitats of NW Europe (height of colony 4cm, 1.6in). (**7**) *Sertularia operculata*, a hydroid; a colony of polyps (height of colony 45cm, 17.7in). (**8**) *Eunicella verrucosa*, a sea fan; Mediterranean and N Atlantic. (**9**) *Peachia hastata*, a "sit-and-wait" burrowing anemone; Mediterranean and N Atlantic (length 10cm, 3.9in). (**10**) *Corynactis viridis*, an anemone-like animal; N Atlantic (diameter 5cm, 2in). (**11**) *Alcyonium digitatum*, or dead man's fingers, a colony of polyps; Mediterranean and N Atlantic (height of colony 20cm, 7.9in).

together by the jelly-like mesogloea which in the jellyfish forms the bulk of the animal. The mesogloea contains a matrix of elastic collagen fibers which aid both the change and maintenance of body shape. This is particularly obvious in the pulsating swimming movements characteristic of jellyfish, during which contractions of the swimming bell brought about by radial and circular muscles are counteracted by vertically running, elastic fibers.

Muscle contraction results in an increase in bell depth and hence fiber stress; fiber shortening subsequently restores the bell to its original shape. In the medusae of hydrozoans the resulting water jets are concentrated and directed by the shelf-like velum projecting inwards from the rim of the bell, where there are tentacles, towards the mouth. Structural support in the relatively large anthozoan polyps is also provided by septa (mesenteries) which contain retractor muscles. When mobile, polyp locomotion may be brought about in a number of ways: by creeping upon the pedal disk, by looping or, rarely, by swimming (for example the anemones, *Stomphia*, *Boloceroides*).

The **hydras and their allies** (class Hydrozoa) are considered to be the group that exhibits the most primitive medley of features. The class contains a plethora of medusa and polyp forms which are, for the most part, relatively small. We can plausibly imagine the early hydrozoan life cycle as being similar to that of the hydrozoan order Trachylina. Here the medusae have a relatively simple form and the typical cnidarian

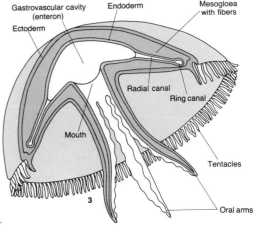

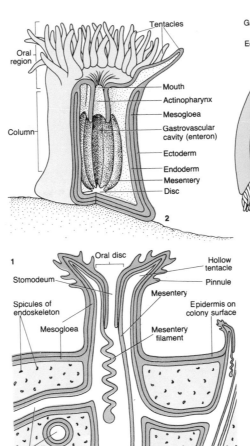

larva, the planula, gives rise in turn to a hydra-like stage which buds-off the next generation of medusae. Significantly this stage is predominantly free-swimming (pelagic), but in other hydrozoan orders subsequent polyp elaboration has resulted in the interpolation of bottom-living, hydra-

◀▲ **The three main forms of cnidarians.** (1) A colonial polyp (soft coral). (2) A solitary polyp (sea anemone). (3) A medusa (jellyfish).

▶ **Sheltering among the tentacles** of a large anemone, this clownfish (genus *Amphiprion*) probably gains shelter and protection. In return it grooms the anemone's tentacles and brings in scraps of food. Mucus from the anemone probably protects the fish from the stinging cells.

▶▶ **Paralyzed** OVERLEAF by batteries of stinging cells (shown as fine yellow dots on the tentacles), this worm lies trapped amongst the polyps of a coral (genus *Tubastrea*).

The 3 Classes of the Phylum Cnidaria

Hydras and their allies
Class: Hydrozoa
About 2,700 species in 6 orders (or more).

Fossil record: some hydroids; many hydrocorals.

Features: of 4- (tetramerous) or many- (polymerous) fold symmetry; solitary or colonial; life cycle can include polyp and medusa or exclusively one or other; mesogloea without cells; digestive (gastrovascular) system lacks a stomodeum (gullet); stinging cells (nematocysts) and internal septa absent; sexes separate or individuals bisexual; gametes mature in the ectodermis which frequently secretes a chitinous or calcareous external skeleton; medusa has shelf-to-bell rim (velum); tentacles generally solid.
Orders: Actinulida; Hydroida

(**hydroids, sea firs**); Milleporina; Siphonophora; Stylasterina; Trachylina.

Jellyfishes
Class: Scyphozoa
About 200 species in 5 orders.

Fossil record: minimal.

Features: dominant medusa form with 4-fold (tetramerous) symmetry; polyp phase produces medusae by transverse fission; solitary (either swimming or attached to substrate by stalk); mesogloea partly cellular; digestive (gastrovascular) system has gastric tentacles (no stomodeum) and is usually subdivided by partitions (septa); sexes usually separate; gonads in endodermis; complex marginal sense organs; skeleton absent; tentacles generally solid; exclusively marine.

Sea anemones, corals
Class: Anthozoa
About 6,500 species in probably 12 orders in 2–4 subclasses.

Fossil record: several thousand species known.

Features: exclusively polyps; predominantly with 6-fold (hexamerous) or 8-fold (octomerous) symmetry; pronounced additional tendency to bilateral symmetry; solitary or colonial; have flattened mouth (oral) disk with an inturned stomodeum; cellular mesogloea; sexes separate or hermaphrodite; gonads in endodermis; digestive (gastrovascular) cavity divided by partitions (septa) bearing gastric

filaments; skeleton (when present) is either a calcareous external skeleton or a mesogloeal internal skeleton of either calcareous or horny construction; some forms specialized for brackish water; tentacles generally hollow.

Subclass: Alcyonaria (or Octocorallia)
Orders: Alcyonacea (**soft corals**); Coenothecalia (**blue coral**); Gorgonacea (**horny corals**); Pennatulacea (**sea pens**); Stolonifera; Telestacea.

Subclass: Zoantharia (or Hexacorallia)
Orders: Actiniaria (**anemones**); Antipatharia (**thorny corals**); Cerianthidia; Corallimorpharia; Madreporaria (**hard** or **stony corals**); Zoanthidea.

Associations and Interdependence in Anemones, Crabs and Fishes

Cnidarians are involved in a variety of associations with other animals, ranging from obtaining food or other benefits from another animal (commensalism) to being interdependent (symbiosis).

An example of commensalism occurs between the hydroid genus *Hydractinia* and hermit crabs, particularly in regions deficient in suitable polyp attachment sites. This is understandable since the shells inhabited by such crabs provide substitute sites and, moreover, the relationship provides *Hydractinia* with the opportunity for scavenging food morsels. Whether any benefit is gained by the crab is unclear, though the development of defensive dactylozooids in *H. echinata* specifically in response to a chemical stimulus emanating from crabs suggests that it does and that there is therefore a mutually beneficial coexistence (mutualism). The association between the cloak anemone *Adamsia palliata* and the crab *Pagurus prideauxi* is, to the contrary, far more intimate. These species normally form a partnership when small. With time the crab comes to outgrow its shelly refuge, but by secreting a horny foot (pedal) membrane the anemone progressively enlarges the shell lip, thus obviating any need for the crab to change shells. A crab lacking an anemone will, upon contact with *Adamsia*, recognize it and attempt to transfer it to its shell.

A range of intermediate degrees of association is provided by the anemone *Calliactis parasitica*, which associates with hermit crabs but also frequently lives independently. An interesting, one-sided association is that of the Hawaiian crab genus *Melia* which has the remarkable habit of carrying an anemone in each of its two chaelae, thereby enhancing its aggressive armament; the crab even raids their food.

Clown fish (*Amphiprion* species) live within the tentacles of sea anemones. Their hosts provide protection to these fish and the fish protect the anemones from would-be predators (primarily other fish which are chased away by the territorial behavior of the clown fish). Additionally the clown fish apparently act as anemone cleaners. Although it has been suggested that inhibitory substances are secreted by these fish to reduce the discharge of stinging cells, recent work has failed to confirm this. It is more likely that stinging is avoided by the secretion of a particularly thick mucous coat, which during acclimatization of the fish to its host anemone may possibly become modified, with the levels of certain excitatory (acidic) components being reduced.

like colonies. Further evolution in specialized niches where dispersal is not at a premium has led in turn to the secondary reduction of medusae; indeed most hydroids lack or almost totally lack a medusoid phase.

Early hydroids were probably solitary inhabitants of soft substrates. Subsequent evolution produced types living in sand (Actinulida) and fresh water (Hydridae—hydras). Most colonial types, however, occur on hard surfaces, anchored by rooting structures. The interconnecting stems (stolons) are protected and supported by a chitinous casing (perisarc) which may or may not enclose the polyp heads. The functional interconnection of members of these colonies permits a division of labor between polyps and an associated variety in form (polymorphism). While one form (gastrozooid) retains both tentacles and a digestive cavity, the form that defends the colony (dactylozooid) has lost the cavity. Another form, gonozooid, is dedicated solely to the budding-off of medusae or, in species lacking medusae, to producing gametes. The delicate branching of hydroid colonies is highly variable, but universally serves to space out member polyps and to raise them well above the substrate, thereby reducing the chance of clogging by silt and sediments.

The evolution of various forms in the class Hydrozoa has culminated in the formation of the complex floating siphonophore colonies (oceanic hydrozoans), each colony composed of a diverse array of both medusae and polyps; they are characteristic of warmer waters. Essentially each individual within the colony is interlinked by a central stolon. In addition to the three polyp forms found in hydroids, there can be up to four forms of medusae: (1) muscular swimming bells that propel the entire colony (for example, *Muggiaea, Nectalia*); (2) gas-filled flotation bells (for example, the Portuguese man-of-war—*Physalia*); (3) bracts which play either a supportive or protective role, or both; (4) medusa buds. Freed from the substrate, these colonies are able to reach large sizes with, for example, the trailing colonial stemwork of the Portuguese man-of-war often extending for several meters below the apical float. Such colonies are capable of paralyzing and ingesting relatively large prey items such as fish. Recent research indicates that some species (for example those of the genus *Agalma*) may attract large prey by moving tentacle-like structures, which are replete with stinging cells (nematocysts) and which bear a remarkable resemblance to small zooplankton (copepods).

At one time it was thought that the pinnacle of this evolutionary line was illustrated by animals such as the by-the-wind-sailor (*Vellela*), which has a disk-like, apical float bearing a sail which catches the wind, thus facilitating drifting. It is now thought, however, that these organisms simply consist of one massive polyp floating upside down, and that they are related to gigantic bottom-living hydroids. The large size of these bottom-living giants—up to 10cm (4in) in *Corymorpha* and 3m (10ft) in *Branchiocerianthus*—has been permitted by their adoption of a deposit-feeding life style, often at great depth in still water.

Finally, two groups of hydrozoans produce a calcareous external skeleton: these are the tropical milleporine and stylasterine hydrocorals.

Among cnidarians, it is the **jellyfishes** (class Scyphozoa) that have most fully exploited the free-swimming mode of life though the members of one scyphozoan order (the Stauromedusae) are bottom-living, with an attached, polyp-like existence. Jellyfish medusae have a similar though more complex structure than the medusae of hydroids with the disk around the mouth prolonged into four arms, a digestive system comprising a complex set of radiating canals linking the central portion (stomach) to a peripheral ring element, and a relatively more voluminous mesogloea. The mesogloea in some genera (for example *Aequoria*, *Pelagia*) helps buoyancy by selectively expelling heavy chemical particles (anions) (such as sulfate ions), which are replaced by lighter ones (such as chloride ions). A wide size range of prey organisms are taken, though many species, including the common Atlantic semaeostomes of the genus *Aurelia*, are feeders on floating particles and thus concentrate on small items. The arms of *Aurelia* periodically sweep around the rim of the bell, gathering up particles which accumulate there following deposition on the animal's upper surface. In contrast the arms around the main mouth of the Rhizostomae have become branched, and have numerous sucking mouths, each capable of ingesting small planktonic organisms such as copepods. Within this group are the essentially bottom-living, suspension-feeding forms of the genus *Cassiopeia*, which lies upside down on sandy bottoms, their frilly arms acting as strainers. The bell shapes of members of two orders are distinctive: coronate medusae have bells with a deep groove and cubomedusae have bells that are cuboid in shape.

The gametes of jellyfish are produced in gonads which lie on the floor of the digestive cavity and are initially discharged into it. Fertilization normally occurs after discharge of the gametes. Many species, however, have so-called brood pouches located on the undersurface where the larvae are retained. After release larvae settle and give rise to polyps which produce additional polyps by budding. These polyps also produce medusae by transverse division (fission), a process which results in the formation of stacks (strobilae) of so-called ephyra larvae. When released the ephyra larvae feed mainly on protozoans and grow and change into the typical jellyfish.

Corals and sea anemones (class Anthozoa) only exist as polyps. Sea anemones (order Actiniaria) always bear more than eight tentacles and usually have both tentacles and internal partitions (mesentaries) arrayed in multiples of six.

Many anemone species, especially the more primitive ones, are burrowers in muds and sands but most dwell on hard substrates, cemented there (permanently or temporarily) by secretions from a well differentiated disk. The disk around the mouth (oral disk) is provided with two grooves (siphonoglyphs) richly endowed with cilia which serve to maintain a water flow through the relatively extensive, digestive cavity. The oral disk extends inwards to produce a tubular gullet or stomodeum which acts as a valve, closing in response to increases in internal pressure. In common with jellyfish some anemones feed on particles suspended in the seawater for which leaf-like tentacles, prodigious mucus production and abundant food tracts lined with cilia are required; a good example is the common plumose anemone, genus *Metridium*. Asexual reproduction occurs by budding, breaking-up or fission, while sexual reproduction may involve either internal or external fertilization of gametes. Some species brood young, either internally or externally at the base of the column.

Members of two other orders are also anemone-like: the cerianthids have greatly elongated bodies adapted for burrowing into sand, but have only one oral groove (siphonoglyp). Zoanthids lack a pedal disk, are frequently colonial and often live attached to other organisms (epizoic).

Also included in the subclass Zoantharia are the hard (stony) corals (order Madreporaria) whose polyps are encased in a rigid, calcium carbonate skeleton. The great majority of hard corals live in colonies

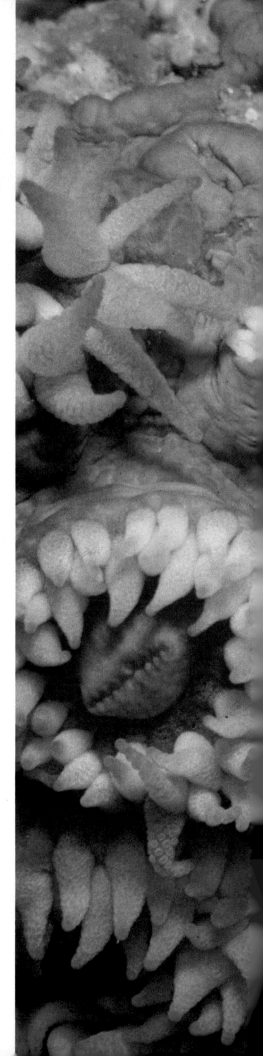

which are composed of vast numbers of small polyps (about 5mm, 0.2in), but the less abundant solitary forms may be large (*Fungia* up to 50cm, 20in, across); most are tropical or subtropical in distribution. In colonial forms the polyps are interconnected laterally; they form a superficial living sheet overlying the skeleton, which is itself secreted from the lower outer (ectodermal) layer.

Corals exhibit a great diversity of growth forms, ranging from delicately branching species to those whose massive skeletal deposits form the building blocks of coral reefs (see pp180–181). An interesting growth variant is exhibited by *Meandrina* and its relatives in which polyps are arranged continuously in rows, resulting in the production of a skeleton with longitudinal fissures, a feature which accounts for its popular name, the brain coral.

Closely related to the hard corals are the members of the order Corallimorpharia which lack a skeleton. Included in this group is the jewel anemone (genus *Corynactis*), so named because of its vivid and highly variable coloration. Since it reproduces asexually, rock faces can become covered by a multicolored quiltwork of anemones. The black or thorny corals (order Antipatharia) form slender, plant-like colonies bearing polyps arranged around a horny axial skeleton; they possess numerous thorns.

Octocorallian corals comprise a varied assemblage of forms, but all possess eight feather-like (pinnate) tentacles. The polyps project above and are linked together by a mass of skeletal tissue called coenenchyme, which consists of mesogloea permeated by digestive tubules. Thus in contrast to hard corals, the octocorallians have an internal skeleton. This assemblage includes the familiar gorgonian (horny) corals, sea whips and fans, and the precious red coral, genus *Corallium*. Most of these have a central rod composed of organic material (gorgonin) around which is draped the coenenchyme and polyps, the former frequently containing spicules which may impart a vivid coloration. Such is the case with *Corallium*, whose central axis consists of a fused mass of deep red calcareous spicules; this material

is used in jewelry. The tropical organ pipe coral, genus *Tubipora* (order Stolonifera) produces tubes or tubules of fused spicules which are cross-connected by a regular series of transverse bars. In contrast, the soft corals (order Alcyonacea) only contain discrete spicules within the coenenchyme (for example dead man's fingers, genus *Alcyonium*). The order Coenothecalia is solely represented by the Indo-Pacific blue coral, genus *Heliopora*, which has a massive skeleton composed of crystalline aragonite

▲ **A waving mass of tentacles**—this dahlia anemone (*Tealia felina*) traps prey 20m (65ft) below the surface of the North Sea. In the foreground are the arms of a buttle star.

◄ **The stinging cell** (cnidoblast) of a cnidarian: (**a**) before discharge of the nematocyst; (**b**) after discharge of the nematocyst thread.

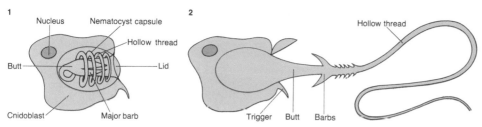

1

Nucleus Nematocyst capsule

Hollow thread

Butt Lid

Cnidoblast Major barb

2

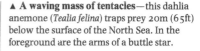

Hollow thread

Trigger Butt Barbs

fibers fused into plates (lamellae): its blue color is imparted by bile salts. Many species in most of these groups have several forms (especially gastrozooids, dactylozooids and gonozooids). This is also true of the sea pens (order Pennatulacea) which are inhabitants of soft bottoms. Each possesses a large, stem-like, primary polyp (as is also the case in the order Telestacea) housing a skeletal rod which becomes embedded in the substrate as a result of waves of contractions. Second-ary polyps are arranged laterally on this stem and exhibit two forms; many polyps (siphonozooids) act as pumps, promoting water circulation through the colonial digestive system. Familiar examples are the sea pansy (genus *Renilla*) and the sea pen (genus *Veretillum*), both of which when dis-turbed exhibit waves of glowing phosphor-escence. These are controlled by the nervous system and are inhibited by light. Their role is not clear, though it is likely that they are a response to intrusion by would-be pred-ators such as fish. A number of other cnidarians display a similar phenomenon, a good example being the hydromedusae of the genus *Aequorea*.

To trap prey cnidarians normally employ stinging cells (nematocysts). The discharge of these is now thought to be under nervous control. Discharge involves a collagenous thread being rapidly shot out, uncoiling and turning inside out in the process, sometimes to expose lateral barbs. Hollow stinging cells frequently contain a toxin which can enter the body of the prey. The released toxins may be extremely potent: especially danger-ous are those released by the cubomedusan sea wasps (for example, genus *Chironex*)—

jellyfish that have been responsible for kill-ing several humans, particularly off Austra-lian coasts. Victims usually succumb rapidly to respiratory paralysis. Nematocysts may be pirated by sea slugs and used for their own protection (see p263).

The cnidarian nervous system shows a certain amount of organization and local specialization. This is especially evident in anemones where nerve tracts accompany the retractor muscles responsible for protec-tive withdrawals. The marginal ganglia of scyphomedusae and the circumferential tracts of hydromedusae have been found to contain pacemaker cells which are respon-sible for initiating and maintaining swim-ming rhythms. In *Polyorchis* it has been found that the giant nerves controlling movement are all coupled together electri-cally, ensuring that they function collec-tively as a giant ring nerve fiber capable of initiating synchronously muscle contrac-tion from all parts of the bell.

Similarly the behaviors of individual polyps in hydroid and coral colonies are integrated by the activities of colonial nerve nets. Additional powers for integrating con-trol are provided by conduction pathways apparently constituted by sheets of electri-cally coupled, epithelial cells. For example, the shell-climbing behavior of anemones of the genus *Calliactis* seems to depend upon the interplay of activities between two epi-thelial systems—one on the outside (ecto-dermal), the other inside (endodermal)—and the nerve net, though conclusive evi-dence as to the exact cellular locations of these additional systems has been difficult to obtain.   RB

Aggression in Anemones

A number of anemones display a well-defined aggressive sequence which, for the most part, is used in confrontation with other anemones. These anemones all possess discrete structures located at the top of the column which contain densely packed batteries of stinging cells (nematocysts): they are called acrorhagi. They can be inflated and directed at opponents.

The common intertidal beadlet anemone (*Actinia equina*), whose distribution encompasses the Atlantic seaboards of Europe and Africa and which also occurs in the Mediterranean, is an example upon which attention has been focused. Although this species can vary considerably in color (red to green), the acrorhagi are always conspicuous thanks to their intense bluish hue. Aggression is triggered by the contact of tentacles. One individual usually displays column extension and bends so that some of its simultaneously

enlarging acrorhagi make contact with the opponent (after 5–10 minutes). There follows a discharge of the stinging cells (nematocysts) which normally results in a rapid withdrawal by the victim.

Experiments have suggested that in common with more advanced animals contest behavior is ritualized, but that in these lowly forms it is dependent upon "simple physiological rules" rather than upon complex behavioral ones. The "rules" used apparently decree that larger anemones should act aggressively more rapidly than smaller ones and, as a result, will subsequently win contests.

The North American anemone *Anthopleura elegantissima* reproduces asexually by fission. In consequence intertidal rocks can become entirely covered by a patchwork of asexually produced anemones. Close inspection reveals that each densely packed clone (ie the mass of

asexually produced offspring) is separated from its neighbors by anemone-free strips, and that these are maintained by aggressive interactions involving acrorhagi. It is clear, therefore, that the aggressive behavior of individuals constituting the boundary of a clone serves to provide territorial defense for the entire clone, the central members of which, significantly, are more concerned with reproduction than aggression. Thus there is, as in hydroid colonies, a functional division of labor, despite the lack of physical interconnection between the clonal units. Individuals at the interclonal border have more and larger acrorhagi than centrally placed members, a difference apparently dependent solely on the former experiencing aggressive contact with nonclonemates. Such a dichotomy is thought to be indicative of the presence of a sophisticated self/nonself recognition system.

The Living and the Dead
The origins and biological organization of coral reefs

Coral reefs are extraordinary oases in the midst of oceanic deserts, for they support immensely rich and diverse faunas and floras but occur primarily in the tropics where the marked clarity of the water indicates a relative dearth of planktonic organisms and other nutrients. "Coral" consists of the skeletons of hard or stony coral.

The success of reef-building (hermatypic) corals in tropical waters, where high light intensities prevail throughout the year, is strictly dependent upon the nutrient-manufacturing (autotrophic) activities of interdependent (symbiotic) algae (zooxanthallae) which live within each polyp. Such dependence also necessarily restricts the algae to these waters. Moreover, since these algae flourish best at temperatures higher than 20°C (68°F), reef development is further limited to depths of less than 70m (230ft), ie where light intensities are greatest. Corals do, however, survive both at higher latitudes and in deeper waters, but where they do their capacity to secrete limestone for reef building is found to be severely curtailed as a result of the reduced metabolic support provided by the algae. Finally, restrictions on the distribution of corals are also imposed by the deposition of silt, freshwater run-off from land and cold, deepwater upwelling. The two former factors, for example, restrict reef development in the Indo-Pacific Ocean towards offshore island sites, while the last hinders coral growth off the west equatorial coast of Africa, where the Guinea current surfaces. It should not be forgotten, though, that in the development of most reefs, encrusting (calcareous) algae (for example, *Lithophyllum*, *Lithothamnion*) normally play an extremely important part. In many cases the limestone they produce acts as a cement.

In 1842 Charles Darwin distinguished three main geomorphological categories of reef which are still in use today: fringing reefs, barrier reefs and atolls. As the name suggests, the first are formed close to shore, on rocky coastlines. Barrier reefs are, on the contrary, separated from land by lagoons or channels which have usually been produced as a result of subsidence. (The best known and largest barrier reef is the Great Barrier Reef off the northeast coast of Australia, though the name is somewhat misleading as along its length (1,900km, 1,200mi) occur a host of different reef configurations—more than 2,500). Finally, atolls are found around subsiding volcanoes.

The continuation of coral growth is heavily dependent upon changes in water level. At present the world is in a period between glaciations and the rising sea level permits vertical growth to continue at about 0.3–1.5cm (0.1–0.6in) per year, a rate that has apparently been maintained over the last 100,000 years. Core drillings taken at Eniwetok atoll in the Pacific have extended downwards for up to 1.6km (1mi) before hitting bed rock: from analyses of both the fauna and flora in cores obtained, it has proved possible to reconstruct past fluctuations in sea level. The majority of the world's coral reefs started development during the Cenozoic era (not later than 65 million years ago) and consist predominantly of corals of the order Madreporaria.

All coral reefs have a similar biological organization with the reef plants and animals, as on rocky shores, lying in zones in accordance with their tolerances to physical factors. This is most evident on the exposed, windward faces of those reefs subject to continuous wave crash where especially prolific growths of both corals and algae develop. However, a reef can only grow outwards if debris accumulates on the reef slope; with increasing water depths, such material tends to slide down the slope and thus becomes unavailable. Below 30–50m (100–165ft), hermatypic corals are replaced by nonhermatypic ones, and by fragile, branching alcyonarians (gorgonians, etc). Above the slope is the reef crest which, in the most exposed situations, is dominated by encrusting calcareous algae (for example, the genus *Porolithon*). These algae form ridges which are full of cavities thereby providing numerous recesses which

▲ **Life around Bermuda**, a coral reef scene from the western Atlantic. The dominant organism here is the Common sea fan (*Gorgonia ventilina*), which is surrounded by hard brain corals and erect sponges.

◄ **Life on a coral surface.** Coral limestone and the skeletons of dead corals provide a substrate for many small encrusting organisms, including other cnidarians. Here *Parazoanthus swiftii* (not itself a coral) is spreading its colonial polyps to capture food suspended in the water.

are colonized by a multitude of invertebrates, including zoanthids, sea urchins and vermetid gastropods. Where wave action is not too severe, windward reef crests are usually dominated by a relatively small number of coral species, notably stout *Acropora* and hydrocoralline *Millepora* species.

Zonation is far less marked on the leeward side of the reef crest, where a different set of problems has to be faced of which sediment accumulation from land run-off is perhaps the most acute. Nevertheless, the relatively sheltered nature of this habitat permits the rapid proliferation of branched corals. In common with the coral faunas of

windward slopes, those of leeward faces display dramatic changes of coral form with increasing depth, changes which can either be attributable to the replacement of species by others or to changes in species forms. For example, on Caribbean reefs, dominant species such as *Monastrea annularis* display both stout (shallow-water) and branching (deepwater) growth forms. Recent work on *M. cavernosa* has indicated that forms found at equivalent depths are distinctive: the polyps of the shallow-water form are open continuously, whilst those of the deep-water form (which house far fewer zooxanthellae) are open only at night.   RB

COMB JELLIES

Phylum: Ctenophora
About 100 species in 5 orders (sometimes grouped into 2 classes).
Distribution: worldwide, marine.
Fossil record: none.
Size: from very small (about 0.4cm, 0.15in) to over 1m (3.3ft) in length.
Features: basically radially symmetrical but masked by superimposed bilateral symmetry; body wall 2-layered (diploblastic) with a thick jelly-like mesogloea and nerve net (these features making them similar to sea anemones and jellyfishes); 8 rows of plates of fused cilia (comb plates) upon whose activity locomotion predominantly depends; tentacles when present help in the capturing of zooplankton; digestive/gastrovascular system with a stomodeum (gullet), stomach and a complex array of canals; one phase in life cycle (not equivalent to polyp or medusa); bisexual.
Orders: Beroida (class Nuda), including *Beroë gracilis*; Cestida, including **Venus's girdle** (genera *Cestum*, *Velamen*); Cydippida, including *Pleurobrachia*; Lobata, including *Mnemiopsis*; Platyctenea, including *Coeloplana*, *Tjalfiella*, *Ctenoplana*, *Gastrodes*.

▼ **Feeding time in the oceans:** a comb jelly (genus *Lampea*) eating a chain of red salps. The comb jelly's coiled white tentacle and the rows of cilia (comb plates) on the ridges of the animal can be seen in this picture.

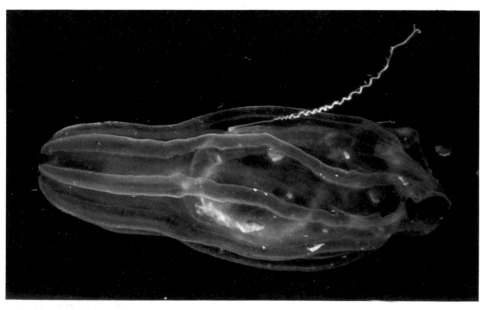

S MALL, translucent, gelatinous globular animals, comb jellies float through the open seas, like ghosts, capturing prey with their whip-like tentacles. The body consists of three zones—a voluminous mesogloea sandwiched, as in sea anemones and jellyfishes, between thin ectodermal (outer) and endodermal (inner) cell layers. Most noticeable, however, are the eight rows of plates of cilia (comb rows) whose activities serve to propel the animal while it is searching for zooplanktonic prey, which it captures by means of the pair of tentacles loaded with adhesive cells.

Most comb jellies resemble species in the common genus *Pleurobrachia* (order Cydippida) which occurs in the colder waters of both the Atlantic and Pacific oceans, and is often found stranded in tidal pools. Its globular body is up to 4cm (1.5in) in diameter, and has two pits into which the tentacles can be retracted. These tentacles function as drift nets, catching passing food items while the animal hovers motionless. When extended these appendages may be up to 50cm (20in) in length. They have lateral filaments and bear numerous adhesive cells (colloblasts), each of which has a hemispherical head fastened to the core of the tentacle by a straight connective fiber and by a contractile spiral one, the latter acting as a lasso. Once caught prey is held by the colloblasts, which produce a sticky secretion, until transferred to the central portion (stomach) of the digestive system following the wiping of the tentacles over the mouth. It has been reported that when feeding upon pipe fishes of a similar size to itself, *Pleurobrachia* will play them in much the same way as an angler tires out

a hooked fish. The stomach, in which digestion commences, leads to a complex array of canals where there is further digestion and intake of small food particles which are subsequently broken down by intracellular digestion. These canals are especially routed alongside those body regions having high energy consumption levels, notably the eight comb rows. The gonads lie in association with the lining of the gastric system; gametes pass out through the mouth. Following external fertilization a larva, which is a miniature version of the adult, is produced.

The common comb jellies of temperate waters are all like *Pleurobrachia*, which has the most primitive body form: shape in the other orders departs from this. The elongate lobate comb jellies are laterally compressed and have six lobes projecting from their narrow mid region, four of which are delicate and two stout. These serve to capture food: the tentacles are small and lack sheaths. *Mnemiopsis*, which is about 3cm (1.2in) in size and occurs in immense swarms, has, in association with the production of these lobes, four long and four short comb rows. Elongation and compression have been carried much further in the Cestida, resulting in organisms resembling thin gelatinous bands. Species in the two genera concerned (*Cestum*, *Velamen*) are collectively referred to as Venus's girdle, and are found in tropical waters and the Mediterranean, only occasionally straying into northern waters. The graceful swimming of these forms is principally dependent upon undulations of the whole body, brought about by muscle fibers embedded in the mesogloea. They feed entirely by means of tentacles set in grooves running along the oral edge. In the order Beroidea, the thimble-shape body is similarly laterally compressed, but mainly occupied by the greatly enlarged stomodeum rather than mesogloea. Species in this group are up to 20cm (8in) in length and often have a pinkish color. There are no tentacles; instead food is caught by lips which curl outwards to reveal a glandular and ciliated (macrocilia) area. Relatively large food items are rapidly taken in by a combination of suction pressure (brought about by the contraction of radial muscles in the mesogloea) and ciliary action. The common North Sea species *Beroë gracilis* feeds exclusively upon *Pleurobrachia pileus*.

The final order, the Platyctenea, is a curious group which, contrary to other flattened ctenophores, are compressed from top to bottom and have, for the most part, assumed a bottom-living (benthic) creeping

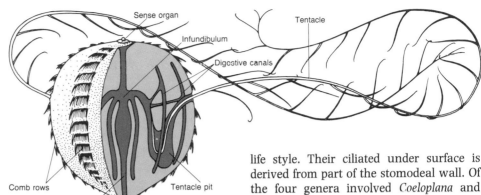

The structure of a comb jelly (*Hormiphora plumosa*). Labels: Sense organ, Infundibulum, Tentacle, Digestive canals, Comb rows, Stomadeum, Mouth, Tentacle pit.

► **The structure of a comb jelly** (*Hormiphora plumosa*).

▼ **A gentle predator.** A comb jelly (genus *Pleurobrachia*) trails its adhesive tentacles—like fishing lines—through the water to ensnare microscopic food organisms. Comb jellies are noted for their luminescence, which is produced by glandular structures—appearing as greenish streaks—lying in association with their eight radial gastrovascular canals.

life style. Their ciliated under surface is derived from part of the stomodeal wall. Of the four genera involved *Coeloplana* and *Tjalfiella* are flattened creeping forms, the latter being practically sessile (ie attached to the substrate). In contrast *Ctenoplana* is partially planktonic, having become adapted to creeping on the water's surface also. The most specialized, though, is *Gastrodes* which is a parasite of the free-swimming sea squirt genus *Salpa*. The flattened adult stage is a free-living, bottom-dwelling form, but the planula-type larva bores into the tunicate test and develops into a bowl-shaped, intermediate parasitic stage.

The waves of movement of the comb plates responsible for swimming are initiated and synchronized by impulses arising within the apical sense organ which functions primarily as a statocyst, detecting tilting, although it is also sensitive to light. It contains a sensory epithelium bearing four groups of elongate sensory "balancer" cilia, upon which a calcareous ball (statolith) perches. The orientation of the animal is controlled very simply: irrespective of whether the comb jelly is swimming upward or downward, deflection of the balancer cilia away from an upright position brought about by the statolith results in a change in the frequency of beating of the cilia which is spontaneously generated. The overall effect exerted through the differential activitation of the four groups of "balancer cilia," and subsequently the four pairs of comb rows to which they are electrically connected, is to ensure that the animal swims vertically rather than obliquely. Since the power stroke of the individual comb plates is opposite to direction of waves of activity passing along the comb rows, the animal normally travels mouth forward. However, this activity is also under control exerted by the nerve net, which upon receipt of a mechanical stimulus can cause either a cessation or reversal of beating. There is also some evidence to suggest that the synchronization of beating of the comb plates within any one row might be partially dependent upon direct mechanical coupling.

RB

ENDOPROCTS

Phylum: Endoprocta (or Entoprocta or Kamptozoa)
Over 130 species in 3 families.
Distribution: worldwide, mainly marine.
Fossil record: none.
Size: minute, 0.5–4mm (0.02–0.16in).

Features: stalked, body cavity a pseudocel; solitary or colonial; body comprises a stalk or peduncle supporting a calyx bearing a circle of partially retractile tentacles within which both mouth and anus open; the calyx encloses paired excretory protonephridia with flame cells; no vascular system; sexes on some or different individuals depending on species; sexual and asexual reproduction.

Family: Pedicellinidae, including *Pedicellina*, *Myosoma*, *Barentsia*, and 3 other genera.

Family: Urnatellidae, sole genus *Urtanella*.

Family: Loxosomatidae, including *Loxosoma*, *Loxocalyx*, *Loxosomella*, *Loxomespilon*, *Loxostemma*.

ENDOPROCTS are small inconspicuous animals that are usually less than 1mm (0.04in) in length. Most inhabit the sea but one genus lives in fresh water. They are sedentary and live attached to hard surfaces or other organisms by a stalk. In the latter case they live as commensals neither receiving benefit from nor giving benefit to the host. Some of the commensal species, such as *Loxosomella phascolosomata*, are catholic and dwell on various marine invertebrates such as the sipunculids *Golfingia* and *Phascolion* as well as on some bivalve shells. Others, like *Loxomespilon perenzi*, are host-specific living on the polychaete worm *Sthenelais*. Some are even specific to certain parts of the host like the exhalent pores or oscula of sponges or the segmental appendages or parapodia of polychaete worms. Preferences for these specific habitats among endoprocts may reduce competition for survival between endoproct species and place individuals where they can most benefit from water currents etc.

Relatively little is known about endoprocts. The European fauna is best known and it is certain that many species remain to be discovered, particularly in the tropics.

Individual endoprocts are called zooids and live, according to species, either as solitary animals or in a colony where many zooids are linked together by creeping root-like stolons. The zooids have a fairly simple structure, each one consisting of a cup-like body called the calyx supported on a stalk-like peduncle. The calyx and peduncle are comprised of body wall tissue, which is soft and flexible, cloaked with a thin protective layer of cuticle under which lies a thin layer of epidermis and a thin layer of muscle. The muscle brings about the nodding movements of the calyx on the peduncle. Towards its top the calyx is slightly constricted by a rim above which radiate about 40 hollow

▶ **The structure of endoprocts.** (**1**) A section showing in detail the internal structure of a zooid (genus *Pedicillina*). (**2**) Part of a colony of *Pedicillina* showing each zooid linked by a creeping stolon. Many endoprocts live in association with other host animals, from whose feeding and respiration currents individual endoproct zooids obtain their own nutrient requirements. (**3a**) One such host is the sea mouse (*Aphrodite aculeata*), upon which lives the endoproct *Loxosemella fauveli*, individuals of which are shown (**3b**) after removal of the scales covering the body of the sea mouse.

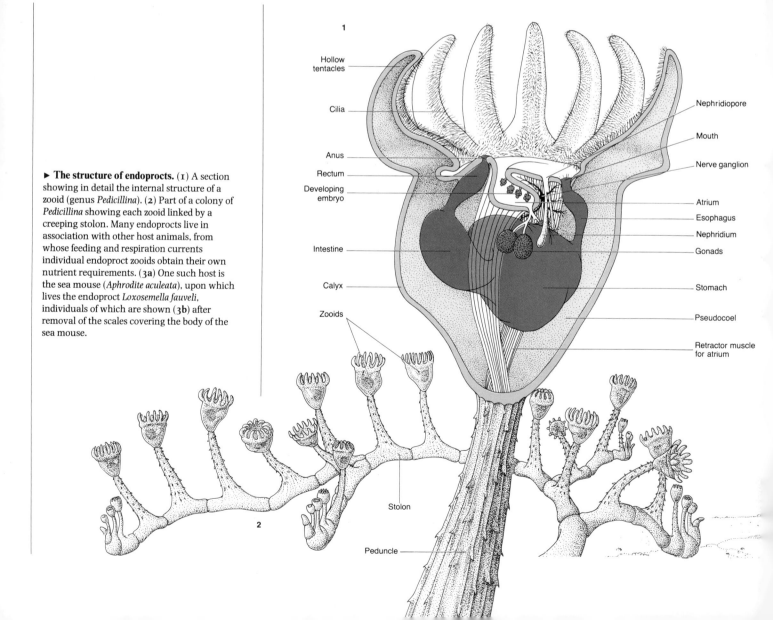

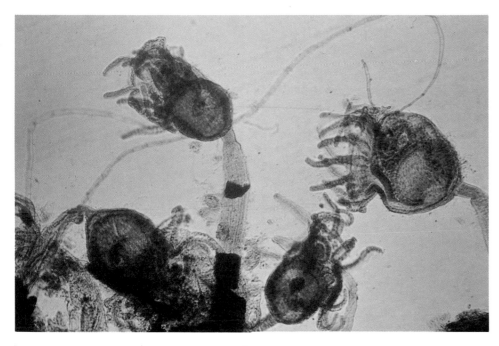

▲ **Animals on stalks,** zooids of the endoproct genus *Pedicellina*.

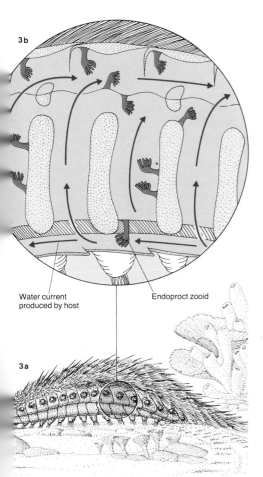

Water current produced by host

Endoproct zooid

3a

3b

water current leaving the tentacle crown.

Many species undergo asexual reproduction by budding. The style of budding, for example directly from the calyx or from the peduncle (as well as the pattern of growth form in colonial types), is characteristic of particular genera. Budding from the calyx is customary in solitary forms like *Loxosomella*. When the buds reach an advanced stage they drop off to occupy a new site and lead an independent life.

Some species are hermaphrodite, others separate-sexed. Fertilization is believed to be within each zooid. The resulting embryos are then brooded in the atrium until they are released as free-swimming larvae. Following the planktonic phase the larvae settle and grow into new zooids.

The three families of endoproct (Pedicellinidae, Urnatellidae and Loxosomatidae) can be distinguished by the form of the zooids and the growth habit. In the first family there are no solitary species. Here the calyx is separated from the peduncle by a diaphragm so that when conditions are unfavorable or when damaged by predators, such as small grazing arthropods, the calyx can be shed. It can be regrown: examples of this family are frequently met with showing a range of regenerating calyces.

The Urtanellidae contains only one genus, *Urtanella*, which is the only freshwater one. Here the stolon is small and disk-like and the calyces may be shed and show regeneration frequently. One species occurs in both western Europe and the eastern USA; a second is known from India.

The Loxosomatidae, which are the most abundant endoproct species, are all solitary, usually with a short peduncle attached by a broad base with a cement gland or a muscular attachment disk. The former type cannot move, being cemented down, but the latter can detach and reattach themselves as conditions require. The peduncle and calyx are continuous and there is no diaphragm. The calyx cannot be shed in the Loxosomatidae.

For many years endoprocts were classified with the ectoprocts (moss animals or sea mats) in one phylum, the Polyzoa, but most authorities now regard them as a separate group. One key difference between the two is that in ectoprocts (phylum Bryozoa or Ectoprocta) the anus opens outside the ring of tentacles, not within the calyx. In the endoprocts the anus (proctodeum in anatomical terms) lies within the ring of tentacles. This explains the meaning of the two names: Endoprocta, anus inside; Ectoprocta, anus outside.   AC

tentacles. These can be folded over the top of the calyx and partially retracted so that the muscular web of tissue which connects their bases affords them some protection in unfavorable circumstances. The tentacles are covered with cilia, minute beating threads projecting from the skin.

On top, inside the circlet of tentacles, the calyx is penetrated by the mouth and anus, between which runs a U-shaped gut. The space inside the calyx is taken up by the body cavity (pseudocoel), the fluid contents of which bathe the gut and other internal organs. The body cavity fluid contains some free cells which wander about. The reproductive organs lie close to the gut and their short ducts discharge into a fold, the atrium, in the top of the calyx; embryos may be brooded in the atrium. Endoprocts lack an internal circulatory system and have no special respiratory structures; oxygen dissolved in the surrounding water simply diffuses into the zooid. A pair of nephridia are responsible for excretion of nitrogenous waste and these discharge through a single nephridiophore on the top of the calyx just behind the mouth.

Endoprocts are suspension feeders utilizing small organic particles and microorganisms borne in the water currents to supply them with food. The cilia on the tentacles drive water in between the tentacle bases and up and out through the central opening in the tentacle crown. The food particles are caught by the cilia on the sides of the tentacles and passed down to the mouth bound up in a string of mucus. Unwanted particles are flicked into the

ROTIFERS, KINORHYNCHS, GASTROTRICHS

Phyla: Rotifera, Kinorhyncha, Gastrotricha

Rotifers or wheel animalcules
Phylum: Rotifera
About 1,700 species in about 100 genera.
Distribution: worldwide, mainly freshwater
and damp soils.
Fossil record: none.
Size: microscopic, 0.04–2mm (0.002–0.08in).

Features: solitary or colonial; body cavity a
pseudocoel; body comprises head section, trunk
and tail piece; head bears a crown of hairs
(cilia); trunk houses internal jaws (mastax);
tailpiece in some forms bears gripping toes;
excretion by means of flame cell
protonephridia; body sometimes encased in a
lorica.

Kinorhynchs
Phylum: Kinorhyncha
About 120 species in the class Echinoderida.
Distribution: probably worldwide in marine
coastal areas.
Fossil record: none.
Size: less than 1mm (0.04in).

Features: free-living; body segmented and
covered in spines (no cilia); excretion via a pair
of protonephridia, each fed by a single flame
cell.

Gastrotrichs
Phylum: Gastrotricha
About 200 species.
Distribution: widespread, marine and fresh
water.
Fossil record: none.
Size: microscopic.

Features: free-living usually with some external
areas covered with cilia; body cavity a
pseudocoel; body of three layers (triploblastic)
often with one or more pairs of adhesive
organs; excretory protonephridia when present
consist of a single flame cell; cuticle covered
with spines, scales or bristles.

Class: Chaetonotoida
Mainly fresh water. Front end of body usually
distinct from trunk; adhesive tubes at rear;
protonephridia present; mainly
parthenogenetic (asexually reproducing)
females, including genus *Chaetonotus*.

Class: Macrodasyoidea
Marine, chiefly reported from Europe. Bodies
straight with adhesive tubes on front, rear and
sides of the body; no protonephridia; bisexual;
including genus *Macrodasys*.

► **Little animals of great endurance.** ABOVE
Antarctic rotifers (*Philodina gregeria*) have
survived being frozen for over a hundred years.
They can also withstand immersion in liquid
nitrogen.

► **The structures of representative genera** of
(1) rotifers (genus *Brachionus*); (2) kinorhynchs
(genus *Echinoderes*); (3) gastrotrichs (genus
Chaetonotus).

Rotifers' most conspicuous structure is their crown of hairs (cilia) borne on a retractable disk: the corona. The cilia beat in such a way that it resembles a wheel spinning, or two wheels spinning in opposite directions when viewed under the microscope. Thus the early microscopists termed rotifers "wheel animalcules."

The three phyla treated here are relatively little-known members of the animal kingdom. They are all aquatic, living mainly in fresh water. Some authorities have grouped them with the roundworms and horsehair worms in one phylum, known as the Aschelminthes, on account of certain developmental and structural similarities. Others believe that they are best considered phyla in their own right because the criteria for grouping them are somewhat debatable.

All these animals show bilateral symmetry and a body cavity, but the latter is a pseudocoel lying between the body wall and the gut (see p148) and not a true coelom. The animals are not segmented and the body is supported by a layer of skin (cuticle). The alimentary tract is usually "straight through," with a mouth, esophagus, intestine, rectum and anus. Excretion is carried out through structures known as protonephridia. Reproductive strategies vary from group to group although the sexes are usually separate. Development is along protostome lines (see p152).

The **rotifers** constitute quite a large phylum with many species living in fresh water and soils all over the world. They inhabit lakes, ponds, rivers and ditches as well as guttters, puddles, the leaf axils of mosses and higher plants and damp soil. Most are free living, but a few form colonies and some live as parasites.

Rotifer bodies can generally be divided into a head at the front, a long middle section or trunk housing most of the viscera, and a tailpiece or foot terminating in a gripping toe. The head as such is not well developed in comparison with those of the higher animals but it does bear a characteristic eye spot sometimes colored red. The most conspicuous structure is the corona. The beating of its cilia draws in a stream of water toward the head which brings with it food in the form of other microorganisms. These are then passed into the muscular esophagus which houses powerful teeth and jaws. The chewing structure is technically termed the mastax and the teeth and jaws are made of strong cuticle for chewing and grinding the food. In some predatory rotifers long teeth may protrude out of the mouth and

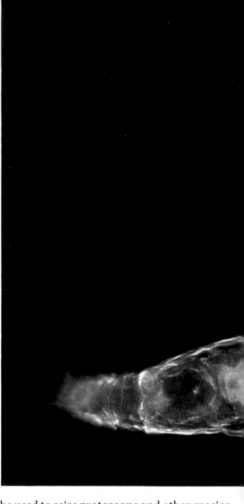

be used to seize protozoans and other species of rotifer.

The trunk houses the internal organs including the mastax. Its contents appear complex: they include elements of the gut, excretory system and reproductive system as well as the musculature for retracting the corona. The tailpiece is important for posture and stance—many rotifers use it to cling to fronds of aquatic plants or other substrates while they feed or rest. The tail may also serve as a sort of rudder.

The breeding strategies of rotifers typically include a phase of reproduction where females lay unfertilized eggs which develop and grow into other females (parthenogenetic). When the sexual season dictates or when adverse environmental conditions prevail some of the females lay eggs which need to be fertilized by a male. At the same time others lay smaller eggs which develop into males for this purpose. The males then fertilize the new females by injecting them with sperm through the body wall. The resulting fertilized eggs are tolerant of harsh conditions and can withstand drought, an

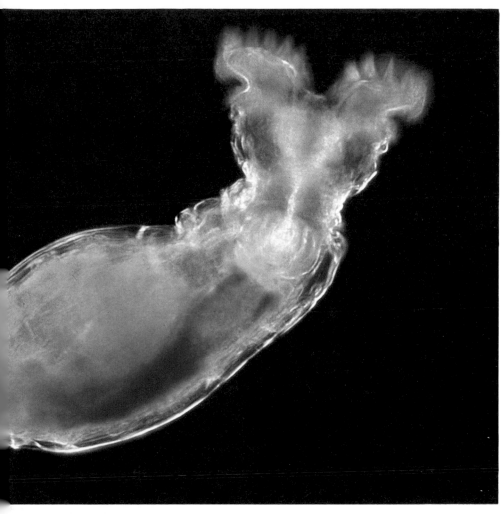

Despite their obscurity rotifers are of ecological and direct economic importance. Millions of them may exist in a small body of water and their populations may rise and fall rapidly. They are important members of numerous aquatic food webs. Soil-dwelling rotifers play a role in soil breakdown.

The **kinorhynchs** are poorly known marine "worms" reaching up to 1mm (0.04in) in length. They probably exist worldwide but most of the 100 or so known species have been recorded from European coastal sands and muds—a reflection probably of where people have searched for them rather than of their actual distribution. Their bodies appear segmented but the divisions are only "skin-deep" and they are quite unrelated to the annelid worms or arthropods. They creep about in sand and mud using the spines of the head to gain a hold while the rear of the body is contracted forward. Then the tail spines dig in and the head is advanced and so the process is repeated. The head is not well developed. Male and female kinorhynchs have a similar appearance and sexual reproduction occurs year round. A larva emerges at hatching which molts several times before becoming an adult.

Gastrotrichs are also minute animals often found in habitats shared with rotifers. They are "worm-like" and live among detrital particles in both fresh and seawater. The genus *Chaetonotus* is a good example. Its head is rounded and attached to the trunk by an elongated neck; the tail is usually forked and the outer surfaces of the body are covered with sticky knobs (papillae). The underside bears hairs (cilia) arranged in bands whose coordinated beating causes the animal to glide over the substrate. The upper surface of the animal is usually armed with spines and scales.

The gastrotrich pharynx generates a sucking action which is employed to draw food into the gut. It takes single-celled algae, bacteria and protozoans. The reproductive strategies vary according to the groups of gastrotrichs. Males are unknown in the class Chaetonotoidea where reproduction is by parthenogenesis among females. Here two types of egg are laid. In one, hatching takes place very quickly and the young mature in about three days. In the other case eggs can lie dormant, surviving desiccation, and will hatch when more favorable circumstances prevail. The members of the class Macrodasyoidea are bisexual. Like the kinorhynchs, gastrotrichs are of little ecological or economic importance.   AC

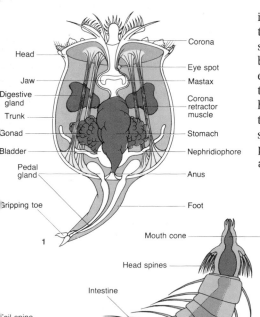

Head
Jaw
Digestive gland
Trunk
Gonad
Bladder
Pedal gland
Gripping toe

Corona
Eye spot
Mastax
Corona retractor muscle
Stomach
Nephridiophore
Anus
Foot

1

important attribute for species dwelling in temporary pools of water. They can also serve as a dispersal phase as they can be blown about by the wind. When they eventually hatch only females develop from them. Survival over difficult times is also helped by the ability of many rotifers to tolerate water loss and to shrivel up into small balls and thus remain in a state of suspended animation. This ability is referred to as cryptobiosis or "hidden life".

Mouth cone
Head spines
Intestine
Tail spine

Mouth

2

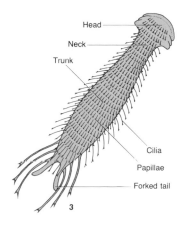

Head
Neck
Trunk

Cilia
Papillae
Forked tail

3

LAMPSHELLS

Phylum: Brachiopoda
About 260 species in 69 genera and 2 classes.
Distribution: worldwide, marine.
Fossil record: very extensive, early Cambrian
(about 600 million years ago) to recent with
greatest profusion in the late Paleozoic (up to
225 million years ago).
Size: shell length from less than 1mm to 4cm
(0.3–1.5in) plus stalk. Some extinct species
reached 30cm (12in).

Features: stalked animals with bilaterally
symmetrical shells in two halves (bivalve)
comprising dorsal and central valves; body
cavity a coelom; circulatory system open with
contractile dorsal vessel; excretion via paired
nephridia which also serve to release the
reproductive cells; conspicuous loop-shaped
ciliated respiratory and filter-feeding tentacle
crown (lophophore).

Class: Articulata
Three living orders, 4 extinct. Valves locked at
rear by a tooth and socket; lophophore has
internal support; anus absent; includes genus
Terebratella.

Class: Inarticulata
Two living orders, 3 extinct. Valves held
together by muscles only; lophophore lacks an
external skeleton; anus present; includes genus
Lingula.

► **A covey of lampshells** (genus *Terebratulina*).
It is just possible to make out the lophophore,
with its many fine tentacles, inside the shells.

▼ **The precursor of a juvenile lampshell:** a
larva of the genus *Lingula*. The large disk-like
structure marks the development of the mantle
and shells while the developing lophophore
crown shows as the group of tentacles. As the
shell gets heavier the lava sinks and takes up
adult life without metamorphosis.

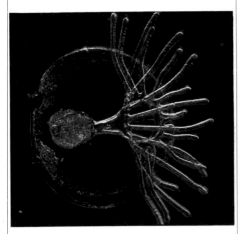

► **Lampshells:** ABOVE exterior view, attached
to the substrate; BELOW the body plan.

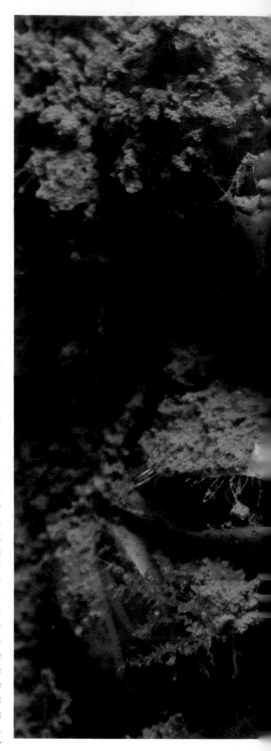

THE common name "lampshell" refers to
the superficial resemblance between the
shells of some of these animals and the
Roman oil lamp. Lampshells are really
animals of the past, for although they are
present in modest numbers in various
marine habitats the world over nowhere
today do they dominate the seas of the world
as they did in the late Paleozoic. Over
30,000 fossil species of lampshells have
been identified between Cambrian and
recent times.

Lampshells resemble bivalve mollusks
(clams, mussels etc) and indeed were classified with them until the middle of the 19th
century. Their bodies are ensheathed in a
mantle and enclosed by two hinged valves
(shells) which protect the soft animal
within. The two lobes of the mantle secrete
the shells and also enclose and protect the
crown (lophophore). However, there are
some startling differences between lampshells and mollusks which begin with the
issues of symmetry and orientation.

Today lampshells are found living mainly
on the continental shelves either attached
to rocks or other shells (for example *Crania*)
or some (for example *Lingula*) living in burrows in mud. Attachment may be direct or
by a cord-like stalk. Some occur in very shallow water, even on shores. Few prosper at
any depth. The distribution of these animals
is sporadic but where they do occur they
may exist in great numbers. The anatomy
of lampshells is variable and quite complex.
A salient feature is that the upper (dorsal)
valve is smaller than the lower (ventral)
one. In most forms the shells are both convex, and the apex of the lower one may be
extended to give the effect of the spout of the
Roman lamp that the common name
alludes to. In some of the burrowing forms
the shells are more flattened. The shells
themselves may be variously decorated with
spines, concentric growth lines fluted or
ridged, and their colors vary from orange
and red to yellow or gray.

The two valves are hinged along the rear
line and the manner of their contact forms
the basis for a division of the phylum into
two classes: Inarticulata and Articulata. In
the inarticulate lampshells the valves are
linked together by muscles in such a way
that they can open widely, but in the articulate lampshells they carry interlocking processes and these limit the extent of the gape.
In inarticulate lampshells as many as five
pairs of muscles control the movements of
the valves while in articulates two or three
sets of muscles are involved.

There are other differences between the
two classes. In inarticulates the shells are
formed from calcium phosphate and the gut
terminates in a blind pouch, there being no
anus. In articulates the shell is made of calcium carbonate and the gut is a "through"
system with an anus. The stalk or pedicle
by which most species are attached to the
substrate is itself attached to the lower
valve. The animals often position themselves so that the lower valve is uppermost,
thus adding confusion to the ideas of orientation and symmetry. In a few species the
pedicle has been completely lost and here
the animals are directly attached to the
substrate by the lower valve with the upper
valve uppermost.

margin of the valves. They tend to be held close to the upper valve. Suspended food material is trapped in mucus and swept to the mouth via a special groove. Special currents carry away rejected particles. The mouth leads to an esophagus which in turn gives on to the stomach.

There is an open circulatory system with a primitive pumping heart situated above the stomach. Blood channels supply the digestive tract, tentacles of the lophophore, the gonads and the nephridia. There is no pigment and there are few blood cells. Circulation of abolished food is likely to be the main function of the blood system and oxygen transport seems to be the responsibility of the coelomic fluid.

Lampshells have one or two pairs of nephridia for excretion and these discharge through pores situated near the mouth. The nervous system is rather simple with a nerve ring around the esophagus and a smaller dorsal and larger ventral ganglion. From these ganglia nerves supply the lophophore, mantle lobes and the various muscles that control the valves.

There are very few hermaphrodite lampshells. There are generally up to four gonads per individual and the ripe gametes are discharged into the coelom and leave the body via the nephridiopores. The eggs are generally fertilized in the sea. In most cases the embryo develops into a free-swimming larva, but in a few species it may be brooded. The larval development and planktonic period vary considerably between species.

AC

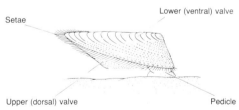

The body of the animal lies within the valves sheathed in mantle tissue which of course are extensions of the body wall. A true body cavity (coelom) lies between the body wall and the gut. The epidermis of the exposed inner mantle surface is hairy (ciliated). The lophophore is suspended in the mantle cavity and consists of a folded crown of hollow tentacles surrounding the mouth. It is supported by the upper valve of the shell and in some species there is actually a special skeletal structure to carry it. For feeding, the valves gape to the front allowing water to flow over the lophophore. The individual tentacles, which are hollow and ciliated on the outside, can reach to the front

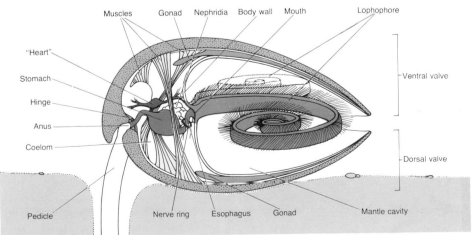

MOSS ANIMALS

Phylum: Bryozoa (or Ectoprocta)
Moss animals or sea mats.
About 4,000 species in about 1,200 genera
and 3 classes.
Distribution: worldwide, some freshwater but
mainly marine; sedentary bottom-dwelling
animals found in all oceans and seas at all
depths but mainly between 50 and 200m
(165–655ft).
Fossil record: from Ordovician era (500–440
years ago) to recent times; about 15,000
species.
Size: individuals 0.25–1.5mm (0.01–0.06in),
colonies 0.5–20cm (0.2–8in) wide.

Features: microscopic colonial animals
attached to a substrate, permanently living
inside a secreted tube or case (the zoecium);
body cavity a coelom; mouth encircled by a
hairy (ciliated) tentacle crown (lophophore);
gut U-shaped; anus outside the lophophore; no
nephridia or circulatory system.

Class: Gymnolaemata
Two orders.
Marine. Including genera *Alcyonidium*,
Memranipora.

Class: Phylactolaemata
One order. Fresh water.

Class: Stenolaemata
Four living orders, three extinct. Marine.
Including genus *Crisia*.

▶ **Colonial conditions:** BELOW individual
zooids of bryozoans (genus *Flustrella*) protrude
their ciliated lophophores from the colony
surface.

▼ **Colonial structure:** an encrusting colony of
bryozoans (genus *Electra*).

THE moss animals are an important group of sedentary aquatic invertebrates. They live attached to the substrate, either on rocks or empty shells, or on tree roots, weeds or other animals where they can find a hold. They are of interest because of their number and diversity. Some types, for example *Flustra*, may dominate conspicuous regions of the sea bed and have particular groups of organisms living alongside them. Others, such as *Zoobotryon*, are important because they can foul man-made structures, for example piers, pilings, buoys and ships' hulls.

Most moss animals are marine, inhabiting all depths of the sea from the shore downwards, in all parts of the world. A smaller group inhabit fresh water, where they are relatively inconspicuous. Moss animals are colonial animals. This means that many individuals termed zooids live together in a common mass. In a colony the first individual (formed from the larva, itself the result of sexual reproduction) is the founder. From the founder all other individuals in the colony are produced by asexual reproduction and thus share a common genetic constitution.

In many colonies of moss animals the individuals are all similar and all participate in feeding and sexual reproduction. Each individual is housed in a protective cup (zoecium), with walls reinforced with gelatinous horny or chalky secretions. The nature of these individual cup walls is conferred on the texture of the colony as a whole, so that some like *Alcyonidium* may feel soft and pliant and some like *Pentapora* feel hard, sharp and stiff. Equally the form of the colonies varies. Some genera, such as *Electra*, adopt a flat encrusting habit and therefore become dependent on currents flowing very

close to the rocks to bring them their food, while others like *Myriapora* grow up from the substrate so as to exploit stronger currents flowing above the substrate.

Each feeding individual in a colony is constructed on a similar plan. There is no head as such. A crown of hollow tentacles, filled by the coelom, forms a food-collecting and respiratory surface. (This parallels the tentacles of lampshells and horseshoe worms.) This crown is termed the lophophore. At the center of the lophophore the mouth opens and leads in to a simple U-shaped gut with esophagus and stomach. The anus lies close to, but *outside* the lophophore—the ectoproct condition (compare the endoprocts, p184, where the anus too lies inside the ring of tentacles). (It should perhaps be added here that opinions vary widely on the relationship of the ectoprocts and endoprocts, but there are good reasons for considering them to be separate phyla.)

The lophophore can be withdrawn for protection by a set of muscles and it can be extended by other muscles which deform one wall of the cup. This is often the upper or frontal wall, which includes a flexible, frontal membrane. When special muscles pull the frontal membrane in, fluid pressure in the coelom increases and the lophophore is protracted. Many interesting evolutionary developments have been explored by moss animals to ensure that a flexible frontal membrane to the cup, or some other means of adjusting the cup volume, is retained while making sure that predators are deterred from gaining entry through what is potentially a weak spot. Clearly rigid protection and flexibility are not easily reconciled. It is the epidermis of the animal that secretes the horny or gelatinous cuticle and in the chalky ectoprocts this in turn is

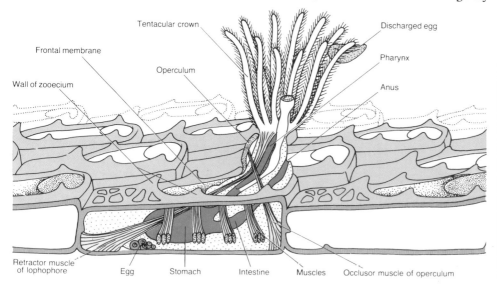

Tentacular crown
Discharged egg
Frontal membrane
Operculum
Pharynx
Wall of zooecium
Anus
Retractor muscle of lophophore
Egg
Stomach
Intestine
Muscles
Occlusor muscle of operculum

► **A foliose hydrozoan** attached to the rocky seabed. The skeletal system of rigid bones protecting the zooids helps this colony maintain its shape in the current. The lophophore can be discerned sieving the seawater.

▼ **The forms of some genera** of colonial moss animals. (1, 2) *Bugula*. (3) *Flustra*. (4) *Flustrellidra*. (5) *Myriapora*. (6) *Cupuladria*. (7) *Sertella*. (8) *Cellaria*. (9) *Pentapora*. (10) *Alcyonidium*.

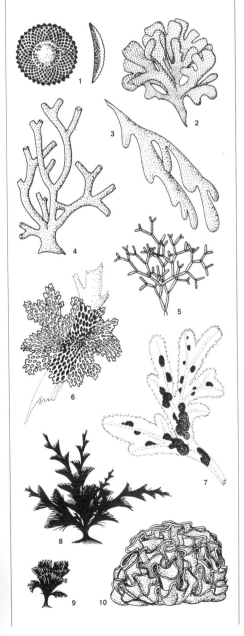

reinforced with calcium salts.

The lophophore tentacles are covered with waving hairs (cilia) to make efficient filter-feeding organs. When the lophophore is withdrawn the hole it goes in by may be covered over with a flap-like operculum or lid, present in some groups, but not in others.

Most animals are hermaphrodite with male and female organs developing from special regions of the inner lining of the body cavity. Ripe sperm and eggs are liberated into the coelom and find their way out of the body by separate pores. In many species the fertilized egg is brooded in a large chamber or ovicell, but in some a free-swimming larva, in many cases called a cyphanautes, spends a period in the plankton feeding on minute algae before settling, metamorphosing and founding a new colony.

The body contains no circulatory or excretory mechanisms; as it is of such small volume these roles are undertaken by diffusion.

In some of the more highly evolved moss animals (class Gymnolaemata, order Cheilostomata) there are several forms in the colonies. Here some zooids have given up the role of feeding and are supplied with food by others. Instead they function to defend the colony against small creeping predators like amphipod crustacea, and against clogging up with silt. There are two such forms: aviculariae and vibraculae. The first are so-called because they resemble minute birds' heads with snapping beak-like jaws. The cup and the operculum have been modified to act as minute jaws with delicate sense organs. They can seize hold of and detain small animals. The others have a large elongated operculum which resembles a bristle and which is vibrated back and forth to act as a current generator sweeping away silt.

AC

HORSESHOE WORMS

Phylum: Phoronida
About 11 species (more known as larvae only) in 2 genera.
Distribution: marine, bottom-dwelling, mainly from shallow temperate and tropical seas.
Fossil record: none.
Size: 0.6–20cm (0.2–8in).

Features: bilaterally symmetrical worm-like animals living permanently in a secreted tube; body cavity a coelom; mouth surrounded by a ciliated tentacle crown (lophophore); gut U-shaped; anus not enclosed by lophophore; closed circulatory system containing red blood corpuscles; paired nephridia also serve as ducts for release of reproductive cells.

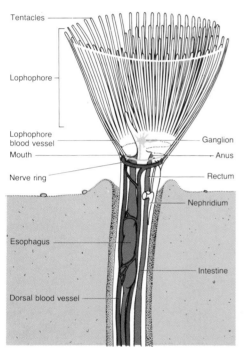

Tentacles
Lophophore
Lophophore blood vessel
Mouth
Nerve ring
Ganglion
Anus
Rectum
Nephridium
Esophagus
Intestine
Dorsal blood vessel

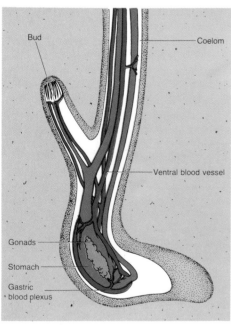

Bud
Coelom
Ventral blood vessel
Gonads
Stomach
Gastric blood plexus

U NLIKE the moss animals, to which they are probably related, the horseshoe worms form a small and relatively insignificant phylum. Only some 11 species are known as adults and they are not of any ecological or commercial importance.

Horseshoe worms live attached to the substrate, in secreted tubes of chitin-like substance which quickly become decorated with fragments of shells and grains of sand. These animals are small and obscure and are normally discovered only by accident when for example items of substrate are being carefully examined. They live in shallow temperate and tropical habitats. None dwell in fresh water. Their known distribution in America, Australia, Europe and Japan is more likely a reflection of where marine biologists are active in discovering them rather than an accurate delineation of where they actually dwell. A number of larvae, technically known as actinotrochs, have been collected, the adults of which are unknown. It is certain that the development of marine biology will see an increase in the number of known species of horseshoe worms.

The horseshoe worm body is essentially worm-like. At the front is a well developed horseshoe-shaped lophophore which in plan resembles two crescents, a smaller one set inside a larger one with their ends touching. The mouth lies between the two crescents. From these crescent-like foundations a number of ciliated tentacles arise to form the lophophore crown, and in some species the crescent tips may be rolled up as spirals thus increasing the extent of the lophophore. The beating of the cilia drives currents of water down the tentacles and out between their bases. Any small food particles are rolled up with secreted mucus and passed to the mouth.

The gut itself is U-shaped, a long esophagus leads back from the mouth to the stomach which is situated near the rear of the animal. From this the intestine leads back to the front and the rectum opens by the anus which is situated on a small protrusion (papilla) just outside the lophophore. The body cavity, a true coelom, houses paired excretory nephridia whose openings lie near the anus, and the gonads. Most species are hermaphrodite, and the sex cells are liberated into the coelom and find their way to the exterior via the excretory ducts.

The body wall consists of an outer tissue (epithelium) covering a thin sheath of circular muscles, which when they contract make the animal long and thin. Inside the circular muscles is a thick layer of longitud-

inal muscle which is responsible for contracting and shortening the body. The muscles enable the animal to move inside the tube and they are controlled by a simple nervous system. There is a nerve ring at the front set in the epidermis at the base of the lophophore. From this nerves arise and innervate the lophophore tentacles and the muscles of the body wall.

One interesting feature of the horseshoe worms is their clearly defined blood system with red pigment (hemoglobin) borne in corpuscles. Blood is carried forward by a dorsal vessel which lies between the limbs of the gut and back to the posterior by a ventral vessel. Small branches penetrate each tentacle of the lophophore. The blood is moved around the system by contractions that sweep along the dorsal and ventral vessels.

▲ **Water fans:** a colony of horseshoe worms feeding. Hemoglobin is visible in the blood vessels of some of the individuals.

◀ **The internal structure** of a horseshoe worm (genus *Phoronis*).

One species, *Phoronis ovalis*, is known to reproduce asexually and to establish large aggregations by budding. In sexual reproduction the eggs are generally fertilized in the sea. Typically an actinotroch larva develops which lives in the plankton for several weeks before undergoing rapid change to settle on the sea bed, form a tube and take up the adult mode of life.

One aspect of interest which zoologists attach to the horseshoe worms is their evolutionary position in general and their relationship in particular with the phyla of moss animals and lampshells. In the course of development embryos of species in all these phyla show a form of development where the body becomes divided into three sections, each with its own region of body cavity (coelom). As the animals develop the front section becomes progressively reduced and all but disappears. The middle region with its own coelom forms the lophophore and the rear section forms the bulk of the body of the adult. Zoologists have described these animals as oligomerous, that is they have few sections or segments. They are thus quite different from those groups like the annelid worms and the arthropod phyla where there are many segments to the body, ie metamerous. The general pattern of development in these groups is inclined toward that of the protostome groups, for example the annelid worms, but occasional features are inconsistent with this suggesting that in evolutionary terms they may occupy a position among the coelomate groups intermediate between the annelids and the arthropod phyla on the one hand, and the echinoderms on the other.

AC

Acoelomate

Flatworms
Phylum: Platyhelminthes

Ribbon worms
Phylum: Nemertea

Pseudocoelemate

Roundworms
Phylum: Nematoda

Spiny-headed worms
Phylum: Acanthocephala

Horsehair worms
Phylum: Nematomorpha

Coelomate – segmented

Segmented worms
Phylum: Annelida

Coelomate – not segmented

Echiurans
Phylum: Echiura

Priapulans
Phylum: Priapula

Sipunculans
Phylum: Sipuncula

Beard worms
Phylum: Poganophora

Arrow worms
Phylum: Chaetognatha

Acorn worms
Phylum: Hemichordata

W HAT is a worm? Everyone thinks they know how to deal with this question, but in reality it is a difficult one to answer. The standard reply will refer to earthworms—after all they are the most familiar invertebrate animals as they occur in our gardens and pasture lands. The sea angler will think of marine worms too, valuable as bait, like lugworms and ragworms. Others will remember the leeches, renowned as parasites, but also living as predators. Lugworms, ragworms, earthworms and leeches are all examples of one of the best known groups of worms, the segmented or annelid worms. This group contains a great variety of forms, marine and freshwater, terrestrial and parasitic, but they are not the only worms in the animal kingdom. The typical worm is a long, thin animal with a variably developed head and tail. It is therefore bilaterally symmetrical, unlike many of the creatures described in the previous sections on sedentary and free-swimming invertebrates which were asymmetrical or radially symmetrical.

Since worms have evolved at many evolutionary levels in the animal kingdom, there are many variations on the worm "theme." It is easiest to understand the different levels to which worms have evolved by looking at their structure in terms of number of body layers, origins and types of body cavities and their disposition, as well as observing whether or not the animals are divided into a number of segments. The simplest structure is found in flatworms and ribbon worms, where there is no internal body space. In roundworms and spiny-headed worms a body cavity is present between the gut and body wall, but this space is not lined on its inner edge by mesoderm (the middle tissue block) and is hence termed a pseudocoel. In the earthworms, ragworms, leeches and various other phyla the body cavity is known as a true coelom, since it is formed within the mesoderm and hence banded by it on both sides. Of those worm-like animals

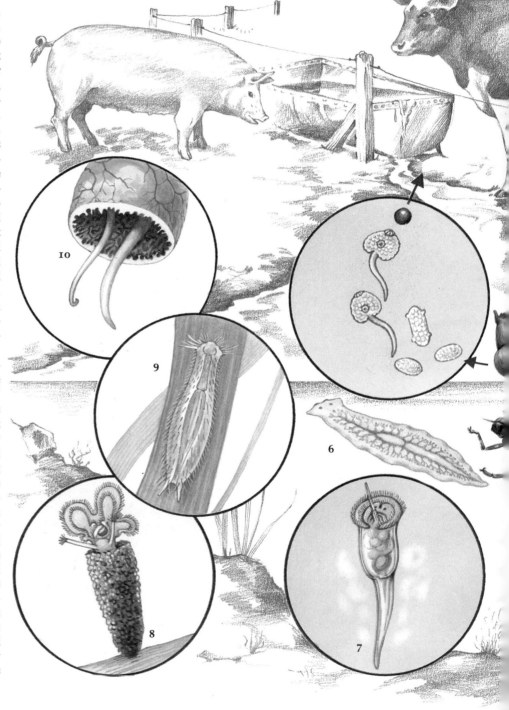

▼ **Worm-like parasites** of farmland and some other minute invertebrates. (1) *Fasciola hepatica*, the common liver fluke, a commercially significant parasite of sheep and cattle in Europe. The adults live in the liver. Eggs pass out of the primary host in feces and hatch into the first larval stage, a miracidium, which swims by cilia in the surface film of water on vegetation. The miracidium detects a secondary host, the fresh water snail *Lymnaea truncatula*, and enters it, forming a cyst in the digestive gland. From this a third larval stage, the rediae, develops, followed by the cercariae which leave the snail. These may encyst prior to reinfection of the primary host, thus completing the life cycle. The liver fluke flourishes in damp pastures where the snails exist freely. Adult size up to about 3cm (1.2in). (2) A species of *Brachionus*, an example of a rotifer protected by a firm outer structure called a lorica. (3) *Stephanoceros fimbriatus*, a sessile rotifer protected by a gelatinous case. (4) *Dugesia subtentaculata*, a free-living flatworm (planarian). (5) A species of *Gordius*, a parasitic worm (5cm, 2in). (6) *Dendrocoelum lacteum*, a free-living flatworm (planarian). Like *Dugesia* it glides through the environment by means of its cilia. (7) *Conochilus hippocrepis*, an example of a free-living rotifer protected by a gelatinous case. (8) *Floscularia ringens*, a rotifer that builds itself a protective tube out of pellets (microscopic). (9) A species of *Chaetonotus*, an example of a gastrotrich. (10) *Ascaris lumbricoides suilla*, a round worm parasite of the gut of the domestic pig, widely distributed (length about 2.5cm, 1in).

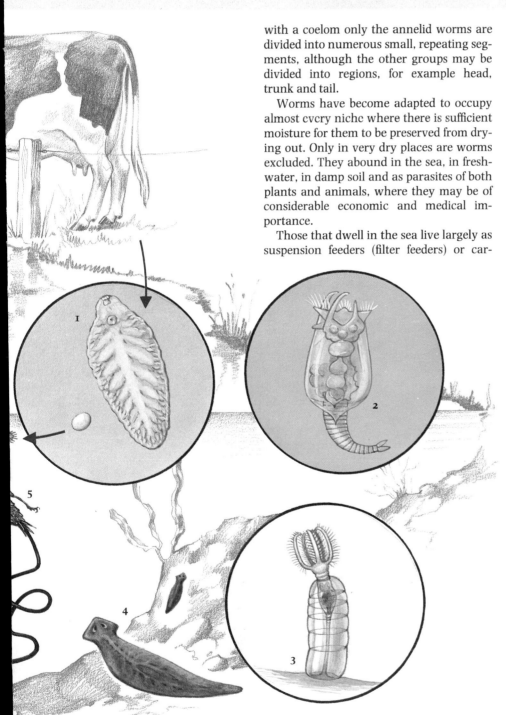

with a coelom only the annelid worms are divided into numerous small, repeating segments, although the other groups may be divided into regions, for example head, trunk and tail.

Worms have become adapted to occupy almost every niche where there is sufficient moisture for them to be preserved from drying out. Only in very dry places are worms excluded. They abound in the sea, in freshwater, in damp soil and as parasites of both plants and animals, where they may be of considerable economic and medical importance.

Those that dwell in the sea live largely as suspension feeders (filter feeders) or carnivores. The suspension feeders are equipped with filters and either depend on currents to sweep food to them, like the polychaetes *Sabella* and *Serpula* or horseshoe worms such as *Phoronis*, or, like the polychaete *Chaetopterus*, use modified limbs to generate a current to pump water through their filters. Carnivores need to subdue and to capture their prey. Ribbon worms have a long proboscis which is sometimes armed with stylets and glands which can be coiled around the victim. The annelids include forms which have powerful jaws like *Nereis* and *Marphysa*. By complete contrast the arrow worms are planktonic and not bottom dwellers. They too are efficient predators, their mouths being armed with seizing teeth which can grip small drifting organisms and readily engulf them. Feeding on detritus and swallowing mud are common ways of life for bottom dwellers in the sea and many annelids as well as the sipunculans and echiurans have adopted such strategies, often simply swallowing the nutrient-rich deposits through which they burrow. The acorn worms too live in muddy deposits and use the gills sited in their pharynx to filter water currents for food. Possibly the most enigmatic life styles of the free-living worms are those of the beard worms which live in tubes buried in the mud of the ocean bed. These animals lack a gut but have a crown of tentacles. They appear to absorb their nutrient requirements from the surrounding sea water.

Many worms have evolved to a parasitic mode of life. Without doubt the parasitic flatworms, flukes and tapeworms, have achieved great success deploying complex life cycles to ensure reinfection of their hosts. The spiny-headed worms are a small, exclusively parasitic group that have no gut, absorbing their nutrients directly from their hosts. Many roundworms feed as external or internal parasites on the bodies of plants and animals using special stylets and teeth to pierce and suck. AC

FLATWORMS AND RIBBON WORMS

▶ **Sliding smoothly** OPPOSITE through a bed of sea squirts, a Banded planarian (*Prosthecereaus vittatus*). Most free-living planarian flatworms such as this marine species avoid strong light and hide under stones or vegetation.

▶ **Body plan of a planarian flatworm** (genus *Planaria*), showing digestive, excretory, reproductive and nervous systems.

Flatworms were the first animals to evolve distinct front and rear ends. They are also the simplest of the animals with a three-layered (triploblastic) arrangement of the body cells. The most remarkable groups are the parasitic flukes and tapeworms, but many species are free living.

The turbellarians (class Turbellaria) are a widely distributed group of mostly free-living flatworms with a ciliated epidermis. They usually possess a gut but have no anus, and are classified by the shape of the digestive tract. Best known are the planarians, which may be black, gray or brightly colored, are aquatic in fresh or sea water, and feed on protozoans and small crustaceans, snails and other worms. *Convoluta* has no gut and depends on symbiotic algae for its food. Some members of the order Rhabdocoela (eg *Temnocephala* species) live attached to crustaceans and to other invertebrates and feed on free-living organisms, but most have become parasitic, such as *Fecampia* species, which live inside crabs and other crustaceans.

Some planarians regenerate complete worms from any piece. The more sophisticated parasitic flatworms, however, are unable to replace lost parts. In general the lower the degree of organization of an animal, the greater its ability to replace lost parts. Even a small piece, cut from such an animal usually retains its original polarity: a regenerated head grows out of the cut end of the piece which faced the front end in the whole animal. The capacity for regeneration decreases from front to rear: pieces from forward regions regenerate faster and form bigger and more normal heads than pieces from further back. In some planarians only pieces from near the front are able to form a head; those further back effect repair but do not regenerate a head.

The head of a planarian is dominant over the rest of the body. Grafts of head pieces reorganize the adjacent tissues into a whole worm in relation to themselves. Grafts from tail regions, on the other hand, are generally absorbed. However, the dominance of the head over the rest of the body is limited by distance, for example, if the animal grows to a sufficient length. This is what happens when planarians reproduce asexually: the rear part starts to act as if it were "physiologically isolated" and then finally constricts off as a separate animal.

True parasitism is universal in the two other classes, the Trematoda and Cestoda. The majority of monogenean trematodes are external parasites living on the outer surface of a larger animal, many on the gills of fishes. A few inhabit various internal organs and are true internal parasites, such as *Polystoma* in the urinary bladder of a frog and *Aspidogaster* in the pericardial cavity of a mussel species. The digenean trematodes, including the flukes, are all internal parasites. The adults inhabit in most cases the gut, liver or lungs of a vertebrate animal, swallowing and absorbing the digested food, blood or various secretions of their host. Among these are *Fasciola* species in the liver of sheep and cattle, *Schistosoma* in the blood of man and cattle, and *Paragonimus* in the lungs of man. The internal parasites are parasitic throughout the greater part of their life. After an initial short period as a free-living ciliated larva known as a miracidium, the young enters a state of parasitism as a sporocyst or redia in a second host, and after a second free interval as a tadpole-shaped cercaria, may enter the body of a third host to become encysted. The second host is often a mollusk, and the cercaria may complete its life cycle by becoming encysted in the same animal or a fish.

The cestodes are the most modified for a parasitic existence, as they remain internal parasites throughout life. They invariably inhabit the gut of a vertebrate. The intermediate host is frequently also a vertebrate—commonly the prey of the final host. As an adult, *Taenia crassicollis* is parasitic in the intestine of the cat; the cysticercus larval stage occurs in the livers of rats and mice. The adult tapeworm *Echinococcus granulosus* inhabits the gut of dogs and foxes; its hydatid cyst larval stage may be found in the liver or lungs of almost any mammal, but especially sheep and occasionally man.

Flatworms are bilaterally symmetrical animals without true segmentation. The

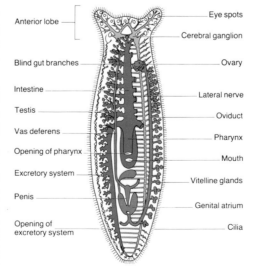

Anterior lobe

Blind gut branches

Intestine

Testis

Vas deferens

Opening of pharynx

Excretory system

Penis

Opening of excretory system

Eye spots

Cerebral ganglion

Ovary

Lateral nerve

Oviduct

Pharynx

Mouth

Vitelline glands

Genital atrium

Cilia

Flatworms and Ribbon Worms

Flatworms
Phylum Platyhelminthes

Free-living flatworms
Class Turbellaria
About 1,600 species.

Order Acoela, eg *Convoluta*.
Order Rhabdocoela, eg *Dalyellia*,
Fecampia, *Mesostoma*, *Temnocephala*.
Order Tricladida (**planarians**),
eg *Dendrocoelum*, *Planaria*, *Polycelis*,
Procerodes, *Rhynchodemus*.
Order Polycladida (**planarians**),
eg *Planocera*, *Thysanozoon*.

Parasitic flatworms
Class Trematoda
About 2,400 species.

Subclass Monogenea
—monogeneans
Order Monopisthocotylea,
eg *Gyrodactylus*.
Order Polyopisthocotylea,
eg *Polystoma*.

Subclass Aspidogastrea
Eg *Aspidogaster*

Subclass Digenea—flukes, digeneans
Order Strigeatoidea, eg *Alaria*,
Schistosoma.
Order Echinostomida, eg *Echinostoma*,
Fasciola.
Order Opisthorchiida, eg *Heterophyes*,
Opisthorchis.
Order Plagiorchiida, eg *Paragonimus*,
Plagiorchis.

Tapeworms
Class Cestoda
About 1,600 species.

Subclass Cestodaria
Order Amphilinidea, eg *Amphilina*.
Order Gyrocotylidea, eg *Gyrocotyle*.

Subclass Eucestoda
Order Tetraphyllidea,
eg *Phyllobothrium*.
Order Protocephala, eg *Proteocephalus*.
Order Trypanorhyncha,
eg *Tetrarhynchus*.
Order Pseudophyllidea,
eg *Dibothriocephalus*, *Ligula*.
Order Cyclophyllidea, eg *Echinococcus*,
Taenia.

Ribbon worms
Phylum Nemertea

Class Anopla
Order Palaeonemertini, eg *Carinella*.
Order Heteronemertini,
eg *Baseodiscus*, *Lineus*.

Class Enopla
Order Metanemertini,
eg *Drepanophorus*, *Prostoma*.

shape is leaf-like or ribboned in the planarians, or cylindrical in some rhabdocoels. While a distinct head is rarely developed, there is often a difference marking the anterior end, eg the presence of eyes, a pair of short tentacles, a slight constriction to form an anterior lobe. The mouth is on the underside, mostly toward the front. In some polycladids there is a small ventral sucker on the underside, and in some rhabdocoels there is an adhesive organ at the front and rear. In *Temnocephala* species a row of tentacles is often present. The trematodes closely related to the Turbellaria in internal organization resemble them in external form, with futher modifications to accommodate the parasitic existence. They are generally leaf-like with a thicker, more solid body. Suckers on the underside fix the parasite to its host. Usually there is a set at the front, or a single sucker surrounding the mouth and a rear set, or a single large rear sucker. Among trematodes, monogeneans often have more numerous suckers. Cestodes are ribbon-like, the anterior end is, in most cases, attached to the host by means of suckers and hooks placed on a rounded head (scolex). The hooks are borne on a retractile process, the rostellum. In the order Pseudophyllidea a pair of grooves takes the place of suckers and there are no hooks. In many cestodes parasitic in fishes the head bears four prominent flaps, the bothridia. In *Tetrarhynchus* species there are four long narrow rostella covered with hooklets. The cestodes are mouthless, and nothing distinguishes upper and lower surfaces. The body or strobila, which is narrower at the front, is made up of a series of segment-like proglottides which become larger toward the rear.

The outer surface of the body wall of parasitic flatworms is differently modified in the three classes and the underlying layers of muscle are also differently arranged. A characteristic of flatworms is the mesenchyme, a form of connective tissue filling the spaces between the organs. There is an alimentary canal in the turbellarians (except the Acoela) and in the trematodes, which also have a muscular pharynx and an intestine. There is no gut in the cestodes, which take in nutrients through the surface of the body wall. Flatworms have a bilateral nervous system involving nerve fibers and nerve cells. The degree of development of the brain varies in the different groups, being greatest in the polycladids and some monogenean trematodes: there is a grouping of nerve cells at the front end into paired cerebral ganglia, especially in the free-living forms. Sense organs in adult free-living turbellarians include light, chemo-sensory and chemo-tactic receptors, and these may also occur in the free-living stages of trematodes and cestodes. An excretory system exists in nearly all flatworms except the Acoela. There is usually a main canal running down either side of the body, with openings to the exterior. Opening into the twin canals are small ciliated branches that finally end in an organ known as a flame cell. There is no circulatory-vascular system.

Both male and female reproductive organs occur in the one animal, which is usually hermaphrodite, one of the exceptions being the trematode genus *Schistosoma*. The reproductive organs are most complex in the parasitic forms. The testes are often numerous, their united ducts leading into a muscular-walled penis resting in the genital opening. The female reproductive organs comprise the ovaries, which supply the ova, and the vitellarium, which supplies the ova with yolk and a shell. The ovaries discharge their ova into an oviduct

► **Gliding through the leaf litter,** a free-living flatworm on the forest floor in Trinidad. Although mostly thought of as either aquatic or parasitic, flatworms are quite common sights in damp tropical forests.

► **Banded ribbon worm on the seabed** (genus *Tubulanus*). The body of many ribbon worms is long and cylindrical and often brightly colored. Most species are carnivorous, usually capturing living invertebrate prey.

▼ **Body plan of a ribbon worm,** showing the proboscis lying outside the body.

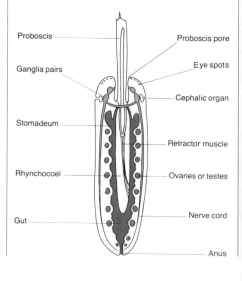

Proboscis
Ganglia pairs
Stomadeum
Rhynchocoel
Gut
Proboscis pore
Eye spots
Cephalic organ
Retractor muscle
Ovaries or testes
Nerve cord
Anus

Memory and Learning in Planarians

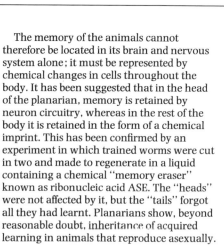

Planarians can learn and have a memory. They acquire a conditioned reflex in response to a bright light followed by an electric shock. If animals trained in this way are cut into two, both halves regenerate into a whole organism which retains its acquired learning, indeed both "heads" and "tails" show as much retention as uncut animals.

Similar results are found in planarians taught to find their way through a simple maze. Further experiments indicate that sexually mature worms fed on trained animals learn quicker than "control" animals kept on a normal diet. Two-headed planarians, produced by surgery, learn more quickly than others, while animals whose brain is removed are incapable of learning.

The memory of the animals cannot therefore be located in its brain and nervous system alone; it must be represented by chemical changes in cells throughout the body. It has been suggested that in the head of the planarian, memory is retained by neuron circuitry, whereas in the rest of the body it is retained in the form of a chemical imprint. This has been confirmed by an experiment in which trained worms were cut in two and made to regenerate in a liquid containing a chemical "memory eraser" known as ribonucleic acid ASE. The "heads" were not affected by it, but the "tails" forgot all they had learnt. Planarians show, beyond reasonable doubt, inheritance of acquired learning in animals that reproduce asexually.

which later forms a receptacle where fertilization occurs. The oviduct next receives the vitalline ducts which lead into the genital atrium. The location and arrangement of the genital opening in relation to the exterior are such to prevent self-fertilization and ensure cross-fertilization. Development in rhabdocoels and monogenean trematodes is direct. In digenean trematodes, cestodes and some planarians a metamorphosis occurs. Asexual reproduction occurs commonly in the turbellarians by a process of budding. Other planarians may fragment into a number of cysts each of which develops into a new individual.

The **ribbon worms** (phylum Nemertea) are almost entirely non-parasitic, marine worms, often highly colored, with only a few forms living on land or in freshwater, and one group as external parasites on other invertebrates. They are commonly looked upon as most closely related to the turbellarian flatworms, but are more highly organized.

Ribbon worms are often found burrowing in sand and mud on the shore or in cracks and crevices in the rocks. Some are able to swim by means of undulating movements of the body. Nearly all are carnivorous, either capturing living prey, mostly small intertebrates, or feed on dead fragments. The body is nearly always long, narrow, cylindrical or flattened, unsegmented and without appendages. The entire surface is covered with cilia and with gland cells secreting mucus which may form a sheath or tube for the creature. Beneath the ectoderm cell layer are two or three layers of muscle. There is no true coelom or body cavity, the space between organs being filled with mesenchyme. The proboscis, the most characteristic organ, lies in a cavity (rhynchocoel) formed by the muscular wall of the proboscis sheath. The proboscis may be everted for feeding. In representatives of the order Metanermertini (eg *Drepanophorus* species) there are stylets on the proboscis, providing formidable weapons. In other nemerteans (eg *Baseodiscus*) there are no stylets but the prehensile proboscis can be coiled round its prey and conveyed to the mouth underneath the front end. The gut is a straight tube from mouth to anus, but various regions are recognizable in some species. There is a blood-vascular system, the blood being generally colorless, with corpuscles. The excretory system resembles that of ribbon worms, but the nervous system is more highly developed. The brain is composed of two pairs of ganglia, above and below, just behind or in front of the mouth. Certain ganglia are probably related to special sense organs on the anterior end of the worm, and most ribbon worms have eyes. From the brain a pair of longitudinal nerve cords runs back down the body and there is a nerve net the complexity of which varies in the three orders. In most species the sexes are separate. The ovaries and testes are situated at intervals between the intestinal ceca and each opens by a short duct to the surface. Most ribbon worms develop directly, but some have a pelagic larva stage (pilidium), ending with a remarkable metamorphosis—the young adult develops inside the larva, from which it emerges.

Although very common, ribbon worms generally are of little economic or ecological importance.   GD

Pathogenic Parasites
Life cycles and medical significance of digenean flukes

Few animals affect humans so adversely as flukes of the subclass Digenea. They include creatures that cause one of the two most prevalent diseases of humans, schistosomiasis or bilharzia, and others that cause serious losses of livestock. These digenean flukes can be divided into blood flukes (eg *Schistosoma* species), lung flukes (eg *Paragonimus westermani*), liver flukes (eg *Fasciola* and *Opisthorchis* species), and intestinal flukes (eg *Fasciolopsis buskii*, *Heterophyes heterophyes*). Liver flukes can cause serious economic loss in sheep and cattle industries.

Other groups of parasitic flatworms contain representatives that, directly or indirectly, are harmful to people. Some monogenean parasites of fish often cause serious losses in fish-farming stocks kept in overcrowded conditions. Human cestode disease caused by adults of the tapeworm species *Taenia solium*, *T. saginata* and *Diphyllobothrum latum* is relatively non-pathogenic, but hydatid disease produced by the hydatid cysts of *Echinococcus granulosus*, cysticercosis caused by the cysticerus larval stage of *T. solium*, and sparganosis caused by the plerocercoid larvae of the genus *Spirometra* can be pathogenic.

The Common liver fluke (*Fasciola hepatica*) of sheep and cattle, almost worldwide in its distribution, is replaced by *F. gigantica* in parts of Africa and the Far East. Young flukes live in the liver tissue feeding mainly on blood and cells, while the adults live in the bile ducts. The flukes are hermaphrodite, with male and female organs in the same worm. Large numbers of eggs are produced and pass out in the feces onto the pasture. After a variable period, depending on the temperature, a miracidium hatches out. Moving on its cilia, this first-stage larva seeks out and penetrates the appropriate snail host, which in Europe and the USA is the Dwarf pond snail (*Lymnaea truncatula*), an inhabitant of temporary pools, ditches and wet meadows. The miracidium develops in its secondary host into a sporocyst which produces two or three generations of multiple rediae. Each redia produces free-living cercariae, the number being determined by temperature. From one miracidium therefore many thousands of infective cercariae are produced. The tadpole-shaped cercariae leave the snail and may encyst on grass as metacercariae until eaten by a suitable primary host, which may be a sheep, cow, donkey, horse, camel, etc and may also be a human eating, for example, infected watercress.

Spectacular losses due to acute fluke infes-

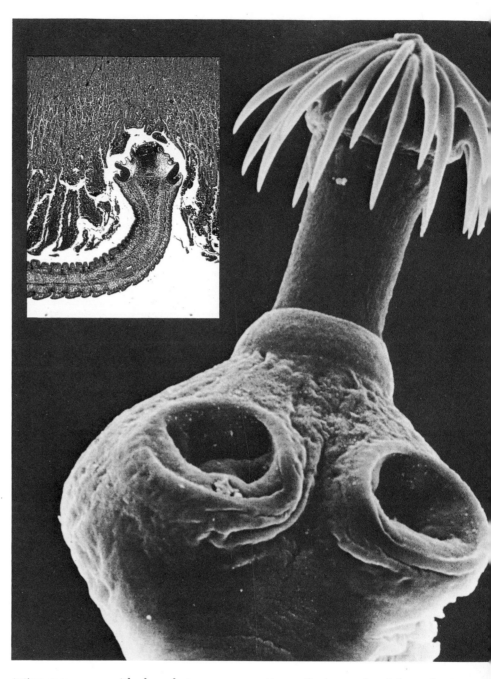

tation may occur with sheep but are rarer in cattle. The annual economic losses from fascioliasis in cattle and sheep in the UK for example are estimated to be approximately £10m and £50m respectively. Annual total losses in Holland may be 16m guilders, and in the USA $7m and in Hungary equivalent to $20m. It is possible in some countries to predict or forecast outbreaks of fascioliasis by using a meteorological system relying on monthly rainfall, evaporation and temperature data. In some countries chemical control of the snail host is possible, but in others reliance is placed on the periodic and strategic dosing of infected animals with such drugs as rafoxanide, oxyclozanide and

nitroxynil. A new breakthrough in treatment is triclabendazole which promises to be effective against early immature and mature *F. hepatica* in sheep and cattle.

Schistosomiasis or bilharzia affects the health of over 250 million people in 74 developing countries and it is estimated that another 600 million people are at risk. There are three principal schistosome species affecting humans. *Schistosoma haematobium* causes disease of the bladder and reproductive organs in various parts of Africa, especially Egypt, and the Middle East; *S. mansoni*, affecting mainly the liver and intestines, occurs in many parts of Africa, the Middle East, Brazil, Venezuela,

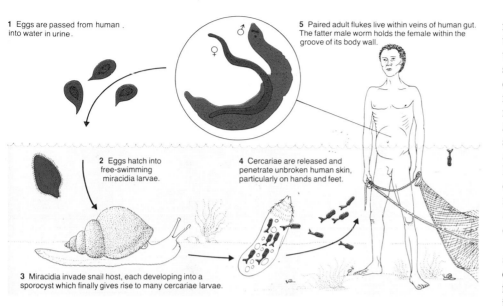

1 Eggs are passed from human into water in urine.

5 Paired adult flukes live within veins of human gut. The fatter male worm holds the female within the groove of its body wall.

2 Eggs hatch into free-swimming miracidia larvae.

4 Cercariae are released and penetrate unbroken human skin, particularly on hands and feet.

3 Miracidia invade snail host, each developing into a sporocyst which finally gives rise to many cercariae larvae.

▲ **Life cycle of the human blood fluke**
Schistosoma haemotobium which causes urinary or vesical schistosomiasis though damage to the bladder wall as its eggs bore through to get out of man. Water snails are the host for the larval stage.

◄ **Surreal palm tree with alien eyes**—the hooked, piston-like armory (rostellum) of the tapeworm *Acanthrocirrus retrirostris* taken in its larval form from a barnacle. Below the shaft on the head is a ring of four suckers, two of which are visible here. These are powered by flexible muscles and attach the tapeworm to the host tissue. The INSET shows the head of a dog tapeworm (*Taenia crassiceps*) embedded in the host's gut wall.

▼ **Life cycle of the liver fluke** *Fasciola hepatica* in sheep. This parasite also infects cattle.

Surinam and some Caribbean islands; *S. japonicum*, also affecting the liver and intestines and sometimes the brain, occurs in China, Japan, the Philippines and parts of Southeast Asia.

Schistosomes live in the blood vessels feeding on the cells and plasma. They do little damage themselves but are prolific egg-layers. The eggs pass through the veins into the surrounding tissues of the bladder, intestines or liver, causing severe damage. The long cylindrical adult females live permanently held in a ventral groove of the more muscular males, laying eggs which eventually pass out in the host's urine or feces. Adult schistosomes possess adaptations which enable them to evade the immunological defenses of the host and they

may live and reproduce in one person for up to 30 years. Should the eggs reach fresh water they hatch, releasing a free-swimming miracidium which actively seeks out and penetrates the appropriate intermediate freshwater snail host—a pulmonate (eg *Bulinus, Physopsis*) or, for *S. japonicum*, a prosobranch snail. The parasite multiplies asexually inside the snail and after a period of between 24 and 40 days has produced a large number of infective cercariae. These escape from the snail and need to burrow through human skin or into another suitable host for the parasite to develop into either a male or female worm.

The geographical range of each species of schistosome is confined to the distribution of suitable snail hosts. The most important factor affecting the spread of schistosomiasis is the implementation of water-resource development projects in developing countries—primarily the construction of hydroelectric dams and also irrigation systems. The Aswan High Dam in Egypt, and Lake Volta in Ghana have already aggravated and increased the spread of a disease already endemic in these countries. The debilitating disease schistosomiasis and malaria are the two most prevalent diseases of the world and have received much attention from the World Health Organization (WHO), which stimulates research into the basic problems related to the spread and control of these important diseases. The WHO claims that schistosomiasis could be eliminated by a combination of clean-water projects and the intensive use of drugs such as metrifonate and a new drug (praziquantel) against the appropriate schistosomes. Some experts believe a vaccine is needed too, but as yet no such vaccine is available, although one may appear in the next few years as a result of applying new technology using monoclonal antibodies.

GD

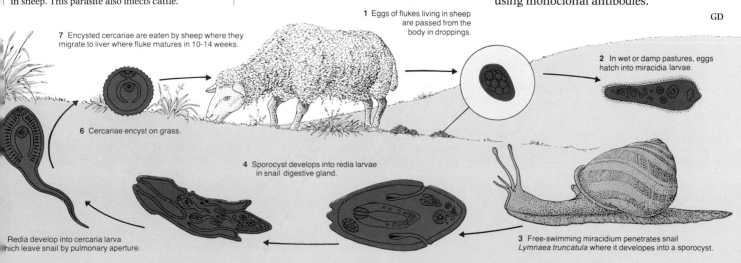

7 Encysted cercariae are eaten by sheep where they migrate to liver where fluke matures in 10-14 weeks.

1 Eggs of flukes living in sheep are passed from the body in droppings.

2 In wet or damp pastures, eggs hatch into miracidia larvae.

6 Cercariae encyst on grass.

4 Sporocyst develops into redia larvae in snail digestive gland.

Redia develop into cercaria larva which leave snail by pulmonary aperture.

3 Free-swimming miracidium penetrates snail *Lymnaea truncatula* where it developes into a sporocyst.

ROUNDWORMS AND SPINY-HEADED WORMS

Phyla: Nematoda, Acanthocephala, Nematomorpha

Roundworms
Phylum: Nematoda
Roundworms, eelworms, threadworms or nematodes
About 12,000 named species, but many may remain to be discovered.
Distribution: worldwide; mostly free living in damp soil, freshwater, marine; some parasites of plants and animals.
Fossil record: sparse; around 10 species recognized in insect hosts from Eocene and Oligocene (54–26 million years ago).
Size: from below 0.05mm to about 1m (0.02in–3.3ft) but mostly between 0.1 and 0.2mm (0.04–0.08in).
Features: round unsegmented worms sheathed in resistant external cuticle; body tapers to head and tail; cavity between gut and body wall a pseudocoel; mouth and anus present; lack circular muscles, cilia, excretory flame cells and a circulatory system; nervous system quite well developed; sexes separate.

Class: Aphasmida
About 14 orders; includes genus *Trichinella*.
Class: Phasmida
About 6 orders; includes genus *Ascaris, Loa, Mermis, Onchocera, Wucheria*.

Spiny-headed worms
Spiny-headed or thorny-headed worms
Phylum: Acanthocephala
About 700 species in 3 classes.
Distribution: widespread as gut parasites of terrestrial, freshwater and marine vertebrates; arthropods are intermediate hosts.
Fossil record: only 1 species from Cambrian (600–500 million years ago).
Size: about 1mm to 1m (0.04in–3.3ft), but mainly below 20mm (0.8in).
Features: worms lacking a mouth, gut and anus; cavity a pseudocoel; peculiar protrusible proboscis ("spiny head") armed with curved hooks; sexes separate.

Class: Arahiacanthocephala
Two orders; includes genus *Gigantorhynchus*.
Class: Palaeacanthocephala
Two orders; includes genus *Polymorphus*.
Class: Eoacanthocephala
Two orders; includes genus *Pallisentis*.

Horsehair worms
Horsehair, gordian or threadworms
Phylum: Nematomorpha
About 80 species.
Distribution: worldwide; adults free living and primarily aquatic in freshwater; juveniles parasitic in arthropods; *Nectonema* marine.
Fossil record: none.
Size: 5–100cm (2–39in) long.
Features: unsegmented worms with thick external cuticle; front end slender; body cavity a pseudocoel; through gut with mouth and anus present, but may be degenerated at one or both ends; lack circular muscles, cilia, excretory system; sexes separate with genital ducts opening into common duct with gut (cloaca).

Genera include: *Gordius, Nectonema, Paragordius*.

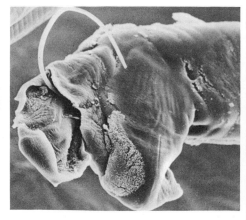

THE roundworm of humans affects some one thousand million people in the world; elephantiasis is a disfiguring disease of the tropics inflicted upon over 250 million individuals. These diseases are just two of the many caused by **roundworms** (Nematoda).

The nematodes are among the most successful groups of animal. They have exploited all forms of aquatic environment and many live in damp soil. Some are even hardy enough to survive in hostile environments like hot springs, deserts and cider vinegar. Furthermore, by parasitizing both animals and plants they have widely extended their habitat ranges. Thus they are significant food and crop pests and the agents of disease in plants, animals and man.

There are many types of nematode, but these animals are all quite similar. They are typically worm-shaped. Their bodies are not divided into segments or regions and generally taper gradually toward both the front and back. The mouth lies at the front and the anus almost at the tail. In cross section, they are perfectly circular and their outer skin is protected by a highly impermeable and resistant cuticle—the secret of the success of this group. The cuticle is complex in structure and is made up of several layers of different chemical composition, each with a different structural layout. The precise form of the cuticle varies from species to species, but in the common gut parasite of man and domestic animals, *Ascaris*, it comprises an outermost keratinized layer, a thick middle layer and a basal layer of three strata of collagen fibers which cross each other obliquely. These flex with respect to each other and allow the animal to make its typical wave-like movements, brought about by longitudinal muscle fibers. The high pressure of fluid maintained in the body cavity (pseudocoel) counteracts the muscles and keeps the worms' cross section round at all times.

Both free-living and parasitic forms may have elaborate mouth structures associated with their food-procuring activities and ways of life. The gut has to function against the fluid pressure of the pseudocoel and in order to do so may depend on a system of pumping bulbs and valves.

In primitive forms excretion is probably carried out by gland cells on the lower surface located near the junction of the pharynx and the intestine. In the more advanced types, there is an H-shaped system of tubular canals with the connection to the excretory pore situated in the middle of the transverse canal.

The nervous system provides a very simple brain encircling the foregut and supplying nerves forward to the lips around the mouth. Other nerves are supplied down the length of the body. The body bears external sensory bristles and papillae.

Free-living nematodes are mostly carnivores feeding on small invertebrates

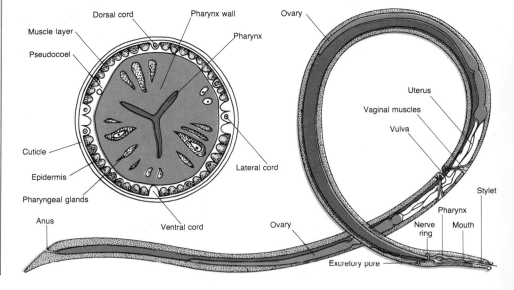

◄ **Pernicious parasite** of man, the hookworm. Shown here is the mating aparatus at the rear end of a male hookworm of the genus *Ancylostoma*. The thread-like protrusion is one of a pair of spicules used to guide sperm into the vulva of the female.

► **Squirming worm**—a free-living nematode worm found in fresh water. The extremely flexible body is bounded by the strong transparent cuticle that is key to the group's success.

▼ **Elephant-like skin** of advanced infections gives elephantiasis its name. The disease, of warm, humid regions of the world, is caused by roundworms (*Wucheria* species). The adult male (1) and female (2) worm parasitize the lymph ducts of their victim, to which they attach themselves, the female by papillae (3) and the male by a spine (4). The adults produce larvae called microfilariae (6) which are sucked up from the blood of an infected person by mosquitos (5) when they feed. Inside the insect the microfiliariae develop into infective larvae (7) which are injected into the blood of another person when the mosquito feeds. There they move to the lymphatic system, completing the cycle. Adults form the tangled masses of worms that cause the build-up of lymphatic fluids resulting in the characteristic swellings (8).

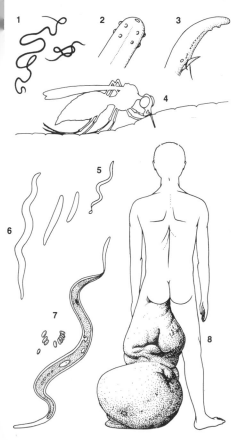

◄ **Anatomy of a roundworm.** (1) Whole animal. (2) Cross section in the region of the phorynx.

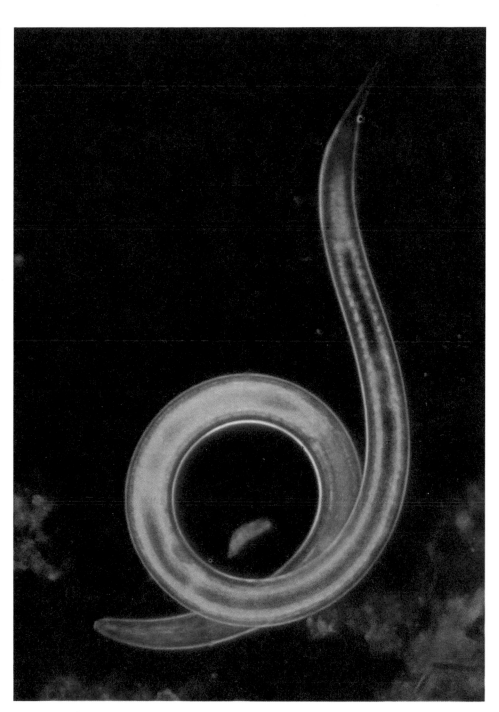

including other nematodes. Some aquatic forms feed on algae and some are specialized for piercing the cells of plant tissue and sucking out the contents. Others are specialized to consume decomposing organic matter such as dung and detritus and/or the bacteria that are feeding on these substances. Nematodes parasitic in man and animals are responsible for many diseases and are therefore of considerable medical and veterinary significance. In humans among other diseases they cause river blindness (*Onchocera* species), roundworm (*Ascaris lumbricoides*) and elephantiasis (*Wucheria* species). Very spectacular is the eye worm *Loa* which can sit on the cornea. Nematode diseases are more common in the tropics.

Parasitic nematodes feed on a variety of body tissues and fluids which they obtain directly from their hosts.

Roundworm of man is one of the largest and most widely distributed of human parasites. Infections are common in many parts of the world and the incidence of parasitism in populations may exceed 70 percent; children are particularly vulnerable to infection. The adult parasites are stout, cream-colored worms that may reach a length of

30cm (1ft) or more. They live in the small intestine, lying freely in the cavity of the gut and maintaining their position against peristalsis (the rhythmic contraction of the intestine) by active muscular movements.

The adult female has a tremendous reproductive capacity and it has been estimated that one individual can lay 200,000 eggs per day. These have thick, protective shells, do not develop until they have been passed out of the intestine and for successful development of the infective larvae a warm, humid environment is necessary. The infective larva can survive in moist soil for a considerable period of time (perhaps years), protected by the shell. Man becomes infected by accidentally swallowing such eggs, often with contaminated food or from unclean hands. The eggs hatch in the small intestine and the larvae undergo an involved migration around the body before returning to the intestine to mature. After penetrating the wall of the intestine the larvae enter a blood vessel and are carried in the bloodstream to the liver and thence to the heart and lungs. In the latter they break out of the blood capillaries, move through the lungs to the bronchi, are carried up the trachea, swallowed and thus return to the alimentary canal. During this migration, which may take about a week, the larvae molt twice. The final molt is completed in the intestine and the worms become mature in about two months.

An infected person may harbor one or two adults only and, as the worms feed largely on the food present in the intestine, will not be greatly troubled, unless the worms move from the intestine into other parts of the body. Large numbers of adults, however, give rise to a number of symptoms and may physically block the intestine. As in trichinosis, migration of the larvae round the body is a dangerous phase in the life cycle and, where large numbers of eggs are swallowed, severe and possibly fatal damage to the liver and lungs may result. Chronic infection, particularly in children, may retard mental and physical development.

Elephantiasis is a disfiguring disease of man restricted to warm, humid regions of the world and occurs in coastal Africa and Asia, the Pacific and in South America.

The blood of humans infected with the parasite contains the microfilaria stage of the worm, that is, the fully developed embryos still within their thin, flexible eggshells, that have been released from the mature female worms. During the day the microfilariae accumulate in the blood vessels of the lungs, but at night, when the

mosquitoes are feeding, the microfilariae appear in the surface blood vessels of the skin and can be taken up by the insects as they suck blood. The daily appearance and disappearance of the microfilariae in the peripheral blood is controlled by the activity pattern of the infected person and is reversed when the person is active at night and asleep during the day.

It is an impressive example of the evolution of close interrelationships between parasites and their hosts and ensures maximum opportunity for the parasite to complete its life cycle. Microfilariae that are taken up by a mosquito undergo a period of development in the body muscles of the insect before becoming infective to man. As the mosquito feeds, larvae may once again enter the human host.

The adult worms may reach a length of 10cm (4in), but are very slender. They live in the lymphatic system of the body, often forming tangled masses and their presence may cause recurrent fevers and pains.

In long-standing infections, however, far more severe effects may be seen, brought about by a combination of allergic reactions to the worms and the effects of mechanical

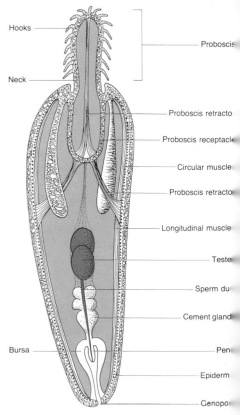

▲ **"Gordian knot"** of massed horsehair worms (*Gordius* species) under a stone in a drying streambed. These worms are often found in drinking troughs used by horses, and for this reason gave rise to the myth that they were horsehair come to life.

◀ **Structure of a spiny-headed worm,** showing the eversible pharynx.

blockage causing accumulation of lymph in the tissues. Certain regions of the body are more commonly affected than others, notably the limbs, breasts, genitals and certain internal organs, which become swollen and enlarged. The skin in these areas becomes thickened and dry and eventually resembles that of an elephant (hence elephantiasis). In severe cases the affected organ may reach an enormous size and thus bring about debilitating or even fatal secondary complications.

Drug treatment for the elimination of the worms is useful in the early stages of infection, but little can be done where chronic disease has produced true elephantiasis. Indeed, the parasites may no longer be present at that time.

Spiny-headed worms are all parasites living in the intestines of various groups of vertebrates. They are particularly successful at parasitizing the bony fish and the birds—they have completely failed to conquer the cartilaginous fishes (skates and rays). Unlike the parasitic flatworms (flukes and tapeworms) and the parasitic roundworms (nematodes), spiny-headed worms are of little medical or economic significance. This is probably because the insects and crustaceans that serve as secondary hosts to the juvenile spiny-headed worms are not eaten by man. In a few species another vertebrate may serve as a secondary host.

Spiny-headed worms completely lack a gut. Food digested by the host's gastric system is absorbed across the body wall and nourishes them. The body wall is composed of a fibrous epidermis which contains channels, sometimes referred to as lacunae, which are not connected to the interior or the exterior of the animal. They probably function to circulate absorbed food materials. The front is equipped with a strange reversible proboscis clad with hooks. The proboscis is extended to attach the worm to the lining of the host's gut. It is retracted by muscles

into a special proboscis sac and everted by fluid pressure by a reduction in the proboscis sac volume. Beside the sac are two bodies called lemnisci, filled with small spaces and thought to be food storage areas. There are excretory nephridia in some spiny-headed worms. The nervous system is simple; there is a ventral mass of nerve cells at the front of the body from which arise longitudinal nerve cords.

The sexes are separate and male spiny-headed worms are often larger than females. Internal fertilization takes place with the male using a penis to transfer sperm to the female. The fertilized eggs develop within the female pseudocoel until they reach a larval stage encased in a shell. These larvae are liberated and pass out with the feces of the host. If eaten by the appropriate secondary host, they emerge from the egg cases, penetrate the secondary host's gut wall and come to lie in its blood space where they remain until the secondary host is eaten by the primary host in a prey/predator relationship or by casual accident, as when water containing a small infected crustacean is drunk. At this point the primary host is reinfected.

Horsehair worms are unusual animals closely related to the nematode worms. Adults typically live in the soil around ponds and streams and lay their eggs in the water, attaching them to water plants. The larvae are parasitic and usually attack insects. One of the puzzles about their life cycle is that many of the insects acting as hosts for the horsehair worm larvae are terrestrial, not aquatic, for example, crickets, grasshoppers and cockroaches. It could be that these animals become infected by drinking water containing the larvae. The development inside the host can take several months and the larvae usually leave the host insect when it is near water to lead a free existence in the moist soil. The adults probably do not feed.   AC

Eelworm Traps

In a reversal of the roles played by certain gnat larvae and the fungal fruiting bodies on which they feed above ground, soil-living nematodes face the hazard of predatory fungi. Eelworms, active forms that thread their way through the soil particles, fall prey to over 50 species of fungi whose hyphal threads penetrate the eelworms' bodies.

There are several ways by which the hyphae penetrate the eelworms. Some fungi form sticky cysts which adhere to the eelworm, then germinate and enter its body.

Other fungi form sticky threads and networks, like a spider's web, which trap the eelworms.

Some fungi form lasso-like traps. These consist of three cells forming a ring on a side branch of a hypha. An eelworm may merely push its way into the ring and become wedged or the trap may be "sprung," the three cells suddenly expanding inward in a fraction of a second, to secure the eelworm (see RIGHT).

It is interesting that these fungi do not need to feed on eelworms: they only develop traps if eelworms are present in the soil.

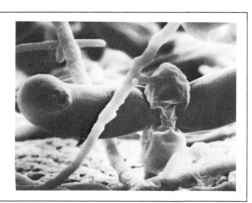

SEGMENTED WORMS

Phylum: Annelida
About 11,500 species in 3 classes.

Distribution: worldwide in land, water and moist habitats.
Fossil record: polychaete tubes (burrows) known from Precambrian (over 600m years ago).
Size: 1mm–3m (0.04in–10ft) long.
Features: body typically elongate, divided into segments and bilaterally symmetrical; body cavity (coelom) present; gut runs full length of body from mouth to anus; nerve cord present in lower part of body cavity; excretion via paired nephridia; appendages, when present, never jointed.

Earthworms
Class: Clitellata

Ragworms and Lugworms
Class: Polychaeta

Leeches
Class: Clitella

▶ **Fan on the sea floor.** Many annelid worms, such as this polychaete tube worm *Serpula vermicularis*, live a sedentary life in a tube, only the fan-like feeding parts protruding to extract food and oxygen from the sea water. In the center of the fan is the round operculum, a type of stopper which is used to close off the tube after the fan is withdrawn.

▼ **Structure of an earthworm,** cut away to show internal structure, particularly the digestive system, reproductive organs and nerve and blood systems.

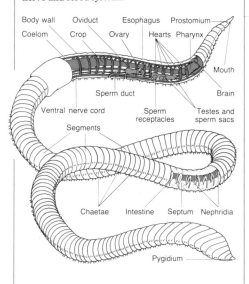

Body wall Oviduct Esophagus Prostomium
Coelom Crop Ovary Hearts Pharynx
Mouth
Sperm duct Brain
Ventral nerve cord Sperm receptacles Testes and sperm sacs
Segments
Chaetae Intestine Septum Nephridia
Pygidium

▶ **Mating earthworms.** Earthworms (*Lumbricus terrestris*) are hermaphrodite and during mating mutually exchange sperm.

WORMS that can be 3m (10ft) long; worms that can suck blood and worms that are vaguely reminiscent of a strip of rag—earthworms, leeches and ragworms. These are representatives of the three groups of worms known collectively as annelids. Annelids are characterized by a long, soft body and cylindrical or somewhat flattened cross-section.

An annelid worm is basically a fluid-filled cylinder, with the body wall comprising two sets of muscles, one circular, the other longitudinal. The fluid-filled cavity (the coelom) is effectively a hydrostatic skeleton (fluid cannot be compressed) upon which the muscles work antagonistically to produce changes in width and length of the animal.

Annelids are divided into a number of segments which in some forms give rise to a ringed appearance of the body—hence the name ringed worms sometimes applied to this group. Although some annelids can be broken up into individual segments and each of these can regenerate from the front or back into a completely new worm, the segments of which the normal adult is composed are not really independent. The gut, vascular system and nervous system run from one end of the segmental chain to the other and coordinate the whole body. The series of segments is bounded by a non-segmental structure at each end: the prostomium and a pygidium. The mouth opens into the gut immediately behind the prostomium, while the gut terminates at the anus on the pygidium. The prostomium and the pygidium are variously adapted: the prostomium, lying at the front, naturally develops organs of special sense such as the eyes and tentacles and hence contains a ganglion or "primitive brain," which connects with the nerve cord that runs from one end of the body to the other. The segments too, although primitively similar, are variously adapted, and the annelids demonstrate the variety of changes that can be rung on this apparently simple and monotonous body plan.

Earthworms and other oligochaete worms have a worldwide distribution. Earthworms may be found in almost any soil, often in very large numbers, and sometimes reaching a great size (*Megascolecides* of Australia, reaches 3m (10ft) or more in length). The

Darwin and Earthworms

Charles Darwin, the founder of modern evolutionary theory, published in 1881 a book entitled *The Formation of Vegetable Mould through the Action of Worms*, summarizing 40 years of his observation and experiment. Having discussed the senses, habits, diet and digestion of earthworms, he concluded that they live in burrow systems, feeding on decaying animal and plant material, in addition to quantities of soil. They produce casts, forming a layer of vegetable mold, thereby enriching the surface soil.

He noted that earthworms frequently plug their burrow entrances with either small stones or leaves, the latter also serving as a food source. Darwin saw signs of intelligence in the way in which leaves are grasped. When the leaf tip is more acutely pointed than the base, most are dragged by the tip, but where the reverse is true, the base is more likely to be grasped. Pine needles are almost invariably pulled by the base, where the two elements of the needle are united, forming the "point".

The habit of producing casts at the burrow entrance results in the gradual accumulation of surface mold, and Darwin estimated that annually some 20–45 tonnes of soil per hectare (8–18 tonnes per acre) may be brought to the surface in pasture. This soon buries objects left on the soil surface, and may aid the preservation of archaeological sites, but under certain conditions cast formation may also accelerate soil erosion. Burrowing affects the aeration and drainage of the soil, which is beneficial to crop production.

Darwin failed to distinguish between earthworm species (only relatively few British earthworms make permanent burrows or produce surface casts), but he was one of the first to recognize the important role that earthworms play in the terrestrial habitat. He firmly established their beneficial effects on soil, which have subsequently been largely confirmed.

commonest European earthworms are *Lumbricus* and *Allolobophora* species. The aquatic oligochaetes, sometimes known as bloodworms because of their deep red color, are smaller than earthworms and generally simpler in structure. Some are found in the intertidal zone, under stones or among seaweeds. But many of the aquatic genera are found in freshwater habitats, living, for example, in mud at the bottom of lakes (*Tubifex*). Such worms commonly have both anatomical and physiological adaptations which equip them to withstand the relatively deoxygenated conditions commonly found in polluted habitats.

Oligochaetes are remarkably uniform. They entirely lack appendages, except for gills in a few species. Some families are restricted to one type of environment, for example where there are fungi and bacteria associated with the breakdown of organic material; some may eat just microflora.

Oligochaetes are exclusively hermaphrodite, with reproductive organs limited to a few front segments. Male and female segments are separate, with segments containing testes always in front of those with ovaries (the reverse is true in leeches). The clitellum, a glandular region of the epidermis (outer skin), is always present in mature animals. It secretes mucus to bind together copulating earthworms, and produces both the egg cocoon and the nutritive fluid it contains. The exact position of the

The 3 Classes of Segmented Worms

Earthworms

Class: Clitellata, order Oligochaeta
About 3,000 species in 284 genera.

Distribution: worldwide in terrestrial, freshwater, estuarine and marine habitats.

Size: 1mm–3m (0.04in–10ft) long.

Features: no head appendages; few bristle-like chaetae usually in 4 bundles per segment; coelom spacious and compartmented; has blood vascular system, well developed body-wall musculature and ventral nerve cord with giant fibers in some; bisexual with reproductive organs confined to a few segments; glandular saddle (clitellum) present which secretes cocoon in which eggs develop; no larval stage.

Families: Aeolosomatidae, 4 genera; Alluroididae, 4 genera; Dorydrilidae, 1 genus; Enchytraeidae, 23 genera, including *Enchytraeus, Marionina, Achaetus*; Eudrilidae, 40 genera; Glossoscolecidae, 34 genera; Haplotaxidae, 1 genus; Lumbricidae, 10 genera, including **earthworms** (*Lumbricus, Allolobophora, Eisenia*); Lumbriculidae, 12 genera; Megascolecidae, 101 genera, including *Megascolex, Pheretima, Dichogaster*; Moniligastridae, 5 genera; Naididae, 20 genera, including *Chaetogaster, Dero, Nais*; Opistocystidae, 1 genus; Phreodrilidae, 1 genus; Tubificidae, 27 genera, including some **bloodworms** (*Tubifex* species), *Peloscolex*.

Ragworms and lugworms

Class: Polychaeta
About 8,000 species in some 80 families.

Distribution: worldwide, essentially marine.

Size: 1mm–2m (0.04–6.6ft) long.

Features: morphology extremely variable; various feeding and sensory structures may be present at front end; body segments with lateral appendages (parapodia) bearing bristle-like chaetae; coelom spacious; has blood vascular system, nephridia and ventral nerve cord with giant nerve fibers in some; usually sexual reproduction, with or without free-swimming larval phase.

About 80 families including: Aphroditidae (**sea mice**); Arenicolidae, including **lugworm** (*Arenicola marina*); Cirrotulidae; Eunicidae; Glyceridae, including some **bloodworms** (*Glyceris* species); Nephthyidae, including **catworms** (*Nephthys* species); Nereididae, including **ragworm** or **sandworm** (*Nereis virens*); Onuphidae (**beachworms**); Opheliidae; Polynoidae (**scaleworms**); Sabellariidae; Sabellidae (**fanworms**); Serpulidae; Spionidae; Syllidae; Terebellidae (including some **bloodworms**).

Leeches

Class: Clitella, order Hirudinea
About 500 species in 140 genera.

Distribution: worldwide, predominantly freshwater, some terrestrial or marine.

Size: 5mm–12cm (0.2–4.7in) long.

Features: number of segments 33; suckers at front and rear; coelom greatly reduced; blood vascular system tends to be restricted to remaining coelomic spaces (sinuses); hermaphrodite with one of mating pair transferring sperm; glandular saddle (clitellum) produces cocoon; no larval stage.

Families: Acanthobdellidae, genus *Acanthobdella*; Americobdellidae, genus *Americobdella*; Erpobdellidae, including *Erpobdella*; Glossiphoniidae, including *Glossiphonia, Theroemyzon, Placobdella*; Haemadipsidae, including *Haemadipsa*; Hirudidae, including **Medicinal leech** (*Hirudo medicinalis*), *Haemopis*; Piscicolidae, including *Pisciola, Branchellion, Ozobranchus*; Semiscolecidae, including *Semiscolex*; Trematobdellidae, including *Trematobdella*; Xerobdellidae, including *Xerobdella*.

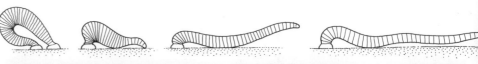

clitellum and the details of the reproductive organs and their exterior openings is of fundamental importance in the classification and identification of oligochaetes.

During copulation two earthworms come together, head to tail, and each exchanges sperm with the other. The sperm are stored by each recipient in pouches called spermathecae until after the worms separate. A slimy tube formed by the clitellum later slips off each worm, collecting eggs and the deposited sperm as it goes, and is left in the soil as a sealed cocoon. The eggs are well supplied with yolk, and are also sustained by the albumenous fluid surrounding them. In earthworms there is a tendency for the eggs to be provided with less yolk and rely more on the albumen. Development of the worms is direct, that is, there are no larval stages.

Parthenogenesis (development of unfertilized eggs) and self-fertilization are known in some species, and the aquatic members of the Aeolosomatidae and Naididae almost exclusively reproduce asexually.

Some earthworms are potentially long-lived, but the majority of oligochaetes probably have one- to two-year life cycles. Cocoon production occupies much of the year, interrupted by unfavorable environmental conditions, such as dryness, or by a seemingly fixed dormancy in some earthworms, that does not appear to be related to outside conditions.

An earthworm moves by waves of muscular contraction and relaxation which pass along the length of the body, so that a particular region is alternately thin and extended or shortened and thickened. A good grip on the walls of the burrow is aided by the spiny outgrowths of the body wall called chaetae which project from the body wall. Chaetae are found in both polychaetes and oligochaetes; but, as their names imply, polychaetes have many chaetae in each segment, whereas oligochaetes have relatively few. Furthermore, the number of chaetae in each segment of an oligochaete is not only small but constant. The earthworm *Lumbricus*, for example, has eight chaetae in each segment. In the aquatic oligochaetes the chaetae may be longer and more slender than those of an earthworm.

The earthworm feeds either on leaves pulled into the burrow with the aid of its suctorial pharynx, or by digesting the organic matter present among the particles of soil which it swallows when burrowing in earth otherwise too firm to penetrate. A muscular gizzard near the front end of the gut serves to break up compacted soil particles into

Bloodsucker—the Medicinal Leech

The most familiar of leeches is the Medicinal leech, which is a native of Europe and parts of Asia. Its use in blood letting in the 18th and 19th centuries led to its introduction to North America. However, it is now believed to be extinct there, and it has also recently become extinct in Ireland. The medicinal value of the practice of "blood letting" for almost all imaginable ills, "vapors and humors" may be questioned, but there is no doubt that the leech's ability to gorge fast for long periods is a perfect adaptation to its life-style.

The Medicinal leech eats blood, usually mammalian but also that of amphibians and even fish. Once attached to a potential blood source by means of its suckers, its three jaws are brought into contact with the skin. Each is shaped like a semicircular saw, bearing numerous small teeth. Sawing action of the teeth results in a Y-shaped incision through which blood is drawn by the sucking action of the muscular pharynx. Glandular secretions are released through each tooth which dilate the host's blood vessels, prevent coagulation of the blood, and act as an anesthetic.

A feeding Medicinal leech may take up to five times its body weight in blood, which passes into its spacious crop. Water and inorganic substances are rapidly extracted and excreted through the excretory cells

(nephridia), but digestion of the organic portion may take some 30 weeks. It was once thought that the Medicinal leech entirely lacked digestive enzymes, relying on bacteria in its own gut for digestion. However, although the involvement of bacteria is still recognized the leech does produce its own enzymes. The time taken to digest this highly nutritious food means that leeches need to feed only infrequently—it has been suggested that not much more than one full meal per year would be sufficient to permit growth.

Although little is known of its life history, the Medicinal leech is thought to take at least three years to mature. Cocoons, containing 5–15 eggs, are laid in damp places 1–9 months after copulation; they hatch 4–10 weeks later.

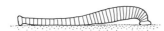

smaller ones, with the result that digestion and absorption of the organic material within the intestine is more efficient. Undigested matter is extruded from the anus on to the surface of the soil as the familiar worm casts. These casts give some indication of the valuable effects earthworms have upon the soil (see box, p206).

The best known of all **leeches** is the Medicinal leech which for centuries has been used in the "medicinal practice" of blood letting (see box). Leeches are easily recognized. Their soft bodies are ringed, usually without external projections and with a prominent often circular sucker at the rear. At the front there is another sucker around the mouth, and although this may be quite prominent as in fish leeches, it frequently is not. The skin is covered with a thin cuticle onto which mucus is liberally secreted. In front there is usually a series of paired eyes.

Leeches lack chaetae (except in the Acanhobdellidae) and have a fixed number (33) of segments. Like oligochaetes, they possess a clitellum, involved in egg cocoon production.

Leeches are divisible into two types: those with an eversible proboscis (Glossiphoniidae and Pisciolidae) and those with a muscular sucking pharynx, which may be unarmed (Erpobdellidae, Trematobdellidae, Americobdellidae, Xerobdellidae) or armed with jaws (Hirudidae, Haemadipsidae).

Not all species feed in the manner of the Medicinal leech. Some species feed on other invertebrates and may either suck their body fluids (eg Glossiphoniidae) or swallow them whole (eg Erpobdellidae). The Hirundidae, including the Medicinal leech, and the terrestrial Haemadipsidae feed on vertebrate blood, often that of mammals. The Pisciolidae are mostly marine, living on fish body fluids.

As well as circular and longitudinal muscles in the body wall, diagonal muscle bundles form two systems spiralling in opposite directions between these layers and opposing both. Muscles running from top to underside are well developed and are used to flatten the body, especially during swimming and movements associated with respiration.

The coelomic space is largely invaded by tissue, leaving only a system of smaller spaces (sinuses). The internal funnels of the excretory nephridia open into this system

and may be closely associated with the blood system, playing a part in blood fluid production. The blood vascular system is intimately connected with the coelomic sinuses, and may be completely replaced by them.

At copulation, one animal normally acts as sperm donor, and in the jawed leeches, a muscular pouch (atrium) opening to the outside acts as a penis for sperm transfer to the female gonadal pore. In other leeches, sperm packets formed in the atrium attach to the body wall of the recipient, and the spermatozoa make their way through the body tissues by a poorly understood mechanism. Fertilization is internal or occurs as gametes are released into the cocoon, which is secreted by the clitellum and slipped over the leech's head. A small number of zygotes are put in each cocoon, which is full of nutrient fluid secreted by the clitellum. In the Glossiphoniidae, cocoons are protected by the parent, and the juveniles may spend some time attached to the parent. Most leeches only go through one- or two-breeding periods, with one- to two-year life cycles.

Leeches are generally accepted as having been derived from oligochaetes, specializing as predators or external parasites. The fish parasite *Acanthobdella* shows many features intermediate between the two groups.

In contrast to the other annelids, polychaetes, such as **ragworms**, have extremely diverse forms and biology, although a common pattern is usually recognizable for each family. Polychaetes are almost all marine and they are often a dominant group, from the intertidal zone to the depths of the oceans. Planktonic, commensal and parasitic forms are also found.

Basically a polychaete is composed of a series of body segments, each separated from its neighbor by partitions (septa). Externally, each segment bears a pair of bilobed muscular extensions of the body wall known as parapodia containing internal supports (acicula), two bundles of chaetae, and a pair of sensory tentacles (cirri). The parapodia are best developed in active crawling forms, for example, ragworms.

Many polychaete families have an eversible pharynx which is most conspicuous in those species that are carnivorous, feed on large pieces of plant material or suck body fluids of other organisms. The pharynx in such families may be armed with jaws. Families with feeding tentacles or crowns use

▲ **Leech locomotion.** The suckers are attached alternately to the substrate, lifted and moved forward, thus removing the need for chaetae and a fluid-filled coelom, both of which are characteristic features of other annelid worms.

◄ **Waiting to pounce** upon passing prey, a jawed leech in the leaf litter of the north Australian rain forest. One good meal of blood each year may be enough for nutritional needs.

◄ **Taking a blood meal** OPPOSITE BELOW, a Medicinal leech hangs onto a human arm. A Medicinal leech may take up to five times its body weight in blood at each meal.

▼ **Structure of a leech.** Externally the body is divided into numerous "segments," although anatomically only 33 (34 according to some) are present, including those fused to form the two suckers. Leeches are protandrous hermaphrodites, with one pair of ovaries and 4–10 pairs of testes, each gonad lying in a sac in the body cavity (coelom). The testes are linked via a system of ducts to a muscular chamber, the atrium, opening to the exterior. The ovaries are positioned in front of the first pair of testes.

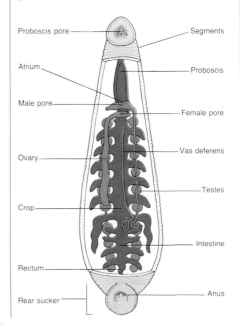

Proboscis pore — Segments
Atrium — Proboscis
Male pore — Female pore
Ovary — Vas deferens
Crop — Testes
Rectum — Intestine
Rear sucker — Anus

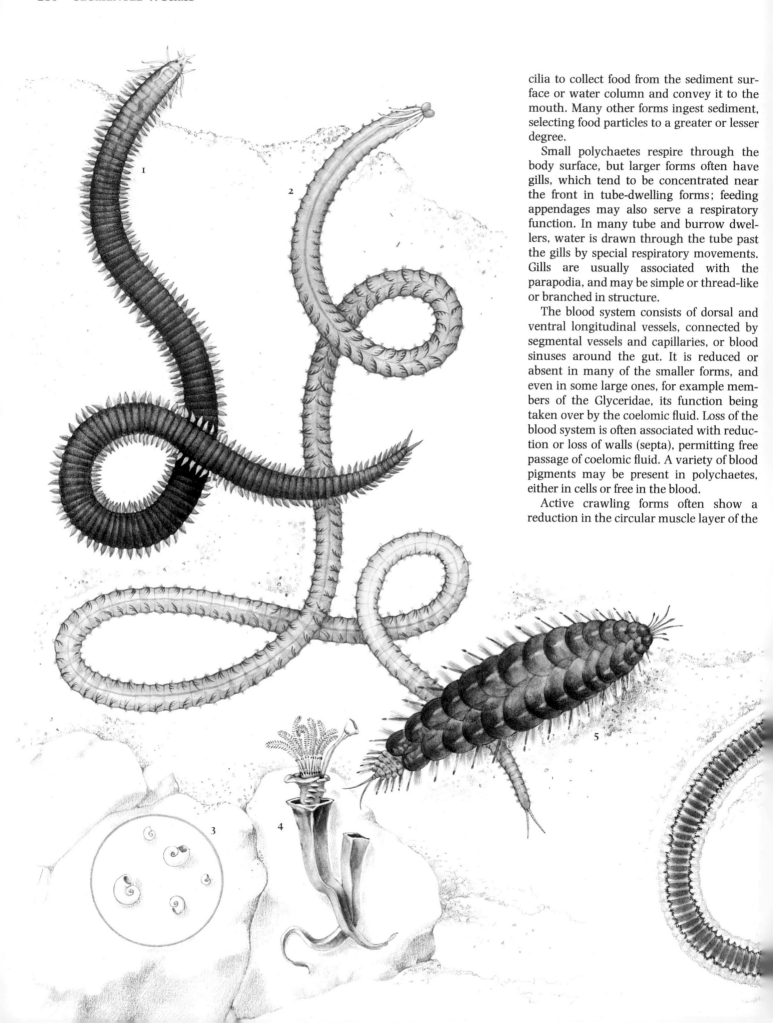

cilia to collect food from the sediment surface or water column and convey it to the mouth. Many other forms ingest sediment, selecting food particles to a greater or lesser degree.

Small polychaetes respire through the body surface, but larger forms often have gills, which tend to be concentrated near the front in tube-dwelling forms; feeding appendages may also serve a respiratory function. In many tube and burrow dwellers, water is drawn through the tube past the gills by special respiratory movements. Gills are usually associated with the parapodia, and may be simple or thread-like or branched in structure.

The blood system consists of dorsal and ventral longitudinal vessels, connected by segmental vessels and capillaries, or blood sinuses around the gut. It is reduced or absent in many of the smaller forms, and even in some large ones, for example members of the Glyceridae, its function being taken over by the coelomic fluid. Loss of the blood system is often associated with reduction or loss of walls (septa), permitting free passage of coelomic fluid. A variety of blood pigments may be present in polychaetes, either in cells or free in the blood.

Active crawling forms often show a reduction in the circular muscle layer of the body wall with corresponding development of the parapodia and their musculature and reduction or loss of the septa. This is correlated with a switch from peristaltic locomotion (as in earthworms) to the use of parapodia as the main propulsive organs.

Connecting the coelomic fluid to the exterior are a pair of excretory nephridia and a pair of genital ducts, the coelomoducts, in each segment. The nephridia pass through a septum, the external pore occurring in the segment behind that in which the nephridium originates. In most polychaetes the coelomoducts and nephridia are fused in fertile segments to produce urinogenital ducts, although often only at sexual maturity. Fertilization is generally external, and may lead to a free-swimming larva which may or may not feed, or development may be entirely on the sea floor (benthic). Adults protect the brood in a number of species, and this is not restricted to those with benthic development. Polychaetes may be hermaphrodite, eggs may develop into larva within the body (viviparous) or internal fertilization may take place. Asexual reproduction, by fragmentation or fission followed by regeneration, occurs in some 30 species.

Ragworms are perhaps the most familiar

◄▼ **Representative species of polychaet worms,** a major element in the benthic fauna of the world's seas. (**1**) *Eulalia viridis*, N Atlantic (15cm, 6in). (**2**) *Marphysa sanguinea*, N Atlantic (60cm, 24in). (**3**) *Spirorbis borealis*, a fanworm, N Atlantic (3.5mm, 0.1in). (**4**) *Pomatoceros triqueter*, a fanworm, N Atlantic (2.5cm, 1in). (**5**) *Harmothoe inibricata*, a scaleworm, N Atlantic (5cm, 2in). (**6**) *Hermodice carunculata*, Mediterranean (30cm, 12in). (**7**) *Perineretis nuntia*, a ragworm, Indo-Pacific (2.5cm, 1in). (**8**) Tentacles of *Reteterebella queenslandica*, Indo-Pacific (3.5cm, 1.4in). (**9**) *Sabellastarte intica*, Indo-Pacific (2.5cm, 1in). (**10**) *Spirobranchus gigantens*, Indo-Pacific (1.5cm, 0.6in).

polychaete worms, mainly through their use by sea anglers. Some species are carnivorous, feeding on dead or dying animals; some are omnivorous; others feed only on weed. Polychaete worms predominate among bait species. In Europe ragworm and lugworm are most commonly used for bait, and the catworms occasionally collected. In North America the sandworm and the bloodworm are extensively used, as are the long beachworms (species of the family Onuphidae) in Australia. All these species live in intertidal soft sediment, although ragworms are often found in muddy gravel.

They live for protection in the mud or muddy sand from which they emerge or partially emerge to feed on the plant and animal debris on the surface. The burrows may be located by the holes on the surface of the mud. The burrows themselves are U-shaped or at least have two openings, for the worms must irrigate them in order to respire. They do this by undulating their bodies, and this serves not only to renew the water within the burrow, but enables them to detect in the incoming water signs of food in the vicinity. Vast populations may occur in suitable habitats. Some species can tolerate the stringent conditions in estuaries and some are found in very low salinities.

Scaleworms also occur on the seashore and below the low water mark. Their backs are covered or partly covered by a series of more or less disk-shaped scales. The scales are commonly dark in color and overlap slightly. They are protective, not only in providing concealment but also in their ability to luminesce. The luminescent organs are under nervous control, so that the animal can "flash" when alarmed. Not all scaleworms do this, however.

Another polychaete worm is the sea mouse which is found below low tide level on sandy bottoms. It is often brought up by fishermen in the dredge. The body is covered with a fine "felt" of silky chaetae, and it is

▲ **Marine mouse.** The fine felt of silky bristles or chaetae gives this polychaete worm a furry appearance from which the common name of sea mouse is derived. Shown here is *Aphrodite aculeata*. Sea mice spend most of their lives in sand or mud in shallow waters.

▶ **Lawn of fanworms** spreading their feeding crowns from the seabed. These representatives of the family Sabellidae are found off Lizard Island, Queensland, Australia at a depth of 30m (100ft).

▼ **Structure of a polychaete worm.** (1) Whole body. (2) Cross section. The basic structure is a series of body segments, each separated from its neighbor by partitions, the septa. Externally, a segment bears a pair of bilobed parapodia containing internal supports, the acicula, two bundles of chaetae, and a pair of sensory cirri. In front of the mouth, which is on a segment that often lacks chaetae, is the prostomium containing the brain.

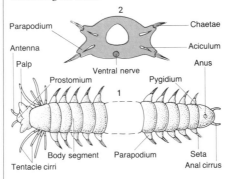

▶ **Varied heads of polychaete worms.** In families where it is well developed, the prostomium may have no appendages or (1) bear a small number of sensory antennae, with or without a pair of sensory/feeding palps; eyes are often present, for example in *Nothria elegans*. In species without such appendages, feeding tentacles may be present behind the prostomium, as in the families Spionidae and Cirratulidae. In other families the prostomium is reduced due to the presence of feeding appendages in the form of extensible tentacles, as in (2) members of the family Terebellidae, or a stiff crown or fan, as in the families Sabellidae (3) and Serpulidae, conditions associated with life in a tube. Specialized chaetae may form part of a lid-like opercular structure in other tube-dwelling families, such as (4) the Sabellariidae.

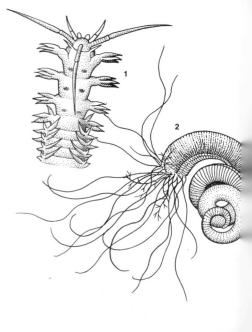

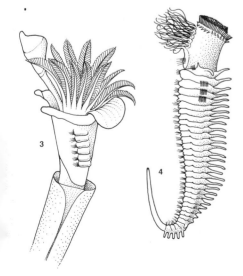

3

4

to this apparently furry appearance that its common name is due.

Sea mice live for the most part just beneath the surface of the sand or mud. They are rather lethargic although they can scuttle quite rapidly for a short distance when disturbed. If the matted "felt" over the back is removed a series of large disk-shaped scales will be seen overlapping the back: the sea mouse is nothing more than a large scaleworm in which the characteristic scales are concealed by the hair-like chaetae.

Fanworms are the most elegant of all the marine polychaete worms. Pinnate or branched filaments radiate from the head to form an almost complete crown of orange, purple, green or a combination of these colors. The crown is developed from the prostomium. It forms a feeding organ and, incidentally, a gill. The remainder of the body is more or less cylindrical.

All fanworms secrete close-fitting tubes which provide protection. Although their often gaily-colored crowns must tempt predatory fish, they can all contract with startling rapidity. These startle responses are made possible by relatively enormous giant nerve fibers which run from one end of the body to the other within the main nerve cord. In *Myxicola* this giant fiber occupies almost the whole of the nerve cord. Almost 1mm across, it is one of the largest nerve fibers known. Giant nerve fibers are associated with particularly well developed longitudinal muscles which enable the worm to retract promptly when danger threatens. Other movements are relatively slow.

PRG

ECHIURANS, PRIAPULANS, SIPUNCULANS

Phyla: Echiura, Priapula, Sipuncula

Echiurans
Phylum: Echiura
About 130 species in 34 genera.
Distribution: worldwide, largely marine but with a few brackish water species; intertidal down to 10,000m (33,000ft).
Size: usually between 3 and 15cm (1.2–6in) long, occasionally up to 75cm (30in).

Features: unsegmented worms with coelom; trunk with muscular proboscis at front; mouth at base of proboscis; gut coiled; anus terminal; blood vascular system usually present; unsegmented ventral nerve cord; usually a pair of bristles (setae) just behind mouth; one to many excretory nephridia and a pair of anal excretory organs; sexes separate with extreme sexual dimorphism in some; produce free-swimming trochophore larva in some others.

Families: Bonelliidae, including genus *Bonellia*; Echiuridae, including genera *Echiurus*, *Thalessema*; Ikedaidae, including genus *Ikeda*; Urechidae, including genus *Urechis*.

Priapulans
Phylum: Priapula
Nine species in 6 genera.
Distribution: exclusively marine, intertidal and subtidal.
Size: 0.2–20cm (0.08–16in).

Features: unsegmented worms with coelom; body divided into 2 or 3 regions; introvert at front with terminal mouth and eversible pharynx; trunk may have tail attached; possess chitinous cuticle which is periodically molted; cilia lacking; sexes separate; produce free-living lorica larva.

Families: Priapulidae (genera *Priapulus*, *Priapulopsis*, *Acanthopriapulus*, *Halicryptus*); Tubiluchidae (genus *Tubiluchus*); Chaetostephanidae (genus *Chaetostephanus*).

Sipunculans
Phylum: Sipuncula
About 320 species in 17 genera.
Distribution: worldwide, exclusively marine; intertidal down to 6,500m (21,000ft).
Size: trunk 0.3–50cm (0.1–20in); introvert half to 10 times as long as trunk.

Features: unsegmented worms with coelom; body divided into introvert and trunk; mouth at tip of retractable introvert; gut U-shaped with anus usually at front of trunk; possess 1–2 excretory nephridia; ventral nerve cord unsegmented; trunk wall of outer circular and inner longitudinal muscles; asexual reproduction rare; usually unisexual; larvae free-swimming.

Families: Sipunculidae, including genus *Sipunculus*; Golfingiidae, including genera *Golfingia*, *Phascolion*, *Themiste*; Phascolosomatidae (genera *Phascolosoma*, *Fisherana*); Aspidosiphonidae, including genus *Aspidosiphon*.

ECHIURANS, priapulans and sipunculans are three groups of marine animals whose existence is only really known to scientists—hence they lack common names! All are worm-like, but their bodies are not segmented as in segmented (annelid) worms. The three groups are not particularly closely related to each other.

Echiurans are delicate soft-bodied animals living in soft sediments or under stones in semi-permanent tubes, or inhabiting crevices in rock or coral. They are exclusively marine, although a few forms penetrate into brackish waters, and many members of the family Bonelliidae are found at very great depth.

Food, usually in the form of surface detritus, is caught up in mucus and transported by cilia down the muscular proboscis to the mouth. In *Urechis* species a net of mucus is used to filter bacteria out of water pumped through the burrow, and the net is then consumed along with the food.

In *Urechis* the rear end of the gut is modified for respiration, rhythmic contractions of the hind gut drawing in and expelling water. This may be linked to the absence of a blood circulatory system in this genus, the coelomic fluid containing blood cells having taken over this function. In all other echiurans the blood circulatory system is separate from the coelom.

The musculature of the trunk wall includes an outer circular layer and an inner longitudinal layer. An additional oblique layer may be present, its position being an important feature in defining families.

Reproduction is always sexual, the sexes being separate. Mature gametes are collected into the excretory nephridia just before spawning, in some cases by complex coiled collecting organs. Where fertilization occurs in the water (families Echiuridae, Urechidae) a free-swimming trochophore larva results, gradually changing to the adult morphology. In the Bonelliidae, where the male lives in or on the female, fertilization is presumed to be internal and larval development is essentially on the sea floor. In *Bonellia*, if the larva makes contact with an adult female, it tends to become male, and if not, female.

Priapulans comprise a small phylum of uncertain affinities, composed of exclusively marine species inhabiting soft sediments. The more familiar forms are the relatively large family Priapulidae, but recently the small sand-living genus *Tubiluchus* and the small tube-dwelling *Chaetostephanus* have been described.

The trunk of priapulans bears warts and spines, and although it may have rings, no true segmentation is present. A tail consisting of one or two projections may be present, although there is no tail in *Halicryptus* or *Chaetostephanus*. In *Priapulus* the tail is in the form of a series of vesicles which constantly change shape and volume and may be involved in respiration. In *Acanthopriapulus* the tail is muscular, bearing numerous hooks, and may serve as an anchor during burrowing. In *Tubiluchus* the tail is a long tubular structure.

The pharynx is eversible and muscular, armed with numerous teeth, or, in *Chaetostephanus*, bristles (setae). In the Priapulidae, it is used to capture living prey, but in *Tubiluchus* a scraping function is more likely. The intestine is in the trunk, the anus opening at its rear, with no part of the gut in the tail.

The sexes are separate, with differences in size between the sexes in *Tubiluchus*, probably associated with copulation.

The free-living larval stage may be extremely long, the larva feeding in surface sediment layers for two years or more before changing to the adult condition.

The Priapula were formerly considered to be related to the Kinorhyncha (p186) because of the body armature, lack of cilia and the molting of the cuticle. Now, however, they are regarded as a coelomate group.

Sipunculans occupy a variety of marine habitats. Many live in temporary burrows in soft sediment, while others can bore into chalky (calcareous) rocks, as in the genera *Phascolosoma* and *Aspidosiphon*, or corals, as in *Cloeosiphon* and *Lithacrosiphon*. Some, such as *Phascolion*, inhabit empty gastropod shells.

All sipunculans have an extensible introvert up to 10 times the length of the trunk, with the mouth at its tip, which can be withdrawn completely into the trunk. It allows feeding at the substrate surface while the body remains protected, and can be readily regenerated if lost. The introvert is extended by fluid pressure brought about by contraction of the musculature of the trunk wall. In some cases, for example *Sipunculus* species, the circular and longitudinal muscle layers may be arranged in separate bundles. Up to four special muscles retract the introvert. Most sipunculans consume sediment, although some are thought to filter material from the water using cilia on their tentacles.

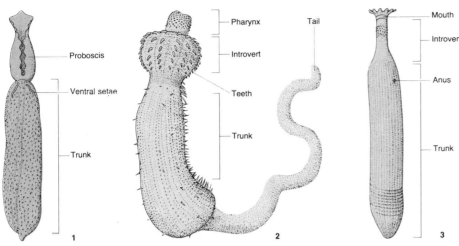

▲ Body forms. (1) In echiurans the proboscis is sited in front of the mouth and is not entirely retractable into the trunk. When fully extended, the muscular proboscis may be several times the length of the trunk. It may be forked and is well supplied with cilia and mucus glands.

(2) The priapulan body is divided into 2–3 regions—an introvert at the front, a trunk and a tail (which may be absent). The introvert bears the mouth at its tip and is a large barrel-shaped structure armed with rows of teeth or, in *Chaetostephanus*, with complex setae, and is a characteristic feature of the group. It can be withdrawn into the trunk by a series of muscles and is used in burrowing.

All sipunculans have **(3)** an extensible introvert up to 10 times the length of the trunk, with the mouth at its tip, which can be withdrawn completely into the trunk. It allows the sipunculan to feed at the surface of the seabed while the body remains below it protected. The introvert of a sipunculan can readily be regenerated if it is lost.

▼▶ Representative species of minor worms.
(1) *Thalassema neptuni*, an echiuran (7cm, 2.8in). (2) *Priapulus caudatus*, a priapulan; N Atlantic (8cm, 3in). (3) *Chaetopterus variopedatus*; N Atlantic and Mediterranean (23cm, 9in). (4) *Phascolion strombi*, a sipunculan which lines empty shells (eg that of *Turitella*) with mud and forms a burrow inside; N Atlantic.

Excretion occurs through the one or two nephridia, which connect the large fluid-filled coelomic space to the exterior, opening on the front of the trunk. They also serve as storage organs for gametes immediately before spawning. Sipunculans have no blood vascular system, the coelomic fluid performing the circulatory function.

Externally, sipunculans may have hooks or spines on the introvert and small protrusions (papillae) and glandular openings on the introvert and trunk. In many boring forms a horny or calcareous shield is present at the front end of the trunk, protecting the animal as it lies retracted in its burrow.

Asexual reproduction is known in only two species, sexual reproduction with separate sexes being the rule. Some populations of *Golfingia minuta* are hermaphrodite. In all sipunculans studied, much of the germ cell development occurs free in the coelomic fluid. At spawning, eggs and sperm are released via the nephridia, and fertilization occurs in the sea. The mode of larval development, however, varies between species and may include a free-swimming trochophore type larva, or may be direct, with juvenile worms emerging from egg masses. In several species, the floating stage occupies a period of months, and long distance transportation of such larvae across the Atlantic has been suggested.   PRG

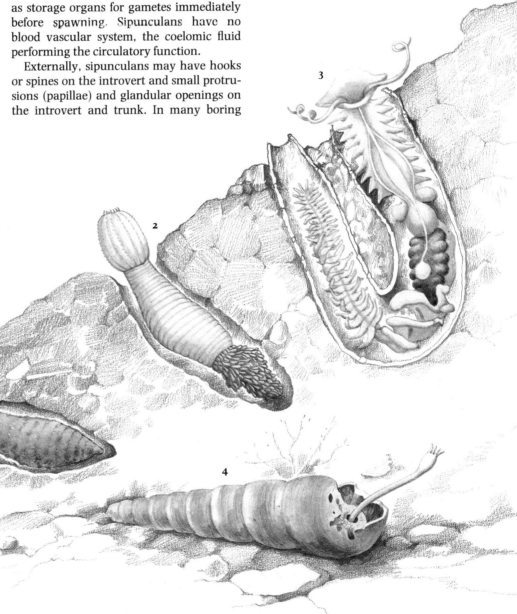

Phyla: Pogonophora and Chaetognatha

Beard worms

Phylum: Pogonophora
About 150 living species in 2 orders and 7 families.
Distribution: all oceans, usually at considerable depths.
Fossil record: none.
Size: length 5cm to 1.5m (2in–5ft), usually 8–15cm (3.2–6in); diameter up to 3mm (0.1in).

Features: bilaterally symmetrical, solitary, tube-dwelling worms; body divided into 3 zones—head (cephalic) lobe bearing tentacles, trunk, and posterior opisthosoma; gill slits, digestive tract and anus all lacking; protostome development.

Order: Athecanephria
Families: Oligobrachiidae (including genus *Oligobrachia*) and Siboglinidae (including genus *Siboglinum*).

Order: Thecanephria
Families: Lamellibrachiidae (including genus *Lamellibrachia*); Lamellisabellidae (including genus *Lamellisabella*); Polybrachiidae; Sclerolinidae (including genus *Sclerolinum*); Spirobrachiidae (including genus *Spirobrachia*).

Arrow worms

Phylum: Chaetognatha
About 70 living species in 7 genera.
Distribution: all oceans, planktonic apart from 1 benthic genus.
Fossil record: one dubious record.
Size: between 0.4cm (0.16in) and 10cm (4in), most less than 2cm (0.8in) long.

Features: small bilaterally symmetrical free-living animals with deuterostome development; lack circulatory and excretory systems; body torpedo-shaped with paired lateral fins and tail fin; mouth at front and armed with strong grasping spines.

Seven genera: *Bathyspadella, Eukrohnia, Heterokrohnia, Krohnitta, Pterosagitta, Sagitta, Spadella.*

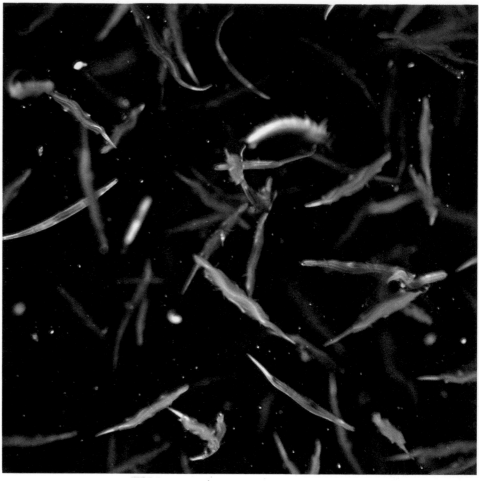

B EARD worms are the most unlikely of animals. They are a zoological curiosity of comparatively recent discovery. Their anatomy was not fully appreciated until 1963 when the first whole specimens were obtained. The name Pogonophora comes from the Greek word *Pogon* meaning a beard—this being a reference to the shaggy group of tentacles carried on the front of the body.

These long, thin animals—often 500 times as long as broad—live in tubes which are generally completely buried in the mud or ooze of the ocean floor. There are two exceptions to this: *Sclerolinum brattstromi* lives inside rotten organic matter like paper, wood and leather in Norwegian fjords; and *Lamellibrachia barhami* builds tubes which project from the sediment.

Beard worms have a front head (cephalic) lobe bearing between one and 100 or more tentacles. Immediately behind the head is the forepart, or bridle, which appears to be important in tube building. Behind this lies the trunk, which comprises the greatest part of the body. It is covered with minute tentacles (papillae) which toward the front are arranged in pairs in quite a regular fashion. They become more irregular further back along the trunk and some may be enlarged. These papillae are thought to enable the worm to move inside its tube, and they are associated with plaques, hardened plates which probably assist in this respect. Further along the trunk there is a girdle where the skin is ridged and rows of bristles, similar to those of annelid worms, occur on the ridges. These probably help the worm maintain its position inside the tube.

The distinct rear region of the body, the opisthosoma, is easily broken from the rest of the body, and because of this its presence was not appreciated for many years. It comprises between five and 23 segments, each of which carries bristles larger than the ones on the girdle. The opisthosoma probably

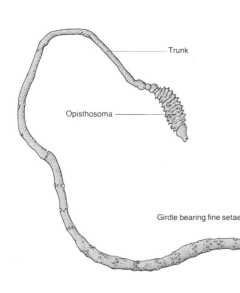

Trunk

Opisthosoma

Girdle bearing fine setae

Papillae

Trunk

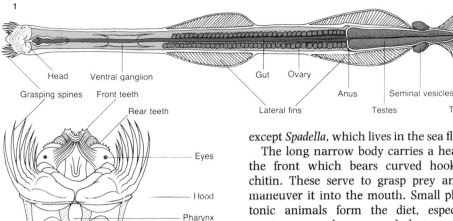

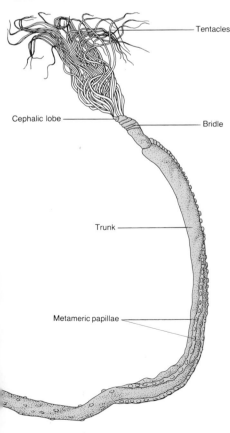

▶ **Body plan of an arrow worm.** (1) Whole body, showing the main internal organs. (2) Detail of head, showing spines and teeth used in feeding.

◀ **Arrow worms** in marine plankton. The existence of this group of animals was not discovered until 1829, when nets capable of sieving out plankton from the sea were developed.

▼ **Beard worm body plan.** A typical pogonophoran, showing the whole worm, most of which is normally buried in its narrow tube in the seabed.

protrudes from the bottom of the tube in life and acts as a burrowing organ.

The tube itself is made from chitin, and according to the species the lengths range from several centimeters to about 1.5m (5ft). Beard worm tubes are, however, very narrow, never exceeding 3mm (0.1m) in diameter. They are quite stiff and may appear banded.

One of the most peculiar features of the beard worms is the lack of a gut within the body. It is probable that nutrients are absorbed directly from the environment as organic molecules. The large surface-area-to-volume ratio in these animals favors such uptake of nutrients.

The sexes are separate and the sperms are released in packets (spermatophores) which are trapped by the tentacles of the females. The packets gradually disintegrate, and the freed sperms fertilize the eggs, which develop into larvae while still inside the female tube. When large enough they move out and settle nearby.

There has been much discussion about the evolutionary position of the beard worms. It seems likely that the possession of a coelom and the segmentation of the opisthosoma places them near the annelid worms.

Arrow worms are small inconspicuous marine animals that were unknown until collected in 1829 after the invention of the towed plankton net which strained minute life from the sea. Their torpedo-shaped bodies, combined with their ability to dart rapidly forward through the water for short distances, justify the name arrow worms. They are transparent and almost all are colorless. All known species are similar in appearance and all live in the plankton,

except *Spadella*, which lives in the sea floor.

The long narrow body carries a head at the front which bears curved hooks of chitin. These serve to grasp prey and to maneuver it into the mouth. Small planktonic animals form the diet, especially crustaceans such as copepods, but occasionally slightly larger food such as other arrow worms and fish larvae may be taken. The mouth carries small teeth which assist in the act of swallowing. A pair of small simple eyes is also borne on the head. These may or may not be pigmented and serve mainly as light detectors rather than as organs that actually form an image.

The trunk carries two pairs of lateral fins and the body terminates in a fish-like tail segment. The gut runs straight through from mouth to anus and has two small pouches or diverticula at the front.

Arrow worms are hermaphrodite and their sex organs are relatively large for the overall body size. The ovaries lie in the trunk region and the testes and seminal vesicles lie toward the tail separated by a partition. The form of the seminal vesicles, where sperm is stored, is used in identifying the various species. The breeding cycle varies in length according to distribution. It is completed rapidly in tropical waters, in about six weeks in temperate latitudes, and it may take up to two years in polar regions. Fertilization takes place internally and fertilized eggs are released into the sea where they develop as larvae before maturing into adults. The larvae resemble small adults. In the genus *Eukrohnia* the larvae are brooded in special pouches.

The outside of the body is equipped with small sensory tufts which respond to vibrations in the water such as are generated by certain suitable prey species. This facility allows the arrow worms to detect prey and to successfully engulf them. If they themselves suffer an accident, for example in an attack by a predator, lost parts, including the head, can be regenerated.

The evolutionary affinities of the arrow worms remain unclear despite various attempts to show relationships with other phyla. Today they are generally supposed to be without close affinities.   AC

ACORN WORMS

Phylum: Hemichordata
About 90 species in 2 classes.
Distribution: marine; acorn worms worldwide, pterobranchs mainly European and N American waters.
Fossil record: acorn worms, none; pterobranchs possibly rare from Ordovician (500–440 million years ago) to present.
Size: acorn worms 2cm to 2.5m (0.8in–8.2ft); pterobranchs as colonies up to 10m (33ft) in diameter with individuals up to 1.4cm (0.6in) although often less.

Features: essentially worm-like with a coelom developed along deuterostome lines; may be solitary (acorn worms) or colonial (pterobranchs) with a body divided into three zones: proboscis, collar and trunk; nervous system quite well developed; blood circulatory system present; with or without gill slits and tentacles; nephridia lacking; sexes separate.

Acorn worms
Class: Enteropneusta
One order containing, for example, the genera *Balanoglossus, Glossabalanus, Saccoglossus.*

Pterobranchs
Class: Pterobranchia
Two orders containing, for example, the genera, *Cephalodiscus, Rhabdopleura.*

HEMICHORDATES may be described as a minor phylum of marine invertebrates because they are not abundant and relatively few species are known. For the zoologist, however, they present a fascinating group of animals since they display a number of features which indicate similarities to the chordates (lancelets, fish, mammals etc), a group with which they were once classified. These characters include gill slits and a nerve cord situated on the upper (dorsal) side of the body. In some species the nerve cord is hollow and like those of vertebrates (fish, mammals etc). At no stage of their development, however, does a rudimentary backbone (notochord) appear, and they are excluded from the phylum Chordata on this technicality. Hemichordates are divided into the two classes Enteropneusta (acorn worms) and Pterobranchia, which have no common name. The pterobranchs are regarded as being more primitive than the acorn worms and some of these lack gill slits and have solid nerve cords. Hemichordates also show similarities with the echinoderms. The larva is a tornaria, which has features in common with certain echinoderm larvae, for example the asteroid bipinnaria. Still more interesting, the pterobranchs also, with their appendages clothed with tentacles, show similarities to the sea mats (Bryozoa), horseshoe worms (Phoronida) and lampshells (Brachiopoda).

The worm-like acorn worms are the only hemichordates likely to be encountered other than by a scientist, and then only rarely. They grow to 2.5m (8.2ft) in length but are often smaller. Their bodies are made up of three sections, the proboscis at the front, the collar and the trunk at the rear which forms the bulk of the body. It is the way in which the proboscis joins the collar, resembling an acorn sitting in its cup, which gives the group its common name. The proboscis is a small conical structure connected to the collar by a short stalk. The collar itself is cylindrical and runs forward to ensheath the proboscis stalk. The collar bears the mouth on its underside. The trunk makes up most of the body length and at the front end this contains a row of gill slits on each side. A ridge runs down the middle of the back of the trunk. The reproductive organs are borne outside the gill slits; in some forms the trunk is extended as wing-like genital flaps to contain them.

The body cavity (coelom) is present in all three parts of the body. There is a single

▶ **Like a monster from outer space** OPPOSITE, the front region of the acorn worm *Balanoglossus australiensis*, a species found in waters around Australia living in sand under stones.

▶▼ **Body plans of hemichordates.** (1) A colony of individuals of the pterobranch genus *Rhabdopleura*, showing individuals within tubes that are connected by a stolon. (2) Close-up of the head of *Rhabdopleura* protruding from its tube. (3) General body form of the acorn worm *Saccoglossus*. (4) Front end of the acorn worm *Protoglossus*, showing water currents carrying food particles. (5) Burrow system of the acorn worm *Balanoglossus*.

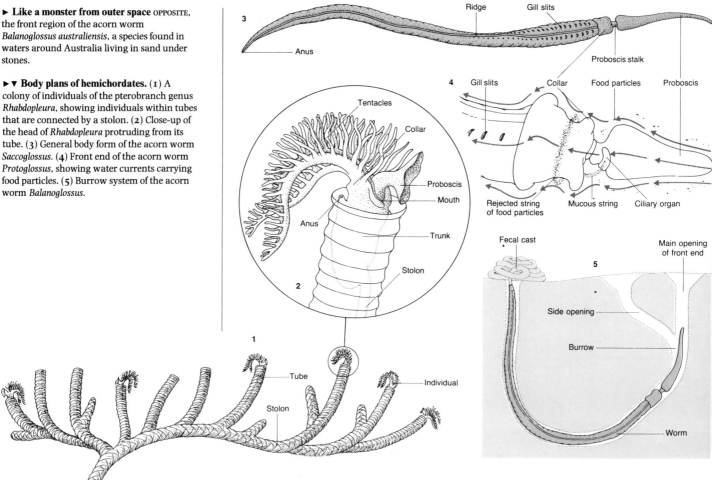
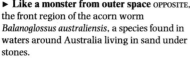

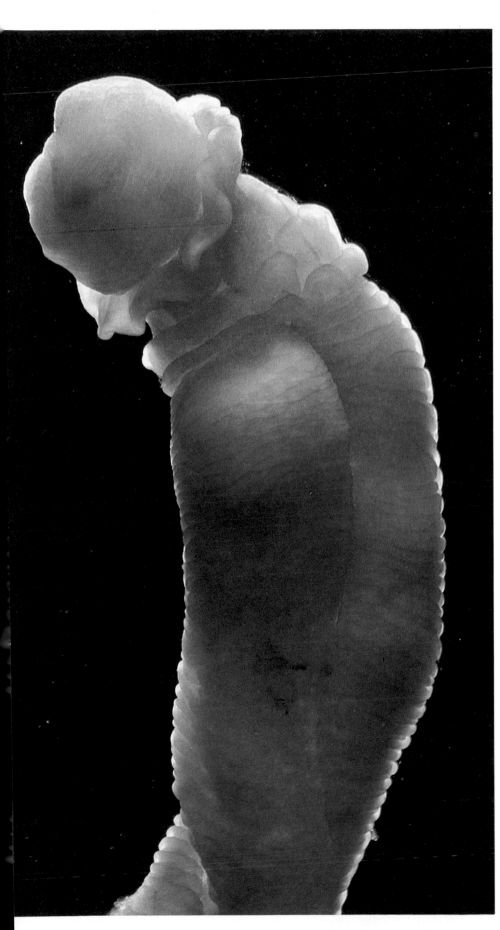

cavity in the proboscis, a pair of cavities in the collar and a pair in the trunk.

Acorn worms live in shallow water and some species may be found burrowing in sandy and muddy shores where they may be recognized by their characteristic coiled fecal casts. The burrows may be more or less permanent. Other species can live under stones and pebbles.

Most of the burrowing acorn worms feed on the organic material in the sand or mud in which they live, simply by ingesting the sediment as an earthworm ingests soil. Some feed by trapping suspended plankton and particles of detritus in the mucus covering on the proboscis. Cilia then pass these particles back to the mouth and the collar can play a role in rejecting unsuitable particles.

The gut runs from the mouth on the collar, via the pharynx of gill slits in the front trunk, to the rear of the trunk where the anus is situated. There is a long, thin blind-ended branch (diverticulum) running from the gut near the mouth into the proboscis, which once was mistaken for a notochord (hence the earlier taxonomic association with the chordates).

The gill slits were originally probably feeding mechanisms, which have subsequently become involved in gas exchange. The cilia they bear pump in water through the mouth and out through the gills.

The nervous system of the acorn worms is fairly primitive by comparison with the chordates, and of course the animals are not highly active. The sexes are separate and the eggs are fertilized in the sea. Initial development resembles that seen in the echinoderms. In some species it proceeds to a tornaria larva which lives in the plankton before it metamorphoses. Other types develop directly, with a juvenile worm appearing.

The pterobranchs are very different, although their bodies still show the three zones. The proboscis is smaller and shield- or plate-like, while the collar is extensively developed with its outgrowths of tentacles. According to the group, there may be two backward-curving arms bearing tentacles (as in *Rhabdopleura* species) or between five and nine (as in *Cephalodiscus*). It is thought that the tentacles function in food gathering.

The gill slits are few and inconspicuous and none is present in *Rhabdopleura*. In this group the gut is U-shaped with an anus opening on the top side of the collar. Again the sexes are separate and many individuals may live grouped together.   AC

"ARTHROPODS"

Crustaceans
Phylum: Crustacea.

Chelicerates
Phylum: Chelicerata.

Uniramians
Phylum: Uniramia.

Water bears
Phylum: Tardigrada.

Tongue worms
Phylum: Pentastomida.

JOINTED-LIMBED animals or arthropods (*Greek arthron*, joint; *pous, podos*, foot) include the commonest animals in the world today—the insects—as well as such other familiar creatures as the land-dwelling centipedes and spiders and the aquatic crustaceans, including lobsters and crabs.

The most obvious characteristics of an arthropod are the paired jointed limbs or other appendages along the body, and the hard outer covering (exoskeleton) containing chitin which makes their presence possible. Chitin is made up of long fibrous molecules with loose similarities to plant cellulose, and can be formed into many shapes, being truly plastic. For growth to take place, the skeleton must be shed, so arthropods grow in a series of stages of increasing size, the new cuticle hardening after being expanded to the new size.

Arthropods are typically mobile, show bilateral symmetry (right and left hand sides of the body being counterparts) and have a distinct head. As in annelid worms, the body is made up of a series of segments, each of which in primitive forms resembles the next and bears a pair of appendages. However, the segments are usually grouped into larger, efficient functional units such as the head, thorax and abdomen of insects and crustaceans, or the prosoma and opisthosoma of spiders. The brain of arthropods consists of dorsal (back, or upper) and ventral (underside, or lower) parts connected by a ring of nervous tissue round the esophagus, and, unless otherwise modified, leads to a double nerve cord on the underside with segmental swellings (ganglia). The body organs are bathed in blood, for the blood system is open, comprising blood-filled spaces with few arteries and veins. Arthropods do have a coelom (the extensive body cavity of earthworms and of ourselves), but it is usually reduced to a number of blind sacs leading via tubules to the outside, and is used for regulation of osmotic pressures within the body and/or for excretion.

Until recently, these and other shared characteristics were believed to indicate that the arthropods should be considered as a single, closely related, evolutionary group—a monophyletic phylum. However, detailed work, particularly in the fields of comparative functional morphology by Dr Sidnie Manton and of comparative embryology by Professor D. T. Anderson, has now shown that there are at least three distinct evolutionary lineages among living arthropods: a chitinous exoskeleton and segmental jointed appendages (along with other common arthropod features) have been evolved together more than once, from unrelated ancestors. Thus compound eyes and biting jaws, apparently so alike in insects and crustaceans, have been evolved separately. Such superficial similarities are to be expected: the same functions are carried out in the two groups, and only certain structural features are capable of fulfilling them.

Dr Manton's comparative study of form and function centered on jaw mechanisms and their associated head structure, and on the nature of body limbs and the locomotory habit in many arthropods. She concluded that three independent living arthropod phyla should be differentiated—the Crustacea, the Chelicerata (including the terrestrial arachnids such as spiders, scorpions and their marine relatives the horseshoe or king crabs and sea spiders) and the Uniramia, which includes the insects. Evidence drawn from the developmental patterns of arthropod embryos, using fate maps to illustrate which original embryonic cells give rise to which adult organ systems, gives strong support to Manton's thesis of three independent living arthropod phyla. The extinct trilobites are also clearly arthropods, and although they do not fall into any of the living arthropod groups, they do show some similarities to the chelicerates. Indeed, trilobites, crustaceans and chelicerates may be more closely related to each other than to uniramians.

Crustaceans are enormously successful in aquatic, particularly marine, environments.

▲ **Always hungry,** the Shore or Green crab investigates a possible food source (here a periwinkle) using its two pairs of antennae and its pincers or chelae. This almost cosmopolitan crab is abundant on rocky shores and in estuaries from Brazil to North America, Europe, North Africa, Sri Lanka, Australia and Hawaii. It is most active at night and at high tide, moving upshore with the tide.

Most living chelicerates are terrestrial arachnids, but the origin of these forms lay in the sea. Marine horseshoe crabs are indeed living fossils, providing a clue as to the nature of the marine ancestors of scorpions and spiders, and are themselves closely related to a large group of fossil chelicerates—the eurypterids or sea scorpions. Sea spiders (Pycnogonida) may well be more recent marine relatives of the arachnids.

Two other living groups of aquatic organisms, the water bears and the tongue worms, show arthropod affinities, particularly the possession of a cuticle (albeit thin) and structures describable as limbs. Various authorities have considered that water bears (tardigrades) should be considered as uniramians and that tongue worms may have crustacean affinities.

Easily the most numerous of arthropods (and of all animals) are the one million or more terrestrial species of Uniramians. They are so called because their limbs consist of a single branch, as opposed to the originally two-branched limbs of crustaceans and chelicerates, although most of the latter have subsequently evolved one-branched limbs as in spiders. In addition to the winged insects (Pterygota), uniramians include the worm-like velvet worms (genus *Peripatus*), those close relatives of the winged insects with six legs, such as the springtails (Collembola) and silverfish (Thysanura), and those multi-legged forms (myriapods), the millipedes and centipedes.

In spite of their names, millipedes do not have as many as a thousand legs nor do centipedes quite have a hundred. Millipedes are typically slow-moving herbivores living under stones or wood. They appear to have two pairs of appendages per segment but each "segment" is strictly formed by fusion of two original ones. Centipedes are typically faster-moving, flattened carnivores with poison claws delivering death to their invertebrate prey.

PSR

CRUSTACEANS

Phylum: Crustacea
About 39,000 species in 10 classes.
Distribution: worldwide, primarily aquatic, in
seas, some in fresh water, woodlice on land.
Fossil record: crustaceans appear in the Lower
Cambrian, over 530 million years ago.
Size: from microscopic (0.15mm/0.006in)
parasites to goose barnacles 75cm (30in) and
lobsters 60cm (24in) long.

Features: 2 pairs of antennae; typically
segmented body covered by plates of chitinous
cuticle subject to molting for growth; body
divided into head, thorax and abdomen; head
and front of thorax may fuse to form
cephalothorax often covered by shield-like
carapace; segments from 20 in decapods to 40
in some smaller species; compound eyes; paired
appendages on each segment typically include
the antennae, main and 2 accessory pairs of
jaws (mandibles, maxillules, maxillae), and
typically 2-branched limbs on thorax (called
pereiopods) and abdomen (generally called
pleopods) usually jointed, variously adapted for
feeding, swimming, walking, burrowing,
respiration, reproduction, defense (pincers);
breathing typically by gills or through body
surface; body organs in blood-filled hemocoel;
double nerve cord on underside; foregut,
midgut, hindgut; sexes typically separate;
development of young via series of larval
stages.

IN addition to many familiar animals, such as crabs and lobsters, the phylum Crustacea includes a myriad of little-known smaller relatives, some of which are major components of plankton. Crustaceans are clearly arthropods, with their tough, chitin-rich exoskeletons, which require molting for growth to take place, and their paired jointed appendages, or limbs, arranged on segments down the body. They are primarily aquatic, mostly marine, but include freshwater representatives. One group, the woodlice, has successfully colonized land.

The numerous segments of crustaceans are usually grouped into three functional units (tagmata)—the head, thorax and abdomen. The head is made up of six segments fused with a presegmental region (the acron) while the thorax and abdomen (which often ends in a non-segmental telson) vary in number of segments. Some of the smaller crustaceans contain up to 40 segments, whereas the decapods (shrimps, crabs, lobsters), the biggest crustaceans, have 20—six in the head, eight in the thorax and six in the abdomen. Some thoracic appendages may fuse to the back of the head, forming a cephalothorax, and a shield-like carapace may cover the head and part or all of the thorax: crustacean evolution is the story of reduction in the number of segments and their grouping into tagmata, with consequent specialization of the appendages for specialized roles. Although the fossil record goes back well over 500 million years, it tells us little about the origin of crustaceans (see pp246–253 for the basis of deductions concerning evolutionary changes within the crustaceans).

The "typical" crustacean head bears five pairs of appendages. There are none on the first segment, but the following two segments both bear a pair of antennae—the first antennae (antennules) and the second antennae (antennae). All crustaceans have two pairs of antennae. The antennae have probably shifted in evolution, for they now lie in front of the mouth, which may have been terminal in a pre-crustacean ancestor. They are typically sensory, although in some crustaceans they are employed in locomotion or even used by a male to clasp the female in reproduction.

Behind the mouth is the fourth segment bearing the mandibles, the main jaws developed from jaw-like extensions (gnathobases) at the base of the appendages, while the remainder of the limb has been lost, or reduced to a sensory palp. The fifth and sixth head segments bear the first maxillae (maxillules) and second, or true, maxillae respectively; these accessory jaws assist

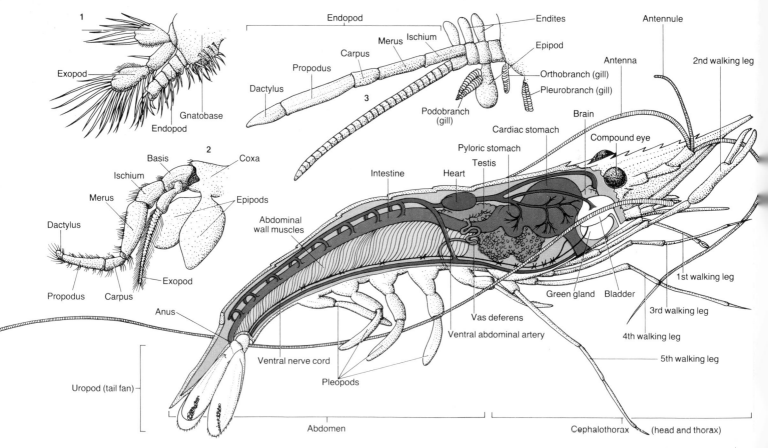

the order Branchiopoda (eg fairy shrimps and brine shrimps). Such limbs are usually moved in rhythm to create swimming and feeding currents, food being passed along the underside of the body to the mouth—so promoting the evolution of jaws from limb gnathobases. Extensions from the outer side of the limb bases often act as respiratory surfaces.

One major line of crustacean evolution has produced large walking animals. The two-branched swimming and filtering leg has evolved into an apparently one-branched (uniramous) walking leg (stenopodium)—strictly the endopod alone, the exopod being reduced and lost during larval development or previously in evolution. Such cylindrical legs have a reduced surface area and are not suitable as respiratory surfaces, which are particularly necessary in large crustaceans. Exites or extensions of the body wall at the base of the legs are therefore used as gills. The decapods (shrimps, crabs, lobsters) also show increased development of the head (cephalization). The first three pairs of thoracic limbs are adapted as accessory mouthparts (maxillipeds) and a carapace shields the cephalothorax.

Crustaceans typically have separate sexes and the fertilized egg as it divides shows a characteristic modified spiral cleavage pattern. Crustaceans usually develop through a series of larval stages of increasing size and numbers of segments with their associated limbs. The simplest larval stage is the nauplius larva, with three pairs of appendages—the first and second antennae and the mandibles. It occurs in many living crustaceans, often as a pelagic dispersal stage that swims and suspension-feeds with the three pairs of appendages. These limbs therefore have different functions in larva and adult, usually relinquishing their larval roles to other appendages which develop further down the body. Nauplius larvae are followed by a variety of larger larvae, according to the type of crustacean, although many crustaceans bypass the nauplius equivalent by developing within the egg.

The chitin-rich exoskeleton of crustaceans is divided into a series of plates which may be hardened by calcification or by tanning. The body organs lie in a blood-filled space—the hemocoel, and the blood often contains hemocyanin, a copper-based respiratory pigment which bears oxygen and is equivalent to hemoglobin in vertebrates. The coelom, the major body cavity of many animals, is restricted to the inner

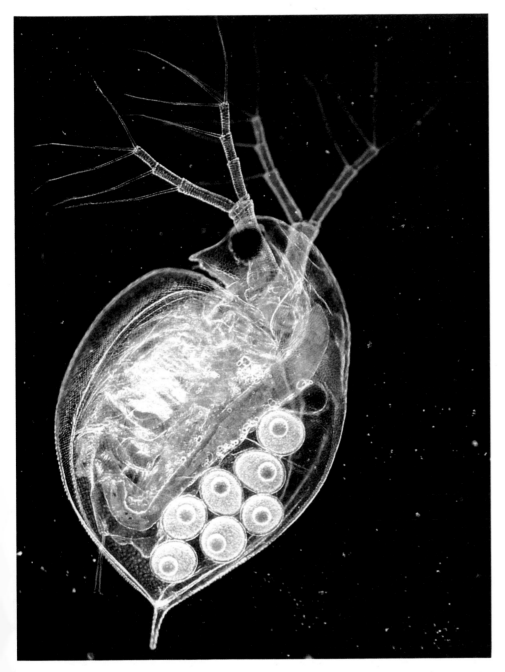

▲ **Using its head to swim:** the powerful two-branched antennae of the tiny freshwater flea *Daphnia* provide the jerking power-stroke to lift the animal through the water. This female bears seven embryos in the brood pouch between her abdomen and carapace.

◀ **Body plan of a crustacean,** showing the main typical external features. Illustrated is *Palaemon*, a genus of free-swimming prawns of the order Decapoda.
Crustacean limbs range from the primitive and leaf-like "phyllopods" to advanced walking legs as in lobsters. Shown here are thoracic appendages of (1) *Hutchinsoniella* (class Cephalocarida); (2) a malacostracan of the superorder Syncarida; (3) a decapod malacostracan (superorder Eucarida).

mastication and are similarly derived by partial or complete reduction of the limb to a gnathobase.

The ancestors of crustaceans were probably small marine organisms living on the sea bottom but able to swim, with a series of similar appendages down the length of a body not divided into thorax and abdomen; all appendages would have been used in locomotion, feeding and respiration. The limbs of early crustaceans were two-branched (biramous), with an endopod (inner limb) and exopod (outer limb) branching from the base (protopodite). The ancestral limb would have had two flat leaf-like lobes, as in several living crustaceans of

cavity of the coxal glands—paired excretory and osmoregulatory organs each consisting of an end sac and a tubule that opens out at the base of the second antenna (antennal gland) or the second maxilla (maxillary gland).

The alimentary canal consists of a cuticle-lined foregut and hindgut, with an intermediate midgut giving rise to blind pockets (ceca), perhaps modified as a hepatopancreas combining the functions of digestion, absorption and storage. The nervous system consists of a double ventral nerve cord, primitively with concentrations of the cord (ganglia) in each segment, but often further concentrated into a few large ganglia.

The following account covers the more important crustacean groups. For a complete list of classes within the phylum Crustacea, see overleaf.

Cephalocarids

Cephalocarids, first described only in 1955, are often considered to be the most primitive of living crustaceans. The nine species of the class Cephalocarida so far discovered are all minute inhabitants of marine sediments, feeding on bottom detritus. The body, under 4mm (0.2in) in length, is not divided into thorax and abdomen, and the leaf-shaped body appendages are very similar to each other and indeed to the second maxillae. Many experts believe the cephalocarid limb to represent an ancestral type from which the limbs of other living crustaceans have evolved.

Fairy shrimps and water fleas

Fairy shrimps and water fleas are among the branchiopods, small, mostly freshwater, crustaceans. They have leaf-like trunk appendages used for swimming and filter-feeding. Flattened extensions (epipodites) from the first segment (coxa) of these limbs act as gills, giving the class the name of "gill legs" (Branchio-poda).

Fairy shrimps (order Anostraca) live in temporary freshwater pools and springs, (some, the brine shrimps, in salt lakes), typi-

cally in the absence of fish. Unlike other branchiopods they lack a carapace (hence An-ostraca). They have elongated bodies with 20 or more trunk segments, many bearing appendages of the one type. They are usually about 1cm (0.4in) long, but some giants reach 10cm (4in). Fairy shrimps swim upside down, beating the trunk appendages in rhythm, simultaneously filtering small particles with fine slender spines (setae) on the legs. Collected food particles are then passed along a groove on the underside to the mouth.

When mating, a male fairy shrimp clasps the female with its large second antennae. The female lays her eggs into a brood sac. The eggs on release are extremely resistant to drying out (desiccation). Some eggs will hatch when wetted, but others require more than one inundation—a successful evolutionary strategy, ensuring that populations are not wiped out when insufficient rain falls to maintain the pool long enough for completion of the fairy shrimp life cycle. Fairy shrimps hatch as nauplii and grow to maturity in as little as one week.

The brine shrimps are found in salt lakes, the brackish nature of which eliminates possible predators. The eggs similarly resist drying out and are sold as aquarium food. *Artemia salina* has a remarkable resistance to salt, surviving immersion in saturated salt solutions. The gills absorb or excrete ions as appropriate and this brine shrimp can produce a concentrated urine from the maxillary glands.

Water fleas (order Cladocera) are laterally compressed with a carapace enclosing the trunk but not the head, which projects on the underside as a beak. Overall length is just 1–5mm (0.04–0.2in). The powerful second antennae are used for swimming. The trunk is usually reduced to about five segments, of which two may bear filtering appendages. Most water fleas, including *Daphnia* species, live in freshwater and filter small particles, but some marine cladocerans are carnivorous.

Water fleas brood their eggs in a dorsal

▲ ► **Representative species of planktonic crustaceans.** (1) A species of *Sagitta*, a genus of arrow worms which feed on copepods (1–4cm, 0.4–1.6in). (2) A species of *Oikopleura*, a genus of larvaceans, dwelling inside a gelatinous protective "house" (3cm, 1.2in). (3) A free-living copepod. It swims by using its well-developed antennae. (4) A phyllosoma larva of a crawfish. This delicate planktonic larva develops into the massive bottom-dwelling crawfish shown on p232. (5) A megalopa, the late larval stage of crabs which has head, thoracic and abdominal appendages. (6) A caprellid, a minute bottom-dwelling crustacean. These are often less than 1mm (0.04in) long and often live in association with other invertebrates or plants. (7) *Salpa fusiformis*, a planktonic relative of the sea squirts. Individuals form chains of varying lengths and drift in surface waters. (8) A zoea larva, ie the early larval stage of a crab. It has a full complement of head appendages but only the first two pairs of thoracic appendages. After several molts it develops into a megalopa. (9) *Pycnogonum littorale*, a sea spider which lives on the lower shore and in shallow water under stones etc in NW Europe (2cm, 0.8in).

chamber and asexual reproduction (parthenogenesis) is common. Populations may consist only of females reproducing parthenogenetically for several generations until a temperature change or limitation of food supply induces the production of males. The brood chamber, which now encloses fertilized eggs, may be cast off at the next molt and can withstand drying, freezing and even passage through the guts of vertebrates. Some *Daphnia* species show cyclomorphosis—cyclic seasonal changes in morphology, often involving a change in head shape.

In the clam shrimps (order Conchostraca), most of which are some 1cm (0.4in) in length, the remarkably clam-like bivalve carapace encloses the whole body, but in tadpole shrimps (order Notostraca) part of the trunk extends behind the large shield-like carapace and the overall body length may reach 5cm (2in). Either may be found with fairy shrimps in temporary rain pools.

Copepods

The class Copepoda includes over 180 families of mostly minute sea- and fresh-water inhabitants which provide a major source of food for fish, mollusks, crustaceans and other aquatic animals. As dominant members of the marine plankton, copepods of the order Calanoida may be among the most abundant animals in the world. On the other hand, members of the order Harpacticoida are common inhabitants of usually marine sediments, and Cyclopoida species may be either planktonic or bottom dwelling (benthic) in the sea or freshwater. Some copepods are parasitic, infesting other invertebrates, fish (often as fish lice on the gills) and even whales. Some of these attain 30cm (12in) in length.

Free-living copepods are small and capable of rapid population turnover. The large first antennae may be used for a quick escape, but more commonly they act as parachutes against sinking, while the thoracic limbs are the major swimming organs.

Many calanoid copepods can filter feed on phytoplankton, using the feathery second maxillae to sieve a current driven by the second antennae which beat at about 1,000 strokes per minute. The setae on the second maxillae are adapted to trap a particular size of microscopic plant organisms (phytoplankton) and the first pair of thoracic limbs (maxillipeds) scrape off the filtered material.

In fact calanoid copepods cannot survive in the marine planktonic ecosystem by filtering alone. Filter feeding is an energy-

▲ **Conspicuous in many food webs,** copepods occur even in the ocean depths. *Megacalanus princeps* is a large, pigmented species found at depths of 500–1,000m (1,640–3,280ft).

Feeding largely on microscopic plants, especially during the spring bloom, planktonic copepods are themselves a major food item for fishes, mollusks and other sea creatures.

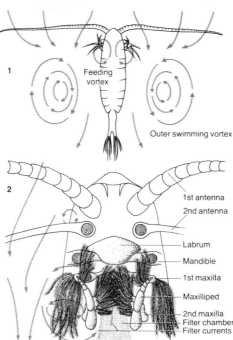

1 Feeding vortex

Outer swimming vortex

2

1st antenna
2nd antenna

Labrum
Mandible
1st maxilla
Maxilliped
2nd maxilla
Filter chamber
Filter currents

◄ **Techniques of filter feeding** in a copepod (*Calanus* species): (1) swimming and feeding currents created by the feathery second antennae; (2) detail of filter currents and filter apparatus.

▶ **Opportunistic goose barnacles** OVERLEAF hang down below floating pumice stone in the warm waters of the Great Barrier Reef off Australia. From the shell plates, six thoracic cirri protrude to filter out food. The species illustrated is *Lepas anserifera*.

sapping process that requires the driving of large volumes of water through a fine sieve. There must therefore be a minimum level of phytoplankton in the sea simply to repay the cost of filtering. Phytoplankton populations in temperate oceans only reach such concentrations for brief periods in the year—in the North Atlantic, during the spring bloom and perhaps again in the fall. For much of the year, therefore, the calanoid copepod feeds raptorially—seizing large

Class Cephalocarida

One order, 2 families, 4 genera, 9 species (eg *Hutchinsoniella macracantha*). Primitive bottom dwellers feeding on deposited detritus.

Branchiopods
Class Branchiopoda

Mostly small freshwater filter feeders. Comprises the **tadpole shrimps** (order Notostraca), 11 species in 1 family, eg *Triops* species; **clam shrimps** (order Conchostraca), 180 species in 5 families, eg *Cyzicus* species; **water fleas** (order Cladocera), 450 species in 11 families, eg *Daphnia* species; and **fairy shrimps** (order Anostraca) 180 species in 7 families, eg **brine shrimps**, *Artemia* species.

Class Remipedia

One species (*Speleonectes lucayensis*), 5mm (0.4in) long, a blind swimmer in marine caves.

Class Tantulocarida

Four species (eg *Basipodella harpacticola*) in 1 family, microscopic (0.15mm/0.006in) ectoparasites on deepsea bottom-dwelling crustaceans.

Class Mystacocarida

Nine species in 1 family, minute (0.5mm/0.02in) in sediments of seabed feeding on detritus. Eg *Derocheilocaris typicus*.

Branchiurans
Class Branchiura

One order (150 species) of blood-sucking **fish lice** some 7mm (0.3in) long, ectoparasitic on marine or freshwater fish. Eg *Argulus foliaceus*.

Copepods
Class Copepoda

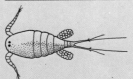

Over 8,400 species in 181 families and 8 orders of mostly minute components of plankton and sediment fauna. Includes: order Calanoida (2,300 species, eg *Calanus tinmarchicus*, in 40 families); order Harpacticoida (2,800 species, eg *Harpacticus* species, in 34 families); order Cyclopoida (450 species, eg *Cyclops* species, in 14 families); and the orders Poecilostomatoida (1,320 species, eg *Ergasilus sieboldi*, in 41 families) and Siphonostomatoida (1,430 species, eg *Caligus* species, in 44 families) of **fish lice**.

Barnacles
Class Cirripedia

Four orders comprising 1,025 species, adults parasitic or permanently attached to substrate, most with long, curved filter-feeding "legs." Comprising order Ascothoracica (45 species, eg *Ascothorax* species, in 4 families), very small parasites on echinoderms and soft corals; order Acrothoracica (50 species, eg *Trypetesa* species, in 3 families), small (0.4mm/0.16in) filter feeders boring into calcareous substrates (eg shells); order Thoracica (700 species in 17 families), filter-feeding barnacles on rocks etc, up to 75cm (30in) long, including **goose barnacles** (*Lepas* species), **acorn** or **rock barnacles** (*Balanus* species) and **whale barnacles** (*Conchoderma* species); and order Rhizocephala (230 species, eg *Sacculina carcini*, in 6 families), internal parasites of decapods (lobsters, shrimps, crabs).

Mussel shrimps and seed shrimps
Class Ostracoda

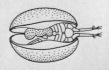

Five orders comprising 54 families and 5,650 species. Small, 1–3mm (0.04–0.12in), mostly bottom dwelling, with two-valved carapace. Eg *Cypris* species.

Malacostracans
Class Malacostraca

Fourteen orders, 359 families, 23,000 species.

Subclass Phyllocarida
Order Leptostraca
One family containing 25 species of pelagic or bottom-dwelling feeders on detritus. Eg *Nebalia bipes*.

Subclass Hoplocarida
Order Stomatopoda
Twelve families containing 350 species of bottom-dwelling carnivorous **mantis shrimps** (eg *Squilla empusa*).

Subclass Eumalacostraca
Four superorders.

Superorder Syncarida
Comprises the orders Anaspidacea (15 species, eg *Anaspides tasmaniae*, in 4 families), S Hemisphere freshwater bottom-dwelling detritus feeders up to 5cm (2in) long; and Bathynellacea (100 species, eg *Parabathynella neotropica*, in 3 families), inhabiting freshwater sediments feeding on detritus, blind, 0.5–1.2mm (0.02–0.04in) long.

Superorder Pancarida
Order Thermosbaenacea, 1 family of 9 species, eg *Thermosbaena mirabilis*, of sediment-dwellers in marine and fresh water, and in hot-spring ground-water, about 3mm (0.12 in) long.

Superorder Peracarida
Six orders, including: the **opossum shrimps** (order Mysidacea, 780 species, eg *Neomysis integer*, in 6 families), free-swimming and bottom-dwelling filter feeders 1–3cm (0.4–1.2in) long; **sea slaters, woodlice, pill bugs** and **sow bugs** (order Isopoda, 4,000 species in 100 families), eg *Armadillidium, Oniscus, Porcellio*, marine, freshwater and terrestrial, typically omnivorous crawlers but some parasites (eg *Bopyrus* species) and wood-boring **gribbles** (*Limnoria* species), 0.5–1.5cm (0.2–0.6in) long; **sand hoppers** and **beach fleas** (order Amphipoda, 6,000 species in 96 families), mostly marine bottom-dwelling scavengers, 0.1–1.5cm (0.04–0.6in) long, but also **skeleton shrimps** (*Caprella* species), **whale lice** (*Cyamus* species) and the "**Freshwater shrimp**" (*Gammarus pulex*).

Superoder Eucarida
Three orders: **krill** (order Euphausiacea), 85 species, eg **Whale krill** (*Euphausia superba*), in 2 families of marine pelagic filter feeders 0.5–5cm (0.2–2in) long; order Amphionidacea, 1 marine free-swimming species, *Amphionides reynaudii*, to 3cm (1.2in) long; and the **decapods** (shrimps, lobsters, crabs) (order Decapoda), 10,000 species in 105 families.

See pp 234–235 for Table of Decapods.

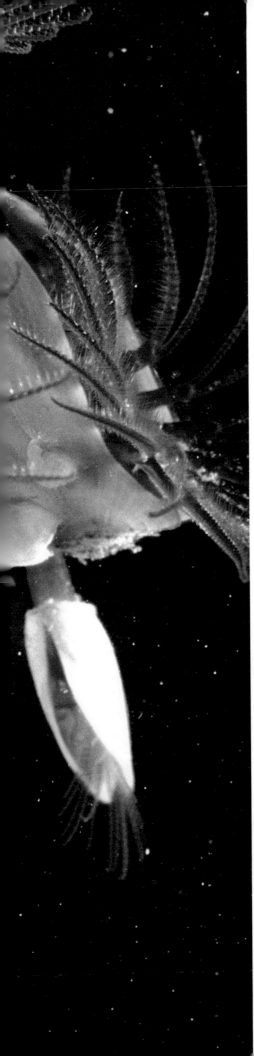

prey items including other members of the zooplankton. During the rich spring "bloom," filtering becomes worthwhile and the copepods can pass through several generations, multiplying very quickly, before returning to low metabolic tickover, often using fat reserves for the rest of the year. The number of copepod generations per year is related to the time period of the phytoplankton bloom.

Copepods pair, rather than copulate directly, the male transferring a packet of sperm (spermatophore) to the female, perhaps in response to a "chemical message" (pheromone) from her. The eggs are fertilized as they are laid into egg sacs carried by the female. Eggs hatch as nauplii and pass through further nauplius and characteristic copepodite stages before adulthood.

It is the larval stages that ensure the dispersal of parasitic copepods. Copepods that are parasitic on the exterior of their host (exoparasites) may show little anatomical modification, but endoparasites often consist of litle more than an attachment organ and a grossly enlarged genital segment with large attendant egg sacs.

Barnacles

Barnacles are sedentary marine crustaceans, permanently attached to the substrate. For protection barnacles have carried calcification of the cuticle to an extreme and have a shell resembling that of a mollusk. The shell is a derivative of the cuticle of the barnacle head and encloses the rest of the body. Stalked goose barnacles, which commonly hang down from floating logs, are more primitive than the acorn barnacles that abut directly against the rock and dominate temperate shores. (For barnacles' economic importance, see pp242–243.)

Barnacles (class Cirripedia) feed with six pairs of thoracic legs (cirri) which can protrude through the shell plates to filter food suspended in the seawater. Goose barnacles trap animal prey, but most acorn barnacles have also evolved the ability to filter fine material, including phytoplankton and even bacteria, with the anterior cirri.

Barnacles are hermaphrodites. They usually carry out cross-fertilization between neighbors. Fertilized egg masses are held in the shell until release as first-stage nauplii. Indeed it was J. Vaughan Thompson's observations (1829) of barnacle nauplius larvae of undoubted crustacean pedigree that removed lingering suspicions regarding the possible molluskan nature of barnacles. There are six nauplius stages of increasing size which swim and filter phytoplankton over a period of a month or so, before giving rise to a non-feeding larva—the cypris larva, named for its similarity to the mussel shrimp genus *Cypris*. This is the settlement stage of the life cycle, able to drift and swim in the plankton before alighting, and choosing a settlement site in response to environmental

---

Barnacles

Barnacles are the most successful marine fouling organisms and more than 20 percent of all known species have been recorded living on man-made objects, including ships, buoys and cables. Goose barnacles (*Lepas* species) attach themselves in enormous clumps to slow-moving sailing ships. They have been replaced today by acorn (or rock) barnacles (*Balanus* species) on motor-powered vessels, although stalked whale barnacles (*Conchoderma* species, named for one of their most important living hosts) are also important on very large crude oil carriers traveling between the Gulf and Europe.

The minute planktonic (nauplius) larvae of barnacles, feeding on phytoplankton, are dispersed in the sea and build up fat reserves to support the non-feeding settlement (cyprid) stage through further dispersal, site selection and metamorphosis. The cyprid, a motive pupa according to Darwin in the 1850s, has a low metabolic rate and lasts as long as a month before alighting on a chosen substrate. The cyprid then walks, using sticky secretions on the adhesive disks of the antennules, responding to current strength and direction, light direction, contour, surface roughness and the presence of other barnacles, as it monitors the suitability of a site for future growth. Finally, cypris cement is secreted from paired cement glands down ducts to the adhesive disks so that they become embedded permanently, whereupon the cypris metamorphoses to the juvenile barnacle which feeds and grows, developing its own cement system.

Barnacles slow down vessels, costing fuel. To counter this, antifouling paint is used which releases toxin in sufficient concentration to kill settling larvae or spores. Copper, the most common antifouling agent, is very toxic and must be released at a rate of 10 micrograms per sq cm per day to prevent barnacle settlement and growth.

Barnacles accumulate heavy metals, specifically zinc as detoxified granules of zinc phosphate. Barnacles are therefore suitable monitors of zinc availability in the marine environment, their high concentrations being easily measurable. Thames estuary barnacles, for example, may contain the fantastic zinc concentration of 15 percent of their dry weight.

factors which the larva detects by an array of sense organs.

Barnacles colonize a variety of substrates, including living animals such as crabs, turtles, sea snakes and whales—with a moving host the barnacle does not need to use energy in beating its cirri. Some barnacles have evolved to become parasites, which bear little similarity to their free-living relatives, except as larvae. Members of the order Rhizocephala (literally "root-headed") parasitize decapod crustaceans.

Mussel shrimps

Mussel shrimps or ostracods (class Ostracoda) are small bivalved crustaceans widespread in sea- and fresh water. Some are planktonic, but most live near the bottom, plowing through the detritus on which they may feed. Algae are another favored food.

The rounded valves of the carapace completely enclose the body, which consists of a large head and a reduced trunk, usually with only two pairs of thoracic limbs. The antennae are the major locomotory organs. Both pairs may be endowed with long bristles (setae) to aid propulsion when swimming, or the first antennae may be stout for digging, or even hooked for climbing aquatic vegetation.

Malacostracans

Including more than half of all living crustacean species, the class Malacostraca is very important in marine ecology and has also successfully invaded freshwater and land habitats.

Malacostracans are modifications of a shrimp-like body plan. The thorax consists of eight segments bearing limbs, of which up to three of the front pairs are accessory mouthparts or maxillipeds. A carapace typically covers head and thorax, though this may have been lost in some peracarids (see below). Members of the primitive superorder Syncarida also lack a carapace but this absence may be of more ancient ancestry: the anaspidaceans are bottom-dwelling feeders on detritus in Southern Hemisphere freshwaters, and the bathynellaceans are minute, blind detritivores living in freshwater sediments.

Most adult malacostracans are bottom dwellers (benthic), and the single-branched (uniramous) thoracic legs are adapted for walking. Some malacostracans can swim using pleopods, the limbs of the first five abdominal segments. The appendages of the sixth abdominal segment are directed back as uropods to flank the terminal telson and form a tail fan.

The gills of malacostracans are usually situated at the base of the thoracic legs in a chamber formed by the carapace, and are aerated by a current driven forward by a paddle on the second maxillae. Malacostracans typically ingest food in relatively large pieces. The anterior part of the stomach is where the large food particles are masticated before they pass to the pyloric stomach, where small particles are filtered and diverted to the hepatopancreas for digestion and absorption. Remaining large particles pass down the midgut and hindgut to be voided.

The most primitive malacostracans are to be found in the order Leptostraca. They are marine feeders on detritus, with a bivalved carapace, leaf-like thoracic limbs and eight abdominal segments. Mantis shrimps (order Stomatopoda) are marine carnivores which wait at their burrow entrances for unsuspecting prey, including fish, before striking rapidly with the enormous claws on the second thoracic appendages. The small sediment-dwelling crustaceans of the order Thermosbaenacea inhabit thermal springs, fresh and brackish lakes, and coastal ground water.

The vast majority of the remaining malacostracans are grouped within the two superorders Eucarida (see p233) and Peracarida. The peracarids are a superorder of malacostracans containing six orders of which the isopods (woodlice or pill bugs and others) and amphipods (sand hoppers and beach fleas) contain 10,000 species between them. Peracarids hold their fertilized eggs in a brood pouch formed on the underside of the body by extensions from the first segments (coxae) of the thoracic legs, the eggs hatching directly as miniature versions of their parents. The carapace, when present, is not fused to all the thoracic segments. The first thoracic segment is joined to the head.

Opossum shrimps (order Mysidacea), although typically marine, are common also in estuaries, where they may be found swimming in large swarms and may constitute the major food of many fish. They filter small food particles with their two-branched thoracic limbs, and on many species a balancing organ (statocyst) is clearly visible on each inner branch of the pair of uropods.

Their common name comes from the large brood pouch on the underside of the female's thorax.

Woodlice (pill bugs) and other isopods

Woodlice (or pill bugs) are the crustacean

success story on land. They are the most familiar members of the order Isopoda. Like other isopods, they are flattened top-to-bottom (dorsoventrally) and lack a carapace covering the segments. The first pair of thoracic limbs is adapted as a pair of maxillipeds, leaving seven pairs of single-branched thoracic walking legs (pereiopods). The five pairs of abdominal pleopods are adapted as respiratory surfaces. Isopods typically molt in two halves, the exoskeleton being shed in separate front and rear portions.

Most isopods walk on the sea bed, but some swim, using the pleopods, and others may burrow. The wood-boring gribble used to destroy wooden piers along the coasts of the North Atlantic by rasping with its file-like mandibles, a major pest until concrete replaced wood. Isopods are also to be found in freshwater, and sea slaters live at the top of the intertidal zone, indicating the evolutionary route of the better known woodlice to life on land.

The direct development characteristic of reproduction in the superorder Peracarida (see above) avoids the release of planktonic larvae, and is a major adaptive preparation

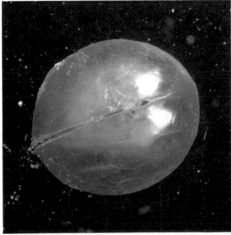

for terrestrial life. Woodlice have behavioral responses, for example to changes in humidity, to avoid desiccatory conditions and often therefore select damp micro-habitats. Members of the pill bug genus *Armadillidium* are able to roll up as the generic name suggests.

Woodlice have adapted their respiratory organs to take up oxygen from air. Members of the genus *Oniscus* show only little change from the ancestral aquatic isopod arrangement of pleopods as gills. Each of the five pairs of pleopods is two-branched and overlaps the one behind. The exopods of the first pair are extensive enough to cover all the remaining pleopods. The innermost fifth pleopods therefore lie in a humid microchamber and the endopods are well supplied with blood to act as the respiratory surface. *Porcellio* and *Armadillidium* species tolerate dryer conditions than can *Oniscus* species and use the outlying exopods of the first pair of pleopods for respiration. The danger of desiccation is reduced, for these exopods have in-tuckings of the cuticle (pseudotracheae) as sites of respiratory exchange.

Most isopods are scavenging omnivores, some tending to a diet of plant matter, especially the woodlice which contain bacteria in the gut to digest cellulose. Of the more carnivorous, *Cirolana* species can be an extensive nuisance to lobster fishermen, devouring bait in lobster pots.

Isopods have also evolved into parasites. Ectoparasitic isopods attach to fish with hooks, and pierce the skin with their mandibles to draw the blood on which they feed. Those isopods (eg *Bopyrus* species) that live in the gill chambers of decapod crustaceans (crabs, shrimps, lobsters) are more highly modified and may cause galls. Some isopods even hyperparasitize rhizocephalan barnacles, themselves parasitic on decapods.

▲ **Hopping for their lives:** Sand hoppers (*Orchestia gammarella*) can use their appendages to jump and thus escape predators, but they are also good swimmers.

▶ **Encased in two shells** ABOVE, the soft body of the mussel shrimp *Gigantocypris* is protected from predators.

◀▶ **Terrestrial wood louse and shoreline sea slater** are related members of the order Isopoda. Common woodlice LEFT (*Armadillidium vulgare*) are familiar land crustaceans able to feed on plant matter and rotting wood. The flattened bodies RIGHT of this male and this female sea slater (*Ligia oceanica*) enable them to shelter in crevices.

Beach fleas and other amphipods

Amphipods (order Amphipoda) are laterally compressed peracarids which often lie on their flattened sides. The most familiar are the beach fleas or sand hoppers found on sandy shores, and the misnamed "Freshwater shrimp" common in European streams. Amphipods are mostly marine. Although they are typically bottom dwellers (benthic), some are free swimming (pelagic).

Like the isopods, amphipods have no carapace, their eyes are stalkless, they have one pair of maxillipeds, seven pairs of single-branched walking legs (pereiopods) and a brooding pouch on the underside. Unlike the isopods, they have gills on the thoracic legs, and on the six abdominal segments there are usually three pairs of pleopods and three pairs of backward-directed uropods.

Most amphipods are scavengers of detritus, able to both creep using the thoracic pereiopods and swim with the abdominal pleopods. Many burrow or construct tubes and may feed by scraping sand grains or filtering plankton with bristle-covered limbs. Beach fleas, which have a remarkable ability to jump, burrow at the top of sandy shores or live in the strand line. This proximity to dry land has facilitated the evolution of terrestrial amphipods in moist forest litter though to a more limited extent than in isopods.

Amphipods of the family Hyperiidae are pelagic carnivores that live in association with gelatinous organisms, such as jellyfish, on which they prey. *Phronima* species are often reported to construct a house from remains of salp tunicates in which the animal rears its young. Skeleton shrimps (family Caprellidae) are predators of hydroid polyps, and the atypical, dorso-ventrally flattened whale lice are ectoparasites of whales.

Krill and decapods—the eucarids

The planktonic krill and the decapods

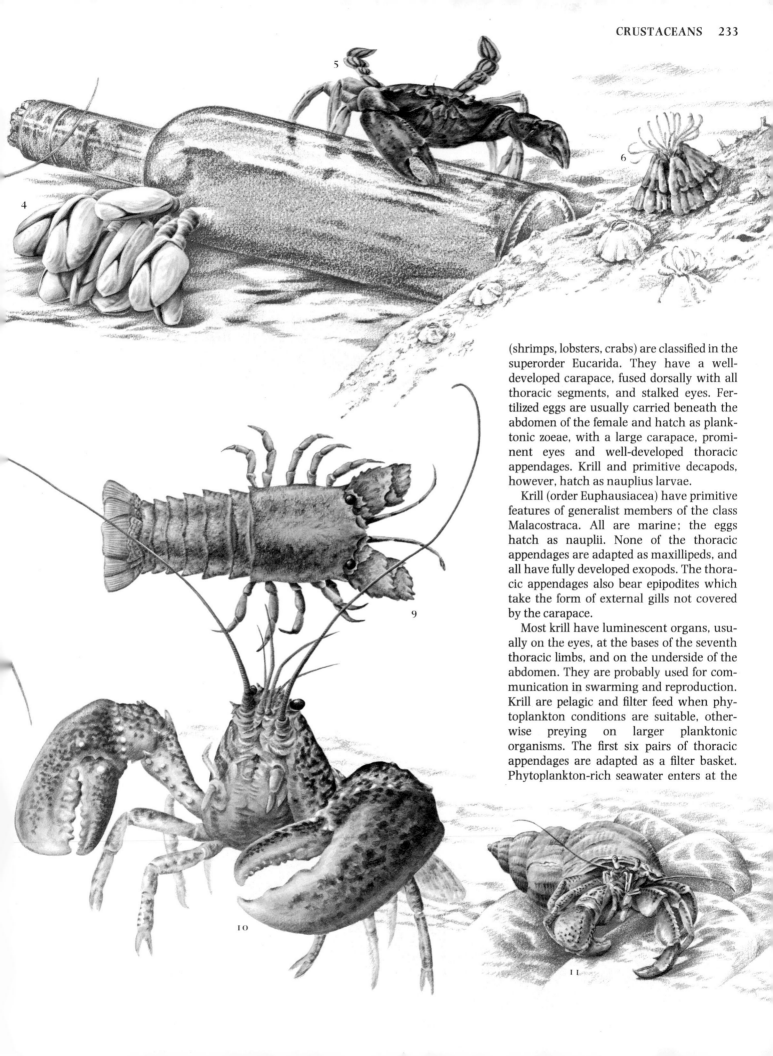

(shrimps, lobsters, crabs) are classified in the superorder Eucarida. They have a well-developed carapace, fused dorsally with all thoracic segments, and stalked eyes. Fertilized eggs are usually carried beneath the abdomen of the female and hatch as planktonic zoeae, with a large carapace, prominent eyes and well-developed thoracic appendages. Krill and primitive decapods, however, hatch as nauplius larvae.

Krill (order Euphausiacea) have primitive features of generalist members of the class Malacostraca. All are marine; the eggs hatch as nauplii. None of the thoracic appendages are adapted as maxillipeds, and all have fully developed exopods. The thoracic appendages also bear epipodites which take the form of external gills not covered by the carapace.

Most krill have luminescent organs, usually on the eyes, at the bases of the seventh thoracic limbs, and on the underside of the abdomen. They are probably used for communication in swarming and reproduction. Krill are pelagic and filter feed when phytoplankton conditions are suitable, otherwise preying on larger planktonic organisms. The first six pairs of thoracic appendages are adapted as a filter basket. Phytoplankton-rich seawater enters at the tips of the legs and is strained as it passes between the leg bases. Whale krill reach about 5cm (2in) long, dominate the zooplankton of the Antarctic Ocean and are the chief food of many baleen whales (see pp154–155).

In decapods, the first three pairs of thoracic appendages are adapted as auxilliary mouthparts (maxillipeds), theoretically leaving five pairs as legs (pereiopods)—hence deca-poda "ten legs." (In fact the first pair of pereiopods is often adapted as claws.) Historically decapods have been divided into swimmers (natantians) and crawlers (reptantians)—essentially the shrimps and prawns on the one hand and the lobsters, crayfish and crabs on the other. More modern divisions rely on morphological characteristics, and decapods are now divided between the suborders Dendrobranchiata and Pleocyemata.

Members of the Dendrobranchiata are all shrimp-like, characterized by their laterally compressed body and many-branched (dendrobranchiate) gills. Their eggs are planktonic and hatch as nauplius larvae. The Pleocyemata have gills which lack secondary branches, being plate-like (lamellate) or thread-like (filamentous) and their eggs are carried on the pleopods of the female before hatching as zoeae.

Prawns and shrimps
The terms shrimp and prawn have no exact zoological definition, and they are often interchangeable.

Prawns, Shrimps, Lobsters, Crabs (Order Decapoda)

Suborder Dendrobranchiata

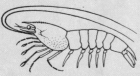

Five families of free-swimmers with many-branched gills and free-floating eggs hatching as nauplius larvae; carnivorous, 0.5–20cm (0.2–8in) long. The family Penaeidae includes **Banana prawn** (*Penaeus merguiensis*), **Brown shrimp** (*P. aztecus*), **Giant tiger prawn** (*P. monodon*), **Green tiger prawn** (*P. semisulcatus*), **Indian prawn** (*P. indicus*), **Kuruma shrimp** (*P. japonicus*), **Pink shrimp** (*P. duorarum*), **White shrimp** (*P. setiferus*), and **Yellow prawn** (*Metapenaeus brevicornis*).

Suborder Pleocyemata
23 families; gills plate-like (lamellate) or thread-like (filamentous), not many-branched; eggs carried by female before hatching as zoeae.

Banded or cleaner shrimps
Infraorder Stenopodidea

One family (Stenopodidae) of bottom-dwelling cleaners of fish, about 5cm (2in) long. Eg *Stenopus* species.

Shrimps, prawns, pistol shrimps
Infraorder Caridea

Twenty-two families of marine, brackish and freshwater swimmers and walkers; predatory scavengers, dominant shrimps in N oceans; 0.5–20cm (0.2–8in) long. Includes **Brown shrimp** (*Crangon crangon*) and **Pink shrimp** (*Pandalus montagui*).

Lobsters, freshwater crayfish, scampi
Infraorder Astacidea

Five families of mostly marine, some freshwater, bottom walkers, hole dwellers; predatory scavengers; up to 60cm (2ft) long. Includes **American lobster** (*Homarus americanus*), **European lobster** (*H. gammarus*) and **Dublin Bay** or **Norway lobster** (*Nephrops norvegicus*).

Spiny and Spanish lobsters
Infraorder Palinura

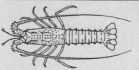

Five families, of marine bottom walkers or hole dwellers; often spiny but lack permanent rostrum (frontal spine) of Astacidea; predatory scavengers up to 60cm (2ft) long. Includes **American spiny lobsters** (*Panulirus argus* and *Panulirus interruptus*) and **European spiny lobster** (*Palinurus elephas*).

◀ **Banded shrimps** (here *Stenopus hispidus*) provide a cleaning service for fish with wounds or parasites. The characteristic coloration may help the "patient" to locate them.

▼ **In a strange procession,** East coast American spiny lobsters migrate annually on the seabed, perhaps a behavioral relic of seasonal movements in an ice age.

The most important "shrimp" families of the suborder Dendrobranchiata are the penaeids and sergestids, the Penaeidae including the most commercially important shrimps in the world (genus *Penaeus*), particularly dominating seas of southern latitudes.

Among the "shrimp" families in the much larger suborder Pleocyemata, the Stenopodidae comprise cleaner shrimps, which remove ectoparasites from the bodies of fish. Carideans (infraorder Caridea) are typified by possessing a second abdominal segment of which the lateral edges overlap the segments to either side. They are the dominant "shrimps" of northern latitudes although present throughout the world's oceans, from the intertidal zone down to the deep sea. Some shrimps (eg *Macrobrachium* species) complete their entire life cycles in freshwater, but many shrimps living in rivers return to estuaries to breed and release their zoea larvae.

In addition to the totally pelagic species, many shrimps are essentially bottom dwellers, only swimming intermittently. In adult

Migrations Along the Seabed

The spiny lobsters (also occasionally called rock lobsters or marine crayfish) lack claws but have defensive spines on the carapace and antennae. Members of the family (Palinuridae) spend their days in crevices in rock or coral, emerging by night to forage for invertebrate prey. They return from their nightly wanderings to one of several dens within a feeding range of hundreds of meters, and after several weeks may move several kilometers to a new location. Among the best known are two American species, *Panulirus argus*, found in the shallow seas off Florida and the Caribbean, and *P. interruptus*, which lives off California. The European spiny lobster is *Palinurus elephas*, the generic name a curious anagram of that of the American species.

Many spiny lobsters take part in spectacular mass migrations. *Panulirus argus*, for example, will abandon its normal behavior in the fall and as many as 100,000 individuals move south, by both day and night, in single files of up to 60 individuals. They may cover 15km (9.3mi) a day and travel for 50km (31mi) at depths between 3 and 30m (10–100ft). The lobsters may be primed to migrate by annual changes in temperature and daylight period, but the immediate trigger is a sharp temperature drop associated with a fall storm—usually the first strong squall of the winter. The spiny lobsters maintain alignment in the queue by touch but may respond initially by sight, recognizing the rows of white spots along the abdomens of their companions.

Mud shrimps, mud lobsters
Infraorder Thalassinidea

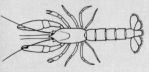

Seven families; bear a carapace up to 9cm (3.5in) long. Live in shallow water in deep burrows in sand or mud. Eg *Callianassa subterranea*.

Anomurans
Infraorder Anomura
Thirteen families.

Hermit crabs
Superfamily Paguroidea

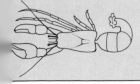

Three families; carapace to 4cm (1.6in); inhabiting marine gastropod shells; scavengers feeding on detritus. Includes *Pagurus bernhardus*, *Lithodes* species (**stone crabs**).

Land hermit crabs
Superfamily Coenobitoidea
Four families with carapace to 19cm (7.5in); on shore, land, in shells; scavenging detritivores. Includes **Coconut crab** (*Birgus latro*).

Squat lobsters, porcelain crabs
Superfamily Galatheoidea

Four families with carapace to 4cm (1.6in); marine, in holes, under stones; scavenging detritivores. Eg **Porcelain crab** (*Porcellana platycheles*).

Mole crabs
Superfamily Hippoidea

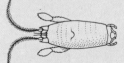

Two families with carapace to 5cm (2in), marine sand burrowers, filter feeders. Eg **Mole crab** (*Emerita talpoida*).

Crabs, spider crabs
Infraorder Brachyura

Forty-seven families with carapace up to 45cm (18in); mostly bottom dwellers, walking on seabed, but some parasitic/commensal with fish, some burrowers, some swimmers; predaceous scavengers. Includes 4 families of **spider crabs**, eg *Macrocheira kaempferi* (family Majidae); **Edible crab** (*Cancer pagurus*) and **Dungeness crab** (*C. magister*) (family Cancridae); **helmet crabs** (family Corystidae), eg *Corystes cassivelaunus*; **swimming crabs** (family Portunidae), eg **Blue crab** (*Callinectus sapidus*), **Henslow's swimming crab** (*Polybius henslowi*), and **Shore** or **Green crab** (*Carcinus maenas*); **Chinese mitten crab** (*Eriocheir sinensis*), *Sesarma* species (family Grapsidae); **pea crabs** (family Pinnotheridae), eg *Pinnotheres* species; **ghost crabs** (*Ocypode* species) and **fiddler crabs** (*Uca* species) (family Ocypodidae); and **coral gall crabs** (family Hapalocarcinidae), eg *Hapalocarcinus* species.

shrimps the thoracic pereiopods are responsible for walking (and/or feeding) and the five pairs of abdominal pleopods for swimming. Flexion of the abdomen is occasionally used for rapid escape. Shrimp zoea larvae have two-branched thoracic appendages, the exopods being used for swimming; the pleopods take over the swimming function in post-larval and adult stages as the exopods are reduced. By the adult stage the "walking" pereiopods are single-branched.

Most pelagic shrimps are active predators feeding on crustaceans of the zooplankton, such as krill and copepods. Bottom-dwelling species are usually scavengers, but range from catholic carnivores to specialist herbivores.

Shrimps usually have distinct sexes, although in some species, including *Pandalus borealis*, some of the females pass through an earlier male stage (protandrous hermaphroditism). Successful copulation usually requires molting by the female immediately before mating, when spermatophores are transferred from the male. Eggs are spawned between two and 48 hours later and are fertilized by sperm from the spermatophore. The eggs of penaeids are shed directly into the water, but in most shrimps the eggs are attached to bristles on the inner branches (endopods) of pleopods 1–4 of the female. The incubation period usually lasts between one and four months, during which time the female does not molt. Most shrimp eggs hatch as zoeae and pass through several molts over a few weeks, to post-larval and adult stages.

▲ **The squat lobster** *Galathea strigosa*, not a close relation of true lobsters, is a colorful crustacean from the lower shore and shallow water.

◄ **"House hunting,"** an important operation for this growing hermit crab seeking a larger home.

▶ **On the attack** OVERLEAF a Costa Rican ghost crab (*Ocypode* species) tackles a hatchling turtle, exposed in its efforts to put to sea for the first time.

House-hunting Hermits

Hermit crabs typically occupy the empty shells of dead sea snails, thereby gaining protection while at the same time retaining their mobility. They are able to discriminate if offered a selection of shells, and will differentiate between shells of different sizes and species, to choose the one that fits the body most closely. Hermit crabs (family Paguridae) change shells as they grow, although in some marine environments there may be a shortage of available shells, and a hermit crab may be restricted to a shell smaller than would be ideal. Some hermit crabs are aggressive and will fight fellows of their own species to effect a shell exchange; aggression is often increased if the shell is particularly inadequate.

Hermit crabs may encounter empty shells in the course of their day-to-day activity but the vacant shell is usually "spotted" by sight; the hermit crab's visual response increases with the size of an object and its contrast against the background. The hermit crab then takes hold of the shell with its walking legs and will climb onto it, monitoring its texture. Exploration may cease at any time, but if the shell is suitable the hermit crab will explore the shell's shape and texture by rolling it over between the walking legs and running its opened claws over the surface. Once the shell aperture has been located, the hermit crab will explore it by inserting its claws one at a time, occasionally also using its first walking legs. Any foreign material will be removed before the crab rises above the aperture, flexes its abdomen and enters the shell backward. The shell interior is monitored by the abdomen as the crab repeatedly enters and withdraws. The crab will then emerge, turn the shell over and re-enter finally.

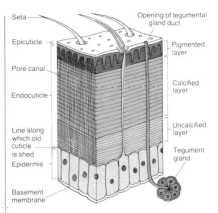

▲▼ **Cuticle of crustaceans** provides a tough protective layer that serves also as an external skeleton (exoskeleton). The cuticle is rich in chitin, a polysaccharide similar in structure to cellulose of plants, and is secreted by the underlying single layer of living cells, the epidermis. The endocuticle of a decapod comprises an outer pigmented, and inner calcified and uncalcified layers. The thin outermost epicuticle is secreted by tegumental glands.

The cuticle, at first soft and flexible, is progressively hardened inward from the epicuticle by the deposition of calcium salts and by sclerotization (tanning), involving the chemical cross-linking of cuticular proteins to form an impenetrable meshwork.

A rigid exoskeleton prevents a gradual increase in size, and so a crustacean grows, stepwise, in a series of molts involving the secretion of a new cuticle by the epidermis and the shedding of the old. Before a molt, food reserves are accumulated and calcium is removed from the old cuticle and much of its organic material resorbed. It is then split along lines of weakness as the crustacean swells (usually by an intake of water), before hardening of the new cuticle. The rest of the old exoskeleton may then be eaten to reduce energy losses BELOW (*Geograpsus grayii*).

During and immediately after molting the temporarily defenseless crustacean hides to avoid predators. Some adult crabs cease molting, but many adult crustaceans molt throughout life.

Most shrimps are mature by the end of their first year and typically live two to three years. (See also pp242–243.)

Lobsters and freshwater crayfish

Lobsters and freshwater crayfish belong to a group known historically as the macrurans ("large-tails") but now divided into three infraorders—the Astacidea (lobsters, freshwater crayfish, scampi), the Palinura (spiny and Spanish lobsters) and the Thalassinidea (mud lobsters and mud shrimps). All have well-developed abdomens, compared with, for example, anomurans or brachyurans. For the group's importance as a source of food for humans, see pp242–243.

Lobsters and freshwater crayfish walk along the substrate on their four back pairs of single-branched thoracic legs (pereiopods). The first pair of pereiopods is adapted as a pair of formidable claws for both offense and defense. The abdomen bears pleopods, but these cannot move the heavy body and have become variously adapted for functions which include copulation or egg-bearing.

Lobsters are marine carnivorous scavengers, usually living in holes on rocky bottoms. The commercially important American lobster reaches 60cm (2ft) in length and weighs up to 22kg (48lb). Lobsters are very long lived and they may survive 100 years.

Freshwater crayfish, however, are more omnivorous. There are more than 500 species of freshwater crayfish, mostly about 10cm (4in) long, and since they typically

require calcium they are often restricted to calcareous waters. Because of their marine ancestry, their internal fluids have an osmotic pressure higher than that of the surrounding medium and water will enter by osmosis across gill and gut membranes. The cuticle covering the rest of the body is made impermeable by tanning and calcification. The convoluted tubule of each antennal (green) gland is very long and resorbs ions from the primary urine (filtered from the blood at the end sac of the gland). A dilute urine, one tenth the osmotic pressure of the blood, may be produced, eliminating water that has entered osmotically. Any salts still lost with the urine are replaced by ions actively taken up by the gills.

Reproduction involves pairing, in which either sperm from the male flows down along the grooves on the first pleopods into a seminal receptacle in the female, or a spermatophore is transferred. Fertilized eggs are incubated on the pleopods of the female and hatch as mysis larvae bypassing in the egg the stages equivalent to nauplius and zoea.

Squat lobsters and hermit crabs

Squat lobsters, hermit crabs and mole crabs are part of a collection of probably unrelated groups intermediate in structure and habit between lobsters and crabs. The abdomen is variable in structure—often being asymmetrical or reduced and held in a flexed position (anom-ura: odd-tailed—hence the scientific name Anomura given to these crustaceans). The fifth pair of pereiopods is turned up or reduced in size.

Hermit crabs have probably evolved from ancestors which regularly used crevices for protection, eventually specializing in the use of the discarded spiral shells of gastropod mollusks. The abdomen is adapted to occupy the typically right-handed spiral of such shells although rarer left-handed shells may also be used. The pleopods are lost at least from the short side of the asymmetric abdomen, but those on the long side of females are adapted to carry fertilized eggs. At the tip of the abdomen, the uropods are modified to grip the interior of the shell posteriorly, and toward the front of the animal thoracic legs may be used for purchase. One or both claws can block up the shell opening. The shell offers excellent protection and still affords mobility (see box, p236).

Not all species of hermit crabs live in gastropod shells. Some occupy tusk shells, coral, or holes in wood or stone. The hermit crab *Pagurus prideauxi* lives in association with a sea anemone, *Adamsia carciniopados*,

whose horny base surrounds the crab's abdomen and greatly overflows the originally occupied shell. This hermit crab avoids the risk of being eaten when transferring from one shell to another, for the protecting anemone moves simultaneously with its associate. The crab is protected by the stinging tentacles of the anemone, which in turn profits from food particles which its crustacean partner releases into the water during feeding.

Hermit crabs live as carnivorous scavengers on sea bottoms ranging from the deep sea to the seashore. They have also taken up an essentially terrestrial existence in the tropics. Land hermit crabs range inland from the upper shore, often occupying the shells of land snails. The Coconut crab has abandoned the typical hermit crab form, and appears somewhat crab-like, but with a flexed abdomen. It lives in burrows and holes in trees, which it can climb, feeding on carrion and vegetation, and can drink. Land hermit crabs have reduced the number of gills that would tend to dry out in the air and collapse under surface tension. The walls of the branchial chamber are richly supplied with blood, which enables them to act as lungs, and some species have accessory respiratory areas on the abdomen enclosed in the humid microclimate provided by the shell. Land hermit crabs are not fully terrestrial, for they have planktonic zoea larvae. The adults must return therefore to the sea for reproduction.

Squat lobsters are well named, for they have relatively large symmetrical abdomens flexed beneath the body. They typically retreat into crevices. Porcelain crabs are anomuran relatives of squat lobsters which look remarkably like true crabs. Mole crabs are anomurans that by flexing the abdomen burrow into the sand when waves break low on warm sandy shores. The first antennae form a siphon channeling a ventilation current to the gills and the setose (bristly) second antennae filter plankton.

True crabs

There are 4,500 species of true crabs which all possess a very reduced symmetrical abdomen held permanently flexed beneath the combined cephalothorax (brachyuran: short-tailed). The terminal uropods are lost in both sexes; four pairs of pleopods are retained on the female's abdomen to brood eggs and in the male only the front two pairs remain to act as copulatory organs. The crabs have a massive carapace extended at the sides, and the first of the five pairs of thoracic walking legs is adapted as large

claws. Typically crabs are carnivorous walkers on the sea bottom.

The reduction of the abdomen has brought the center of gravity of the body directly over the walking legs, making locomotion very efficient and potentially rapid. The sideways gait assists to this end. The shape of a crab is therefore the ultimate shape in efficient crustacean walking. Mainly as a result, the brachyurans have enjoyed an explosive adaptive radiation since their origin in the Jurassic era (195–135 million years ago) and various anomurans (for example mole crabs and porcelain crabs) approach or duplicate the crab-like form.

Crabs live in the deep sea and extend up to and beyond the top of the shore. The blind crab *Bythograea thermydron* is a predator in the unique faunal communities surrounding deep-sea hydrothermal vents, regions of activity in the earth's crust which emit hot sulfurous material 2.5km (1.6mi) below the surface of the sea. On the other hand, crabs of the family Ocypodidae, such as the burrowing ghost and fiddler crabs, live at the top of tropical sandy and muddy shores and the distribution of the genus *Sesarma* extends well inland. Crabs have also invaded the rivers—the common British Shore crab extends up estuaries—and in the tropics crabs of the family Potamidae and some Grapsidae are truly freshwater in habit.

Most crabs burrow to escape predators—descending back-first into the sediment, and several typically remain burrowed for long periods. Of these the Helmet crab has long second antennae which interconnect with bristles to form a tube for the passage of a ventilation current down into the gill chamber of the buried crab. Some crabs have become specialist swimmers, with the last pair of thoracic legs adapted as paddles. The more terrestrial crabs like the ghost crabs are very rapid runners—their speed and nighttime activity contributing to their common name. Other crabs, particularly spider crabs (see box), cover themselves with small plants and stationary animals (eg sponges, anemones, sea mats) for protective camouflage. Pea crabs may live in the mantle cavities of bivalve mollusks, feeding on food collected by the gills of the host, and female coral gall crabs become imprisoned by surrounding coral growth. The female crab is left with just a hole to allow entry of plankton for food and the tiny male for reproduction.

Most crabs are carnivorous scavengers although the more terrestrial ones may eat

▲ **Quest in the trees:** the Blue land crab *Cardisoma guanhumi* can climb trees in search of prey.

▶ **Getting to grips:** the large nippers of the fiddler crab are used for signaling and ritual combat. Here two males of the genus *Annulipes* test their strength.

▼ **Buried alive.** The ghost crab *Ocypode ceratophthalma* seeks refuge from predators by burrowing in the sand. Only its large eyes protrude to keep watch.

Japanese Spider Crabs

Spider crabs belong to a family of true (brachyuran) crabs with long legs and a superficial resemblance to spiders. Included in their number is the world's largest crustacean, the giant spider crab (*Macrocheira kaempferi*), whose Japanese name is Takaashigani—the tall-leg crab.

This remarkable animal may measure up to 8m (26.5ft) between the tips of its legs when these are splayed out on either side of the body. The claws of such a beast may be 3m (10ft) apart when held in an offensive posture. The main body (cephalothorax) of the crab is, however, relatively small and would not usually exceed 45cm (18in) in length or 30cm (12in) in breadth.

Giant spider crabs have a restricted distribution off the southeast coast of Japan. They live on sandy or muddy bottoms between 30–50m (100–165ft) deep. They have poor balance and so live in still waters, hunting slow-moving prey that includes other crustaceans as well as echinoderms, worms and mollusks. Little detail is known of their life history, but they probably pass through zoea and megalopa larval stages before they metamorphose into juvenile crabs. Large specimens are probably more than 20 years old.

The crabs are caught relatively infrequently, but they are used for food. Because of their size they command respect, and their nippers can inflict a nasty wound.

The existence of the giant crabs was first reported in Europe by Engelbert Kaempfer, a physician for the Dutch East India Company who visited Japan in 1690, in his *History of Japan* published in English in 1727 (although he himself died in 1716). The crabs were given their Latin name in honor of Kaempfer in the 19th century by C. J. Temminck, the director of the Leiden Museum, which received much of the natural history collections of the Dutch East India Company.

plant matter. The fiddler crabs process sand or mud in their mouthparts, scraping off the nutritious microorganisms with specialized spoon- or bristle-shaped setae.

Reproduction involves copulation. The second pair of pleopods on the male's abdomen acts like pistons within the first pair to transfer sperm to the female for storage. The eggs are fertilized as they are laid and are held under the broad abdomen of the female. They hatch as pre-zoea larvae, molting immediately to the first of several planktonic zoea stages. Zoea larvae have a full complement of head appendages, but only the first two pairs of thoracic appendages (destined to become the first two or three pairs of adult maxillipeds), used by the zoea for swimming. After several molts the megalopa stage is reached, with abdominal as well as thoracic appendages. This settles on the sea bottom and metamorphoses into a crab. PSR

Economic Importance of Crustaceans
Human food and foulers of vessels

The economic importance of crustaceans is twofold—their commercial value as food items and the costs caused by their effects as foulers of ships or coastal structures.

Crustaceans, especially the decapods (shrimps, lobsters, crabs), are important as food products, whether cropped from the wild or reared by aquacultural processes. Whale krill of the Antarctic Ocean are now being harvested and processed as food, not only for man, but more particularly for agricultural livestock. In the Gulf of Mexico and off the southeastern USA, fishing boats trawl for the Brown shrimp which, with the White shrimp and the Pink shrimp, make up the world's largest shrimp fishery. In Southeast Asia, in Indonesia and the Philippines and in Taiwan, shrimps of the same genus *Penaeus* have been reared for food in brackish ponds for 500 years. Popular species are the Banana prawn, the Indian prawn, the Giant tiger prawn, the Green tiger prawn and the Yellow prawn. The shrimps are trapped in pools at high tide and cultured for several months, often with mullet or milkfish. In Singapore ponds are usually constructed in mangrove swamps and in India rice paddies are used. Typical yields vary from 300 to 1,600kg of edible shrimp per hectare (1,650–8,700lb/acre). In Japan there is a recently developed intensive culture of the Kuruma shrimp. Egg-bearing females are supplied by fishermen and shrimps are reared over two weeks, from eggs through successive larval stages with differing food requirements, to postlarvae before transfer to production ponds. Harvesting takes place 6–9 months later at yields which may attain 6,000kg per ha (32,600lb/acre).

In British waters the Caridean shrimp (also called the Pink shrimp) is fished by beam trawl in spring and summer in the Thames estuary, the Wash, Solway Firth and Morecambe Bay—the Thames fishery dating back to the 13th century. The Brown shrimp is fished off Britain, Germany, Holland and Belgium; boats work a pair of beam trawls and total landings reach 30,000 tons a year. *Palaemon adspersus* is taken off Denmark and off south and southwest Britain.

Another caridean, the tropical Indo-Pacific freshwater prawn *Macrobrachium rosenbergi*, is a giant reaching 25cm (10in) long; it is therefore attractive to aquaculturists, but its larvae are difficult to rear and dense populations do not occur naturally.

Lobsters and crabs are usually caught in traps, enticed by bait to enter via a tunnel of decreasing diameter protruding into a wider chamber. Most are predators or scavengers: lobsters prefer stinking bait but crabs are attracted to fresh fish pieces. Crab or lobster meat usually consists of muscle (white meat) extracted from the claws and legs, and lobster abdominal muscle is also used. The hepatopancreas (brown meat) may be taken in some species. The meat is processed as a meat paste or canned, or the crustacean may be sold fresh or frozen.

The Dublin Bay prawn or Norway lobster supplies scampi—strictly the abdominal muscle. It burrows in bottom mud and is collected by trawling. Most lobsters, however, are caught in lobster pots. American lobsters and European lobsters are of commercial value in temperate seas but are replaced by spiny lobsters in warmer waters: *Panulirus argus* and *P. interruptus* are trapped off Florida and California respectively, *P. versicolor* throughout the Indo-Pacific and *Jasus* species off Australia.

▲ **Four different species** of *Pandalus* make up the principal elements of the shrimp catch from southwest Alaska.

◄ **Harvest of the sea:** commercial shrimp boat unloading at Kachemak Bay, southwest Alaska.

Freshwater crayfish are consumed with enthusiasm in France. The habit has transferred to Louisiana, USA, where between 400,000 and 800,000kg (from 880,000 to 1,960,000lb) of wild Red and White crayfish are trapped each year, and a further 1.2 million kg (2.6 million lb) reared in artificial impoundments.

The Edible crab (*Cancer pagurus*), is taken in pots off European coasts. The British annual catch reaches 6,500 tonnes, boats laying out 200–500 pots daily in strings of 20–70 buoyed at each end. The crabs are usually sold to processing factories, to be killed by immersion in freshwater or by spiking, boiled in brackish water and then cooled to room temperature to set the meat, which is extracted by hand picking or with compressed air. The related Dungeness crab is fished off the west coast of North America, and the Blue crab is taken off the east coast by trap or line or by fishing with a trawl net.

On the negative side, one year's growth of fouling organisms can increase the fuel costs of ships by 40 percent, and barnacles are the most important foulers. Their presence impedes the smooth flow of water over a ship's hull and their shell plates disrupt paint films, so enhancing corrosion. Barnacles are cemented to the substrate and necessitate expensive dry docking for removal. Barnacles have swimming larvae that disperse widely, ready to alight on passing ships. Barnacles and tube-building crustaceans may also clog water-cooling intake pipes of industrial installations by the coast.

The commercial importance of wood-boring crustaceans, such as the gribbles has decreased since concrete has replaced wood as a major pier construction material.

PSR

Phylum: Chelicerata (part)

Horseshoe or king crabs

Class: Merostomata.
Order: Xiphosura.
Four species in 3 genera.
Distribution: marine, Atlantic coast of
N America (genus *Limulus*), coasts of SE Asia
(*Tachypleus*, *Carcinoscorpius*).
Fossil record: first appear in the Devonian
period, 400–350 million years ago.
Size: larvae about 1cm (0.4in) long, adults up
to 60cm (2ft) long.

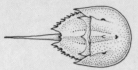

Features: pair of pincer-like mouthparts
(chelicerae) in front of mouth; carapace covers
the back of the prosoma (front portion of body)
and part of the opisthosoma (rear portion of
body); prosoma has 6 pairs of appendages:
chelicerae and 5 pairs of walking legs;
opisthosoma bears 6 pairs of flattened limbs (a
modified genital operculum or lid, and 5 pairs
of leaf-like or lamellate gillbooks for respiration)
and a rear spine; excretion is by coxal glands
at limb bases; circulatory system well
developed; sexes distinct, with external
fertilization; larva has three divisions, like a
trilobite.

Sea spiders

Class: Pycnogonida.
Order: Pantopoda.
Five hundred species in 70 genera and 8
families.
Distribution: marine, worldwide, from shores to
ocean depths.
Fossil record: appear first in the Devonian
period 400–350 million years ago.
Size: narrow, short body 0.1–6cm (0.04–2.5in)
long.

Features: cephalon has a tubular proboscis
with mouth at tip, paired chelicerae and palps;
usually 4 pairs of long legs on protuberances at
the side; adult heart tubular, within blood-
containing body cavity (hemocoel); there are
no excretory, respiratory or osmoregulatory
organs; sexes distinct, with external
fertilization; extra, egg-carrying, legs well
developed on males; males brood eggs which
hatch into protonymphon larvae.

Families: Ammotheidae (eg *Achelia* species);
Colessendeidae; Endeidae; Nymphonidae (deep-
sea, including 100 species of *Nymphon*);
Pallenidae; Phoxichilidiidae; Pycnogonidae;
(eg *Pycnogonum* species); Tanystylidae.

MOST chelicerates are land creatures,
such as spiders, scorpions, harvestmen
and the parasitic mites and ticks. The
aquatic chelicerates, however, are by no
means so well known. They consist of two
apparently dissimilar groups—the horseshoe or king crabs, "living fossils," and the
sea spiders, "all legs and no body."

All chelicerates are typified by having a
pair of pincer-like mouthparts, known as
chelicerae, in front of the mouth opening,
where insects and crustaceans have one and
two pairs of antennae respectively. In addition chelicerates have no biting jaws and in
most (including the horseshoe crabs but not
sea spiders) two distinct parts of the body are
recognizable, the prosoma (front part) and
opisthosoma (rear part). Since they fit this
description and in spite of their common
names and sea-dwelling habit, horseshoe
crabs are therefore more closely related to
terrestrial arachnids than to crustaceans.
They have remained more or less
unchanged for over 300 million years. The
affiliations of the sea spiders are less clear,
their idiosyncratic body shape probably
having been much altered during evolution.
It is probable that sea spiders should be
included in the chelicerates, and some
authorities believe them actually to be
arachnids, and close to their terrestrial
counterparts.

Horseshoe crabs or king crabs have a protective hinged carapace whose domed shield
of horseshoe form covers the prosoma and
the first part of the opisthosoma; a long
caudal spine protrudes behind. They have
compound eyes on the carapace and median
simple eyes. Beneath the dark brown
carapace lie the chelicerae and five pairs of
walking legs, comparable in evolution to the
chelicerae, pedipalps and four pairs of walking legs of spiders.

Horseshoe crabs live on sandy or muddy
bottoms in the sea, plowing their way
through the upper surface of the sediment.
During burrowing, the caudal spine levers
the body down while the fifth pair of walking
legs acts as shovels, the form of the carapace
facilitating passage through the sand. The
animal also uses the caudal spine to right
itself if accidentally turned over.

Horseshoe crabs are essentially scavenging carnivores. There are jaw-like extensions on the bases of the walking legs, used
to trap and macerate prey, such as clams
and worms, before it is passed forward to the
mouth. The stout bases of the sixth pair of
legs can also act like nutcrackers to open the
shells of bivalves which are seized during
burrowing.

The appendages on the rear part (opisthosoma) are much modified. The first pair
forms a protective cover over the remainder,
each of which is expanded into about 150
delicate gill lamellae resembling the leaves
of a book, in an adaptation unique to horseshoe crabs. The movement of the appendages maintains a current over the respiratory
surfaces of the gill books which are well supplied with blood, which in turn drains back
to the heart for pumping around the body.
Small horseshoe crabs can swim along
upside down, using the gill books as paddles.

At night at particular seasons of the year,
male and female horseshoe crabs congregate at the intertidal zone for reproduction.
The female lays 200–300 eggs in a depression in the sand, to be fertilized by an
attendant clasping male. The eggs hatch
after several months as trilobite larvae, so
called because of their similarity to the
trilobites of the fossil record of the Paleozoic
era (600–225 million years ago). Initially
1cm (0.4in) long, the larvae develop to
reach maturity in their third year.

Sea spiders are exclusively marine, and are
to be found from the intertidal zone to the
deep sea. They have an exaggerated hanging stance, like larger examples of their terrestrial namesakes, and are typically to be
found straddling their stationary prey,
which includes hydroid polyps, sponges and
sea mats. Sea spiders grip the substrate with
claws as they sway from one individual of
their colonial prey to another without shift-
ing leg positions. Although they mostly
move slowly over the seabed, many sea
spiders are capable swimmers.

Sea spiders typically have four pairs of
long legs arising from lateral protuberances

along the sides of the small, narrow trunk. There are some ten- and twelve-legged species. The paired chelicerae and palps border the proboscis. Sea spiders usually feed either by sucking up their prey's body tissues through the proboscis or by cutting off pieces of tissue from the prey with the chelicerae, then transferring them to the mouth at the tip of the proboscis. A few sea spiders feed on detritus.

In addition to the walking legs, there is a further pair of small legs, which is particularly well developed in males. As eggs are laid by the female they are fertilized and collected by the male, which cements them on to the fourth joint of each of its small egg-bearing (ovigerous) legs, where they are brooded. The eggs hatch later as protonymphon larvae and develop through a series of molts, adding appendages to the original three larval pairs, the forerunners of the chelicerae, palps and ovigerous legs.

Body colors are variable, generally white/transparent, but red in deep-sea species. The high surface-area-to-volume ratio of the narrow body means that it is only a short distance to the outside from anywhere in the body and respiratory gases and other dissolved substances can be moved efficiently by diffusion. There is therefore no necessity for specialized excretory, osmoregulatory or respiratory organs. The lack of storage space in the body does, however, mean that reproductive organs and gut diverticula have to be partly accommodated in the relatively large legs.   PSR

◄ **Nursing father:** the male sea spider *Nymphon gracile* has special ovigerous legs to carry the fertile eggs.

▼ **Not from outer space** but from distant time, "living fossil" horseshoe crabs (*Limulus* species) mating on the shore.

WATER BEARS
AND TONGUE WORMS

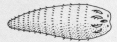
W̲A̲T̲E̲R̲ B̲E̲A̲R̲S̲ and tongue worms are relatively obscure animals which have similarities to arthropods, particularly in having a chitinous cuticle. Some zoological authorities go as far as to include each group in one of the three main arthropod phyla, water bears in the uniramians and tongue worms in the crustaceans. Here water bears and tongue worms are assigned to their own phyla, the Tardigrada and the Pentastomida.

The **water bears** or **tardigrades** (literally "slow steppers") are minute fat-bodied animals said to resemble bears, and often referred to as water bears, moss bears or bear animalcules. They are to be found in the water film on damp moss and also occur in soil, freshwater and marine sediments.

Water bears move by slow crawling, attaching with claws at the end of each leg. In addition to the four pairs of short unjointed legs, there are further signs of segmentation in the muscles and the nervous system. In *Echiniscus* species the cuticle is arranged in segmental plates. Most water bears feed by piercing plant cells with two sharp stylets and sucking out the contents via a muscular pharynx. Some feed on detritus and others are predatory. There are two, probably salivary, glands leading into the mouth. Defecation may be associated with molting, as in *Echiniscus* which leaves its feces in its cast cuticle.

The three "Malpighian" glands leading into the gut are believed to serve an excretory function. There are no respiratory organs, blood vessels, nor heart, because diffusion is sufficient for transport of essential foods in such small animals. The nerve cord along the underside is well organized with ganglia, and the sense organs include two eyespots and sensory bristles.

Water bears have separate sexes, the females usually being the more numerous. However, males have yet to be discovered in certain species, and in other species the females breed asexually (parthenogenesis). In general, reproduction involves copulation with subsequent internal fertilization; some females store transferred sperm in seminal receptacles. Between one and 30 eggs are laid at a time. These may be thin-shelled and hatch soon after laying, or thick-shelled to resist hazardous environmental conditions. Newly hatched young resemble adults and grow by increasing the size, but not the apparently fixed number of their constituent cells—a feature perhaps associated with their very small size. Legs increase from two pairs to four before adulthood. Maturity is reached in about 14 days and they live

between three and 30 months, passing through up to 12 molts.

Water bears are remarkable in withstanding extreme conditions. They can survive desiccation and, experimentally at the other extreme, immersion into chemicals or liquid helium at −227°C. The tardigrade enters a state of dormancy that may last years, reviving when conditions improve.

The phyletic position of the water bears is controversial, for they resemble other groups as well as arthropods. The **tongue worms** or **pentastomids**, however, are clearly arthropodan—disagreement existing over whether they should be placed in their own phylum or included in the crustaceans. Tongue worms are worm-like parasites in the respiratory tracts of carnivorous vertebrates, typically reptiles, although six species are found in mammals including man and two in birds. The life cycle usually involves a herbivorous vertebrate intermediate host for the developing larvae, such as a fish or a rabbit.

Blown-up water bear. This highly enlarged photograph shows clearly the external morphology of the water bear *Macrobiotus richtersi* (× 700).

Digesting its last meal. The plant food in the gut of this tardigrade can be clearly seen.

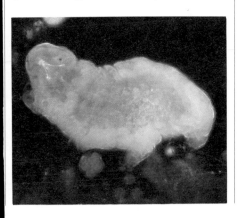

The front end of a tongue worm has five short protuberances bearing, respectively, the mouth and four hooks used by the animal to anchor itself and in feeding. (Once it was thought that there was a mouth on each of the five—hence the name pentastomid.) The mouth is modified for sucking the blood of the host. Some species have superficial ring-like markings along the body, enhancing the worm-like appearance. The worm shape is an adaptation to life in confined passages which must not be blocked; otherwise the host will be suffocated.

The body is covered in a chitinous cuticle which is molted during larval development. Most larvae have two pairs of appendages, and these are retained in adults in the primitive genus *Cephalobaena*. The adults move little, and the nervous system is reduced, as befits a parasite. There is no heart, and respiratory and excretory organs are also absent.

There are distinct sexes, and fertilization is internal. Fertilized eggs are released in the host's feces or by the host sneezing, and lie in vegetation before being ingested by a herbivore. In the case of *Linguatula serrata* the eggs are eaten by a rabbit or hare and the larvae hatch out under the action of digestive juices. The larva bores through the gut wall and is carried in the blood to the liver. Here it forms a cyst and grows through a series of molts to approach adult form. If the rabbit is eaten by a dog or wolf, the larva is released and passes up the dog's esophagus to the pharynx, and so to its adult location in the host's nasal cavity.

Porocephalus crotali lives in snakes with mice as intermediate hosts, and other tongue worms live in crocodiles with fish intermediate hosts. It has been suggested that pentastomids may be modified crustaceans originally parasitic on fish in the manner of the modern genus *Argulus* of crustacean fish lice. It is a plausible step for the parasite to have been stimulated to leave this host when eaten by a crocodile, to take up residence in the mouth of the predator. PSR

Mollusks
Phylum: Mollusca

Spiny-skinned animals
Phylum: Echinodermata

Chordates
Phylum: Chordata

Sea squirts
Subphylum: Urochordata (or Tunicata)

Lancelets
Subphylum: Cephalochordata (or Acrania)

Vertebrates (not in this volume)
Suphylum: Craniata (or Vertebrata)

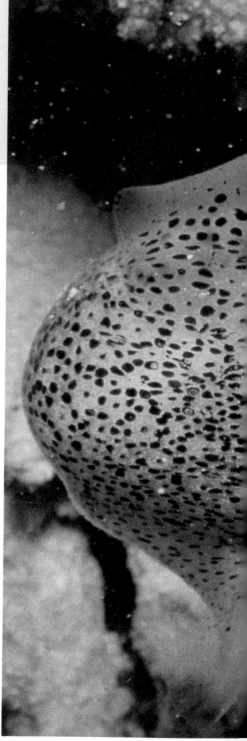

WE think of most animals as having distinct front and rear ends, a definite head, and left and right sides. Not all animal phyla fit this pattern, and some have a body plan and architecture which zoologists still do not fully understand. Among these phyla are the mollusks, echinoderms and urochordates.

In the largest class of mollusks, the gastropods, the advantages of coiling and twisting (torsion) of the body in the snails and slugs are not understood. If torsion confers advantages to the larvae, as has been suggested, why does it persist in many adults, and why have some mollusks abandoned coiling and torsion in the course of evolution? Quite different in appearance, the sophisticated squids and octopuses have produced a further variation on the mol-luskan theme; in the cephalopod ("head-foot") body plan the fleshy foot has migrated to surround the head in the form of tentacles.

The spiny-skinned echinoderms, unlike the mollusks, have no head and many lack even a recognizable front end, being able to move in all directions with similar ease. Their mostly five-sided bodies have no right and left sides. Again the advantages conferred by this form have not really been satisfactorily explained. The fact that echinoderm larvae, unlike the adults, have a distinct front end, as well as left and right sides, makes matters more confusing.

The urochordates, which include the sea squirts and lancelets, are a subphylum of the great phylum Chordata. Their swimming larvae resemble primitive tadpoles, and

▲ **Sophisticate among invertebrates,** this Little cuttle displays the large head and eyes, suckered arms and the pigment spots (chromatophores) of cephalopod mollusks.

◄ **Not all starfishes have five arms,** as this *Solaster endeca* shows. Generally the five-pointed symmetry (pentamery) is superimposed on a radial body plan, as in the brittle stars that surround their relative.

have features which were crucial in identifying the chordate affinities of the group—a distinct head and tail and a rudimentary stiffening structure in the body, the notochord. However, the adults look nothing like chordates at all. The humble-looking sea squirts have a weird shape with no head.

Perhaps all these forms have at some time in their evolution served their owners to advantage, in either their larval or adult lives. Alternatively, some may not be as highly evolved as has been supposed, and the groups in question may have managed to get by in spite, rather than because of, their strange shape. In any event, today the mollusks are a major group of marine animals, significant also in freshwater and on land, where they may carry human disease or be crop pests. The echinoderms are restricted to the sea; a few are economically important as pests in shellfisheries and coral areas or as food for commercial fishes. The urochordates are a group of great interest, meriting further research that may tell us much about animal embryology and development.

AC

MOLLUSKS

Phylum: Mollusca
About 80,000–100,000 species in 7 classes.
Distribution: worldwide, primarily aquatic, in
seas mainly, some in fresh and brackish water,
some on land.
Fossil record: appear in Cambrian rocks about
530 million years ago, modern classes distinct
by about 500 million years ago; abundant
extinct nautiloids, ammonites, belemites.
Size: often smaller than eg crustaceans, but
ranging from tiny 1mm (0.04in) gastropods to
Giant squid up to 20m (60ft) long overall.

Features: body soft, typically divided into head
(lost in bivalves), muscular foot and visceral
hump containing body organs; protected by
hard calcareous shell in most species; no paired
jointed appendages; a fold of skin (mantle)
forms a cavity that protects soft parts, and may
act in defense and locomotion; mouthparts
include usually a toothed tongue (radula);
breathing usually by gills which may serve also
in filter feeding; heart usually present;
cephalopods have veins; nervous system of
paired ganglia—brain and eye of cephalopods
most highly developed of all invertebrates;
sexes typically separate (not land species),
fertilization external or by copulation; young
develop via larval stages including often a free-
floating trochophore and/or veliger before
settling to an often bottom-dwelling life.

▼▶ **Some representative species of mollusks.**
(**1**) A species of *Dentalium* or elephant's tusk
shell; (5cm, 2in). (**2**) *Solen marginatus*, a razor
shell; (12.5cm, 4.9in). (**3**) A chiton (class
Polyplacophora). Chitons are flattened
sedentary mollusks with eight overlapping
protective shell plates. (**4**) *Nucella lapillus*, a
dog whelk; rocky shores of NW Europe (3cm,
1.2in high). (**5**) *Buccinum undatum*, the large
European whelk. It lives on sand and mud
down to about 100m (330ft) (8cm, 3in).
(**6**) A species of *Neopilina*, a genus of limpet-like
mollusks (4cm, 1.6in). (**7**) A species of *Aplysia*
or sea hare (15cm, 5.9in). (**8**) *Tridacna gigas*,
the largest living mollusk or giant clam (1.35m,
4.4ft). (**9**) A common octopus (genus *Octopus*):
the zenith of molluskan organization.
(**10**) *Falcidens gutterosus*, a chaetoderm. It lives
in mud at 40m (130ft) and below;
Mediterranean (1.5cm, 0.6in).

THE diversity of mollusks encompasses
food, pests, dyes, pathogens, parasites
and pearls. This variety is reflected in the
range of body forms and ways of life. Mollusks include coat-of-mail shells or chitons,
marine, land and freshwater snails, shell-
less sea slugs and terrestrial slugs, tusk
shells, clams or mussels, octopuses, squids,
cuttlefishes and nautiluses. In addition to
the 80,000–100,000 living species, the
many extinct species include the ammonites
and belemites. Today's mollusks occur
throughout the world, living in the sea,
fresh and brackish water, and on land.
Apart from those that float, swim weakly (eg
sea butterflies), or powerfully (eg squids), or
burrow (eg clams), mollusks live either
attached to or creeping over the substrate,
whether seabed or ground, or vegetation.

The molluskan body is soft and is typically
divided into head (lost in the bivalves),
muscular foot and visceral hump containing the body organs. There are no paired
jointed appendages or legs, a feature dis-
tinguishing mollusks from the arthropods.
Most species have their soft parts protected
by a hard calcareous shell.

Two notable features of the molluskan
body are the mantle, an intucking of skin
tissue that produces a protective pocket, and
the toothed tongue or radula.

Mollusks have a gut with both mouth and
anus, associated feeding apparatus, a blood
system (generally with a heart), nervous
system with ganglia, reproductive system
(which in some is very complex), and
excretory system with kidneys. The epi-
dermal (skin) tissues of mollusks are generally moist and thin, and liable to drying out.
Gills are present in most aquatic species
which are used to extract oxygen from
water. In the majority of bivalves and some
gastropods, however, the gills are also used
in feeding, when they strain out organisms
and detritus from the water or bottom mud
with minute flickering cilia on the gills.
These particles are then conveyed by tracts
of cilia to the mouth.

In land and some freshwater snails, the
walls of the mantle cavity act as lungs,
exchanging respiratory gases between the
air and body. Many mollusks have a free-
floating (pelagic) larval stage, but this is
absent in land and some freshwater
examples.

Lack of an internal skeleton has kept most
mollusks to a relatively small size.
Cephalopods have achieved the greatest size

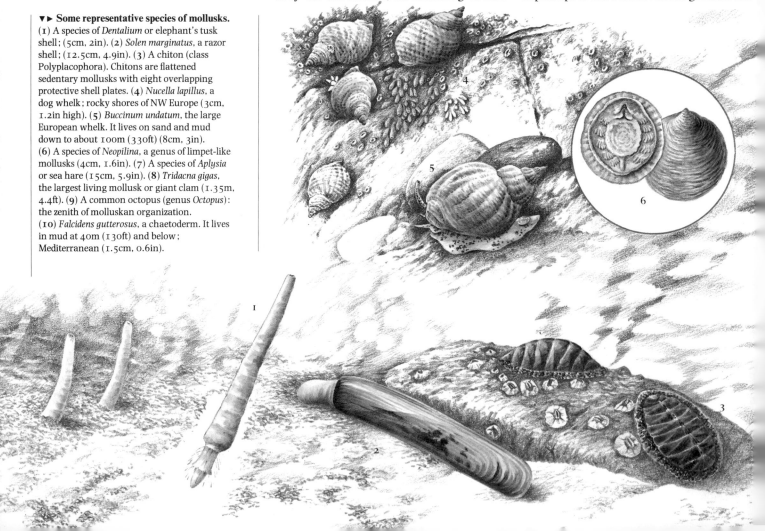

in the Giant squid, which can be 20m (60ft) long, including tentacles. Giant ammonites with shells up to 2m (6.6ft) across existed in the Jurassic period 195–135 million years ago. The largest living bivalves are the tropical giant clams, which can reach 1.5m (4.5ft) in shell length. A substantial number of species measure less than 1cm (0·4in). Some of the smallest, like the tiny gastropod *Ammonicera rota*, are only 1mm (0.04in) long when fully grown.

The mantle

The back of a mollusk is covered by a fold of skin, the mantle, which forms a pocket housing the gills, osphradium (a chemical sensory organ), hypobranchial gland (secreting mucus), anus, excretory pore and sometimes the reproductive opening. This special feature of mollusks has been adapted in many different ways and is present in all molluskan classes.

The cells of the mantle, particularly at the

thickened edge of the mantle skirt, secrete the shell and may also produce slime, acids and ink for defense. Mucus for protection and for cohesion of food particles is secreted by the gill and the hypobranchial glands. Products of the mantle can be defensive, acting to deter predators. The purple gland in the mantle of the sea hare expels a purple secretion when the animal is disturbed. Several species of dorid sea slugs or sea lemons can expel acid from glands in the mantle, while on land the Garlic snail gives off a strong aroma of garlic from cells near the breathing pore.

The mantle wall may be visible and in some sea slugs it is brightly colored and patterned—acting as either warning coloration or camouflage. The glossy and colored shells of cowries are usually hidden by a pair of flaps from the mantle.

Protection of the delicate internal organs was probably an early function of the mantle, which also provides a space into which the head and foot can be retracted when the animal withdraws into its shell. Within the mantle cavity, the gills are protected from mechanical damage from rocks, coral etc as well as from silting. At the same time the gills must have ready access to sea water from which to extract oxygen. In some mollusks, special strips of mantle tissue are developed as tubular siphons which help to separate two currents of water that pass over the gills—the inhalant and exhalant water currents. Fleshy lobes to the mantle, as in freshwater bladder snails of the genus *Physa*, may function as extra respiratory surfaces.

Fertilization of eggs may take place within the mantle cavity of bivalves, and eggs are brooded there in, for example, the small pea shells of freshwater habitats.

The versatile molluskan mantle can become muscular and serve in locomotion. Some sea slugs employ their leaf-like mantle lobes in swimming. Some scallops, such as the Queen scallop, swim by expelling water from the mantle cavity.

The radula

The radula, a toothed tongue, is typically present in all classes of mollusk except the bivalves. It is secreted continuously in a radula sac and is composed of chitin, the polysaccharide also found in arthropod exoskeletons. The oldest teeth are toward the tip: when a row of teeth becomes worn, they detach and are often passed out with the feces. A new row of teeth then moves into position. Inside the mouth is an organ, the buccal mass, which contains and operates the radula during feeding. The radula is carried on a rod of muscle and cartilage (odontophore) that projects into the mouth cavity, while further complexes of muscles and cartilage in the walls of the buccal mass operate the radula, usually in a circular motion.

The form of the radula depends on feeding habits and is used in identifying and classifying individual species. Herbivores, such as land snails and slugs, have a broad radula with many small teeth, while carnivores, including whelks, have a narrow radula with a few teeth bearing long pointed cusps. Limpets, which browse algae off rocks, have an especially hard rasping radula and a few very strong teeth in each row; they leave

▲ **Tiny black eyes** and sensory tentacles fringe the mantle of the Giant scallop (*Pecten maximus*). The eyes detect sudden changes in light caused by arrival of would-be predators. In the free-swimming scallops the mantle margin is also used to direct jets of water propelling the mollusk to safety.

▶ **Rows of rasping teeth** on radula of a herbivorous top shell (*Trochus* species). Carnivorous mollusks have fewer, larger, more pointed teeth on a narrower radula.

◀▼ **Iridescence** LEFT inside a Pearly nautilus shell is caused by reflected light being refracted as it passes calcite back out of the shell through calcite crystals. The mother-of-pearl inside nautilus, top, turban, ormer and other shells, and the pearls formed in the mantle cavity of oysters and mussels are made of the same material. Four shell layers are now recognized BELOW: (1) an inner nacreous layer of calcite crystals in flat layers parallel to the surface; (2) a cross-lamellar component of oblique crystals; (3) a chalky, often pigmented, prismatic layer, usually the thickest, of crystals set at right angles to the surface; and (4) a shiny, horny outer layer, the periostracum, of proteins and polysaccharides.

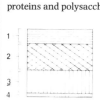

ink along the outside of the lip. After a few days new shell will be seen in front of the ink mark. Newly secreted shell is thin, but gradually attains the same thickness as the rest. Although, in the event of damage, a repair can be made further back from the mantle edge, it will be a rough patch and not contain all the layers of normal shell. When growth stops for a while during cold weather, in drought, or a time of starvation, a line forms on the shell which continues to be visible after resumption of growth. A number of mollusks (eg cockles) normally have regular marks recording interruptions of their growth.

The cross-lamellar component of the shell, revealed by high magnification, consists of different layers of oblique crystals, each layer with a different orientation, rather like the structure of radial car tyres. This is thought to give greater strength without extra weight or bulk.

Shell is mostly composed of calcium carbonate, in calcite form (in the prismatic layer) and in argonite form (in the cross-lamellar layers), together with some sodium phosphate and magnesium carbonate. The mineral component of the shell is laid down in organic crystalline bodies in a matrix of fibrous protein and polysaccharides (conchiolin) secreted by the mantle. Snails can store calcium salts in cells of the digestive gland (hepatopancreas). When needed for growth or shell repair the salts are transported to the mantle by migratory cells.

Molluskan shells show great variation in shape, size, thickness, sculpture, surface texture and shine. Marine examples are often thick and heavy, while land snails, lacking the support of water, tend to have thinner shells.

The spirally coiled shells of gastropods range from tall and spindle-shaped to flat and disk-like; the body whorl containing the animal itself may be small, or enlarged to occupy most of the shell; and likewise the aperture or mouth of the shell can be open or constricted and armed with a range of teeth or ribs – in whelks and other carnivorous sea snails there is a groove (siphonal canal) to house the siphon. With the shell apex uppermost, the mouth of the shell in most mollusks is on the right hand (dextral) side, but some species normally have the mouth on the left (sinistral). Some genera, such as Hawaiian tree snails and *Amphidromus* (both tropical land snails), and the temperate-zone whorl snails, have both dextral and sinistral examples. In some normally dextral species an occasional sinistral species may be found, but this is unusual.

scratch marks on the surface of rock.

In the carnivorous cone shells each tooth is separated from the membrane and is a harpoon-like structure which is delivered into the body of the prey (often a fish or a worm) to facilitate penetration of an accompanying nerve poison. The tiny sacoglossan sea slugs feed on thread-like algae—their radula teeth are adapted to pierce individual algal cells.

The shell

Mollusks usually hatch from the egg complete with a tiny shell (protoconch) that is often retained at the apex of the adult shell. This calcareous shell, into which the animal can withdraw, is often regarded as a hallmark of a mollusk. For the living mollusk, the shell provides protection from predators and mechanical damage, while on land and on the shore it helps to prevent loss of body water. Empty shells have long been a source of fascination in themselves, and many people collect them.

New growth occurs at the shell lip in gastropods and along the lower or ventral margin in bivalves. Shell is secreted by glandular cells, particularly along the edge of the mantle. This is easy to demonstrate in young land snails, by painting waterproof

THE 7 CLASSES OF MOLLUSKS

Includes species, genera and families mentioned in the text. For reasons of space, divisions such as suborders and superfamilies, important in some groups, are omitted. The sequence of families reflects relationships.

Monoplacophorans
Class: Monoplacophora

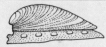

Five species of deep-sea segmented limpets, eg *Neopilina galathea*.

Solenogasters, chaetoderms
Class: Aplacophora

About 200 worm-like marine species in subclasses Solenogastres (eg *Epimenia verrucosa*) and Caudofoveata.

Chitons or coat-of-mail shells
Class: Polyplacophora

About 500 species including the **Giant Pacific chiton** (*Amicula stelleri*) and *Mopalia* species.

Slugs and snails (gastropods)
Class: Gastropoda

About 60,000–75,000 species in 3 subclasses.

Prosobranchs or operculates
Subclass: Prosobranchia

Order Diotocardia
Slit shells (Pleurotomariidae), eg *Pleurotomaria* species. **Ormers** and **abalones** (Haliotidae), eg *Haliotis*, *Ormer* species. **Slit and keyhole limpets** (Fissurellidae), eg *Diodora*, *Fissurella* species, **Great keyhole limpet** (*Megathura crenulata*). **True limpets** (Patellidae), eg **Common limpet** (*Patella vulgata*), **Blue-rayed limpet** (*Patina pellucida*). **Top shells** (Trochidae), eg **Thick top shell** (*Monodonta lineata*). **Turban shells** (Turbinidae), eg **Tapestry turban** (*Turbo petholatus*). **Pheasant shells** (Phasianellidae), eg **Pheasant shell** (*Tricolia pullus*), **Australian pheasant shell** (*Phasianella australis*).

Order Monotocardia
Mesogastropods
Apple snails (Ampullariidae), eg *Pila*, *Pomacea* species. **River snails** (Viviparidae), eg *Viviparus viviparus*. **Winkles** or **periwinkles** (Littorinidae), eg **Dwarf winkle** (*Littorina neritoides*), **Flat winkle** (*L. littoralis*), **Edible winkle** (*L. littorea*). **Round-mouthed snails** (Pomatiidae), eg *Pomatias elegans*. **Spire snails** (Hydrobiidae), eg **Jenkins' spire shell** (*Potamopyrgus jenkinsi*). **Family Omalogyridae**, eg *Ammonicera rota*. **Vermetids** (Vermetidae), eg *Vermetus* species. **Sea snails** (Janthinidae), eg **Violet sea snail** (*Janthina janthina*). **Family Styliferidae** (parasites), eg *Stylifer*, *Gasterosiphon* species. **Family Eulimidae** (parasites), eg *Eulima*, *Balcis* species. **Family Entoconchidae** (parasites), eg *Entoconcha*, *Entocolax*, *Enteroxenos* species. **Slipper limpets, cup-and-saucer** and **hat shells** (Calyptraeidae), eg **Atlantic slipper limpet** (*Crepidula fornicata*). **Family Capulidae** (parasites), eg *Thyca* species. **Ostrich foot shells** (Struthiolariidae), eg *Struthiolaria* species. **Conch shells** (Strombidae), eg **Pink conch shell** (*Strombus gigas*). **Cowries** (Cypraeidae), eg **Money cowrie** (*Cypraea moneta*), **Gold ringer** (*C. annulus*). **Necklace** and **moon shells** (Naticidae), eg *Natica* species.

Neogastropods
Whelks (Buccinidae), eg **Edible whelk** (*Buccinum undatum*). **Dog whelks** (Nassariidae), eg *Bullia tahitensis*. **Spindle shells** (Fasciolariidae), eg *Fasciolaria* species. **Rock shells** or **murexes** (Muricidae), eg *Murex* species, **Common dog whelk** (*Nucella lapillus*), **Oyster drill** (*Urosalpinx cinerea*). **Volutes** (Volutidae), eg *Voluta* species. **Olives** (Olividae), eg *Oliva* species. **Turret shells** (Turridae). **Cones** (Conidae), eg **Courtly cone** (*Conus aulicus*), **Geographer cone** (*C. geographicus*), **Marbled cone** (*C. marmoreus*), **Textile cone** (*C. textile*), **Tulip cone** (*C. tulipa*). **Auger shells** (Terebridae), eg *Terebra* species.

Sea slugs and bubble shells
Subclass: Opisthobranchia

Order Bullomorpha—bubble shells
Acteon shells (Acteonidae), *Acteon* species. **Bubble shells** (Hydatinidae), eg *Hydatina* species. **Cylindrical bubble shells** (Retusidae), eg *Retusa* species. **Bubble shells** (Bullidae), eg *Bullaria* species. **Lobe shells** (Philinidae) eg *Philine* species. **Canoe shells** (Scaphandridae), eg *Scaphander* species.

Order Pyramidellomorpha
Pyramid shells (Pyramidellidae).

Order Thecosomata
Sea butterflies or pteropods (Spiratellidae), eg *Limacina* species.

Order Gymnosomata
Sea butterflies or pteropods (Clionidae), eg *Clione* species.

Order Aplysiomorpha
Sea hares (Aplysiidae), eg *Aplysia* species.

Order Pleurobranchomorpha

Order Acochlidiacea
Family Hedylopsidae, eg *Hedylopsis* species. Family Microhedylidae, eg *Microhedyle* species.

Order Sacoglossa
Bivalve gastropods (Julidae), eg *Berthelinia limax*. Family Elysiidae, eg *Elysia*, *Tridachia* species. Family Stiligeridae, eg *Hermaea* species. Family Limapontiidae, eg *Limapontia* species.

Order Nudibranchia—shell-less sea slugs
Sea slugs (Dendronotidae), eg *Dendronotus* species. **Sea lemons** (suborder Doridacea). Family Tethyidae, eg *Melibe leonina*. Family Aeolidiidae, eg **Common gray sea slug** (*Aeolidia papillosa*). **Floating sea slugs** (Glaucidae), eg *Glaucus atlanticus*, *G. marginata*.

Lung-breathers or pulmonates
Subclass: Pulmonata

Order Systellommatophora—tropical slugs
Eg *Veronicella* species.

Order Basommatophora—pond and marsh snails
Operculate pulmonates (Amphibolidae), eg *Salinator* species. **Dwarf pond snails** (Lymnaeidae), eg *Lymnaea truncatula* **Bladder snails** (Physidae), eg *Physa* species, **Moss bladder snail** (*Aplexa hypnorum*). **Ramshorn snails** (Planorbiidae), eg *Bulinus*. **Freshwater limpets** (Ancylidae).

Order Stylommatophora—land snails and slugs
Hawaiian tree snails (Achatinellidae), eg *Achatinella* species. **Whorl snails** (Vertiginidae), eg *Vertigo* species. **African land snails** (Achatinidae), eg *Archachatina marginata*. Family Oleacinidae, eg *Euglandina* species. **Shelled slugs** (Testacellidae), eg *Testacella* species. **Glass snails** (Zonitidae), eg **Garlic snail** (*Oxychilus alliarius*), *Aegopis verticillus*. Family Limacidae, eg **Gray field slug** (*Deroceras reticulatum*), **Great gray slug** (*Limax maximus*). Family Chamaemidae, eg *Amphidromus* species. Family Helicidae, eg **Brown-lipped snail** (*Cepaea nemoralis*), **Common garden snail** (*Helix lucorum*), **Roman snail** (*H. aspersa*), **Desert snail** (*Eremina desertorum*), *Eobania vermiculata*, *Otala lactea*. **Carnivorous snails** (Streptaxidae).

Tusk or tooth shells
Class Scaphopoda

About 350 species of marine sand burrowers.
Family Dentaliidae, eg **Elephant tusk shell** (*Dentalium elephantinum*). Family Cadulidae, eg *Cadulus* species.

Clams and mussels (bivalves)
Class: Bivalvia (or Pelecypoda)

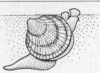

About 15,000–20,000 species of sea, brackish and freshwater.

Order Protobranchia
Includes nut shells (Nuculidae), eg *Nucula* species.

Order Taxodonta—ark shells, dog cockles
Includes dog cockles (Glycimeridae), eg *Glycimeris* species.

Order Anisomyaria
Mussels (Mytilidae), eg **Date mussels** (*Lithophaga* species), **Edible mussel** (*Mytilus edulis*), *Botulus*, *Fungiacava* species. **Pearl oysters** (Pteriidae), eg *Pinctada martensii*. **Scallops** (Pectinidae), eg *Pecten* species, **Queen scallop** (*Aequipecten opercularis*). **File shells** (Limidae), eg *Lima* species. **Saddle oysters** (Anomiidae), eg **Window oyster** (*Placuna placenta*). **Oysters** (Ostreidae), eg *Ostrea*, *Crassostrea* species.

Order Schizodonta
Freshwater pearl mussel (Margaritiferidae), *Margaritifera margaritifera*. **Freshwater** or **river mussels** (Unionidae), eg *Unio*, *Anodonta*, *Lampsilis* species. **Tropical freshwater mussels** (Aetheriidae), eg *Aetheria* species.

Order Heterodonta
Cockles (Cardiidae). **Giant clams** (Tridacnidae), eg *Tridacna gigas*,

T. crocea. **Pea shells** (Sphaeriidae), *Pisidium* species. **Fingernail clam** (Corbiculidae), *Corbicula manilensis.* **Atlantic hard-shell clam** (Arctidae), *Arctica islandica.* **Zebra mussel** (Dreissenidae), *Dreissena polymorpha.* **Ruddy lasaea** (Erycinidae), *Lasaea rubra.* Family Galleomatidae, eg *Devonia perrieri.* **Montagu shells** (Montacutidae), eg *Montacuta* species, *Mysella bidentata.* **Venus** and **carpet shells** (Veneriidae), include **Venus shells** (*Venus* species), eg **quahog** or **Hard-shell clam** (*V. mercenaria*), **smooth Venus** (*Callista* species), **carpet shells** (*Venerupis* species). **False piddock** (Petricolidae), *Petricola pholadiformis*, **oval piddocks** (*Zirfaea* species). **Wedge shells** or **bean clams** (Donacidae), eg *Donax* species. **Tellins** (Tellinidae), eg *Tellina folinacea.*

Order Adepedonta
Trough shells (Mactridae), eg *Spisula* species. **Razor shells** (Solenidae), eg *Ensis* species. **Gapers** (Myidae), eg *Mya, Platyodon* species. **Rock borer** or **Red nose** (Hiatellidae), (*Hiatella arctica.* **Flask shells** (Gastrochaenidae), eg *Gastrochaena* species. **Piddocks** (Pholadidae), eg *Pholas* species, **wood piddocks** (*Xylophaga* species), **shipworms** (Teredinidae), eg *Teredo* species.

Order Anomalodesmata —septibranchs
Dipper clams (Cuspidariidae), eg *Cuspidaria* species.

Cephalopods
Class: Cephalopoda

About 650 marine, mostly pelagic, species.

Nautiloids
Subclass: Nautiloidea
Eg **Pearly nautilus**, *Nautilus* species

Ammonites (extinct)
Subclass: Ammonoidea
Eg **Giant ammonite** (*Titanites titan*).

Subclass: Coleoidea
Order Decapoda—cuttlefishes
Cuttlefishes (eg *Sepia, Sepiola* species), **squids** (eg *Loligo* species), **flying squid** (*Onycoteuthis* species), **Giant squid** (*Architeuthis harveyi*); also extinct **belemites**.

Order Vampyromorpha
Eg *Vampyroteuthis infernalis.*

Order Octopoda—octopuses
Eg *Octopus* species, **Paper nautilus** (*Argonauta* species).

The nautilus, an exception among living cephalopods, has a light, brittle, spiral shell. In section this is seen to be divided, behind the outermost body chamber, by thin walls into progressively smaller earlier body chambers. Each wall (septum) has a central perforation which in the living animal is traversed by a thread-like extension of the body, the siphuncle, which extends to the shell apex. The pressure of gas in the chambers affects buoyancy.

Many shells are strongly sculpted into ribs, lines, beading, knobs or spectacular spines. Such detail, much admired by shell collectors, is also used in the identification of species. The surface of the shells is sometimes rough, but in certain examples, such as cowries and olive shells, it may be smooth and glossy. Many tropical shells are very colorful and may also have attractive patterns and markings.

A number of mollusks in different groups have a reduced shell or none at all. Some sea slugs retain external shells, others have thin internal shells and some (the nudibranchs) like land slugs are without any shell. Shell-less sea slugs have evolved to swim as well as crawl, to squeeze into small crannies, and to develop secretions and body color as means of defense. An external shell is also lacking in some parasitic gastropods, in the worm-like aplacophorans and in most cephalopods.

The evolution of mollusks
The success of the mollusks has been due to their adaptability of structure, function and behavior. Mollusks are thought to be derived from either an ancestor of the Platyhelminthes (planarians, flukes, tapeworms) or from the arthropod annelid line, the latter having trochophore larvae, as do most mollusks.

There is fossil evidence of mollusks in some of the oldest rocks bearing fossils, dating back over 530 million years to the Cambrian period. Mollusks soon evolved in different directions and the modern classes had largely separated out by the end of the Cambrian, 500 million years ago. The early fossils were all marine. Land snails appeared in the coal-measure forests of 300 million years ago, but land snail fossils are rare until deposits of the Tertiary (65–2 million years ago).

The original molluskan shell was probably cap-like. Many families, such as limpets, have reverted to that form from the spiral coiling that was widespread in gastropods (and still is) and in the few remaining shelled cephalopods. Nautiloids

(now represented by only six living species) gave rise to some 3,000 known fossil species, many of which flourished in the Paleozoic seas, but they dwindled in the Mesozoic (225–65 million years ago) when ammonites expanded. After a successful period, ammonites suddenly vanished in their turn at the end of the Cretaceous, along with the dinosaurs.

There is no central theme to molluskan evolution. Different groups of mollusks have adapted to similar habitats, often adopting similar characteristics (convergent evolution). Among bivalves, for example, both true piddocks and the False piddock bore into mudstone but, although the outsides of their shells are similar, the latter is more closely related to venus shells and its shell-hinge teeth are quite different. The bivalve shell of members of the class Bivalvia even has its counterpart in a different class, the bivalve gastropods (see p266).

Isolated islands as in the Pacific show a high level of speciation (evolution of new species) and forms that are endemic (limited to that island), because of the separation of the snail population from other larger populations. In Hawaii there are even local color forms of tree snails in isolated valleys.

Respiration and circulation
Mollusks originated in the sea—and their basic method of breathing is by gills which extract dissolved oxygen from the surrounding water. The typical molluskan gill (ctenidium) consists of a central axis from which rows of gill filaments project on either side (bipectinate). Blood vessels enter the filaments and the surface of each filament is covered with cells bearing cilia; some of these cells create a current in the water with their long cilia, and others pick up food particles. An inhalant current of water enters the gill on one side, passes between the filaments and goes out as an exhalant current on the other side.

In some mollusks the gill serves only for respiration, while in others (eg most bivalves) the two enlarged gills have a dual role of feeding as well as respiration. Gills are delicate structures, which need to be kept clear of clogging particles: the ciliary devices evolved for cleaning the gills later became adapted for feeding. The mantle is often developed at the rear end into a tube (siphon) which projects and takes in water, testing it with sensory cells and tentacles on the way; in some bivalves (eg the tellinids) there is a separate siphon for the exhalant current.

Blood is pumped around the molluskan

body by a heart, and is distributed to the tissues by arteries but, in all except cephalopods, it has to make a slow passage back through blood spaces (hemocoel).

Monoplacophorans and chitons have several pairs of gills in the mantle cavity. In prosobranch or operculate gastropods there is typically one pair of gills, but in the more highly evolved groups the gill is reduced and lost on one side, so winkles and whelks have only one gill in the mantle cavity. The more primitive prosobranchs (eg slit shells, ormers, slit and keyhole limpets) still have two gills. These limpets have lost the typical molluskan ctenidium, replacing it by numerous secondary gills of different structure which hang down around the mantle cavity. Most prosobranchs are aquatic, breathing by gills, although some have adapted to life on land and breathe by a vascularized mantle cavity or lung.

During drought, land prosobranchs can spend long periods of time inactive inside the shell, sealed off by an operculum. Some tropical land species of the family Annulariidae have developed a small tube of shell material behind the operculum which enables the snail to obtain air when the operculum is enclosed. The tropical apple snails live in stagnant water, and these have long siphons which reach above the surface of water to breathe air.

Among the sea slugs and bubble shells, the primitive shelled species such as *Acteon* and the sea hares have a gill in the mantle cavity, but in the shell-less forms there are either secondary gills, as in the sea lemons, or respiration takes place directly through the skin, which may have its surface area increased by numerous papillae.

The pulmonates (land snails and slugs, pond snails), as the name implies, are essentially lung breathers, having lost the gill, and breathe air from a highly vascularized mantle wall. The finely divided blood vessels of the respiratory surface can often be seen as silvery lines through the thin shell. The entrance to the mantle cavity is sealed off and opens by a breathing pore (pneumostome) whenever an exchange of air is needed. The pulmonate pond snails usually breathe air and come up to the surface to open the breathing pore above water.

Some deepwater pond snails of lakes have reverted to filling the mantle cavity with water and no longer use air. While freshwater prosobranchs breathe by gills and are more often found in oxygenated waters of rivers and streams, the pulmonates, breathing air, are better adapted for living in the still stagnant water of ponds and ditches.

▲ **An active predator,** the Common cuttlefish (*Sepia officinalis*) has grabbed a Common prawn (*Palaemon serratus*) by shooting out two long arms. These draw the prey into the grasp of the eight short arms which secure it so that the cuttlefish's sharp beak-like jaws can get to work on the prawn's shell.

◄ **One-way system.** In bivalves such as the Edible mussel, water enters the mantle cavity by an inhalant siphon (the one with the frilly tentacles). After passing over the gills, where food is filtered off as well as respiratory gases exchanged, the water is exhaled by the smaller siphon.

▶ **Naked sea slugs** (*Polycera quadrilineata*) feeding on a sea mat (*Electra* species). The sea slugs and bubble shells (opisthobranchs) include herbivores, suspension feeders, specialist carnivores and parasites, but the majority are carnivores feeding on encrusting marine animals.

Bivalves are all aquatic and breathe by gills. In primitive families, including the nut shells, the gills are relatively small, being only used for breathing, not feeding.

Cephalopods breathe by gills in the mantle cavity. The system is efficient, for there is a greater flow of water through the mantle cavity due to its use in jet propulsion, and faster blood circulation through a closed blood system with veins as well as arteries. The branchial hearts of most cephalopods are not present in nautiluses, which rely instead on duplication of gills to meet the respiratory needs.

Feeding

The primitive mollusk probably fed on small particles, the macrophagous habit (eating large particles) developing later. Most mollusks, except bivalves, feed using the radula (see above). The bivalves and some gastropods (eg slipper limpets of the USA, ostrich foot of New Zealand, and river snails of Europe) are ciliary feeders, either straining food from seawater (filter feeding) or sucking in sludge off the bottom (deposit feeding).

Another typical molluskan feature, found in some prosobranchs and bivalves, is the stomach, its wall protected by chitinous plates from a pointed, forward-projecting style which winds round and brings the string of food from the esophagus into the stomach. The style also secretes digestive enzymes.

The gastropods include browsing herbivores, ciliary feeders, detritus feeders, carnivores and parasites. The muscular mouthparts (buccal mass) and gut show adaptions reflecting this variety. Limpets, top shells and winkles browse on algae and other encrustations on rocks, while the Flat periwinkle eats brown seaweed and the Blue-rayed limpet rasps at the fronds and stems of oarweed (*Laminaria* species). Some will eat carrion while others attack live animals. The whelks, murexes or rock shells, volutes, olives, cones, auger and turret shells (neogastropods) are specialized for a carnivorous diet—the shell has a siphonal canal housing a siphon which directs water over taste cells in the osphradium (a chemoreceptor in the mantle cavity), which helps in the detection of food. Certain carnivorous gastropods, like the Dog whelk and the necklace shells, drill holes in the shells of other mollusks which are then consumed. The murex or rock shells bore mechanically, but necklace shells use acid to soften the shell before excavating with the radula.

The sea slugs and bubble shells include herbivores, suspension feeders, carnivores and parasites, but the majority are carnivores feeding on encrusting marine animals. The sea slug *Melibe leonina*, from the west coast of the USA, is an active swimmer and adapted to feeding on crustaceans, which it catches with the aid of a large cephalic hood: the radula is absent. Shelled opisthobranchs, like species of canoe shell, lobe shell and cylindrical bubble shell, feed on animals in the sand, including mollusks. The inside of the gizzard is lined with special

---

Cones—The Venomous Snails

The 400–500 species of cone shells are found mostly in tropical and subtropical waters. These great favorites among collectors are unusual exceptions among mollusks, being directly harmful to humans.

Cones are carnivores, taking a range of prey, from marine worms to sizeable fish, and those feeding on fish are most dangerous. In common with other carnivorous gastropods, such as whelks, cones have a proboscis. This muscular retractable extension of the gut carries the mouth, radula and salivary gland forward to reach food in confined spaces or for other reasons at a point distinct from the animal. When the probing, extended proboscis of a cone touches a fish, it embeds one of the harpoon-like teeth on the tip of the radula into the prey, accompanied by a nerve poison which paralyzes the fish; the cone swallows the fish whole.

The Geographer cone, Textile cone and Tulip cone have been known to kill humans, while others like the Courtly cone, Marble cone and *Conus striatus* can cause an unpleasant, although not fatal, sting.

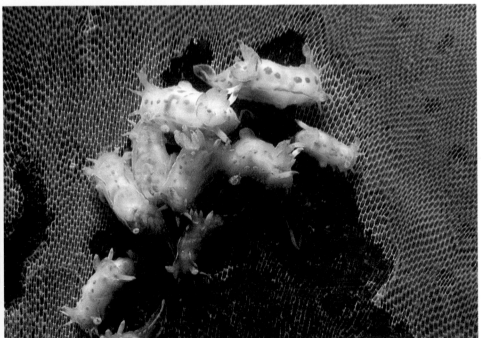

plates which they use to crush the prey. The curious pyramidellids with small coiled shells are parasitic in a range of marine animals.

Pulmonates are chiefly herbivorous, although many of them feed on dead rather than living plant material, and pond snails often consume detritus and mud on the bottom. In these animals the radula is broad, with large numbers of small teeth. There is no style nor chitinous gizzard plates as found in bubble shells. The smaller land snails retain a microphagous diet, while a few species from different families have become carnivorous. These include the shelled slugs that eat earthworms, and a number of tropical land snails, such as the family Streptaxidae and genus *Euglandina*, which eat other snails. Some glass snails (family Zonitidae) have carnivorous tendencies and the large glass snail *Aegopis verticillus* of eastern Europe readily eats land snails.

The more primitive bivalves feed on detritus, which is pushed into the mouth by labial palps, but most modern bivalves are ciliary feeders, making use of phytoplankton, while others take in detritus from the surface of the substrate with siphons. In the more adanced bivalves the gills are used for filtering food and conveying it to the mouth by wrapping it in mucus and passing it along food grooves to the mouth. The crystalline style and stomach plates are well developed in bivalves. The woodboring shipworms have cellulase enzymes used to digest wood shavings. The curious group of bivalve septibranchs have lost the gill and have reduced labial palps and style. They are scavengers, sucking juices of dead animals. Species of the bivalve genus *Entovalva* are parasites inside sea cucumbers, and some of the freshwater mussels are parasitic in fish in the early stages of their life history.

The carnivorous cephalopods mostly catch fish, although the slower-moving octopus takes crustaceans. The radula is relatively small, but the prey is seized by jaws with a hard beak. In cephalopods, enzymes are secreted by gland cells into the tubules of the digestive gland where extracellular digestion takes place: this contrasts with the ingestion of food particles by cells of the digestive gland (intracellular digestion) in other mollusks. Extracellular digestion is a feature in which cephalopod body organization is in advance of the rest of the mollusks and parallels the situation in vertebrates.

Excretion

In mollusks there is a kidney next to the heart which extracts nitrogenous waste from the blood. The excretory duct runs alongside the rectum to the pore at the mantle edge in pulmonates, but in prosobranchs there is a simple opening on the side of the kidney directly into the mantle cavity. In some mollusks certain minerals are selectively resorbed. There is little water regulation in marine mollusks but considerable activity in those of freshwater. Land mollusks conserve their water and little goes out with the excreta, nitrogenous waste being in a insoluble crystalline form and often stored in the kidney. Bivalves give off their nitrogenous waste as ammonia or its derivatives. In some opisthobranchs excretory waste is discharged into the gut.

Breeding

Eggs of mollusks vary considerably. Some are shed into water before fertilization as in bivalves. Many mollusk eggs are very small but those of cephalopods are large and yolky. When fertilization is internal, elaborate egg cases may be secreted and very often gastropods lay eggs in large clutches. Some winkles and water snails deposit eggs in a jelly-like matrix often attached to vegetation. Necklace shells form stiff collars of egg cases which are large for the size of the mollusk, while whelks and murexes deposit eggs in leathery capsules attached to rocks and weed. Land snail eggs tend to be buried in soil: some are contained in a transparent envelope but others have a limy eggshell, and the eggs of one of the large African land snails, *Archachatina marginata*, are of the size and appearance of a small bird's egg. Egg masses of sea slugs can be quite spectacular when found in rock pools.

In aquatic mollusks there is usually a planktonic larva, the primitive trochophore, of short duration, and/or the characteristic veliger larva that develops from it (often within the egg capsule), with shell and ciliary lobes. Some retain the egg inside the body of the female, or in the capsule, from which the young emerge as miniature adults. The veliger lives in the plankton, feeding on algae for a day to several months before settling. Vast numbers of molluskan larvae occur in the plankton and many of them are eaten by other members of the zooplankton or by filterfeeders, or perish when unable to find a suitable habitat to settle.

In land and freshwater prosobranchs the veliger stage is supressed, and a snail hatches direct from the egg. Pulmonates also lack the veliger stage, and floating larvae occur in only some freshwater bivalves. The pelagic larva was important

▶ **Mating in a shoal,** squid (*Loligo* species) off Catalina Island, California. Male squids and other male cephalopods parcel up their sperm into a packet (spermatophore) carried on a specially modified arm called a hectocotylus. This is used to transfer the spermatophore to the mantle cavity of the female, where fertilization occurs. The hectocotylus may be broken off there.

Gastropod snails and slugs more often copulate. This safe transfer of the male sperm by internal fertilization was a prerequisite for gastropods' unique success story among mollusks—the colonization of land.

Some prosobranch snails, such as Jenkins' spire shell, are found in all-female populations that breed parthenogenetically, without the need of males. This has enabled Jenkins' spire shell to colonize streams and ponds over a large part of northwestern Europe within a century.

Other mollusks—monoplacophorans, chitons, bivalves and some gastropods—fertilize eggs externally: both female and male gametes are shed into the sea, or male gametes are brought into the female's mantle cavity on the inhalent current.

▼ **Seen from above, a floating sea slug,** *Glaucus atlanticus*, feeds on a Portuguese man-of-war, a colony of specialized hydrozoan polyps, in the surface waters of the ocean.

in establishing the freshwater Zebra mussel that first came to Britain in the first half of the 19th century. Freshwater or river mussels brood the eggs in the gills and release them as glochidia larvae parasitic on fish.

It is in the sea snails and limpets (prosobranchs) and sea slugs and bubble shells (opisthobranchs) that the veliger is most varied. Chitons and more primitive prosobranchs such as slit shells and ormers have a trochophore larva, with a horizontal band of ciliated cells, which only lasts for a few days. Mollusks with veligers in the plankton for several weeks have a better opportunity for dispersal. At metamorphosis the ciliated lobes (velum) by which the veliger swims and feeds are engulfed, the mollusk ends its planktonic life and sinks to the bottom. Bivalve veliger larvae also occur in the plankton but in some, such as the small midshore *Lasaea rubra*, the young hatch as bivalves and establish themselves near the parent colony. After the bivalve veliger stage is an intermediate pediveliger when the larva searches for a suitable place to settle; if none is found, the velum can be reinflated and the larva is carried to other sites.

Nervous system and sense organs
The brain and eye of the cephalopod are the most highly developed of any invertebrate. The molluskan nervous system essentially consists of pairs of ganglia (masses of nerve tissue), each ganglion linked by nerve fibers. In the more primitive groups the individual ganglia are well separated, but in more highly evolved mollusks, such as land snails and whelks, there is both a shortening of the connectives, bringing the ganglia into closer association, and a concentration of most of the ganglia in the head. The ring of ganglia round the front part of the gut (esophagus) in mollusks, compares with the nervous system of annelids and arthropods. The main pairs of ganglia in mollusks are the cerebral, pleural, pedal, parietal, visceral and buccal (receiving impulses from the head, mantle, foot, body wall and internal organs), and there is a pedal "ladder" arrangeement in monoplacophorans, chitons and the more primitive prosobranchs such as slit shells. In prosobranch gastropods there is the further complication of torsion (see p262).

Most mollusks are sensitive to light, which can be detected by sensors in the shell plates of chitons, the black eyespots associated with the tentacles of most gastropods, and the very elaborate cephalopod eye. The balance of the animal is maintained by special sense organs called statocysts. Prosobranchs have a special chemosensory organ, the osphradium, but there are less specialized patches of chemonsensory cells in other mollusks. The terrestrial slugs have a sense of smell used to find food.

Mollusks are also sensitive to touch: the suckers of octopuses can discriminate texture and pattern (see box p270), while the lower pair of tentacles of land pulmonates are largely tactile and function in feeling the way ahead.

Movement in mollusks
Some mollusks (eg mussels and oysters) anchor themselves to one place, but most move around in pursuit of food, for mating and to escape enemies. The octopus crawls, using suckers on its arms, modifications of the foot. Despite the mollusks' slow image, some, like cephalopods, can swim surprisingly fast.

Usually the foot is the organ involved in locomotion, which involves gliding over the surface of seabed, rock or plant. Land snails and slugs, particularly, lay down a lubricating and protective film of slime or mucus— the silvery trails seen on garden paths and walls. Some species "leap" with long stretches of the foot; the head lunges forward and attaches to ground ahead. Lobes of the foot (parapodia), are often developed for swimming in the sea slugs, and the pelagic thin-shelled *Limacina* and slug-like *Clione* species also have swimming "wings." The sea slugs, freed from the restrictions of a shell, can swim by lateral movements using muscles in the body wall.

Planktonic larvae and some small gastropods move primarily by the beating action of cilia. Veliger larvae of gastropods and bivalves have minute, hair-like cilia on the lobes of the enlarged velum which are used for feeding, respiration and locomotion. They are also able to adjust their depth by retreating into the shell to sink, then re-expanding the velum to halt the descent. Many pond snails move by cilia on the sole of the foot and they can glide along the surface film of water by this method. On land, where the body weight is not supported by water, snails moving by ciliary means are more likely to be the smaller species. Bivalves may swim by shell-flapping (scallops), they may "leap" across the surface (cockles), or burrow. Cockles use the pointed foot for moving the shell across the surface of the sand. Most bivalves, however, use the foot (which can be of considerable size) for burrowing: it is pushed forward and expanded by blood entering the pedal hemocoel

▶ **Jet swimming** by a scallop (*Pecten maximus*), as it escapes from a starfish (here *Marthasterias glacialis*). In flapping the two valves of its shell, the scallop expels jets of water from its mantle cavity. Cockles use their pointed foot for "leaping" across the surface of the sand, but most bivalves use the foot (which may be of considerable size) for burrowing.

▼ **A helmet shell** (*Phalium labiatum*) on the move. On top of the large, muscular foot, at the rear, can be seen the lid-like operculum that protects the animal when it withdraws into its shell. The siphon, and tentacles with eyes at their bases, can also be seen.

As in most mollusks, the muscular foot provides locomotion, by means of wave-like movements that in this case are longitudinal, as in winkles and top shells, for example, while in the terrestrial pulmonates the muscle waves are transverse, traveling from head to tail or the reverse.

(blood space); the muscles of the foot also contract and it then changes shape, often forming an anchor. Further contraction brings the shell down into the sand; as the shell closes, water jetted out of the mantle cavity can help to loosen the sand ahead. The digging cycle then starts again.

Solenogasters and chaetoderms

The shell-less aplacophorans are curious worm-like creatures found in the mud of marine deposits, usually offshore. This small and little-understood group was once classified with the chitons, but is now placed in a group of its own. Indeed, recent research suggests that the class Aplacophora should be divided into two classes, the solenogasters (class Solenogastres) and the chaetoderms (class Caudofoveata).

Aplacophorans do not have an external shell, although there may be tiny, pointed calcareous spicules in the skin, sometimes of a silvery, "fur-like" appearance. In common with other mollusks, most possess a radula, a mantle, mantle cavity, a foot and a molluskan-type pelagic larva. There are dorso-ventral muscles crossing the body, which are reminiscent of similar structures in flatworms, flukes and tapeworms.

Most of the 200 or so species are small, but the solenogaster *Epimenia verrucosa* can reach 30cm (12in) in length. Solenogasters are fairly mobile and can twist their bodies around other objects; the foot is reduced to a ventral groove. Chaetoderms (named for their spiny skin), live in mud, moving rather like earthworms but spending much of their time in a burrow. The radula is often reduced in this group.

Monoplacophorans

First of the seven classes within the plylum, the Monoplacophora is a small group of primitive mollusks that were originally thought to have become extinct about 400 million years ago. The flattish cap-like shell of monoplacophorans resembles that of limpets, and the groups used to be classified with the gastropods.

In 1952 living monoplacophorans were collected 3,750m (11,700ft) down in a Pacific Ocean deep-sea trench off South America, and a new species of monoplacophoran, *Neopilina galathea*, was described. The shell is pale, fairly thin, cap-like in shape and about 2.5cm (1in) long. On the inner surface of the shell, instead of the single horseshoe-shaped muscle scar of limpets, there are several pairs of muscle scars in a row on either side.

The particularly interesting feature of

Neopilina galathea is the repetition of pairs of body organs: there are eight pairs of retractor muscles attaching the animal to the shell, 5–6 pairs of gills, 6–7 pairs of excretory organs, a primitive ladder-like pedal-nervous system with 10 connectives across, and two pairs of gonads. Although there is a parallel in chitons (see below), in monoplacophorans this repetition is taken much further. In consequence, it has been suggested that mollusks evolved from an annelid/arthropod ancestor, rather than from an unsegmented flatworm, and that the original segmentation of the body was lost during early molluskan evolution. Some other researchers disagree, considering the repetition of body organs to be a more recent, secondary character, rather than a primitive one.

In *Neopilina galathea* there is a molluskan radula and posterior mantle cavity and anus, showing that torsion (the twisting of body organs found in gastropods) did not occur in the monoplacophorans. Since the original discovery, further living species of monoplacophorans have been found in deepwater trenches in other parts of the world such as Aden, and now five living species are described.

Chitons

Chitons, or coat-of-mail shells, have a distinctive oval shell consisting of eight plates bounded by a girdle. They are exclusively marine and, with the exception of a few deepwater species, mostly limited to shores and continental shelves. Beneath the shell with its low profile and stable shape, the animal attaches to the rock by a sucker-like foot. The plates of the shell are well articulated—chitons can roll up into a ball when disturbed. The articulations are also an advantage when moving over the uneven surface of rocks.

Although there are no obvious eyes, chitons are sensitive to light through light

receptors in the shell. Chitons are found mostly on rocky shores. When a boulder is turned over, chitons on the underside quickly move down again out of the light.

Chitons have remained substantially unchanged since the Cambrian period 600–500 million years ago. The different species are identified by the relative width of the girdle protecting the mantle, by the sculpturing of the shell valves and the bristles they bear, and also the teeth and the surfaces of the joins between the valves. Most chitons are 1–3cm (0.4–1.2in) long, but some, like the Giant Pacific chiton can reach 20–30cm (8–12in). The larger and more spectacular species are found on the Pacific coast of the USA and off Australia.

Chitons are browsers, rasping algae and other encrusting organisms off the rock with the hard teeth of the radula. One family, the Mopaliidae, is carnivorous, feeding on crustaceans and worms.

The anatomy of the chiton is closer to that

▲ **Delicate colors and patterns** of the mantle that envelops the shell of this Australian cowrie bear little relation to the colorful markings of the gleaming shell beneath.

▼ **Body plan of a gastropod,** here a lung-breathing freshwater snail. Gastropods have adapted to life on land and also to freshwater habitats. Characteristic of gastropods are the flat, creeping foot sole, distinct head with tentacles, mantle, coiled shell made of one piece (univalve), and well developed alimentary, reproductive, circulatory, nervous and excretory systems.

In most gastropods, rotation (torsion) of the body in the developing embryo (1–3) has brought the opening of the mantle cavity, anus and other organs to the front, and the nervous system is twisted. Sea slugs and bubble shells lose torsion in adult life, and the nervous system of land gastropods is not twisted. It has been suggested that torsion provides the larva with a space (the mantle cavity) into which it can quickly contract its head. This may enable it to sink out of danger quickly or help it to settle on the seabed.

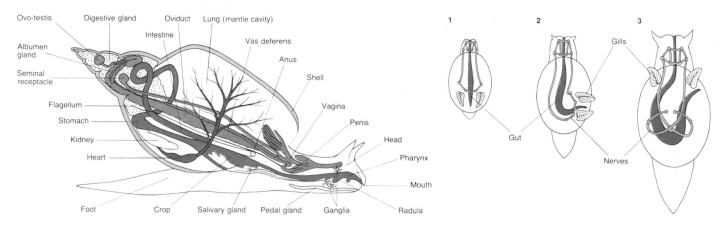

of the primitive ancestor than the more highly evolved gastropods. The mantle cavity and anus are situated toward the rear. The mantle extends forward and houses several pairs of gills, rather more than in most other classes of mollusks. Chitons resemble monoplacophorans and the more primitive groups of prosobranchs, such as the ormers and the top shells, in the ladder-like nervous system with paired ventral nerves and cross-connections. Sexes are separate in the chitons and the eggs are fertilized externally.

Slugs, snails and whelks (gastropods)

This largest class of mollusks contains three-quarters of the living species and shows the greatest variation in body and shell form, function and way of life. Unlike the other classes, which are all aquatic, gastropods have also adapted very successfully to life on land and they have achieved greater diversity in freshwater than the bivalves. Gastropods occur in all climatic zones of the world, colonizing the sea, brackish and freshwater and land. They are among the earliest molluskan fossils.

In the more primitive prosobranchs, gastropods with a cap-like shell, such as slit shells and ormers, which have a trochophore larva with only a short pelagic phase, torsion occurs after the larva has settled. In the more advanced prosobranchs, such as winkles and whelks, the newly hatched veliger larva already has the mantle cavity to the front. Sea slugs and bubble shells do not retain torsion in adult life; loss of shell was influential in the development of this trend. Land and pond snails and slugs have retained torsion, but their nervous system is not twisted.

For the larva, torsion may provide a space into which the animal can quickly contract, enabling it to sink out of danger, or to reach the bottom for settling when it metamorphoses. Advantages of torsion for the adult may include the use of mantle cavity sensors for testing the water ahead or possibly the intake of cleaner water not stirred up by the foot, and providing a space into which the head can be withdrawn. Besides the looping of the alimentary canal and reorientation of the reproductive organs, torsion causes the twisting of the prosobranch nervous system (streptoneury).

The spiral coiling of the gastropod shell is a separate phenomenon from torsion. Coiling occurs also in some of the cephalopods (eg nautiluses), which do not exhibit torsion, and is a way of making a

shell compact. If the primitive cap-like shell had become tall it would have been unstable as the animal moved along. Spiral coiling is brought about by different rates of growth on the two sides of the body. During their evolutionary history, various spirally coiled mollusks in the gastropods, extinct ammonites and nautiloids have uncoiled, producing loosely coiled shells, tubular forms or, in for example limpets, a return to the cap-like shape.

The **prosobranchs** (subclass Prosobranchia) include most of the gastropod seashells—limpets, top shells, winkles, cowries, cones and whelks—as well as a number of land and freshwater species. The prosobranchs, or operculates, have separate sexes, unlike the two other subclasses of gastropod which are hermaphrodite, and the mouth of the

▶ **A dorid sea slug glides** OVERLEAF through a zoological garden. Some members of the family Doridae, called sea lemons, can defend themselves against predators by emitting acid from glands in the mantle.

◀ **Hermaphrodites' courtship embrace.** These two Roman snails will fertilize one another's eggs with an exchange of sperm packets or spermatophores.

▼ **Aeolid sea slug** (*Hermissenda* species) in a Californian tidepool. Members of the family Aeolidiidae use spare parts from their prey, which include jellyfishes and sea anemones. The tentacles of such cnidarians carry nematocysts, stinging structures responsible for the venomous sting of, among others, the Portuguese man-of-war. When a sea slug eats a cnidarian, the victim discharges up to half its nematocysts, but the remainder may reach the papillae secondary organs of breathing on the sea slug's back, via sacs opening off the digestive gland. Later the "live" nematocysts reach the outer skin layer, where the sea slug uses them "second hand" to defend itself against predators.

shell is usually, except in limpet forms, protected by a lid or operculum (see below). Aquatic prosobranchs have a gill in the mantle cavity, together sometimes, especially in the carnivorous groups, with a chemical sense organ (osphradium) and slime-secreting hypobranchial gland. The mantle and associated structures exhibit torsion and as a result of this the principal nerves are twisted into a figure-8 in the more highly evolved orders (Mesogastropoda and Neogastropoda).

In marine species there is usually a planktonic trochophore or veliger larval stage which helps to distribute the species.

Prosobranchs live in sea, brackish and fresh water and on land and have an ancient fossil history going back to the Cambrian. Terrestrial prosobranchs are more abundant and varied in the tropical regions than in temperate zones. Shells vary in shape from the typical coiled snail shell to the cap-like shells of limpets and the tubular form of the warm-water vermetids (family Vermetidae) that look from the shell more like marine worm-tubes than mollusks. This diverse group exploits most opportunities in feeding from a diet of algal slime, seaweed, detritus, suspended matter and plankton (ciliary feeders), to terrestrial plants, dead animal matter and other living animals.

The lid or operculum is secreted by glands on the upperside of the back of the foot. It is the last part of the animal to be withdrawn and therefore acts as a protective trapdoor. It keeps out predators and also prevents water loss in land prosobranchs and intertidal species. The operculum is present in the veliger larvae, even in limpets and slipper limpets which later lose the operculum.

The opercula of most prosobranchs are horny, but hard calcareous ones are found in some of the turban and pheasant shells. The thick operculum of the Tapestry turban shell is green and can be about 2.5cm (1in) across: this is the "cat's eye" used for jewelry. The much smaller pheasant shell *Tricolia pullus* from northern Europe and the Australian pheasant shell and others have conspicuous white calcareous opercula.

The different shapes of opercula usually fit the form of the mouth of the shell. Shells with a narrow aperture like cones, for example, have a tall narrow operculum. In some species, such as the whelk *Bullia tahitensis*, there are teeth on the operculum. In the Pink conch shell these teeth are thought to be defenses against attack by predators that include tulip shells. In some species, particularly the land prosobranchs, the operculum may indeed be small, enabling the animal to retreat further inside its shell.

Sea slugs and bubble shells (opisthobranchs) are hermaphrodite—both male and female reproductive systems function in the same individual—and usually have a reduced shell, or none at all. Bubble shells do have a normal external shell, which in the Acteon shell looks very like that of a prosobranch, as it is fairly solid with a distinct spire. Most of the bubble shells, such as *Hydatina*, *Bullaria* and canoe shell species, have an inflated shell which consists mostly of body whorl with little spire, is rather brittle and houses a large animal. Other opisthobranchs have a reduced shell that is internal, for example the thin bubble shells of *Philine* and *Retusa* species. The sea hares have a simple internal shell plate in the mantle that is largely horny. A few species have a bivalve shell (see below). The rest of the sea slugs have lost their shell altogether. They include *Hedylopsis* species with hard spicules in the skin, *Hermaea* species and other sacoglossans such as *Limapontia*, and the large group of the nudibranchs (meaning "exposed-gills") or sea slugs, including the sea lemons, the family Aeolidiidae and Dendronotidae, and many others.

The bodies of opisthobranchs, particularly nudibranchs, can be very colorful. Although they may function as warning coloration or camouflage, little is known of the function of such bright colors, which are less vivid at depth under water. Some sea lemons emit acid from glands in the mantle as a defense against predators.

Opisthobranchs reproduce by laying eggs, often in conspicuous egg masses. The eggs hatch to veliger larvae.

Some species of bubble shells, such as *Retusa* and Acteon shells, have a blunt foot which they use to plow through surface layers of mud or sand. The round shell-less sea lemons creep slowly on the bottom with the flat foot, but many of the sea slugs are agile and beautiful swimmers, capable of speed. Sea butterflies or pteropods swim in surface waters of the oceans.

About 25 years ago a malacological surprise came to light—an animal with a typical bivalve shell but a gastropod body, complete with flat creeping sole and tentacles. This was *Berthelinia limax*, found living on the seaweed *Caulerpa* in Japan. Other bivalve gastropods have since been discovered, also on seaweed, in the Indo-Pacific and Caribbean as well as Japanese waters. They are classed with the sea slugs of the order Ascoglossa.

Bivalve gastropods have a single-coiled

shell in the veliger larva. In mature shells this is sometimes retained at the prominent point (umbone) of the left-hand valve. This development of a bivalve shell in gastropods is an example of convergent evolution rather than evidence of an ancestor shared with the class Bivalvia.

In **pond snails, land snails and slugs** (subclass Pulmonata) the mantle wall is well supplied with blood vessels and acts as a lung. In parallel with this specialization, the lung-breathing snails have specialized in colonizing land and freshwater, although a few continue to live in marine habitats. Like the sea slugs and bubble shells, pulmonates are hermaphrodite, with a complex reproductive system: the free-swimming larval stage is lost in land and freshwater species and, except in the marine genus *Salinator*, there is no operculum.

The shell is usually coiled, although the varied shapes include the limpet form and in several unrelated families the shell is reduced or lost altogether, leading to the highly successful design of slugs. The thin shell of land snails still offers protection against drying out but is more portable and demands less calcium than do the shells of marine gastropods. Both snails and slugs further conserve body water by being active chiefly at night and by their tendency to seek out crevices. The shell-less slugs are freed from restriction to calcareous soils and can also retreat into deeper crevices.

The body is differentiated into a head with tentacles (one or two pairs), foot and visceral mass. The mantle and mantle cavity are at the front (still showing signs of torsion) but the entrance is sealed off except at the breathing pore (pneumostome), which can open and close. The mouthparts and their muscles (buccal mass) may incorporate both a radula and a jaw. Pulmonates are predominantly plant feeders although there are a few carnivores.

The subclass may be divided into three superorders. In the mostly tropical Systellommatophora the mantle envelops the body. The pond snails (superorder Basommatophora) have eyes at the base of their two tentacles, while the land snails and slugs have two pairs of tentacles and eyes at the tips of the hind pair.

Pulmonates succeed in less stable environments than the sea by their opportunistic behavior and the fact that they can enter a dormant state during adverse periods of cold (hibernation) or drought (aestivation). A solidified plug of hardened mucus (epiphragm) can seal off the mouth

of the shell and in some species, such as the Roman snail becomes hardened. Unlike the operculum of a prosobranch, the epiphragm is neither permanent nor attached by muscle tissue to the animal.

Tusk shells

The tusk shells are a small group (Scaphopoda) of around 350 species which are entirely marine and live buried in sand or mud of fairly deep waters. Only their empty shells are to be found on the beach. Tusk shells or scaphopods occur in temperate as well as tropical waters: the large Elephant tusk shell can be up to 10–13cm (4–5in) long. There are two families, the Dentaliidae, which include the large examples more commonly found, and the Siphonodentaliidae (eg *Cadulus* species), which are shorter, smaller and less tubular in shape. The oldest fossil tusk shells known from the Ordovician period 500–440 million years ago. Like the chitons, they have changed little and show very little diversity of body form and way of life.

The shell is tubular, tapering, curved and open at either end. Scaphopods position themselves in the sand with the narrow end protruding above the surface, and through this pass the inhalant and exhalant currents of seawater, usually in bursts rather than as a continuous flow. The broader end of the shell is buried in the sand. From it the head and foot emerge: the foot creates a space in front into which the animal extends the tentacles of the head that pick up detritus, foraminiferans and other microorganisms from the sand. The tentacles are sensory as well as collecting food and conveying it to the mouth. Food can be broken up by the radula, and the shells of forminiferans are further crushed by plates in the gizzard.

▶ **Wedged in rocks and corals** of shallow Red Sea waters, this giant clam (*Tridacna* species) filter feeds like other bivalves but also houses algae in its brightly colored mantle edge. The algae photosynthesize sugars and starches that are consumed by the clam, which in turn provides shelter and mineral nutrients for its symbiotic partners.

▲ **Mediterranean Fan mussel** or Pen shell (*Pinna nobilis*) is anchored to gravel by strong byssus threads secreted by a gland in the foot. This golden thread was once used in the manufacture of Cloth of Gold, items of which are still to be seen in museums.

▼ **Groups of sensory tentacles** and tiny eyes alternate along the margin of bivalves such as this scallop (*Pecten* species) in the Galapagos Islands. The mantle margin takes over many of the functions fulfilled by the developed head present in other mollusks but lacking in bivalves.

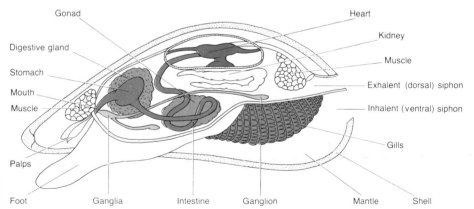

Body plan of a bivalve. Gonad, Digestive gland, Stomach, Mouth, Muscle, Palps, Foot, Ganglia, Intestine, Ganglion, Mantle, Shell, Heart, Kidney, Muscle, Exhalent (dorsal) siphon, Inhalent (ventral) siphon, Gills

▲ **Body plan of a bivalve.** The bivalve body consists of the mantle, often extended into one or two siphons, the visceral mass and relatively small foot. It lacks a developed head. Usually the siphons and foot can be seen protruding from the shell, but in mussels (*Mytilus*, illustrated above) they largely remain inside.

The anatomy of scaphopods is rather simple. The tube is lined by the mantle. There are no gills, oxygen being taken up by the mantle itself, which may have a few ridges with cells bearing tiny hair-like cilia that help to create a current. Oxygen may also be taken in through the skin of other parts of the body. There are separate sexes,

fertilization takes place externally in the sea, and the egg hatches into a pelagic trochophore larva.

Clams, mussels, scallops (bivalves)
Members of this, the second largest class of mollusks with around 15–20,000 living species, are recognized by their shell of two valves which articulate through a hinge plate of teeth and a horny ligament which may be inside or outside the shell.

The bivalve shell can vary considerably in shape, from circular, as in dog cockles, to elongate, as in razor shells. It can be swollen (eg cockles) or flat (eg tellins) and can have radial or concentric shell sculpture (ridges, knobs and spines), bright colors and patterns.

The shell is closed by adductor muscles passing from one valve to the other. Where these attach to the shell, distinctive muscle scars are formed on the inside of the shell. The muscle scars are very important in the

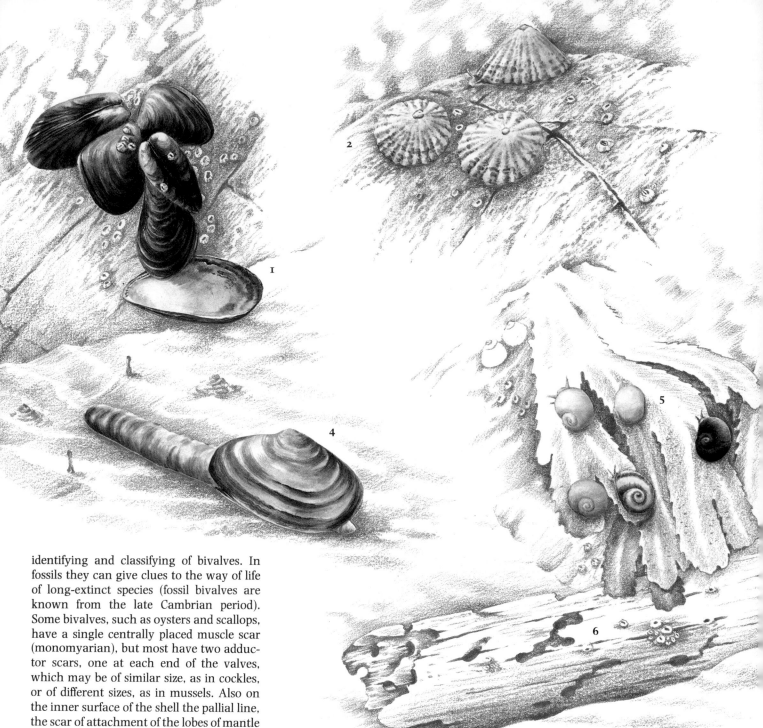

identifying and classifying of bivalves. In fossils they can give clues to the way of life of long-extinct species (fossil bivalves are known from the late Cambrian period). Some bivalves, such as oysters and scallops, have a single centrally placed muscle scar (monomyarian), but most have two adductor scars, one at each end of the valves, which may be of similar size, as in cockles, or of different sizes, as in mussels. Also on the inner surface of the shell the pallial line, the scar of attachment of the lobes of mantle lining the shell, runs from one adductor muscle scar to the other. In those bivalves with small projecting mantle tubes (siphons), living on the surface or in shallow burrows, the pallial line is unbroken and parallel to the ventral margin of the shell, opposite the hinge. In the shells of burrowers (eg venus shells and tellins) which have a long siphon, the pallial line is indented to provide an extra area for attachment of the muscles involved in contracting the siphon when the animal withdraws.

In most bivalves the pair of gills is large and fills the mantle cavity, performing a dual role of respiration and feeding. The primitive nut shells, however, have small gills which are respiratory only—nut shells shovel detritus into the mouth with a pair of labial palps. The carnivorous and more highly evolved dipper clams (septibranchs, eg *Cuspidaria* species) have replaced gills with a wall which controls water flow into the mantle cavity. They feed on very small crustaceans and worms drawn in with water.

The bivalve reproductive system is very simple. The sexes are usually separate, although some, like oysters, do alter sex during their lives. The eggs are fertilized externally, in the sea or in the mantle cavity, by sperm taken in with surrounding water. There is a pelagic bivalve veliger larva in most species.

▲ **Some further representative species of mollusks.** (**1**) *Mytilus edulis*, the Edible mussel. It lives attached to rocks on the lower shore (10cm, 3.9in). (**2**) *Patella vulgata*, a limpet; N Atlantic (7cm, 2.8in). (**3**) *Janthina exigua*, a violet sea snail; N Atlantic (1.5cm, 0.6in high). (**4**) *Mya arenaria*, a clam or gaper (15cm, 5.9in). (**5**) *Littorina obtusata*, the flat periwinkle; NW Europe (1cm, 0.4in). (**6**) Wood bored by *Teredo navalis*, the Common ship worm. (**7**) *Nautilus pompilus*, the pearly nautilus (20cm, 7.9in).

▶ **European Common piddock** (*Pholas dactylus*) in fossilized wood, showing edge of ridged shell and rounded tip of foot.

3

7

Most species live in the sea, but some have colonized brackish and freshwater. The adults lead a relatively inactive life buried in the substrate, or firmly attached to rock by cement or byssus threads, or boring into stone and wood. A few, like scallops and file shells, flap the valves and swim by jet propulsion.

Octopuses, squids, cuttlefishes, nautiluses

Cephalopods are quite different from the rest of the mollusks in their appearance and their specializations for life as active carnivores. The estimated 650 living species are all marine and include pelagic forms,

swimmers of the open sea, and bottom-dwelling octopuses and cuttlefish. While octopuses can be found in rock crevices on the lower shore, most cephalopods usually ocur further out and some penetrate deep abyssal waters, like *Vampyroteuthis infernalis*, which lives 0.5–5km (0.3–3mi) below the surface.

The cephalopods that flourished in the seas of the Mesozoic period over 65 million years ago included nautiloids, ammonites and belemnites. With the exception of nautiluses, these groups, most of which possessed shells, are now extinct.

Most modern cephalopods are descended from the extinct belemnites, which had internal shells.

Cephalopods are typically good swimmers, catching moving fish, and have evolved various buoyancy mechanisms (see p270). They are very responsive to stimuli, due to special giant nerve fibers (axons) with few nerve cell junctions (synapses). This enables messages to pass quickly to and from the brain. (Giant axons are also found

in annelid worms and some other invertebrate groups.) The well-developed cephelopod eye focuses by moving its position rather than changing the shape of the lens. The high metabolic rate of cephalopods is also aided by a particularly efficient blood system with arteries and veins (other mollusks have arteries only) and extra branchial hearts.

All cephelopods except nautiluses have an ink sac opening off the rectum which contains ink, the original artists' sepia. Discharged as a cloud of dark pigment, this confuses an enemy. The cephalopod can also change its color while escaping. Body color and tone are changed by means of pigment cells (chromatophores) in the skin. These are operated under control of the nervous system by muscles radiating from the edge of the chromatophore which can contract it, concentrating the pigment. Stripes and other patterns appear in the skin of cephalopods under certain circumstances.

The possibility that these are a means of communication, for example in recognizing

---

Bivalve Borers

An important number of bivalves, from seven different superfamilies, have adapted from burrowing into soft sand and mud to boring into hard surfaces including mudstone, limestone, sandstone and wood, the wood-boring habit being the most recent to evolve.

Boring developed from bivalves settling in crevices which they subsequently enlarged—one of the giant clams, *Tridacna crocea*, does this. Rock borers of the genus *Hiatella*, although able to use crevices, also erode tunnels of circular section. They push the shell hard against the wall of the burrow by pressure of water in the mantle cavity. At low tide, the red siphons can be seen protruding from holes in the rock low on the shore—they are sometimes known as red-noses.

Most rock borers make their tunnels mechanically by rotation of the shell, often aided by spikes on the shell surface, which erodes away at the rock. The foot may attach the animal to the end of the burrow, as in piddocks. Closure of the siphons helps to keep up fluid pressure. Rock raspings are passed out from the mantle as pseudofeces. While some borers form a tunnel of even width, flask shells (species of *Gastrochaena*) are surrounded by a jacket of cemented shell fragments and live in a rock tunnel that is narrower at the entrance. The contracted siphons dilate outside the shell to form an anchorage during boring.

The date mussels of warm seas and other bivalves have an elongated smooth shell. They make round burrows in limestone by an acid secretion from mantle tissue which is applied to the end of the burrow. The thick,

shiny brown periostracum protects the shell from the mollusk's own acid.

Among the genera of rock borers, *Botula*, *Platyodon* (gapers) and false piddocks drill in clays and mudstone, *Fungiacava* species in coral, and rock borers, flask shells, piddocks and oval piddocks, in rock. The piddocks are recognized by a projecting tooth (apophysis) inside, to which the foot muscles are attached.

Wood piddocks and shipworms burrow into wood. Wood in seawater is merely a transitory habitat, and boring bivalves have therefore adapted in many ways to make the most of what may be a short stay—high population densities, early maturation, prolific breeding and dispersal by pelagic veliger larvae.

Wood piddocks use the wood only for protection, feeding on plankton by normal filtration of seawater, while shipworms exploit the wood further by ingesting the shavings, from which with the aid of cellulase enzymes, they obtain sugar. Shipworms also feed by filtration, but in some species where most food comes from the wood gills are reduced in size.

sex, is being investigated. In bottom-living species like cuttlefish (*Sepia* species), the chromatophores function as camouflage.

The sexes are separate and the male fertilizes the female by placing a sperm package (spermatophore) inside her mantle cavity, where the sperm are released. The yolky eggs are large and there is often a pelagic stage like a miniature adult. Cuttlefish come to inshore waters to breed and after egg laying the spent bodies may be cast up on beaches.

Nautiluses have a brown and white coiled shell. When the nautilus is active, some 34 tentacles protrude from the brown and white coiled shell, and to one side of these are the funnel and hood. The hood forms a protective flap when the nautilus retreats into its shell.

Cuttlefishes are flattened and usually have a spongy internal shell. They often rest on the sea bottom but also may come up and swim. They have 10 arms, eight short and two longer ones, as in squids, for catching prey. The internal shell or "bone" is often found washed up on the seashore.

Squids are torpedo-shaped, active, and adapted for fast swimming. Unlike cuttlefishes, which are solitary, squids move around in shoals in pursuit of fish. The suckers on the 10 arms may be accompanied by hooks. The internal shell or pen is reduced to a thin membranous structure. Oceanic or flying squids can propel themselves through the surface of the water.

Octopuses have adapted to a more sedentary life-style, emerging from rock crevices in pursuit of prey. They can both swim and crawl, using the eight arms. Female octopuses often brood their eggs. The female Paper nautilus is rather unusual in secreting, from two modified arms, a large, thin shell-like egg-case in which she sits protecting the eggs. The male of the species is very small, only one tenth of the size of the female, and does not produce a shell.

Learning in the Octopus

Octopuses have memory and are capable of learning. A food reward, coupled with a punishment of a mild electric shock, has successfully been used to train octopuses to respond to sight and touch. Sight is important to cephalopods in the recognition of prey. The well-developed eye of cephalopods approaches the acuity of the vertebrate eye more closely than that of any other invertebrate. Presented with two distinct shapes, one leading to a food reward and the other to an electric shock, the octopuses in the above test learned the "right" one after 20–30 trials, although they found some shapes easier to recognize than others, and mirror images of the same shape difficult to separate.

Similar experiments using only the tactile stimulus of cylinders with different patterns of grooves, have shown that octopuses distinguish between and respond differently to rough and smooth objects, different degrees of roughness, and objects with differing proportions of rough and smooth surfaces. Touch is perceived by tactile receptors in the octopus's suckers. Octopuses do not distinguish between objects of different weight, but they can recognize sharp edges and distinguish between inanimate objects and food.

Other researches into the brain and nervous systems of octopuses have led to important advances in knowledge of how nervous systems work in general.

► **Prey's-eye view** of a Pacific octopus (*Octopus dofleini*) off the Oregon coast. A red flush replaces the usual yellowish-gray to mottled brown when the octopus is excited.

▼ **Body plan of a cephalopod** (cuttlefish) typically includes mantle, mantle cavity with funnel opening, gills, radula, sharp parrot-like beak, suckered arms (eight short arms, two longer tentacles in cuttlefishes and squid BELOW) and a rounded or, in fast swimmers, pointed torpedo-shaped body. There is usually a reduced internal quilt-like shell or none. A prominent head region is marked by mouth, eyes, arms and cartilage-protected brain.

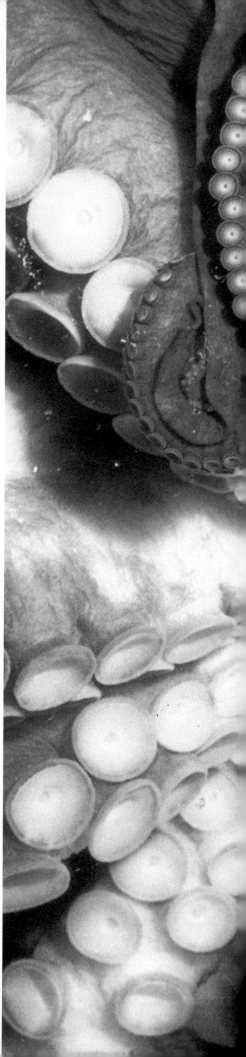

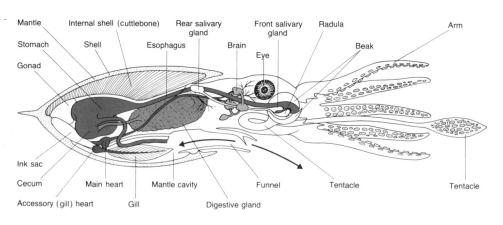

Mantle · Internal shell (cuttlebone) · Rear salivary gland · Front salivary gland · Radula · Arm · Stomach · Shell · Esophagus · Brain · Beak · Gonad · Eye · Ink sac · Cecum · Main heart · Mantle cavity · Funnel · Tentacle · Tentacle · Accessory (gill) heart · Gill · Digestive gland

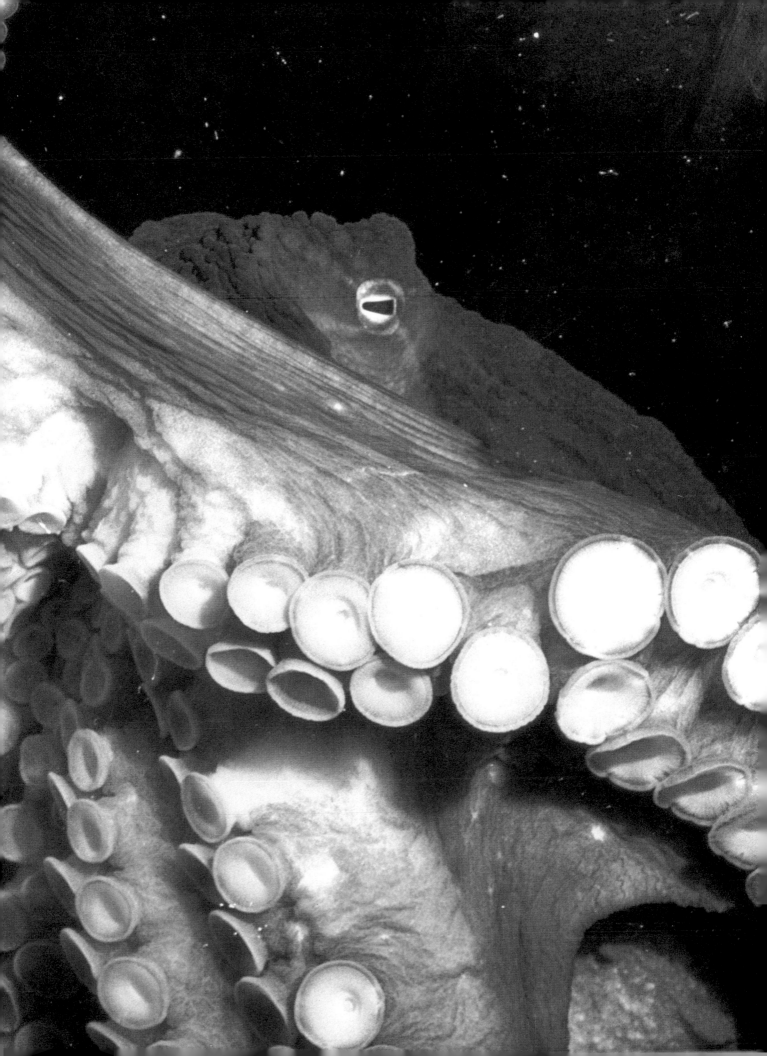

Ecology of mollusks

Mollusks have colonized the sea, fresh water and land. Tropical regions tend to have a more diverse fauna than temperate belts, although temperate New Zealand has one of the richest molluskan land faunas.

Marine habitats can include rocky, coral, sandy, muddy, boulder and shingle shores and also the transitions between freshwater, sea and land found in salt marshes, brackish lagoons, mangrove swamps and estuaries. Beyond the molluskan fauna of the shores is that of the ocean, with communities below the lowtide mark in comparatively shallow water of continental shelves. There is a mosaic of different types of communities on the sea floor, mostly relating to differences in bottom materials. Certain mollusks, like squids, can form part of the free-swimming animal population (nekton) in open water. The veliger larvae of most marine mollusks float passively in the upper waters of the sea, part of the plankton. A few prosobranchs (violet sea snails and heteropods) and opisthobranchs (pteropods or sea butterflies) spend their entire adult lives on or just below the surface as part of the pelagic community. Deep waters of the abyss were once thought to be devoid of life but investigations have revealed a limited but characteristic fauna. With some exceptions (eg squids) abyssal mollusks are small.

Freshwater habitats colonized by mollusks include running waters of streams and rivers, still waters of lakes, ponds, canals, and temporary waters of swamps, all with their own range of species. There are both bivalves and gastropods living in freshwater, the latter including both prosobranchs and aquatic pulmonates. Foreign species can spread dramatically like the freshwater Fingernail clam *Corbicula manilensis*, introduced from Asia to the USA, where it now clogs canals, pipes and pumps. Only the gastropods have successfully colonized land; they include both prosobranchs and pulmonates, although the latter, as both slugs and snails, are the most common in temperate climates.

In the food chains of the sea, mollusks are eaten by other mollusks, as well as by starfish and by bottom-living fish such as rays. Some starfish are notorious predators of commercial mollusks such as oysters and mussels. Some whales eat large quantities of squid. On the shore seabirds probe in mud for mollusks which can form a substantial part of their diet. Such predation by animals which are part of the natural ecosystem can usually be tolerated by mollusks, as they can be prolific breeders.

A few mollusk species have adopted the parasitic way of life. They are nearly all gastropods, with a few bivalves. Among the prosobranchs are parasites on the exterior of the host (ectoparasites), such as needle whelks (eulimids), which are parasites of echinoderms but look like normal gastropods. Internal parasites (endoparasites) are less active and have reduced body organs and less shell. The cap-like genus *Thyca* lives attached to the underside of starfish, in the radial groove under the arms. *Stilifer*, which penetrates skin of echinoderms, has a shell but it is enclosed in fleshy flaps of proboscis (pseudopallium) outside the skin of the host. The further inside the host a parasite is, the more the typical molluskan structure is lost.

Empty gastropod shells are regularly used by hermit crabs (see p238). Commensalism, in which one partner feeds on the food scraps of the other, is shown by the small bivalves *Devonia perrieri*, *Mysella bidentata* and *Montacuta* species which live with sea cucumbers, brittle stars and heart urchins respectively. Symbiosis is demonstrated by the presence of algae (zooxanthellae) in the tissues of sea slugs such as *Elysia* and *Tridachia* species and also in the mantle edge of the Giant clam.

Mollusks are also hosts to their own parasites, many of which may have become established via commensalism. Most parasites of mollusks are larvae of two-suckered flukes. Two commercially and medically important parasites are the liver fluke of sheep and cattle (*Fasciola hepatica*) and blood

▲ **Various oyster species are cultivated** for food in Australia (ABOVE), New Zealand, Japan, the USA and Europe. Japanese pearl oyster culture is also particularly well known. Many natural oysterbeds are all but fished out.

▶ **Mollusk eats echinoderm** in this attack by a Mediterranean Triton shell *Charonia nodifera*. In the food chains of the sea, starfishes are in turn noted predators of bivalves such as mussels and scallops.

▼ **Bird-like flapping** of lateral fins is the characteristic swimming stroke of the Atlantic and Mediterranean cuttlefishes (*Sepiola* species). Much smaller than the torpedo-shaped *Loligo* squids, they may escape predators either by burrowing into the substrate, using the water jet from the mantle funnel, or else by emitting a distracting puff of ink to cover a retreat. When fleeing, the squid abruptly becomes nearly transparent.

flukes or bilharzia of humans (see p200). Mollusks can also be parasitized by arthropods. Familiar examples are the small white mites *Riccardoella limacum* found crawling on the skin of slugs and snails and the small pea crab, *Pinnotheres pisum*, living in the mantle cavity of the Edible mussel.

Mollusks have long been used in human culture for food, fishing bait and hooks, currency, dyes, pearl, lime, tools, jewelry and ornament. Mother-of-pearl buttons were once manufactured from the shells of freshwater mussels, particularly in the USA, where these mussels were originally common in the rivers. Most pearls come from marine pearl oysters, but at one time fine pearls were obtained from the Freshwater

pearl mussel which occurred notably in upland rivers of Wales, Scotland and Ireland. Today pearls are cultured commercially by inserting a "seed pearl" inside the mantle skirt of the oyster.

Mediterranean cooks are famous for their seafood dishes which utilize gastropods such as necklace shells, top shells, ormers, murexes and occasionally limpets; bivalves such as mussels, scallops, date mussels, venus shells, carpet shells, wedge shells, and razor shells; and also cephalopods including cuttlefishes, squids and octopuses. The traditional "escargot" of French cuisine is the pulmonate Roman snail. On the east coast of the USA the hard-shell clams used by cooks are quahogs (introduced from transatlantic liners to the Solent in southern England), while soft-shell clams are from a range of genera including gapers, carpet shells and trough shells.

Both slugs and snails can be pests of agriculture and horticulture. The mollusks are controlled by biological, cultural and chemical methods, the latter being the ones most usually employed.

A few marine mollusks also compete with man's activities. Bivalves like the shipworm bore into marine timbers, and gastropods, including the slipper limpet and oyster drill, are pests of oyster beds.

The major threats to mollusks are destruction of habitats and pollution—the latter being more important for aquatic species, which are subject to crude oil spills, heavy metals, detergents, fertilizer in water run-off from the land, and acid rain from distant industry. Native land snails of deciduous woodland in the American Midwest, for example, are often not able to cope with the more rigorous conditions of cleared land. In consequence, much North American farmland has been colonized by European mollusks, especially slugs, introduced with plants.

In addition to the harvesting of natural populations and measures limiting trading in shells and marine curios, others aim to prevent introductions of non-native species. The International Union for the Conservation of Nature is also producing Red Data books, assembling information on endangered species that can be used in their conservation.

Little is known of the molluskan fauna of some of the potentially richest and most threatened habitats, many of which are delicate and intolerant of disturbance. Hundreds of snail species are likely to be exterminated before they are even described and studied. JEC

SPINY-SKINNED INVERTEBRATES

▶ **Closing the gap.** These starfishes (*Archaster typicus*) are not actually mating, as copulation is unknown in echinoderms. By coming close together, the chance of the male's sperm fertilizing the female's eggs in the sea is increased. This behavior is not seen in other echinoderms.

▶ **Brittle and spiny** BELOW, the arms of the brittle star are made up of many ossicles which fit together rather like the vertebrae of the chordate spine. This enables them to be flexible but also makes them liable to fracture at the joints.

▼ **Arms spread, a feather star** (*Tropiometra* species) extends its arms to strain suspended food material from the sea.

THE echinoderms are distinct from all other animal types and easily recognizable. The name echinoderm means spiny-skinned, for most members of the group have defensive spines on the outside of their bodies. They are found only in the sea, never having evolved to cope with the problem of salt balance that life in freshwater would impose on them. As adults they virtually all dwell on the seabed, either, like sea lilies, being attached to it, or, like the starfishes, brittle stars, sea urchins and sea cucumbers, creeping slowly over it. These five groups or classes represent the types of echinoderms found living in the seas and oceans today.

For animals relatively high on the evolutionary scale, it is remarkable that a head has never been developed. Echinoderms show a peculiar body symmetry known as pentamerism. This is effectively a form of radial symmetry with the body arranged around the axis of the mouth. Superimposed on this radial pattern is a five-sided arrangement of the body which is well shown in the starfish. The result is that the echinoderm body generally has five points of symmetry arranged around the axis of the mouth. These points are very often associated with the locomotory organs or tube-feet (see p275).

While five-pointed symmetry or pentamerism is largely displayed by most present-day adult echinoderms, it is interesting to note that their larvae are bilaterally symmetrical (ie symmetrical on either side of a line along the length of the animal), and that their primitive ancestors, which appeared in the pre-Cambrian seas, were also bi-laterally symmetrical. The causes of pentamerism are unclear, but some authorities have suggested that it leads to a stronger skeletal framework.

The body of echinoderms shows a deuterostome coelomate level of organization (see p152). This means that they are relatively highly evolved invertebrates with a body constructed originally from three layers of cells.

The echinoderm skeleton is made of many crystals of calcite (calcium carbonate). It is unusual because these crystals are perforated by many spaces in life (reticulate) so that the tissue which forms the crystals actually invades them. Such a reticulate arrangement leads to a lightening of the crystal structure and hence a reduction in weight of the animal without any loss of strength (see right). One side effect of this crystal structure is that it is easily invaded by mineral after the death of the animal and thus it fossilizes beautifully.

The skeleton supports the body wall or test. This reinforced structure may be soft (as in sea cucumbers) or hard (sea urchins) but it should never be thought of as a shell because it is covered by living tissue.

The exterior of each class of echinoderm appears different, and so too is the way in which the skeleton has been deployed. In the sea cucumbers the calcite crystals are embedded in the body wall and linked by flexible connective tissue in a way that does not occur in the other classes. In the starfish there is sometimes a flexible body wall, but more often the crystals are grouped close together, sometimes being "stitched" together by fibers of connective tissue running through the crystal perforations. Individual crystals may be extensively developed to form spines or marginal plates. The sea

urchins have carried the skeletal process further, for in almost all the skeleton is rigid, being composed of many interlocking crystals. At the same time there is a reduction in the soft tissue of the body wall. The sea urchins have some of the most complex arrangements of muscle and skeleton in the phylum, for example the chewing teeth or "lantern teeth," as Aristotle called them, and the pedicellariae.

In the sea lilies and feather stars and the brittle stars the skeleton is massive and arranged as a series of plates, ossicles and spines with a minimum of soft tissue. In both these classes the major internal organs or viscera are contained in a reduced area, the cup-like body (theca) of the crinoids and the disk-like central body of the brittle stars. Here the skeleton reinforces the body wall, which remains flexible; but in the arms the ossicles become massive, operating with muscles and connective tissue in a way rather reminiscent of the vertebrae of the human backbone. In the arms of both types there is relatively little soft tissue. In the sea lilies the arms branch near their bases into two or more main axes, each bearing lateral branches called pinnacles. The arms of brittle stars branch only in the basket stars.

The drifting echinoderm larvae also have a skeleton, which serves to support their delicate swimming processes.

Another unique feature of echinoderms is their water vascular system. This probably arose in the primitive echinoderms as a respiratory system pointing away from the substrate which could be withdrawn inside the heavily armored test. As the echinoderms became more advanced it was arranged around the mouth, but still held away from the substrate. Branching processes developed, forming a system of tentacles that became useful for suspension-feeding as well as respiration. It is in this state that the water vascular system is seen in present-day sea lilies and feather stars. Their branched tentacles, also called tube-feet (although in the crinoids they have no locomotory role), are arranged in a double row along the upper side of each arm, bounding a food groove, and along the branches of the arms (pinnules). The tube-feet can be extended by hydraulic pressure from within the animal, and much of the water vascular system is internal. They are supplied with fluid from a radial water canal which runs down the center of each arm, just below the food groove, and which sends a branch into each pinnule. The radial water canal of each arm connects with that of its fellows via a circular canal running around the gullet of the animal. Pressure is generated inside the system by the contraction of some of the tube-feet, and also by special muscles in the canal itself which generate local pressure increases to distend the neighboring tube-feet. The water vascular system in crinoids is associated with several other tubular networks, notably the hemal and peri-hemal systems (whose role is less easy to define) and the radial water canal also runs close to the radial nerve cord which controls the tube-feet.

The activities of the tube-feet relate to gas exchange (respiration) and food gathering. The tube-feet are equipped with mucous glands in crinoids and when a small fragment of drifting food collides with one, the fragment sticks to the tube-feet, is bound in mucus and flicked into the food groove by which it passes down to the central mouth. The tube-feet are arranged in double rows alternating with small non-distendable lappets. This arrangement assures their efficient use in feeding.

Crinoids exploit currents of water in the sea. They do not pump water to get their food, but gather it passively. They "fish" for food particles using the tube-feet and select mainly those in the 0.3–0.5mm (0.01–0.02in) size-range.

In all the remaining groups of echinoderms the orientation of the body is reversed with respect to the substrate. The tube-feet actually make contact with the ground over which the animals are moving and thus take up an additional role in locomotion. This happens in the starfishes, sea urchins and sea cucumbers but not in the basket and brittle stars, where movement is achieved by bending the arms, while the tube-feet are still important in respiration and food gathering. In the basket stars they are well developed for suspension feeding in a way which has interesting parallels with the crinoids. The basket stars, too, exploit currents of water, and arrange their complex branching arms with tube-feet as a parabolic net sieving the water currents for particles in the 10–30μm size range. Thus they do not exactly compete with the crinoids in the same habitat. They are able to withstand stronger currents than the crinoids. In the remaining types of brittle stars there is a range of feeding habits. Some, like *Ophiothrix fragilis*, are suspension feeders, often living in huge beds. Others are detritus or carrion feeders. In many species the tube-feet, which are suckerless, are very important in transferring food to the mouth and have a sticky mucous coating which

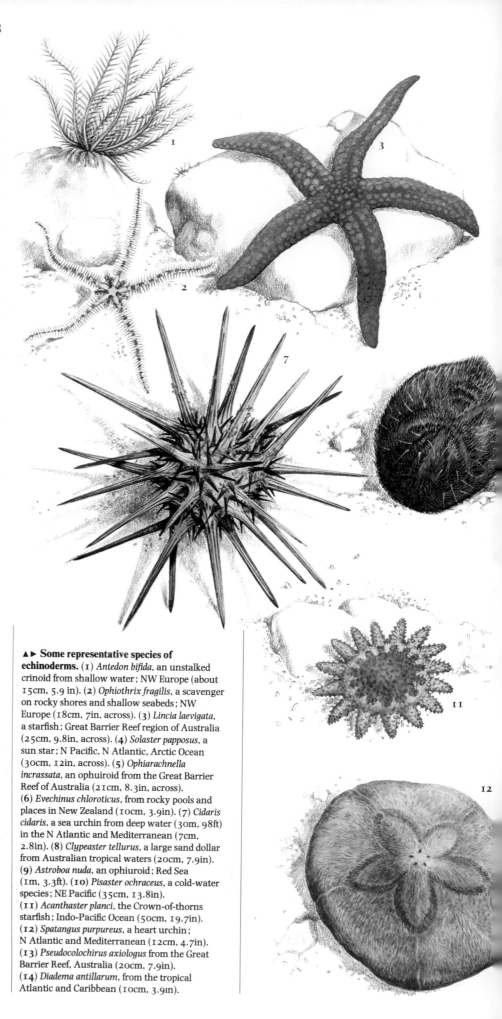

▲▶ **Some representative species of echinoderms.** (**1**) *Antedon bifida*, an unstalked crinoid from shallow water; NW Europe (about 15cm, 5.9 in). (**2**) *Ophiothrix fragilis*, a scavenger on rocky shores and shallow seabeds; NW Europe (18cm, 7in, across). (**3**) *Lincia laevigata*, a starfish; Great Barrier Reef region of Australia (25cm, 9.8in, across). (**4**) *Solaster papposus*, a sun star; N Pacific, N Atlantic, Arctic Ocean (30cm, 12in, across). (**5**) *Ophiarachnella incrassata*, an ophiuroid from the Great Barrier Reef of Australia (21cm, 8.3in, across). (**6**) *Evechinus chloroticus*, from rocky pools and places in New Zealand (10cm, 3.9in). (**7**) *Cidaris cidaris*, a sea urchin from deep water (30m, 98ft) in the N Atlantic and Mediterranean (7cm, 2.8in). (**8**) *Clypeaster tellurus*, a large sand dollar from Australian tropical waters (20cm, 7.9in). (**9**) *Astroboa nuda*, an ophiuroid; Red Sea (1m, 3.3ft). (**10**) *Pisaster ochraceus*, a cold-water species; NE Pacific (35cm, 13.8in). (**11**) *Acanthaster planci*, the Crown-of-thorns starfish; Indo-Pacific Ocean (50cm, 19.7in). (**12**) *Spatangus purpureus*, a heart urchin; N Atlantic and Mediterranean (12cm, 4.7in). (**13**) *Pseudocolochirus axiologus* from the Great Barrier Reef, Australia (20cm, 7.9in). (**14**) *Diadema antillarum*, from the tropical Atlantic and Caribbean (10cm, 3.9in).

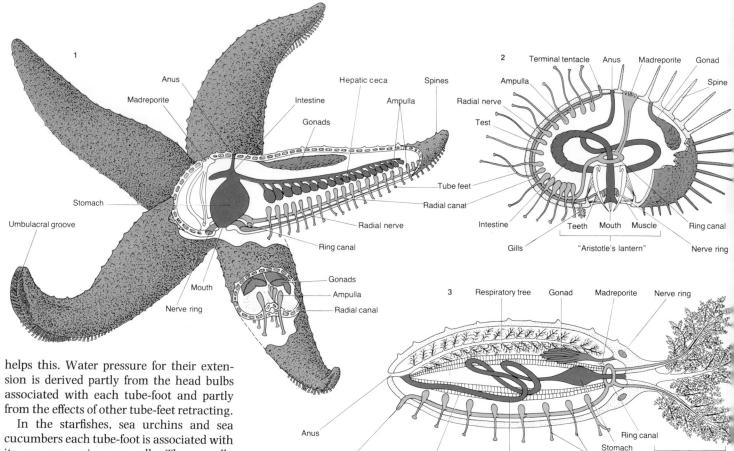

1

Anus
Madreporite
Intestine
Gonads
Hepatic ceca
Ampulla
Spines
Stomach
Umbulacral groove
Tube feet
Radial canal
Radial nerve
Ring canal
Mouth
Nerve ring
Gonads
Ampulla
Radial canal

2

Terminal tentacle
Anus
Madreporite
Gonad
Ampulla
Spine
Radial nerve
Test
Tube feet
Radial canal
Intestine
Teeth   Mouth   Muscle
Ring canal
Gills
"Aristotle's lantern"
Nerve ring

3

Respiratory tree
Gonad
Madreporite
Nerve ring
Anus
Terminal tentacle
Muscle bands
Intestine
Tube feet
Ring canal
Stomach
Tentacles

helps this. Water pressure for their extension is derived partly from the head bulbs
associated with each tube-foot and partly
from the effects of other tube-feet retracting.

In the starfishes, sea urchins and sea
cucumbers each tube-foot is associated with
its own reservoir or ampulla. The ampulla
is thought to play a role in filling the tube-
foot with water vascular fluid. It has its own
muscle system and connects to the foot by
valves to control the flow. However it seems
certain that fluid pressure within the water
vascular system is also important. The
shafts of the tube-feet are equipped with
muscles for retraction and for stepping
movements.

Suckered tube-feet occur in all sea urchins and many sea cucumbers. Some
asteroids, eg *Luidia* and *Astropecten* species,
lack suckers on the tube-feet, and most of
these burrow in sand. Other starfishes,
inhabiting hard substrates, have suckered
tube-feet and use them for locomotion and
for seizing prey. In the burrowing sea urchins, eg *Echinocardium* species, some of the
tube-feet are highly specialized for tunnel-
building and for ventilating the burrow. In
the sea cucumbers the ambulacral tube-feet
may be used for locomotion and respiration,
while those surrounding the mouth have
become well developed for suspension or
deposit feeding and form the characteristic
oral tentacles. There are many closely
related sea cucumbers which feed in slightly
different ways, each having slightly modified oral tentacles so they can exploit food
deposits of detritus particles of different sizes.

The fluid within the water vascular
system is essentially seawater with added
cellular and organic material. Water

vascular fluid is responsible for other tasks
apart from driving the tube-feet. It transports food and waste material and conveys
oxygen and carbon dioxide to and from the
tissues of the body. It contains many cells.
These are mainly amoeboid coelomocytes
which have a role to play in excretion,
wound healing, repair and regeneration. No
excretory organs have been identified in the
echinoderms.

The water vascular system of starfishes,
basket and brittle stars, and sea urchins
appears to communicate to the exterior of

▲ **Body plans of starfish, sea urchin and sea
cucumber.**

▼ **Sucker power:** close-up of the tube-feet and
spines of the Crown-of-thorns starfish.

▷ **Social stars.** OVERLEAF Brittle stars often gather on the seabed. Here *Ophiothrix quinquemaculata* is grouped with other invertebrates on the floor of the Mediterranean.

▶ **Smothering to death:** the spiny starfish *Marthasterias glacialis* feeds on a sea squirt.

▼ **Test appendages,** a selection of the microscopic tong- or forceps-like organs found between the spines of all sea urchins and most starfishes.

Grooming Tools of Echinoderms

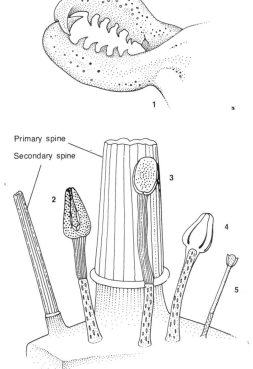

Between their spines, most starfishes and all sea urchins carry unique small grooming organs like microscopic tongs or forceps. These intriguing organs were once thought to be parasites on the tests of the animals which bear them but are an integral part of the animal. Each consists of two or more jaws supported by skeletal ossicles called valves. Some starfish types are directly attached to the test. Others, like those of sea urchins, are carried on stalks, so that the jaws can reach down to the surface of the test.

The jaws are caused to open or close by muscles attached to nerves from both the base of the epithelium of the test and from special receptors responsive to touch and certain chemicals. Most of these receptors are situated on the inside of the jaw blades, but some lie on the outside.

In an undisturbed animal the epithelial nerves may close down most of the pedicellariae (except some on the sea urchins, see below), so that they are inactive with the jaws closed. If an intruding organism strays onto the test, such as a small crustacean or a barnacle larva seeking a place to settle, the resultant tactile stimuli cause the pedicellariae to gape open and thus expose the special touch receptors. If these are stimulated, the jaws rapidly snap shut, trapping the intruder. The pedicellariae of starfish (**1**, skeletal parts only) and the tridentate (in fact, three-jawed)

Primary spine

Secondary spine

(**2**) and ophiocephalous (snake-head) ones (**3**) of sea urchins are all specialized for such activities and often have fearsome teeth to grip their victims.

In the globiferous (round-headed) pedicellariae (**4**) of the sea urchin class Echinoidea, there are venom sacs. Here the jaws close only on objects which carry certain chemicals. The venom is injected into the victim via a hollow tooth in many echinoids and in species such as *Toxopneustes pileosus* it has a powerful effect. Globiferous pedicellariae detach after the venom is injected and remain embedded in the tissue of the intruder. They seem to be mainly deployed by the sea-urchins as defenses against larger predators such as starfish.

In the sea urchins, sand dollars and heart urchins the smallest class of pedicellariae are known as trifoliate (three-leaved) (**5**). They differ from all the others in having spontaneous jaw movements, "mouthing" over the surface of the test in grooming and cleaning activities. This persists even when these pedicellariae are removed from the test.

Far from being the parasites they were once believed to be, pedicellariae therefore serve to keep the surface of the echinoderm free from other animal, or plant, organisms. Their complex structure is a good example of the intricacies of echinoderm biology, but their various roles are not yet all understood.

the animal via a special sieve-like plate, the madreporite. In sea cucumbers the madreporite is internal, while in the crinoids it is lacking altogether. It used to be thought that seawater entered and left the water vascular system via the madreporite, but more recent research in sea urchins shows that in fact very little water actually moves across this special structure. In starfish and sea urchins the madreporite may be associated with orientation during locomotion (see below).

The nervous system of echinoderms is peculiar to the group. Because of the absence of a head there is no brain and no aggregation of nerve organs in one part of the body. In fact, with the exception of the rudimentary eyes (optic cushions) of starfishes, the balance organs (statocysts) of some sea cucumbers and the chemosensory receptors of pedicellariae of sea urchins (see box p279), there are no complex sense organs in echinoderms. Instead there are simple receptor cells responding to touch and chemicals in solution. These appear to be widely spread over the surface of the animals. Some authorities even suggest that all the external epithelial cells of starfish and sea urchins may have a sensory function.

In all living echinoderms the main part of the nervous system comprises the nerve cords which run along the axis of each arm close to the radial water canal. These radial nerve cords are linked together around the esophagus by a circumesophageal nerve cord so that the activities of one arm or ray may be integrated with the activities of the others.

The control of the tube-feet and body-wall muscles is under the command of each radial nerve cord. The responsibilities for coordinated locomotion and direction of movement lie here too. In directional terms echinoderms may move with one arm or ray taking the lead, or even with the space between two acting as a leading edge. Where there is a need to back away, the animal may either go into reverse or actually turn around. In the starfish and sea urchins there is some evidence to suggest that the space between two rays which contains the madreporite may frequently act as the leading edge, possibly because the madreporite has some, as yet unknown sensory function.

Echinoderms are all very sensitive to gravity and generally show a well-defined righting response if they are turned upside down. It has been suggested that in the sea urchins, small club-like organs, sphaeridia, act as organs of balance. All other echinoderms, apart from a few sea cucumbers, lack the balance organs (statocysts) that are frequently found in mollusks and crustaceans.

In starfishes and sea urchins the outer surface of the body is covered by a well-developed epithelium at the base of which lies a network of nerves. This nerve plexus controls the external appendages of these two groups which are richly developed in many species. The appendages include various effector organs, movable spines, pedicellariae (minute tong-like grasping organs), sphaeridia in sea urchins and paxillae and papulae in starfishes. These organs are concerned with defending the animal against intruders and keeping the delicate skin of the test clean from deposits of silt and detritrus.

The basi-epithelial nerve plexus connects with the radial nerve cords of each arm or ray and forms a system of fine nerves linking the receptor site of the epithelium with the various effector organs.

The various groups of existing echinoderms show characteristic patterns of behavior. All echinoderms are sensitive to touch, and to waterborne chemicals which signal the presence of desirable prey or of potential predators which must be avoided. Most starfish species are efficient predators, feeding on other invertebrates, such as worms, mollusks and other echinoderms. They are able to "smell" the presence of suitable prey in the water and move efficiently towards it.

For the common European starfish, *Asterias rubens*, mussels and oysters are significant prey items. The sun stars *Solaster endeca* and *Crossaster papposus* also feed on bivalves, but may attack *Asterias* too. In tropical waters starfishes show a variety of tastes, but the Crown-of-thorns, *Acanthaster planci* is well known for its selection of certain species of reef-building coral as prey. All these echinoderm predators will move efficiently toward the source of chemical scent in the water. When they have arrived at it they will commence attack.

Some of the burrowing starfish (eg *Astropecten* species) will ingest their prey of gastropods whole. *Crossaster* species may attack *Asterias* by hanging on to one ray with the mouth and eating it while the prey drags the predator about. *Acanthaster* feeds on objects too large to be ingested whole, so it everts the stomach membranes through its mouth and smothers the prey with these, digesting the victim outside its body. When the process of digestion is complete the stomach membranes are withdrawn.

Members of the genus *Asterias*, like other starfishes which prey on bivalves, are able to use their tube-feet with their suckers to prize open the valves of a mussel or oyster. They do this by climbing on to the prey and attaching some tube-feet to each valve. The two valves are then pulled apart by the persistent actions of the tube-feet. The muscles which keep the shells closed eventually tire, so that they gape ever so slightly. A gap of one or two millimetres is all that is needed for the starfish to insert some of its stomach folds passed out through its mouth. Once this has been done, digestion of the victim will begin and in the end only the cleaned empty shell will remain.

It is interesting to note that in both tropical starfish (eg *Acanthaster* species) and temperate species (eg *Asterias*), solitary individuals display a different type of feeding behavior from that of individuals feeding together in groups. In these starfishes, regular feeding tends to be solitary and at night, the individuals being well spaced one from the next. In some populations, individuals will gather periodically in large numbers at a superabundant food source and feed by day as well as by night; as a result of such social feeding, the growth rate of individuals considerably surpasses the

The 3 Subphyla and 5 Classes of Living Echinoderms

Sea lilies, feather stars
Subphylum: Crinozoa

About 650 species; sedentary, mostly stalked, at least in young even if adults free-living; branching main nervous system; anal opening on same surface as mouth; 6 classes and most species known from fossils. One surviving class, Crinoidea (order Articulata), lacking madreporite, spines and pedicellariae appendages. Includes genus *Antedon*.

Starfishes
Subphylum: Asterozoa
Stemless, mobile, free-living; mouth surface faces down, nervous system on mouth surface; usually have arms (rays).

Class: Asteroidea —**starfishes**

About 2,000 species; flattened, star-shaped with 5 arms (a few with more); endoskeleton flexible; arms contain digestive ceca; branching madreporite.
Subclasses: Somasteroidea (including genus *Platasterias*); and Euasteroidea, 5 orders including Platyasterida (eg genus *Luidia*).

Class: Ophiuroidea —**brittle stars, basket stars**

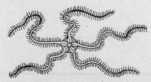

About 1,600 species; flattened, 5 sided with long flexible arms rarely branched, clearly demarked from central "control" disk; madrepore on underside; no anus or intestine; tube-feet lack suckers.
Orders: Oegophiurida (including genus *Ophiocanops*); Phrynophiurida **basket stars** (including genus *Gorgonocephalus*); and Ophiurida **brittle stars** (including genus *Ophiura*).

Sea urchins
Subphylum: Echinozoa
Stemless, mobile, free-living; mouth surface downward or to side; main nervous system on oral surface; without arms or rays.
Six fossil classes, of which 2 are living.

Class: Echinoidea —**sea urchins, sand dollars, heart urchins**

About 750 species; mainly globular or disk-shaped, without arms; covered with numerous spines and pedicellariae; tube-feet usually ending in suckers; endoskeleton comprises close-fitting plates.
Subclass Perischoechinoidea: 1 living order Cidaroida, including genus *Cidaris*.
Subclass Euechinoida: comprises superorders Diadematacea (eg *Asthenosoma, Diadema*); Echinacea (eg *Echinus, Psammechinus, Paracentrotus, Toxopneustes*); Gnathostomata with orders Holectypoida (eg *Echinoneus, Micropetalon*) and Clypeasteroida (eg *Rotula, Clypeaster, Echinocyamus*); and Atelostomata with suborders Holasteroidea (eg *Pourtalesia, Echinosigria*), and Spatangoida (eg *Spatangus, Echinocardium, Brissopis*).

Class: Holothuroidea —**sea cucumbers**

About 500 species; long, sac-like, without arms; bilaterally symmetrical; mouth surrounded by tentacles; no spines or pedicellariae; endoskeleton reduced to microscopic spicules or plates, or absent.
Subclass Dendrochirotacea: orders Dendrochirotida (eg *Cucumaria, Thyone, Psolus*); and Dactylochirotida (eg *Rhopalodina*).
Subclass Aspidochirotacea: orders Aspidochirotida (eg *Holothuria*); and Elasipodia (*Pelagothuria, Psychropotes*).
Subclass Apodacea: orders Molpadiida (eg *Molpadia, Caudina*); and Apodida (eg *Synapta, Leptosynapta*).

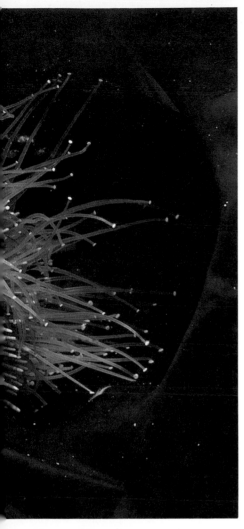

▲ **Sea urchin under pressure.** *Echinus esculentus* was taken for its edible roes and regarded as a delicacy in Tudor England. Now it is fished for its ornamental test.

▼ **Cleaning up.** The sea cucumber *Holothuria forskali* creeps across the seabed using modified oral tube-feet to collect the sediment and the food it contains.

norm. To what extent these differences are acquired or inherited is not yet clear.

In the sea urchins there is a range of feeding behavior. Many of the round (or regular) echinoids (eg the genera *Strongylocentrotus*, *Arbacia* and *Echinus*) are omnivores. They browse on algae and encrusting animals such as hydrozoans and barnacles, using their Aristotle's lantern teeth. In many places echinoids are important at limiting the growth of marine plants and compete very successfully with other types of algal grazers, including gastropod mollusks and fish.

The irregular sea urchins, including the sand dollars and heart urchins, are more specialized feeders. The sand dollars live partly buried in the sand and use their modified spines and tube-feet to collect particles of detritus for food. These are then passed to the mouth along ciliated tracts. The heart urchins burrow quite deeply in sand and gravels and ingest the substrate entire. They have lost the Aristotle's lantern. As the substrate particles pass along their guts any organic material is digested and "clean" substrate is passed out from the anus.

The sea cucumbers have diversified to exploit a number of food sources. In virtually every case the form of their oral tentacles, the specialized tube-feet arranged round the mouth, is adapted to gathering food.

Some groups sweep the surface of sand and mud for particles of detritus and thus live as deposit feeders. Others are suspension feeders relying on currents of seawater to sweep suspended particles of food into their oral tentacles. In both cases the size of the sweeping or filtering fronds and the gap between will dictate the sizes of the particles collected.

The fascinating and often beautiful echinoderms inhabit all the world's seas from the intertidal zone down to the ocean abyss. They are also present in all latitudes from the tropics to the poles. In temperate intertidal zones such as the North Atlantic and North Pacific starfishes and sea urchins are familiar organisms. In some places sea urchins may be harvested for use as food, eg *Paracentrotus lividus* in Ireland and the Mediterranean. Brittle stars, and sea cucumbers, though present, are less conspicuous intertidally. The shallow seas overlying the continental shelves are particularly good habitats for echinoderms where coastal currents, rich in nutrients, sweep detritus and plankton for suspension feeders such as the crinoids and sea cucumbers and nourish prey suitable for starfishes. In tropical areas the development of reefs allows a great diversity of echinoderm species to develop because of the variety of niches. Although all groups of echinoderms inhabit the ocean abyss it is here that the sea cucumbers flourish, often in great densities, moving over the benthic ooze in search of detrital food. Here too some highly unusual epibenthic sea cucumbers have taken to a swimming life, moving along in deep currents and collecting food as they go.

The majority of echinoderm species are dioecious—the sexes are separate. A few are hermaphrodite, passing through a male phase before becoming functional females. In one genus, *Archaster* from the West Pacific, pseudo-copulatory activity occurs, with the one partner climbing on top of the other, but even here fertilization is external. Sperm and eggs are released into the seawater via short gonoducts. In many species this is almost a casual affair, the partners not coming close together. However, synchrony of spawning is essential and this is usually governed by water temperature, and chemical stimuli operating between participants.

Antarctic and abyssal echinoderms often brood their eggs, and a direct development of juveniles occurs in brood pouches or between spines. In the remaining vast majority of species the fertilized eggs drift in the plankton and develop through characteristic larval stages, usually feeding on minute planktonic plants such as diatoms and dinoflagellates. After a period of larval life that ranges from a few days up to several weeks in different species, metamorphosis occurs, and the juvenile echinoderms settle on the sea bed.   AC

SEA SQUIRTS AND LANCELETS

Phylum: Chordata

Sea squirts
Subphylum: Urochordata (or Tunicata)
About 2,000 species in 3 classes and 7 families.
Distribution: worldwide; bottom-dwelling and pelagic at all depths and in all oceans.
Fossil record: very few, from the Cambrian to the Quaternary (600–500 million years ago to recent).
Size: individuals from less than 1cm (0.4in) to about 20cm (8in) long.

Features: chordate characteristics—hollow, dorsal nerve cord, enterocoelic body cavity, gill slits, tail behind anus, and notocord—are all present in the larvae; adults show no segmentation, lack hollow dorsal nerve cord and notochord, and most have the gills surrounded by a large cavity (atrium) and are ensheathed in a test or tunic of tunicin, a substance related to cellulose; hermaphrodite; reproduction often involves budding and an asexual phase; life cycles complex in Thaliacea and Larvacea.

Sea squirts
Class: Ascidiacea
Includes genera *Aplidium, Botryllus, Clavelina, Gonia*.

Salps or pelagic turnicates
Class: Thaliacea
Includes genera *Pyrosoma, Salpa*.

Class: Larvacea
Sexually mature adults resemble larvae.
Includes genus *Oikopleura*

Lancelets
Subphylum: Cephalocordata (or Acrania)
Fewer than 20 species in 2 families.
Distribution: temperate to tropical shallow sea waters, adults bottom-dwellers, burrowing in sand.
Fossil record: none.
Size: up to 5cm (2in) long.

Features: simple fish-like chordates lacking a recognizable head, with hollow dorsal nerve cord similar to vertebrates, but no vertebrae, instead a notochord, muscle blocks segmented; enterocoelic coelom and well-developed gills; can swim; sexes separate; planktonic tornaria larva produced.

Family: Branchiostomidae, genus *Branchiostoma* (or *Amphioxus*).

Family: Asymmetronidae, genus *Asymmetron*.

▶ **Lowly chordates.** Sea squirts can be single individuals like *Ciona intestinalis* with its translucent body and yellow-fringed siphonal openings, or colonial animals like *Botryllus schlosseri* where a few individuals are grouped in star-like patterns around a common exhalant opening.

CHORDATES are "higher animals" possessing a single, hollow, dorsal nerve cord and a body cavity that is a true coelom. The most familiar chordates are the vertebrates, fishes, amphibia, reptiles and mammals, where a definite backbone and bony braincase are to be found (subphylum Craniata or Vertebrata). However, there are a number of lowly chordates which display the phylum's characteristics at a simple level. All aquatic, they are included in the two other subphyla of the chordates, the Urochordata (sea squirts) and Cephalochordata (lancelets).

The adult **sea squirts** look nothing like vertebrates. They are bottom-dwellers growing attached to rocks or other organisms. Their bodies are encased in a thick tunic made of material which resembles cellulose, the main constituent of plant cell walls. At the top of the body lies the inhalant siphon, and on the side is the exhalant siphon. There is no head. Water is pumped in via the inhalant siphon and passes through the pharynx and out between the gills into the sleeve-like atrium. From the atrium it is discharged via the exhalant siphon. The gills serve as a respiratory surface and also act as a filter, extracting suspended particles of food. They are ciliated, the tiny hair-like cilia providing the pumping force to maintain the respiratory and filter current. Acceptable food particles are collected in sticky mucus secreted by the endostyle, a sort of glandular gutter running down one side of the pharynx, and they are then passed into the gut for digestion. Waste products are liberated from the anus which opens inside the atrium near the exhalant siphon.

Sea squirts may be solitary or colonial. In a colony the exhalant siphons of individuals open into a common cloaca, and each individual retains its own inhalant siphon. Colonial squirts may be arranged in masses, as in *Aplidium* species, where the individuals are hard to recognize, or in encrusting plate-like growth, eg *Botryllus* species, where the individuals can easily be made out.

The heart lies in a loop of the gut and services the very simple blood system. Sea squirts are hermaphrodite, each having male and female organs. In some species the eggs are retained in the atrium, where the sex ducts open, and where they are fertilized by sperm drawn in with the feeding and respiratory currents. The embryos can develop here in a protected environment. In other species the eggs and sperm are both liberated into the sea where fertilization occurs. The embryo quickly develops into a

▶ **Like a row of glass bottles,** these colonial salps drift through the deep waters of the ocean.

▼ **Colonial sea squirts** on the Great Barrier Reef. In some species the green coloring is due to commensal algae growing in their tissues.

tadpole-like larva. This process may take from a few hours to a few days, a time-scale which makes these animals ideal for observation and experimental embryology.

The "tadpoles" of sea squirts are small, independent animals like miniature frog's tadpoles. They are sensitive to light and gravity, enabling the animal to select an appropriate substrate for settlement, attach-

ment and metamorphosis. These tadpoles obviously serve to distribute the species too. It is the ascidian tadpole with its chordate characters of dorsal hollow nerve cord, stiffening notochord supporting the muscular tail, and features of the head which tells us much more of the likely evolutionary position of the sea squirts than does the adult.

Some sea squirts are of economic importance because they act as fouling organisms encrusting the hulls of ships and other marine structures. They are also of considerable evolutionary and zoological interest.

The other urochordates, thaliaceans (salps and others), and the larva-like larvaceans, have characters fundamental to the group but have evolved along pelagic and not bottom-dwelling lines. Members of one thaliacean order, the Pyrosomidae, are commonly found in warmer waters and emit phosphorescent light in response to tactile stimulation. These pelagic animals (eg *Pyrosoma* species) live in colonies, but some other types exist as solitary individuals. They have complex life-histories. *Oikopleura* is a larvacean genus of small animals like the tadpoles of sea squirts. Rhythmic movements of the tail draw water through the openings of the gelatinous "house" in which they live and which is secreted by the body.

The **lancelets** are a minor group of chordates containing only two genera and under 20 species. These small, apparently fish-like animals are known as lancelets because of their elongated blade-like form. They show primitive chordate conditions. The notochord, which in sea squirt larvae merely supports the tail, extends into the head (hence the term cephalochordata). There is no cranium as in the craniates or

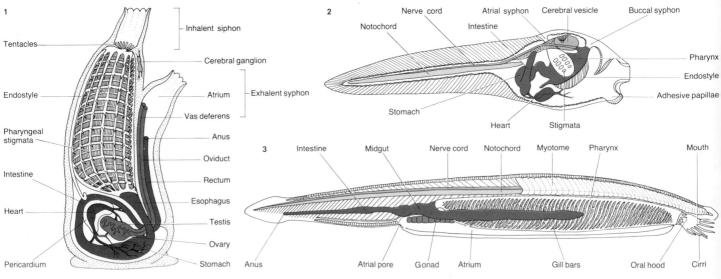

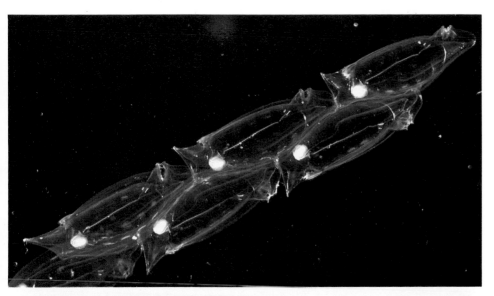

▲ **Verging on the vertebrate condition,** a lancelet. Here *Branchiostoma lanceolatum* lies with its tail in the gravel and its head protruding to maintain a respiratory and filter-feeding current.

◄ **Body plans of** (1) a sea squirt and (2) its "tadpole" larva, and (3) a lancelet.

vertebrates, and the front end of the animal, although quite distinct, lacks the well-developed brain, eyes and other sense organs as well as the jaws associated with vertebrates.

In species of *Branchiostoma*, a hood extends over the mouth equipped with slender, tentacle-like cirri instead of jaws. These form a sieve which assists in rejecting particles of food too large for the lancelet's suspension-feeding habit. The oral hood leads to the extensive pharynx via a thin flap of tissue, the velum. The pharyngeal wall is composed of many gill bars and it is the action of the cilia situated on those bars that draws the water and suspended food paticles into the body by way of the oral hood. The gill bars act as a filter for food as well as providing a surface for the absorption of oxygen. When water passes through the gill bars, food particles are trapped and passed down to the floor of the pharynx, where they

are swept into the endostyle. In this ciliated gutter trapped food is collected up, bound into mucus strands, and passed into the mid-gut. The filtered water passes out into the atrium and leaves the body via the atrial pore. The midgut has a blind diverticulum leading forward. Backward the midgut leads to the intestine and then eventually to the anus.

The anatomy of these coelomates is complex in comparison with most invertebrates and there are some unusual features to it. The excretory system consists of sac-like nephridia lying above the gill bars. Each lancelet nephridium has a number of flame cells reminiscent of those seen in the flatworms (see p198), annelids and mollusks. In evolutionary terms the nephridia (which do not occur in other primitive chordates) are a far cry from the vertebrate kidneys, which must have evolved via a different route.

The muscle blocks are arranged in segments (myotomes) along either side of the body. A hollow dorsal nerve cord runs the length of the animal and shows very little anterior specialization or brain. As it passes back along the body it branches to supply muscles and other organs.

There is a simple circulatory system with blood vessels passing through the gills to collect oxygen and distribute it to the body. The sexes are separate. In *Branchiostoma* the gonads are arranged on both sides of the body at the bases of the muscle blocks, but in *Asymmetron*, whose name speaks for itself, they lie only on the right side. Sperms and eggs are released into the atrium by rupture of the gonad wall. They pass out to the sea through the atriopore. Fertilization occurs in the open water and a swimming larva or tornaria develops. This lives a dual life for several months as it feeds and matures. In the daytime it lies on the seabed but when darkness comes it swims and joins the plankton. When it has attained a length of about 5mm (0.2in) it metamorphoses and becomes bottom-dwelling and more sedentary. The tornaria is the chief distributive phase.

Adult lancelets live in shallow water inshore and occur from temperate to tropical regions. Although they are really burrowers, inhabiting sands and gravels, they can emerge from their burrows and swim actively for short periods. Because of its characters *Branchiostoma* provides an important lesson in our understanding of vertebrate organization, and its simple structure makes it quite easy to identify the basic chordate features, many of which are seen in humans themselves. AC

Bibliography

The following list of titles indicates key reference works used in the preparation of this volume and those recommended for further reading. The list is divided into categories corresponding to those of the volume.

FISHES

Fish Classification

Berg, L. S. (trans. Edwards, J. W.) (1947) *Classification of Fishes both Recent and Fossil*, Ann Arbor, Michigan.

Greenwood, P. H., Miles, R. S., Patterson, C. (eds) (1973) *Interrelationships of Fishes*, Linnean Society of London.

Greenwood, P. H., Rosen, D. E., Weitzman, S H. and Myers, G. S. (1966) "Phyletic Studies of Teleostean Fishes with a Provisional Classification of Living Forms," *Bulletin of the American Museum of Natural History*, vol 131 (4), pp339–456.

Jordan, D. S. (1963) *The Genera of Fishes and a Classification of Fishes*, Stanford University Press, Stanford, California.

Lindberg, D. U. (1974) *Fishes of the World*, John Wiley and Sons, New York.

Moy-Thomas, J. A. and Miles, R. S. (1971) *Palaeozoic Fishes*, Chapman and Hall, London.

Nelson, J. S. (1984) *Fishes of the World* (2nd edn), John Wiley and Sons, New York.

Norman, J. R. (1966) *A Draft Synopsis of the Orders, Families, and Genera of Recent Fishes and Fish-like Vertebrates*, British Museum (Natural History), London.

General Fish Biology

Alexander, R. McN. (1967) *Functional Designs in Fishes*, Hutchinson, London.

Berra, T. M. (1981) *An Atlas of the Distribution of the Freshwater Fish Families of the World*, University of Nebraska Press, Lincoln, Nebraska, and London.

Boulenger, G. A. and Bridge, T. W. (1910) *Fishes* (vol VII of *The Cambridge Natural History*), Cambridge University Press, London.

Goodrich, E. S. (1909) *Cyclostomes and Fishes* (part IX, fascicle I of Lankaster, R. (ed) *A Treatise on Zoology*), London.

Hoar, W. S. and Randall, D. J. (eds) (1969–) *Fish Physiology*, Academic Press, London and New York.

Marshall, N. B. (1954) *Aspects of Deep Sea Biology*, Hutchinson, London.

Marshall, N. B. (1979) *Development in Deep Sea Biology*, Blandford Press, Poole, Dorset.

Marshall, N. B. (1971) *Explorations in the Life of Fishes*, Harvard University Press, Cambridge, Massachusetts.

Marshall, N. B. (1965) *The Life of Fishes*, Weidenfeld and Nicolson, London.

Norman, J. R. (1975) *A History of Fishes* (3rd edn, revised by Greenwood, P. H.), Ernest Benn, London.

Norman, J. R. and Fraser, F. C. (1937) *Giant Fishes, Whales and Dolphins*, Putnam, London.

Nikolsky, G. V. (1963) *The Ecology of Fishes*, Academic Press, London.

Regional Fish Field Guides and Identification

Arnoult, J. (1959) *Faune de Madagascar: Poissons des eaux douces*, Institute de Récherche Scientifique, Tananarive.

Beaufort, L. F. de *et al* (1913–62) *Fishes of the Indo-Australian Archipelago* (10 vols), E. J. Brill, Leiden.

Boulenger, G. A. (1908–16) *Catalogue of the Freshwater Fishes of Africa* (4 vols), British Museum (Natural History), London.

Clemens, W. A. and Wilby, G. V. (1961) "Fishes of the Pacific Coast of Canada," *Bulletin of the Fisheries Research Board of Canada*, vol 68.

Day, F. (1875–78) *Fishes of India*, Dawson, London.

Fowler, H. W. (1936) "The Marine Fishes of West Africa," *Bulletin of the American Museum of Natural History*, vol 70, pp1–1,493.

Hoese, H. D. and Moore, H. D. (1977) *Fishes of the Gulf of Mexico, Texas, Louisiana and Adjacent Waters*, Texas A. & M. University Press, College Station, Texas.

Jayaram, K. C. (1981) *The Freshwater Fishes of India, Pakistan, Bangladesh and Sri Lanka*, Sri Aurobindo Press, Zoological Survey of India, Calcutta.

Khalaf, K. T. (1962) *The Marine and Freshwater Fishes of Iraq*, Ar-Rabitta Press, Baghdad.

Kumada, T. (ed) (1937) *Marine Fishes of the Pacific Coast of Mexico*, Nissan Fisheries Institute, Odawara, Japan.

Lake, J. S. (1971) *Freshwater Fishes and Rivers of Australia*, Thomas Nelson, Melbourne.

Leim, A. H. and Scott, W. B. (1966) "Fishes of the Atlantic Coast of Canada," *Bulletin of the Fisheries Research Board of Canada*, vol 155, pp1–485.

Munro, I. S. R. (1967) *The Fishes of New Guinea*, Department of Agriculture, Stock and Fisheries, Port Moresby, Papua New Guinea.

Nichols, J. T. (1943) *Freshwater Fishes of China*, American Museum of Natural History, New York.

Smith, J. L. B. and Smith, M. M. (1963) *The Fishes of Seychelles*, Department of Ichthyology, Rhodes University, Grahamstown, S.A.

Wheeler, A. C. (1969) *The Fishes of the British Isles and Northwest Europe*, Macmillan, London.

AQUATIC INVERTEBRATES

General Invertebrate Biology

Alexander, R. McN. (1979) *The Invertebrates*, Cambridge University Press, Cambridge.

Barnes, R. D. (1982) *Invertebrate Zoology*, 4th edn, Holt-Saunders, Philadelphia.

Barrington, E. J. W. (1982) *Invertebrate Structure and Function*, Van Nostrand Reinhold, New York.

Clark, R. B. (1964) *Dynamics in Metazoan Evolution*, Oxford University Press, Oxford.

Fretter, V. and Graham, A. (1976) *A Functional Anatomy of Invertebrates*, Academic Press, London, New York, San Francisco.

Grzimek, B. (ed) (1972) *Grzimek's Animal Life Encyclopedia* vols 1 and 3, Van Nostrand Reinhold, New York.

Hyman, L. H. (1940–67) *The Invertebrates*, vols 1–6, McGraw-Hill, New York.

Marshall, A. J. and Williams W. D. (1982) *Textbook of Zoology: Invertebrates*, Macmillan, London.

Meglitsch, P. A. (1972) *Invertebrate Zoology*, 2nd edn, Oxford University Press, Oxford.

Moore, R. C., Lalicker, C. G. and Fischer, A. G. (1952) *Invertebrate Fossils*, McGraw-Hill, New York.

Russell-Hunter, W. D. (1979) *A Life of Invertebrates*, Macmillan, New York. Collier Macmillan, London.

Invertebrate Groups

Berquist, P. R. (1978) *Sponges*, Hutchinson, London.

Bliss, D. E. (ed) (1982–) *The Biology of the Crustacea*, vols 1–10, Academic Press, London and New York.

Berrill, N. J. (1950) *The Tunicata, with an Account of the British Species*, Ray Society, London.

Cheng, T. C. (1973) *General Parasitology*, Academic Press, New York.

Burton, M. (1963) *Revision of Classification of Calcareous Sponges*, British Museum (Natural History), London.

Crofton, H. D. (1966) *Nematodes*, Hutchinson, London.

Dales, R. P. (1967) *Annelids* (2nd edn), Hutchinson, London.

Darwin, C. R. (1851–54) *A Monograph of the Sub-class Cirripedia*, 2 vols, Ray Society, London.

Fretter, V. and Graham, A. (1962) *British Prosobranch Molluscs*, Ray Society, London.

Gibson, R. (1972) *Nemerteans*, Hutchinson, London.

Ingle, R. (1980) *British Crabs*, British Museum (Natural History), London, and Oxford University Press.

King, P. E. (1973) *Pycnogonids*, Hutchinson, London.

Lapage, G. (1963) *Animals Parasitic in Man*, Dover, New York.

Lincoln, R. J. (1979) *British Marine Amphipoda: Grammaridea*, British Museum (Natural History), London.

Manton, S. M. (1977) *The Arthropoda: Habits, Functional Morphology and Evolution*, Clarendon Press, Oxford.

Morton, J. E. (1968) *Molluscs*, Hutchinson, London.

Nichols, D. (1969) *Echinoderms* (4th edn), Hutchinson, London.

Ramazzotti, G. and Maucci, W. (1983) *Il Phylum Tardigrada* (3rd edn), Memorie dell Istituto Italiano di Idrobiologia Dott. Marco de Marchi, vol 41.

Robin, B., Pétron, C. and Rives, C. (n.d.) *Les coraux de Nouvelle-Calédonie, Tahiti, Réunion, Antilles*, Editions du Pacifique, Tahiti.

Rudwick, M. J. S. (1970) *Living and Fossil Brachiopods*, Hutchinson, London.

Ryland, J. S. (1970) *Bryozoans*, Hutchinson, London.

Schumacher, H. (1976) *Korallenriffe*, BLV Vorlagsgesellschaft, Munich, Berne, Vienna.

Stephenson, T. A. and Stephenson A. (1928 and 1935) *The British Sea Anemones*, 2 vols, Ray Society, London.

Tattersall, W. M. and Tattersall, O. M. (1951) *The British Mysidacea*, Ray Society, London.

Thompson, T. E. (1976) *Biology of the Opisthobranch Molluscs 1*, Ray Society, London.

Thompson, T. E. and Brown, G. H. (1984) *Biology of the Opisthobranch Molluscs 2*, Ray Society, London.

Veron, C., Pichon, M., Wijsman-Best, M. and Wallace, C. C. (1976–84) *Scleractinia of Eastern Australia*, parts I–V, Australian Institute of Marine Sciences Monograph Series.

Warner, G. F. (1977) *The Biology of Crabs*, Elek, London.

Yonge, C. M. and Thompson, T. E. (1976) *Living Marine Molluscs*, Collins, London.

Regional Invertebrate Field Guides and Identification

Barratt, J. and Younge, C. M. (1972) *Collins Pocket Guide to the Seashore*, Collins, London.

Campbell, A. C. (1984) *The Country Life Guide to the Sea Shores and Shallow Seas of Britain and Europe*, Country Life Books, London.

Campbell, A. C. (1982) *The Hamlyn Guide to the Flora and Fauna of the Mediterranean Sea*, Hamlyn, London.

Conseil Permanent International pour l'Exploration de la Mer, *Fiches d'Identification du Zooplankton*, Charlottenlund Slot, Denmark. (A series of papers on plankton identification.)

Dakin, W. J. (1952) *Australian Seashores*, Angus & Robertson, Sydney and London.

Day, J. H. (1974) *Marine Life on South African Shores*, Baklema, Rotterdam.

Linnean Society of London, and the Estuarine and Brackish Water Sciences Association, *Synopses of the British Fauna* (new series), E. J. Brill, Leiden. A periodical dealing with identification and general bionomics of many invertebrate groups.

Riedel, R. (1963) *Fauna und Flora der Adria*, Paul Parey, Hamburg and Berlin.

Rickets, E. F. and Calvin, J. (1960) *Between Pacific Tides* (3rd edn), Stanford University Press, Palo Alto.

Picture Acknowledgements

Key: *t* top. *b* bottom. *c* centre. *l* left. *r* right.

Abbreviations: A Ardea. AN Agence Nature. ANT Australasian Nature Transparencies. BCL Bruce Coleman Ltd. J Jacana. NHPA Natural History Photographic Agency. OSF Oxford Scientific Films. PEP Planet Earth Pictures/Seaphot. SAL Survival Anglia Ltd. SPL Science Photo Library.

2 A/V. Taylor. 3 PEP/P. David. 7 Scala, Firenze. 8–9 S. Middleton. 8c Mary Evans Picture Library. 8b BCL/J. Burton. 9b NHPA/H. Switak. 10 Tony Morrison. 11t PEP/P. Capen. 11bA. 12–13 Wildwood Postcard Co., New Jersey. 13l,13r Mary Evans Picture Library. 14–15,15 BCL/H. Reinhard. 16 OSF/J. Paling. 16–17,20 PEP/K. Lucas. 21t AN. 21b Mary Evans Picture Library. 22t M. Sandford. 22–23 A/J. Mason. 23 PEP/K. Lucas. 24 BCL/L. Marigo. 25 PEP/C. Roessler. 26 PEP/I. Lythgoe. 27 Biofotos/H. Angel. 28–29 PEP/H. Voigtmann. 30t PEP/C. Pétron. 30b BCL/A. Power. 32 Biofotos/I. Took. 32–33 OSF/L. Gould. 34 PEP/C. Roessler. 35 OSF/P. Parks. 38 AN. 39tl Dwight R. Kuhn. 39tr BCL/J. Simon. 39c NHPA/I. Polunin. 40–41 A/V. Taylor. 41 ANT/G. Schmida. 44–45 PEP/K. Cullimore. 45 BCL/H. Reinhard. 46–47,47 ANT/G. Schmida. 48–49 Leonard Lee Rue. 48 SAL/J. Foott. 50–51 PEP/G. van Ryckerorsel. 52tl BCL/H. Reinhard. 52bl NSP/G. Kinns & P. Ward. 52–53,53 BCL. 54t ANT/G. Schmida. 54–55 BCL/H. Reinhard. 56–57 ANT. 56t BCL/L. Rue. 60 PEP. 61,62t OSF/P. Parks. 62–63 PEP/P. David. 63t OSF/P. Parks. 64–65 AN. 65l PEP. 65r NHPA/Chaumeton-Bassot. 66–67 A/V. Taylor. 68–69t PEP/C. Roessler. 68–69c OSF/P. Parks. 69b BCL/D. Bartlett. 70 BCL/J. Kenfield. 71,72–73 BCL/H. Reinhard. 72c,72b BCL/J. Burton. 74–75 SAL/J. Root. 75b ANT/G. Schmida. 76b A/I. Beames. 76–77 PEP/K. Lucas. 78b OSF/R. Kuiter. 78–79t,78–79c BCL/J. Burton. 82–83 A/P. Morris. 85 PEP/R. Waller. 88–89 A/P. Morris. 89t PEP/P. David. 90t OSF/D. Shale. 90b PEP/K. Lucas. 91 PEP/C. Roessler. 92–93 OSF/G. Bernard. 97 PEP/P. David. 98–99,98l,99r SAL/J. Foott. 100l,100–101, 101t ANT/G. Schmida. 102 BCL. 103 PEP/B. Cocker. 104–105 A/V. Taylor. 105 A/P. Morris. 106–107 NSP/N. Fain. 107b AN. 108tl Dwight R. Kuhn. 108l BCL/J. Burton. 108b OSF/R. Kuiter. 109 PEP/H. Jones. 110 BCL/A. Compost. 111 PEP/D. Clarke. 113 BCL/A. Power. 114c PEP/B. Wood. 114b A/V. Taylor. 114–115 PEP/C. Roessler. 115b BCL/J. Kenfield. 116 ANT/G. Schmida. 116–117 PEP/H. Voigtmann. 117b OSF/R. Kuiter. 120t Biofotos/H. Angel. 120–121 BCL/J. Burton. 122b BCL/J. Taylor. 122–123 PEP/R. Chesher. 124–125 PEP/P. Scoones. 125 M. Sandford. 126–127 ANT/G. Schmida. 127t SAL/A. Root. 127b PEP/P. Scoones. 128 A/V. Taylor. 129 PEP/H. Voigtmann. 130–131t A/V. Taylor. 130–131b OSF/R. Kuiter. 132t OSF/G. Bernard. 132–133 A/P. Morris. 134,135 A/ V. Taylor. 138 PEP/B. Cocker. 138–139 BCL/M. P. Kahl. 139t A/K. Fink. 140–141 PEP/R. Johnson. 142–143t Leonard Lee Rue. 143c OSF/P. Parks. 142–143b PEP/K. Lucas. 144–145 A/V. Taylor. 146,147 OSF. 149t A/V. Taylor. 149b SPL. 150t OSF/D. Allan. 150b NHPA/B. Wood. 151 PEP/J. Mackinnon. 152 C. Howson. 152–153 PEP/K. Lucas. 154 Biophoto Associates. 155,156b OSF/P. Parks. 156–157 NHPA/M. Walker. 158–159 OSF/P. Parks. 162 Biophoto Associates. 163 SPL. 165 Biophoto Associates. 166–167 B. Picton. 167b NSP/I. Bennett & F. Myers. 168 OSF. 169 C. Howson. 170–171 NHPA/B. Wood. 175 PEP/B. Wood. 176–177 NHPA/B. Wood. 178–179 PEP/J. Greenfield. 180b BCL. 180–181 OSF/F. Ehrenström. 182 PEP/L. Madin. 183 OSF/P. Parks. 185 Biophoto Associates. 186–187 J. Walsh. 188b OSF/P. Parks. 188–189 B. Picton. 190 OSF/J. Cooke. 191 C. Howson. 192–193 OSF/S. Foote. 197 B. Picton. 198–199 Premaphotos. 199b OSF/P. Parks. 200tl S. Stammers. 200r SPL/C. Ellis. 202,203 SPL. 204–205 Biofotos/H. Angel. 205b D. Weathered. 206 Premaphotos. 207 OSF/A. Kuiter. 208t ANT/R. & D. Keller. 208b SPL/M. Dohrn. 212 A/P. Morris. 212–213 OSF/ R. Kuiter. 216 BCL. 219 OSF/K. Atkinson. 220–221 OSF. 223 NHPA/ G. Bernard. 226–227 PEP/P. David. 228–229 OSF/D. Shale. 230–231 BCL. 230b NHPA/N. Callow. 231c AN/Chaumeton. 231b Premaphotos. 234 A/V. Taylor. 235–236 BCL. 236–237,237b PEP/D. Maitland. 238–239 BCL. 240t NHPA/ J. Carmichael. 240b PEP/D. Maitland. 241 BCL. 242,242–243 NSP/I. Bennett. 244 Biofotos/H. Angel. 245 BCL. 246–247 SPL. 247b OSF/P. Parks. 248b C. Howson. 248–249 PEP/B. Wood. 252 NHPA/J. Carmichael. 252–253 OSF/ G. Bernard. 253b Biophoto Associates. 256–257 OSF/G. Bernard. 256b Biofotos/ H. Angel. 257b B. Picton. 258 PEP/D. Maitland. 259 BCL. 260–261 NSP/ I. Bennett. 261t PEP/J. Lythgoe. 262 OSF/R. Kuiter. 263t OSF/G. Bernard. 263b M. Fogden. 264–265 A/V. Taylor. 266t PEP/C. Pétron. 266b Biofotos/ S. Summerhays. 267 PEP/F. Jackson. 269 Biofotos/H. Angel. 270–271 PEP/ K. Lucas. 272b B. Picton. 272–273 BCL. 273 PEP/C. Pétron. 274b,274–275 NHPA/A. Bannister. 275b PEP/B. Wood. 278 NHPA/B. Wood. 279 C. Howson. 280–281 PEP/A. Svoboda. 282–283 OSF/G. Bernard. 283b B. Picton. 284–285 J. Jamieson. 286 OSF/D. Shale. 287t OSF/P. Parks. 287b OSF/G. Bernard.

Artwork

Abbreviations: SD Simon Driver. ML Mick Loates. DO Denys Ovenden. JL Joe Little. NW Norman Weaver. RL Richard Lewington. RG Roger Gorringe. BC Barbara Cooper. PB Priscilla Barrett. SC Stephen Cocking. MM Malcolm McGregor.

5 SD. 6 ML. 18 DO. 26t SD. 26c ML. 27 JL. 31 ML. 36 DO. 40 JL. 42 DO. 49 ML. 59 DO. 61,63,65,70 ML. 72 SD. 80 DO. 82,83,86,87 NW. 89 ML. 91,94 DO. 100,103 SD. 112 NW. 118 DO. 120,123 ML. 137 RL. 148t,148b,148r SD. 153 RL. 154 BC. 159t,159b SD. 161 RG. 153,164,165,169,170,171 SD. 173 RG. 174,178,183,184,187,189,190, 191,192 SD. 194 RG. 196,198,201t,201b,202,203,204,206 SD. 208 RL. 209 SD. 210 RG. 212 SD. 213 RL. 215t SD. 215b RG. 216,217,218,222 SD. 225 ML. 226 SD. 227 RL. 232 ML. 234 RL. 237 SD. 244,246 RL. 250 ML. 252 SD. 254 RL. 262,267 SD. 268 ML. 270 SD. 277 RG. 278 SD. 279,282 RL. 286 SD.

GLOSSARY

FISHES

Adaptation   features of an animal which adjust it to its environment. Adaptations may be genetic, ie produced by evolution and hence not alterable within the animal's lifetime, or they may be phenotypic, ie produced by adjustment on the behalf of the individual and may be reversible within its lifetime.

Adipose fin   a fatty fin behind the rayed DORSAL FIN, normally rayless (exceptionally provided with a spine or pseudorays in some catfish).

Adult   a fully developed and mature individual, capable of breeding but not necessarily doing so until social and/or ecological conditions allow.

Air bladder   see SWIM BLADDER.

Algae   very primitive plants, eg epilithic algae, algae growing on liths (ie stones).

Ammocoetes   the larval stage of the lamprey.

Anadromous   of fish that run up from the sea to spawn in fresh water.

Brood sac or pouch   a protective device made from fins or plates of one or other parent fish in which the fertilized eggs are placed to hatch in safety.

Cartilage   gristle.

Caudal fin   the "tail" fin.

Caudal peduncle   a narrowing of the body in front of the caudal fin.

Cerebellum   a part of the brain.

Cilia   tiny hair-like protrusions.

Class   a taxonomic level. The main levels (in descending order) are Phylum, Class, Order, Family, Genus, Species.

Cogener   a member of the same genus.

Colonial   living together in a COLONY.

Colony   a group of animals gathered together for breeding.

Conspecific   a member of the same species.

Cryptic   camouflaged and difficult to see.

Ctenoid scales   scales of "advanced" fishes which have a comb-like posterior edge, thereby giving a rough feeling.

Cutaneous respiration   breathing through the skin.

Cycloid scales   scales with a smooth posterior (exposed) edge.

Denticle   literally a small tooth; used of dermal denticles, ie tooth-like scales (all denticles are dermal in origin).

Diatoms   small planktonic plants with silicaceous tests (shells).

Dimorphism   the existence of two distinctive forms.

Disjunct distribution   geographical distribution of taxons that is marked by gaps. Many factors may cause it.

Display   any relatively conspicuous pattern of behavior that conveys specific information to others, usually to members of the same species; often associated with courtship but also in other activities, eg "threat display."

Dorsal fin   the fin on the back.

Endostyle   a complex hairy (ciliated) groove that forms part of the feeding mechanism of the larval lamprey.

Epigean   living on the surface. See also HYPOGEAN.

Esca   the luminous lure at the end of the ILLICIUM (the fishing rod) of the angler fishes.

Family   either a group of closely related species or a pair of animals and their offspring. See CLASS and Introduction.

Feces   excrement from the digestive system passed out through the anus.

Fin   in fishes the equivalent of a leg, arm or wing.

Fin girdles   bony internal supports for paired fins.

Ganoid scales   a primitive type of thick scale.

Gape   the width of the open mouth.

Genus   the lowest taxonomic grouping. See CLASS and Introduction.

Gills   the primary respiratory organs of fish. Basically a vascularized series of slits in the PHARYNX allowing water to pass and effect gas exchange. The gills are the bars that separate the gill slits.

Gill slits   the slits between the gills that allow water through.

Gular plates   bony plates lying in the skin of the "throat" between the two halves of the lower jaw in many primitive and a few living bony fishes.

Heterocercal   a tail shape in which the upper lobe is longer than the lower and into which the upturned backbone continues for a short distance.

Hypogean   living below the surface of the ground, eg in caves.

Illicium   a modified dorsal fin ray in angler fishes which is mobile and acts as a lure to attract prey.

Introduced   of a species which has been brought from lands where it occurs naturally to lands where it has not previously occurred. Some introductions are natural but some are made on purpose for biological control, farming or other economic reasons.

Invertebrate   animals lacking backbones, eg insects, crustacea, coelenterates, "worms" of all varieties, echinoderms etc.

Krill   small shrimp-like marine crustaceans which are an important food for certain species of seabirds, whales, and fish.

Lamellae   plate-like serial structures (eg gill lamellae) usually of an absorbent or semipermeable nature.

Larva   a pre-adult form unlike its parent in appearance.

Lateral line organs   pressure-sensitive organs lying in a perforated canal along the side of the fish and on the head. They as it were feel at a distance.

Maxillary bone   the posterior bone of the upper jaw. Tooth-bearing in primitive fish, it acts as a lever to protrude the tooth-bearing anterior bone (premaxilla) in advanced fish.

Metamorphosis   a dramatic change of shape during the course of ontogeny (growing up). Usually occurs where the adult condition is assumed.

Mollusk   a shellfish.

Monotypic   the sole member of its genus.

Natural selection   the process whereby individuals with the most appropriate ADAPTATIONS are more successful than other individuals, and hence survive to produce more offspring. To the extent that the successful traits are heritable (genetic)

they will therefore spread in the population.

Neoteny   a condition in which a species becomes sexually mature and breeds whilst still in the larval body form, ie the ancestral adult body stage is never reached.

Niche   the position of a species within the community, defined in terms of all aspects of its life-style (eg food, competitors, predators and other resource requirements).

Olfactory sac   the sac below the nostrils containing the olfactory organ.

Opercular bones   the series of bones including the operculum (gill flap) and its supports.

Operculum   the correct name for the bone forming the gill flap.

Order   a level of taxonomic ranking. See CLASS and Introduction.

Osmosis   the tendency for ions to flow through a semipermeable membrane from the side with the greatest concentration to the side with the least. This means that in the sea fish fluids pick up ions and have to get rid of them whereas in fresh water retention of vital ions is essential.

Oviparous   egg-laying.

Ovipositor   a tube by which eggs are inserted into small openings and cracks.

Ovoviviparity   the retention of eggs and hatching within the body of the mother.

Pelvic girdle   the bones forming the support for the pelvic fins.

Perianal organ   an organ around the anus.

pH   a measure of the acidity or alkalinity of water: pH7 is neutral; the lower the number the more acid the water and vice versa.

Pharyngeal teeth   teeth borne on modified bones of the gill arches in the "throat" of the fish.

Pharynx   that part of the alimentary tract that has the gill arches.

Photophore   an organ emitting light.

Piscivore   fish-eater.

Plankton   very small organisms and larvae that drift largely passively in the water.

Predator   an animal that forages for live prey; hence "anti-predator behavior" describes the evasive actions of prey.

Prehensile   capable of being bent and/or moved.

Rostrum   snout.

Scale   a small flat plate forming part of the external covering of a fish; hence deciduous scale, a scale that easily falls off the fish.

Scutes   bony plates on or in the skin of a fish.

Spawning   the laying and fertilizing of eggs, sometimes done in a spawning ground.

Specialist   an animal whose life-style involves highly specialized strategems, eg feeding with one technique on a particular food.

Species   a population, or series of populations which interbreed freely, but not with others. See CLASS and Introduction.

Spiracle   a now largely relict GILL SLIT lying in front of the more functional gill slits.

Subcutaneous canal   a canal passing beneath the skin.

Swim bladder   or air bladder. A gas- or air-filled bladder lying between the gut and the backbone. It may be open via a duct to the PHARYNX so that changes of pressure can be accommodated by exhalation or inhalation of atmospheric air. If closed, gas is secreted or excreted by special glands. Its main function is buoyancy but it can also be used, in some species, for respiration, sound reception, or sound production.

Teleosts   a group of fishes, defined by particular characters. The fishes most familiar to us are almost all teleosts.

Temperate zone   an area of climatic zones in mid latitude, warmer than the northerly areas, but cooler than subtropical areas.

Territory   area that a fish considers its own and defends against intruders.

Tropics   strictly an area lying between 22.5°N and 22.5°S. Often, because of local geography, animals' habitats do not match this area precisely.

Tubercles   small keratized protrusions of unknown and doubtless different functions which are either permanent or seasonally or irregularly present on the skin of fish.

Type species   the species on which the definition of a genus depends.

Vascularized   possessed of many small, usually thin-walled, blood vessels.

Velum   a hood around the mouth of larval lampreys (ammocoetes) that is a feeding adaptation.

Vertebrate   an animal with a backbone primitively consisting of rigidly articulating bones.

Villi   small hair-like processes that often have an absorptive function.

Viviparous   producing live offspring from within the body of the mother.

Vomerine teeth   teeth carried on the vomer, a median bone near the roof of the mouth.

Weberian apparatus   a modification of the anterior few vertebrae in ostariophysan fishes (carps, catfish, characins etc) that transmit sound waves as compression impulses from the SWIM BLADDER to the inner ear thereby enabling the fish to hear.

AQUATIC INVERTEBRATES

Abdomen   a group of up to 10 similar segments, situated behind the THORAX of crustaceans and insects, which in the former group may possess appendages.

Abyss, abyssal   the part, or concerning the part, of the ocean, including the ocean floor, that extends downward from some 4,000m (13,000ft).

Acoelomate   having no COELOM (main body cavity).

Acrorhagi   groups of NEMATOCYSTS.

Adaptation   a characteristic which enhances an organism's chances of survival in the environment in which it lives, in comparison with the chances of a similar organism lacking the same characteristic.

Adult   a fully developed and mature individual capable of breeding, but not necessarily doing so until social and/or ecological conditions allow.

Aestivation   dormancy during the summer or dry season.

Americ having a body not divided into SEGMENTS.

Ampulla a small contractile fluid reservoir associated with the tube-feet of some echinoderms.

Anabiosis suspended animation, with a low metabolic rate enabling an animal to survive adverse environmental conditions, particularly desiccation.

Ancestral stock a group of animals usually showing primitive characteristics, which is believed to have given rise to later more specialized forms.

Anisogamous of reproduction, involving gametes of the same species that are unalike in size or in form.

Antennae the first pair of head appendages of uniramians and the second pair of crustaceans.

Antennules the first pair of head appendages of crustaceans.

Apophysis an outgrowth or process on an organ or bone.

Aquatic associated with, or living in water.

Arthropod an invertebrate, such as an insect, spider or crustacean, which is TRIPLOBLASTIC and COELOMATE, with a chitinous, jointed EXOSKELETON, paired, jointed limbs and a lack of NEPHRIDIA and CILIA; in some classifications, a member of the phylum Arthropoda; here includes the members of the phyla Crustacea, Chelicerata, Uniramia, Tardigrada, Pentastomida, Onychophora.

Arachnid a member of the Arachinda, a class of the phylum Chelicerata which includes spiders, scorpions, mites and ticks.

Asexual reproduction reproduction that does not include fertilization (exchange of GAMETES) or MEIOSIS. See BINARY FISSION; BUDDING; GEMMULE; PARTHENOGENESIS.

Asymmetrical having no plane of symmetry, eg an animal of indeterminate shape that cannot be divided into two halves which are mirror images.

Atrium the volume enclosed by the tentacles of an endoproct; also the chamber through which the water current passes before leaving the body of a sea squirt or lancelet.

Autrophy the synthesis, by an organism, of its own organic constituents from inorganic material, ie independent of organic sources; autotrophic organisms may synthesize food phototrophically (eg green plants) or chemotrophically (eg bacteria) via inorganic oxidations.

Axon a long process of a nerve cell, normally conducting impulses away from nerve cell body.

Axopodium a stiff filament or pseudopodium which radiates outward from the body of a heliozoan or radiolarian.

Bacterium member of a division of unicellular or multi-cellular microscopic PROKARYOTIC organisms, lacking CHLOROPHYLL. Distinct from both plants and animals, rod-like, spherical or spiral in shape, occasionally forming a mycelium.

Benthic associated with the bottom of seas or lakes.

Bilateral symmetry a bilaterally symmetrical animal can be halved in one plane only to give two halves which are mirror images of each other. Most multi-cellular animals are bilaterally symmetrical, a form of symmetry generally associated with a mobile, free-living life style.

Binary fission a form of ASEXUAL REPRODUCTION of a cell in which the nucleus divides, and then the CYTOPLASM divides into two approximately equal parts.

Biomass a measure of the abundance of a life form in terms of its mass.

Bipectinate comb-like, of structures with two branches, particularly the OSPHRADIUM and/or CTENIDIUM in some mollusks.

Biramous of those ARTHROPODS (eg crustaceans) with forked ("two-branched") appendages.

Bivalve a shell or protective covering composed of two parts hinged together and which usually encases the body; also, a member of the molluskan class Bivalvia, which includes most bivalved animals.

Bothria long, narrow grooves of weak muscularity found in Pseudophyllidea (an order of tapeworm); form an efficient sucking organ.

Branchial hearts contractile hearts near the base of each CTENIDIUM in certain mollusks.

Brood sac a thoracic pouch of certain crustaceans into which fertilized eggs are deposited and where they develop.

Buccal mass a muscular structure surrounding the RADULA, horny jaw and ODONTOPHORE of a mollusk (not a bivalve).

Budding a form of ASEXUAL REPRODUCTION in which a new individual develops as a direct growth from the body of the parent.

Calcareous composed of, or containing, calcium carbonate as in the spicules of certain sponges or the shells of mollusks.

Cambrian a geological period some 600–500 million years ago, also the oldest system of rocks in which fossils can be used for dating, containing the first shelled fossil remains.

Carapace the dorsal shield of the exoskeleton covering part of the body (mainly anterior) of most crustaceans; particularly large in eg crabs, it protects the animal from both predation and water loss.

Carboniferous a geological period some 350–280 million years ago.

Carnivore an animal that feeds on other animals.

Catabolic of processes involving breakdown of complex organic molecules by living organisms, typically animals, resulting in the liberation of energy.

Caudal relating to the tail or to the rearmost SEGMENT of an invertebrate.

Cecum a blindly-ending branch of the gut or other hollow organ.

Cellulose the tough, fibrous fundamental constituent of the cell walls of all green plants and some algae and fungi.

Cephalization development of the head during evolution; different organisms show different degrees of cephalization, generally according to their "level of evolution".

Cephalothorax the fusion of head and anterior thoracic segments in certain crustaceans to form a single body region which may be covered by a protective CARAPACE.

Cerata (sing. ceras) projections on the back of some shell-less sea slugs, often brightly colored, which may bear NEMATOCYSTS from cnidarians and may act as secondary respiratory organs.

Cercaria a swimming larval form of flukes; produced asexually by REDIA larvae while parasitic in snails; cercaria infects a new final or intermediate host via food or the skin.

Chaetae the chitinous bristles characteristic of annelid worms.

Chela the pincer-like tip of limbs in some arthropods.

Chelicera one of the first pair of appendages behind the mouth of a chelicerate.

Chelicerate a member of the phylum Chelicerata.

Chitin a complex nitrogen-containing polysaccharide which forms a material of considerable mechanical strength and resistance to chemicals; forms the external "shell" or CUTICLE of arthropods.

Chloroplast a small granule (plastid) found in cells and containing the green pigment chlorophyll, site of PHOTOSYNTHESIS.

Chordate a member or characteristic of the phylum Chordata, animals which possess a NOTOCHORD.

Chromatophore a cell with pigment in its CYTOPLASM.

Cilia (sing. cilium) the only differences between FLAGELLA and cilia is in the former's greater length and the greater number of cilia found on a cell; flagella measure up to $22\mu m$, cilia up to $10\mu m$.

Ciliary feeding feeding by filtering minute organisms from a current of water drawn through or toward the animal by CILIA.

Ciliated having a number of cilia on a surface which beat in a coordinated rhythm; ciliary action is a common method of moving fluids within an animal body or over body surfaces, employed in CILIARY FEEDING, and a common means of locomotion in microscopic and small animals. The ciliated ciliates are a major class (Ciliata) of protozoans.

Cirri (sing. cirrus) in barnacles, paired thoracic feeding appendages; in protozoans, short, spine-like projections in tufts called CILIA; in annelids, broad flattened projections situated dorsally on segments; in flukes and some turbellarian flatworms the cirrus is the male copulatory apparatus.

Class a rank used in the classification of organisms; consists of a number of similar ORDERS (in some cases only one order may be distinguished); similar classes are grouped into a PHYLUM.

Cleavage see RADIAL CLEAVAGE; SPIRAL CLEAVAGE.

Clitellum the saddle-like region of earthworms which is prominent in sexually mature worms.

Coelom the main body cavity of many TRIPLOBLASTIC animals, situated in the middle layer of cells or MESODERM and lined by EPITHELIUM. In many organisms the coelom contains the internal organs of the body and plays an important part in collecting excretions which are removed via NEPHRIDIA or coelomoducts.

Coelomate having a COELOM.

Coelomocyte a free cell in the coelom of some invertebrates which appears to be involved with the excretion of waste material, wound healing and regeneration.

Colony an organism consisting of a number of individual members in a permanent colonial association.

Commensalism a relationship between members of different species in which one species benefits from the relationship, often by access to food; the other species neither benefits nor is harmed.

Community a naturally occurring group of different organisms inhabiting a common environment, sometimes named for one of its members, eg the *Donax* community of sandy beaches named for a genus of bivalve mollusks.

Compound eyes the type of eyes possessed by most crustaceans and insects, composed of many long, cylindrical units (ommatidia) each of which is capable of light reception and image formation.

Conjugation the union of GAMETES, or two cells (in certain bacteria); or the process of sexual reproduction in most ciliates.

Convergent evolution the evolution of two organisms with some increasingly similar characteristics but different ancestry.

Copulation the process by which internal fertilization is accomplished, the transfer of sperm from one member of a species to another via specialized organs.

Corona the characteristic, ciliated, wheel-like organ at the anterior end of rotifers.

Coxa the basal segment of an arthropod appendage which joins the limb to the body.

Cretaceous a geological period extending from some 135 to 65 million years ago.

Cross-fertilization the fusion of male and female GAMETES produced by different individuals of the same species.

Cryptobiosis a form of suspended animation enabling an organism to survive adverse environmental conditions.

Ctenidium one of the pair of gills within the MANTLE CAVITY of some mollusks.

Cuticle the external layer covering certain multicellular animals (eg arthropods), formed from a collagen-like protein or CHITIN, which is secreted by the EPIDERMIS. It acts as a physiological barrier between the animal and its external environment, may reduce water loss, acts as a barrier to the entry of microorganisms and in arthropods acts as an EXOSKELETON.

Cyst a thick-walled protective membrane enclosing a cell, larva or organism.

Cytoplasm all the living matter of a cell excluding the nucleus.

Dactylozooid the specialized defensive polyp of colonial hydrozoans.

Desiccation loss of water, or drying out.

Detritivore an animal that feeds on dead or decaying organic matter.

Detritus organic debris derived from decomposing organisms which provides a food source for a large number of organisms.

Deuterostome a member of a major branch of multicellular animals (the others are PROTOSTOMES). The mouth is formed as a secondary opening and the original embryonic blastopore becomes the anus. The embryo undergoes radial cleavage, the body cavity (ENTEROCOEL) arises as a pouch from the ENDODERM, and the central nervous system is dorsal.

Devonian   a geological period from some 400 to 350 million years ago.

Dextral   of spirally coiled gastropod shells in which, as is usual, the whorls rise to the right and the aperture is on the right where the shell is viewed from the side.

Diatom   a single-celled alga, a component of the PHYTOPLANKTON.

Dimorphism   the presence of two distinct forms, eg in color or size, in a species or population.

Dioecious   having separate sexes.

Diploblastic   a multicellular animal having a body composed of two distinct cellular layers, the ECTODERM and ENDODERM.

Dispersal   the movement of individuals away from their previous home range, often as they approach maturity.

Display   a relatively conspicuous pattern of behavior that conveys specific information to others, usually involving visual elements.

Dinoflagellate   a unicellular member of the PHYTOPLANKTON characterized by the possession of two FLAGELLA, one directed posteriorly, the other lying at right angles to the posterior flagellum.

Diverticulum   (plural -ae) a blind-ending tube forming a side branch of a cavity or passage.

DNA   deoxyribonucleic acid, a complex molecule, found almost exclusively in chromosomes of plants and animals, whose "double helix" structure contains the hereditary information necessary for an organism to replicate itself.

Dorsal   situated at or related to, the back of an animal, ie the side which is generally directed upwards.

Dorso-ventral   a plane running from the top to the bottom of an animal, as in a dorso-ventrally flattened horseshoe crab or sea slater.

Ecology   the study of animal and plant communities in relation to each other and their natural surroundings.

Ecosystem   an intricate community of organisms within a particular environment, interacting with one another and with the environment in which they live.

Ectoderm (is)   the superficial or outer germ layer of a multicellular embryo which mainly develops into the skin, nervous tissue and excretory organs.

Ectoparasite   a parasite which lives on the outside of its host and may be permanently attached or only come into contact with the host when feeding or reproducing.

Elongate   relatively long in comparison with width.

Endemic   confined to a given region, such as an island or country.

Endocuticle   the inner layer of the crustacean CUTICLE which is composed of CHITIN.

Endoderm (is)   the innermost of the three germ layers in the early embryo of most animals, developing into, for example, in jellyfishes the lining of the ENTERON or, in many animals, the GUT lining.

Endoparasite   a parasite which lives permanently within its host's tissues (except for some reproductive or larval stages). Often there are primary and SECONDARY HOSTS for different stages of the life cycle. Endoparasites are typically highly specialized.

Endoskeleton   an internal skeleton, as in echinoderms and vertebrates.

Enterocoelom   a COELOM that is thought to have arisen from cavities in the sacs of the MESODERM of the embryo.

Enteron   the body cavity of cnidarians which is lined with ENDODERM and opens to the exterior via a single opening, the mouth.

Epibenthic   living on the seabed between lowwater mark and some 200m (670ft) depth.

Epicuticle   the outer layer of the crustacean CUTICLE, a thin, non-chitinous protective layer.

Epidermis   the outer tissue layer of the epithelium.

Epithelium   a sheet or tube of cells lining cavities and vessels and covering exposed body surfaces.

Epizoic   a sedentary animal which is attached to the exterior of another animal but is not parasitic, ie is epizoic.

Esophagus   part of the foregut of certain invertebrates connecting the pharynx with the stomach or crop and concerned with the passage of food along the gut.

Eukaryote   a cell, or organism possessing cells, in which the nuclear material is separated from the CYTOPLASM by a nuclear membrane and the genetic material is borne on a number of chromosomes consisting of DNA and protein; the unit of structure in all organisms except bacteria and blue-green algae.

Exoskeleton   the skeleton covering the outside of the body, or situated in the skin.

Extracellular digestion   digestion of food within an organism but not within its constituent cells.

Family   a rank used in the classification of organisms, consisting of a number of similar GENERA (or sometimes only one). Similar families are grouped into an ORDER. In zoological classifications the name of the family usually ends in -idae.

Fibril   a small fiber, or subdivision of a fiber; used as contractile ORGANELLES in protozoans.

Filamentous   a type of structure, eg a crustacean GILL, in which the branches are thread-like, but not sub-branched, and are arranged in several series along the central axis.

Filter feeding   a form of SUSPENSION FEEDING in which food particles are extracted from the surrounding water by filtering. Filtering requires the setting up of a water current usually by means of CILIA, with mucus being used to trap particles and sometimes to filter them from the surrounding water.

Fission   see BINARY FISSION.

Flagellum   (plural flagella) a fine, long thread, moving in a lashing or undulating fashion, projecting from a cell.

Flame cell   the hollow, cup-shaped cell lying at the inner end of a protonephridium, important in the excretory system of some invertebrates. The inner end bears FLAGELLA whose beating causes body fluids to enter the NEPHRIDIUM.

Fragmentation   a form of SEXUAL REPRODUCTION in which an organism produces eggs in SEGMENTS of its body, which then break off and themselves split after leaving the host body, allowing the eggs to develop eventually into new organisms.

Free-living   having an independent life-style, not directly dependent on another organism for survival.

Funnel   part of the molluskan "foot" in cephalopods responsible for respiratory currents to the CTENIDIA and for jet propulsion.

Gamete   a female (ovum) or male (spermatozoan) reproductive cell whose nucleus and often CYTOPLASM fuses with another gamete, so constituting fertilization.

Gametocyte   a cell which undergoes MEIOSIS to form GAMETES; an oocyte forms an ovum (female gamete) and a spermatocyte forms a spermatozoan (male gamete).

Ganglion   a small discrete collection of nervous tissue containing numerous cell bodies. The nervous system of most invertebrates consists largely of such ganglia connected by nerve cords and which may be concentrated into a cerebral ganglion constituting the "brain."

Gastrozooid   a type of individual POLYP in colonial hydrozoans which captures and ingests prey.

Gemmule   a mass of sponge cells which acts as a resting stage under adverse conditions and is composed of amoebocytes surrounded by two membranes in which SPICULES are embedded.

Gene   the unit of the material of inheritance, a short length of chromosome and the set of characters which it influences in a particular way.

Generalist   an animal not specialized, not adapted to a particular niche; may be found in a variety of habitats.

Genus   a rank used in classifying organisms, consisting of similar SPECIES (in some cases only one species). Similar genera are grouped into FAMILIES.

Gill   the respiratory organ of aquatic animals.

Gill book   a type of gill, possessed by eg horseshoe crabs, formed by the five posterior pairs of appendages on the OPISTHOSOMA.

Gizzard   part of the alimentary canal where food is broken up, preceding main digestion. In crustaceans its walls bear hard "teeth."

Gonoduct   the duct through which sperm and eggs are released into the surrounding water.

Gut   the alimentary canal—a tube concerned with digestion and absorption of food. In most animals there are two openings (cnidarians and flatworms have only one), the mouth into which food is taken and the anus from which material is ejected.

Hemal system   a tubular system of undecided function found in echinoderms.

Hemocoel   the major secondary body cavity of arthropods and mollusks which is filled with blood. Unlike the COELOM it does not communicate with the exterior and does not contain germ cells. However, body organs lie within or are suspended in the hemocoel. It functions in the transport and storage of many essential materials.

Herbivore   animal that feeds on plants.

Hermaphrodite   an animal producing both male and female GAMETES; among unisexual animals hermaphrodites may occur as aberrations.

Hermatypic   of corals, reef-building corals with commensal zooanthellae.

Heterotrophic   heterotrophic organisms are unable to synthesize their own food substances from inorganic material, therefore they require a supply of organic material as a food source. They include all animals, all fungi, most bacteria and a few flowering plants.

Hibernation   dormancy in the winter period.

Holoplankton   organisms in which the whole life cycle is spent in the PLANKTON.

Host   see INTERMEDIATE HOST; PRIMARY HOST; SECONDARY HOST.

Hybrid   a plant or animal resulting from a cross between parents that are genetically different, usually from two different species.

Hydrostatic skeleton   a fluid-filled cavity enclosed by a body wall which acts as a skeleton against which the muscles can act.

Hyperparasite   (verb hyperparasitize) an organism which is a parasite upon another parasite.

Infusariiform   a larval stage of mesozoans produced in members of the order Dicyemida by the hermaphrodite RHOMBOGEN generation, and in the order Orthonectida by free-living males and females and reinfecting the host.

Inorganic material   material not derived from living or dead animals or plants, carbon atoms being absent from the molecular structure.

Intermediate host   an organism which plays host to parasitic larvae before they mature sexually in the final or definitive host.

Interstitial   living in the spaces between SUBSTRATE particles.

Intracellular digestion   digestion of food within the cell.

Introduction   a species which has settled in lands where it does not occur naturally as a result of human activities.

Invertebrate   an animal which is not a member of the subphylum vertebrata of the Chordata, ie it lacks a skull surrounding a well-developed brain and does not have a skeleton of bone or cartilage.

Isogamy   (adjective isogamous) a condition in which the GAMETES produced by a species are similar, ie not differentiated into male and female.

Jurassic   a geological period that extended from about 195 to 135 million years ago.

Keratin   a tough, fibrous protein rich in sulfur; the outer layer of the CUTICLE of nematode worms is keratinized.

Kinety   in ciliate protozoans, a row of kinetosomes and FIBRILS; from kinetosomes arise CILIA, the fibrils linking each kinetosome in a longitudinal row.

Kingdom   the uppermost rank of classification dividing bacteria and blue-green algae, algae, plants, fungi, protista and animals into their respective kingdoms.

Lacunae   a minute space, in invertebrate tissue containing fluid.

Larva   a general term for a distinct pre-adult form into which most invertebrates hatch from the egg and which may develop directly into adult form or into another larval form.

Lymph   an intercellular body fluid drained by lymph vessels; contains all the constituents of blood plasma except the protein, and varying numbers of cells.

Macronucleus   one of two nuclei found in ciliate protozoans.

Macrophagous diet   a diet of pieces which are large relative to the size of animal; feeding usually occurs at intervals.

Madreporite   a delicate, perforated sieve plate through which seawater may be drawn into the WATER VASCULAR SYSTEM of echinoderms; may be internal (eg sea cucumbers) or prominent external convex disk (starfishes).

Malpighian tubule/gland   a tubular excretory gland which opens into the front of the hindgut of insects, arachnids, myriapods and water bears.

Mandible   the paired appendages behind the mouth of crustaceans and uniramians, used in biting and chewing, and having grinding and biting surfaces.

Mantle   a fold of skin covering all or part of the body of mollusks; its outer edge secretes the shell.

Mantle cavity   the cavity between the body and MANTLE of a mollusk, containing the feeding and/or respiratory organs.

Maxilla   paired head appendages of crustaceans and uniramians which are located behind the MANDIBLES on the fifth segment. They act as accessory feeding appendages.

Maxilliped   the first one, two or three pairs of thoracic limbs of malocostracan crustaceans which have turned forward and become adapted as accessory feeding appendages rather than being involved in locomotion.

Maxillule   paired head appendages of crustaceans and uniramians which are located on segment six behind the MAXILLAE. They also function in the manipulation of food.

Medusa   the free-swimming sexual stage of the cnidarian life cycle, produced by the asexual BUDDING POLYPS.

Megalopa   a postlarval stage of brachyuran crustaceans in which, unlike the adult (eg crab), the abdomen is large, unflexed and bears the full number of appendages.

Meiosis   cell division whereby the DNA complement is halved in the daughter cells. Compare MITOSIS.

Meroplankton   organisms passing part of their life cycle in the PLANKTON, usually the larval forms of BENTHIC animals. Compare HOLOPLANKTON.

Merozoite   a stage in the life cycle of some parasitic protozoans which enters red blood corpuscles of the host.

Mesenchyme   embryonic connective tissue consisting of scattered, irregularly branching cells in a jelly-like matrix; gives rise to connective tissue, bone, cartilage, blood.

Mesoderm   the cell layer of TRIPLOBLASTIC animals that develops into tissues lying between the ENDODERM and ECTODERM.

Mesogloea   the layer of jelly-like material between the ECTODERM and ENDODERM of cnidarians (jellyfishes etc).

Mesozoic   a geological era ranging from 225 to 65 million years ago, comprising the TRIASSIC, JURASSIC and CRETACEOUS systems.

Metabolic rate   the rate at which the chemical processes within an organism take place.

Metameric   having many similar SEGMENTS constituting the body.

Metamorphosis   the period of rapid transformation of an animal from larval to adult form, often involving destruction of larval tissues and major changes in morphology.

Metazoan   an animal, as in the vast majority of invertebrates, whose body consists of many cells in contrast to PROTOZOANS which are unicellular; a member of the subkingdom Metazoa.

Microfilaria   the larval form of filaroid nematode worm parasites found in the SECONDARY HOST, usually mosquitoes.

Microflora   microscopic bacteria occurring in the soil.

Microhabitat   the particular parts of the habitat that are encountered by an individual in the course of its activities.

Micronucleus   one of two nuclei found in the protozoan ciliates, the smaller micronucleus provides the gametes during conjugation. See MACRONUCLEUS.

Microphagous diet   a diet of pieces of food which are minute relative to the animal's own size; feeding occurs continually.

Microtubule   a very small long, hollow cylindrical vessel conveying liquids within a cell.

Miracidium   a ciliated larva of flukes which emerges from eggs passed out with the feces of the vertebrate host and parasitizes snails, where it reproduces asexually.

Mitosis   cell division in which daughter cells replicate exactly the chromosome pattern of the parent cell, unlike MEIOSIS.

Molt   periodic shedding of the arthropod EXOSKELETON. Possession of a hardened exoskeleton prevents continuous growth until the adult stage is reached. Molting occurs under hormonal control after the secretion of a new and larger CUTICLE. An increase in size occurs during the short period prior to the hardening of the new cuticle, involving water or air uptake into the internal spaces. New tissue then grows into these spaces after the hardening of the new cuticle, ie between molts.

Monoblastic   organisms having a single cell layer (eg sponges).

Monophyletic   descended from a common ancestor. Some scientists hold a monophyletic view of arthropods, while others recognize several phyla of jointed-limbed invertebrates with separate evolutionary origins.

Morphology   the structure and shape of an organism.

Multicellular   composed of a large number of cells.

Myotome   a block of muscle, one of a series along the body of a lancelet, sea squirt larva, or vertebrate.

Natural selection   the mechanism of evolutionary change suggested by Charles Darwin, whereby organisms with characteristics that enhance the chance of survival in the environment in which they live are more likely to survive and produce more offspring with the same characteristics than organisms without those characteristics, or with other less advantageous characteristics.

Nauplius   the first larval stage of some crustaceans which is divided into three segments each possessing a pair of jointed limbs that develop into the adult's two pairs of antennae and the mandibles. The nauplius uses its limbs in feeding and locomotion.

Nekton   aquatic organisms, such as fish, which, unlike the smaller PLANKTON, can maintain their position in the water column and move against local currents.

Nematocyst   the characteristic stinging ORGANELLE of cnidarians (eg jellyfishes) located particularly on the tentacles. A short process at one end of the ovoid cell (cnidoblast) containing the nematocyst acts as a trigger opening the lid-like OPERCULUM. Water entering the cnidoblast swells the nematocyst, a long thread-like tube coiled up inside. The nematocyst discharges, ensnaring prey in its barbed coils or releasing poison down the tube into the victim.

Nematogen   the first and subsequent early generations of certain mesozoans (order Dicyemida), parasites of immature cephalopods.

Nephridiopore   the pore by which a NEPHRIDIUM opens into the external environment.

Nephridium   an excretory tubule opening to the exterior via a pore (nephridiopore). The inner end of the tubule may be blind, ending in FLAME CELLS or it may open into the COELOM via a ciliated funnel.

Nerve cord   a solid strand of nervous tissue forming part of the central nervous system of invertebrates

Niche   the position of a species within the community, defined in terms of all aspects of its life-style.

Nocturnal   awake and active by night, particularly of animals which hunt for food by night.

Notochord   a row of vacuolated cells forming a skeletal rod lying lengthwise between the central nervous system and gut of all CHORDATES.

Oligomeric/ous   having a few segments constituting the body.

Omnivore   an animal that feeds on both plant and animal tissue.

Operculate   the condition of gastropods having an OPERCULUM.

Operculum   a lid-like structure; the calcareous plate on the top surface of the foot of some gastropods, serving to close the aperture when the animal withdraws into the shell.

Opisthosoma   the posterior body region of chelicerates which may be segmented in primitive forms but generally has the segments fused.

Order   a group used in classifying organisms, consisting of a number of similar FAMILIES (sometimes only one family). Similar orders are grouped into a CLASS.

Ordovician   a geological period extending from about 500 to 440 million years ago.

Organ   part of an animal or plant which forms a structural and functional unit, eg spore, lung.

Organelle   a persistent structure forming part of a cell, with a specialized function within it analogous to an ORGAN within the whole organism.

Organic material   material derived from living or dead animals and plants, molecules making up the organisms being based on carbon, the other principle elements being oxygen and hydrogen.

Osmotic pressure   and regulation osmotic pressure is the force that tends to move water in an osmotic system, ie the pressure exerted by a more concentrated solution on one of a lower concentration. The body fluids of a freshwater animal will exert an osmotic pressure on the surrounding aqueous medium, causing water to enter the animal. Osmoregulation is the maintenance of the internal body fluids at a different osmotic pressure from that of the external aqueous environment.

Osphradium   a patch of sensory EPITHELIUM located on gill membranes of mollusks.

Paleozoic   a geological era ranging from 600-225 million years ago, comprising the CAMBRIAN, ORDOVICIAN and SILURIAN systems in the older or lower Paleozoic sub-era and the DEVONIAN, CARBONIFEROUS and PERMIAN systems in the newer or Upper Paleozoic sub-era.

Palp   an appendage, usually near the mouth, which may be sensory, aid in feeding, or be used in locomotion.

Papilla   a small protuberance ("little nipple") above a surface

Parapodium   one of a pair of appendages extending from the sides of the segments of polychaete worms.

Parthenogenesis   the development of a new individual from an unfertilized egg. It occurs when rapid colonization is important under adverse environmental conditions, or when there is an absence or only a small number, of males in the population.

Pathogen   an agent which causes disease, always parasitic.

Pedicellariae   minute, pincer-like grooming and defensive structures on the body of sea urchins and starfishes.

Pedipalp   an appendage borne on the third prosomal segment of chelicerates, sensory or prehensile in horseshoe crabs, adapted for seizing prey in scorpions, and sensory or used by the male in reproduction in spiders.

Peduncle   a narrow part supporting a longer part, eg the muscular stalk by which the body of an endoproct is attached to the SUBSTRATE.

Pelagic   of organisms or life-styles in the water column, as opposed to the bottom SUBSTRATE.

Pentamerism   the fivefold RADIAL SYMMETRY typical of echinoderms.

Pericardial cavity   the cavity within the body containing the heart. In vertebrates a hemocoelic space, which is an expanded part of the blood system, supplying blood to the heart.

Peristalsis   rhythmic waves of contraction passing along tubular organs, particularly the GUT, produced by a layer of smooth muscle.

Permian   a geological period from 270 to 255 million years ago, marking the end of the PALEOZOIC era.

Pharynx   part of the alimentary tract or GUT behind the mouth, often muscular.

Pheromone a chemical substance which when released by an animal influences the behavior or development of other individuals of the same species.

Photosynthesis the synthesis of organic compounds, primarily sugars, from carbon dioxide and water using sunlight as a source of energy and chlorophyll or some other related pigment for trapping the light energy.

Phyletic concerning evolutionary descent.

Phylogeny the evolutionary history or ancestry of a group of organisms.

Phylum a major group used in the classification of animals. Consists of one or more CLASSES. Several (sometimes one) phyla make up a KINGDOM.

Physiology the study of the processes which occur within living organisms.

Phytoplankton microscopic algae that are suspended in surface waters of seas and lakes where there is sufficient light for PHOTOSYNTHESIS to take place.

Pinnate of tentacles, GILLS, resembling a feather or compound leaf in structure, with similar parts arranged either side of a central axis.

Pinnule a jointed appendage found in large numbers on the arms of crinoids giving a feather-like appearance, hence the name feather star.

Plankton drifting or swimming animals and plants, many minute or microscopic, which live freely in the water and are borne by water currents due to their limited powers of locomotion.

Planula the free-swimming ciliated larva of cnidarians (jellyfishes and allies).

Plasma the fluid medium of the blood in which highly specialized cells are suspended; mainly water, containing a variety of dissolved substances which are transported from one part of the body to another.

Plasmodium the asexual stage of orthonectid mesozoans, resembling the protozoan plasmodium, which divide repeatedly by FISSION, filling the hosts' tissue spaces.

Pneumostome the aperture to the lung-like MANTLE CAVITY of pulmonates.

Polyp the stage, the most important in the life cycle of most cnidarians, in which the body is typically tubular or cylindrical, the oral end bearing the mouth and tentacles and the opposite end being attached to the SUBSTRATE.

Polysaccharide a carbohydrate produced by a combination of many simple sugar or monosaccharide molecules, eg starch and CELLULOSE.

Primary host the main host of a parasite in which the adult parasite or the sexually mature form is present.

Proboscis a tubular organ that may be extended from the mouth of many invertebrates such as moths and butterflies; in ribbon worms, the proboscis can be everted.

Proglottides (sing. proglottis) the segments which make up the "body" of a tapeworm. When mature, each proglottis will contain at least one set of reproductive organs.

Prokaryote cell having, or organism made of cells having, genetic material in the form of simple filaments of DNA, not separated from the CYTOPLASM by a nuclear membrane (cf EUKARYOTE). Bacteria and blue-green algae have cells of this type.

Prosoma the anterior body region of CHELICERATES, composed of eight SEGMENTS, analogous to the head and THORAX of other arthropods, or the CEPHALOTHORAX of chelicerates. The segments are generally fused and are only distinguishable in the embryo.

Prostomium the anterior non-segmental region of annelid worms, bearing the eyes, ANTENNAE and a pair of PALPS; comparable to the head of other phyla.

Protein a complex organic compound composed of numerous amino acids joined together by peptide linkages, forming one or more folded chains. The sequence of amino acids is peculiar to a particular protein.

Protoconch the first shell of a gastropod which is laid down by the larva.

Protonephridium a type of excretory organ in which the tubule usually ends in a FLAME CELL.

Protostome a member of one major branch of the muticellular animals (the other, complementary, branch is the DEUTEROSTOMES). The mouth is formed from the embryonic blastopore, the embryo undergoes SPIRAL CLEAVAGE, the body cavity is formed by the MESODERM splitting into two, and the central nervous system is ventral.

Protozoan an organism of the phylum Protozoa, Kingdom Protista, differing from animals of the Kingdom Animalia in consisting of one cell only, but resembling them and plants and differing from bacteria in having at least one well-defined nucleus.

Pseudocoel the secondary body cavity of roundworms, rotifers, gastrotrichs and endoprocts, between an inner MESODERM layer of the body wall and the ENDODERM of the gut, it is not a true COELOM.

Pseudopodium (plural: -a) a temporary projection of the cell when the fluid endoplasm flows forward inside the stiffer ectoplasm. Occurs during locomotion and feeding.

Pseudotrachea a branched TUBULE resulting from intuckings of the CUTICLE of certain terrestrial isopods which acts as a specialized respiratory surface. Pseudotracheae resemble the TRACHEAE of uniramians and certain arachinids, although they have evolved independently.

Pulmonate having a lung, eg certain snails and slugs.

Pygidium the terminal, non-segmental region of some invertebrates which bears the anus.

Radial symmetry a form of symmetry in which the body consists of a central axis around which similar parts are symmetrically arranged.

Radula the "toothed tongue" of mollusks, a horny strip with ridges or "teeth" on its surface which rasp food. Absent in members of the class Bivalvia.

Ray a radial division of an echinoderm, eg a starfish "arm."

Redia a larval type produced asexually by a previous larval stage of flukes (Trematoda). Lives parasitically in snails and reproduces asexually, giving rise to more rediae or to CERCARIAE.

Reticulopodium a type of PSEUDOPODIUM characteristic of the foraminiferans; reticulopodia are thread-like, branched and interconnected.

Rhombogen an hermaphrodite form of dicyenid mesozoan derived from a NEMATOGEN when the cephalopod host has reached maturity. It resembles the nematogen morphologically and gives rise to INFUSARIIFORM larvae.

Rostrum the anterior plate of the crustacean CARAPACE, present in malacostracans, which extends toward the head and ends in a point.

Sclerotization hardening of the arthropod CUTICLE by TANNING.

Scolex the head region of a tapeworm, which attaches to the wall of the host's gut by suckers and/or hooks.

Secondary host the host in which the larval or resting stages of a parasite are present.

Sedentary sedentary organisms, or stages in the life cycle of certain organisms, are permanently attached to a SUBSTRATE; as opposed to FREE-LIVING.

Segment a repeating unit of the body which has a structure fundamentally similar to other segments, although certain segments may be grouped together into TAGMATA to perform certain functions, as in the head, THORAX or ABDOMEN.

Segmentation the repetition of a pattern of segments along the length of the body, or along an appendage. The similarity between different segments of an animal may be imperfect, particularly the segments forming the head.

Septum a portion dividing a tissue or organ into a number of compartments.

Seta (plural setae) a bristle-like projection on the invertebrate EPIDERMIS.

Sexual reproduction reproduction involving MEIOSIS and fertilization, usually fusion of two GAMETES, one female and one male. See also COPULATION, CONJUGATION, FRAGMENTATION.

Siliceous composed of or containing silicate, as in the skeleton of glass sponges (see SPICULES).

Silurian a geological era, 440–400 million years ago.

Sinistral of gastropod shells, with whorls rising to the left and not as usual to the right (compare DEXTRAL).

Sinus a space or cavity in an animal's body.

Siphon a tube through which water enters and/or leaves a cavity within the body of an animal, eg in mollusks and in sea squirts.

Solitary a life-style in which an organism exists by itself and not in permanent association with others of the same species.

Specialist an organism having special adaptations to a particular habitat or mode of life; its range of habitats or variety of modes of life may thus be limited and, as a result, its evolutionary flexibility also.

Speciation the origin of species, the diverging of two like organisms into different forms resulting in new species.

Species a taxonomic rank, the lowest commonly used; reproductively an isolated group of interbreeding organisms. Similar species make up a GENUS.

Spermatophore a package of sperm produced by males, usually of species in which fertilization is internal but does not involve direct COPULATION.

Spermatotheca an organ, usually one of a pair, in a female or hermaphrodite that receives and stores sperm from the male.

Spicule a mineral secretion (calcium carbonate or silica) of sponges which forms part of the skeleton of most species and whose structure is of importance in sponge classification.

Spiral cleavage a form of embryonic division which occurs in PROTOSTOMES; in the spiral arrangement of cells any one cell is located between the two cells above or below it. In all other many-celled animals, ie DEUTEROSTOMES, there is radial CLEAVAGE.

Spore a single- or multi-celled reproductive body that becomes detached from its parent and gives rise directly or indirectly to a new individual.

Sporocyst a sac-like body formed by the MIRACIDIUM larva of a blood fluke while within the snail, the intermediate host; produces numerous CERCARIAE, over 3,000 per day from a single sporocyst.

Sporozoite SPORE produced in certain protozoans which then develops into gametes.

Statocyst the balancing organ of a number of invertebrates consisting of a vesicle containing granules of sand or calcium carbonate. These granules move within the vesicle and stimulate sensory cells as the animal moves, so providing information on its position in relation to gravity.

Stolon the tubular structure of colonial cnidarian POLYPS that anchors them to the SUBSTRATE and from which the polyps arise.

Stomodeum a region of unfolding ECTODERM from which derive the mouth cavity and foregut in many invertebrates.

Strobila the "body" of a tapeworm, consisting of a string of segments, through which food is absorbed from the gut of the host.

Stylet a small, sharp appendage, for example in water bears used to pierce plant cells.

Suborder members of an ORDER forming a group of organisms which differ in some way from the other members but also resemble them in many characteristics.

Substrate the surface or sediment on or in which an organism lives.

Superfamily a division containing a number of families or a single family differing in some way from other families which are included in the same ORDER.

Suspension feeding a feeding mechanism in which small organisms and other matter suspended in the water are removed and consumed.

Symbiosis a close and mutually beneficial relationship between individuals of two species.

Synapse the site at which one nerve cell is connected to another.

Tagmata (sing. tagma) functional body regions of arthropods and annelids consisting of a number of segments; eg the head, THORAX and ABDOMEN of crustaceans.

Tanning hardening of the arthropod CUTICLE achieved by the cross-linking of the protein chains by arthoquinones, involving also polyphenol and polyphenoloxidase catalysts.

Taxon a taxonomic grouping of organisms or the name applied to it.

Taxonomy the study of the classification of organisms acording to resemblances and differences.

Tegumental gland a gland below the EPIDERMIS of the crustacean cuticle. Ducts from the glands convey the constituents of the EPICUTICLE to the cuticle surface during molting, when the new epicuticle is formed.

Telson the posterior segment of the arthropod abdomen which is present only embryonically in insects. In certain crustaceans the telson is flattened to form a tail fin which is used in swimming.

Terrestrial associated with, or living on the earth or ground.

Tertiary the geological period of time from the end of the CRETACEOUS era 65 million years ago to the present time, divided into a number of epochs, the last 2 million years sometimes distinguished as the Quaternary Period.

Test an external covering or "shell" of an invertebrate, especially sea squirts (tunicates), sea urchins etc; is in fact an internal skeleton just below the EPIDERMIS.

Thorax the segmented body region of insects and crustaceans which lies behind the head and which typically bears locomotary appendages. Up to 11 segments are present in crustaceans but only 3 in insects.

Tissue a region consisting mainly of cells of the same sort and performing the same function, associated in large numbers and bound together by cell walls (plants) or by intercellular material (animals).

Torsion the process of twisting of the body in the larval stage of gastropods.

Trachea a cuticle-lined respiratory tubule of uniramians and certain arachnids which is involved in gas exchange. Tracheae open to the exterior via a spiracle which can often be sealed to reduce desiccation. The tracheae are branched and ramify into the tissues, they end in thin-walled, blind-ending tracheoles within the cells.

Triassic a geological period extending from 225 to 195 million years ago, marking the beginning of the MESOZOIC era.

Trichocyst rod-like or oval ORGANELLE in the ECTODERM of protozoans which may discharge a long thread on contact with prey.

Trochophore an oval or pear-shaped, free-swimming, planktonic larval form of organisms from different phyla, including segmented worms and mollusks.

Tube feet or podia—hollow, extensive appendages of echinoderms connected to the WATER VASCULAR SYSTEM that may have suckers, or serve as stilt-like limbs or be ciliated to waft food particles toward the mouth.

Tubule long hollow cylinder within a cell, normally for conveying or holding liquids.

Tunicin, tunic a form of CELLULOSE, the main constituent in the fibrous matrix forming the tunic or TEST of sea squirts.

Unicellular an organism composed of only a single cell.

Uniramian a member of the phylum Uniramia which includes the insects, centipedes and millipedes. They possess a single pair of antennae, and mandibles. The appendages are basically unbranched or UNIRAMOUS in contrast to those of crustaceans.

Uniramous condition describing an arthropod limb which is not branched and is found in insects and myripods (hence the Uniramia) and chelicerates. Some crustacean limbs are secondarily uniramous where one of the branches of the BIRAMOUS limb has been lost.

Urogenital tract ducts and tubules common to the genital and urinary systems voiding via a common aperture.

Uropod flattened extension of the sixth abdominal appendage of malacostracan crustaceans which together with the flattened TELSON form a tail fin used in swimming.

Vacuole a fluid-filled space within the CYTOPLASM of a cell, bounded by a membrane.

Valve in bivalves, one half of the two-valved shell.

Vascular containing vessels which conduct fluid—in animals usually blood, as in the vascularized MANTLE CAVITY of pulmonate snails.

Veliger a free-swimming larval form of mollusks possessing a VELUM; develops from a TROCHOPHORE; foot, mantle, shell and other adult organs are present.

Velum the veil-like ciliated lobe of the VELIGER larva, used in swimming; also the inward-projecting margin of the umbrella in most hydrozoan medusae.

Ventral situated at, or related to, the lower bottom side or surface.

Vermiform a worm-like larval stage of dicyenid mesozoans formed within the axial cells of the NEMATOGEN generation or generally meaning worm-like.

Vertebrate an organism which belongs to the subphylum Vertebrata (Craniata) of the phylum Chordata; differs from other chordates and invertebrates in having a skull which surrounds a well-developed brain, and a skeleton of cartilage and bone.

Water vascular system or ambulacral system, a system of canals and appendages of the body wall that is unique to echinoderms, derived from the COELOM and used, eg, in locomotion in starfishes.

Zoea a planktonic larval form of some decapod crustaceans which possesses a segmented THORAX, a CARAPACE and at least three pairs of BIRAMOUS thoracic appendages. In contrast to the antennal propulsion of the NAUPLIUS these thoracic appendages are used in locomotion. The abdominal pleopods appear but are not functional until the postlarval stage.

Zoochlorella a symbiotic green alga of the Chloroplyceae which occurs in the amoebocytes of certain freshwater sponges, the gastrodermal cells of some hydra species and the jelly-like connective tissue (parenchyme) of certain turbellarian flatworms.

Zooid a member of a colony of animals which are joined together; may be specialized for certain functions.

Zoospore a motile spore which swims by means of a flagellum, is produced by some unicellular animals and algae, and is a means of ASEXUAL REPRODUCTION.

Zooplankton small or minute animals that live freely in the water column of seas and lakes and consists of adult PELAGIC animals or the larval forms of pelagic and some BENTHIC animals; most are motile, but the water movements determine their position in the water column.

Zygote a fertilized ovum before it undergoes cleavage.

INDEX

A **bold number** indicates a major section of the main text, following a heading: a *bold italic* number indicates a fact box on a group of species: a single number in (parentheses) indicates that the animal name or subjects are to be found in a boxed feature and a double number in (parentheses) indicates that the animal name or subject are to be found in a spread special feature. *Italic* numbers refer to illustrations.

N

O

M